Animal Nutrition Science

This page intentionally left blank

Animal Nutrition Science

Gordon McL. Dryden

Senior Lecturer in Animal Nutrition
School of Animal Studies
The University of Queensland, Gatton
Australia

www.cabi.org

CABI is a trading name of CAB International

CABI Head Office
Nosworthy Way
Wallingford
Oxfordshire OX10 8DE
UK

Tel: +44 (0)1491 832111
Fax: +44 (0)1491 833508
E-mail: cabi@cabi.org
Website: www.cabi.org

CABI North American Office
875 Massachusetts Avenue
7th Floor
Cambridge, MA 02139
USA

Tel: +1 617 395 4056
Fax: +1 617 354 6875
E-mail: cabi-nao@cabi.org

A catalogue record for this book is available from the British Library, London, UK.

Library of Congress Cataloging-in-Publication Data
Dryden, Gordon McL.
    Animal nutrition science / Gordon McL. Dryden.
        p. cm.
    Includes bibliographical references and index.
    ISBN 978-1-84593-412-5 (alk. paper)
1. Animal nutrition--Textbooks. I. Title.

SF95.D79 2008
636.08'52--dc22

2007050644

ISBN: 978 1 84593 412 5

Typeset by SPi, Pondicherry, India.
Printed and bound by CPI Group (UK) Ltd, Croydon, CR0 4YY

Every reasonable effort has been made to obtain and evaluate information from a wide variety of reputable sources and to ensure its accuracy at the time of writing. Nevertheless the author and publisher disclaim all responsibility for any liability, loss, injury or damage incurred as a consequence, directly or indirectly, of the use and application of any of the contents of this volume.

Contents

Preface   vii

Acknowledgements   xi

1. Nutritional Ecology   1

2. The Nutritive Value of Animal Foods: Introductory Concepts About Foods, Nutrients and Food Analysis   16

3. Methods of Evaluating the Availability of Nutrients in Foods   28

4. Physico-chemical Composition and Digestibility of Forages and Cereal Grains   40

5. The Nutritive Value of Concentrate Foods   57

6. Secondary Substances in Concentrates and Roughages   74

7. Digestion and the Supply of Nutrients   85

8. Water Use and Requirements   115

9. Minerals: Their Functions and Animal Requirements   130

10. Vitamins   151

11. Voluntary Food Intake   162

12. Quantitative Nutrition: Requirements for Energy and Protein   180

13. Ration Formulation   218

14. Nutritional Investigations: Measures of Nutritional Status   226

15. Nutrition and the Environment   241

16. Nutritional Genomics   248

17. Animal Responses to Stock Food Processing   256

18. Feed Mill Design and Management: an Introduction   273

Index   293

This page intentionally left blank

Preface

The Many Facets of Animal Nutrition

Animal nutrition is one of the most diverse disciplines in the animal sciences. Its sub-disciplines range from the biochemistry of nutrient use and digestive physiology, through nutritional genomics and the mathematical modelling of nutrient requirements, to the ongoing development and refinement of animal feeding systems. There are two reasonably distinct areas of study: the first is the basic biochemistry and physiology of the discipline and the second involves the application of these concepts to the design of feeding systems. These two areas are usually considered separately. This is not because one is 'scientific' and the other 'practical' (an incorrect and sterile distinction), but because the biochemical and physiological topics are general in nature, while studies of feeding systems must, necessarily, be focused on each animal type and the relevant welfare, production and economic considerations.

How This Book Is Structured

This book describes how animals obtain, digest and use food and nutrients, how their nutrient requirements can be quantified and how nutrient use can affect the environment. The discussion is not confined to any particular type of animal – domestic or wild, ruminants or monogastrics. The book is intended for senior undergraduate or postgraduate readers, and so it assumes a basic understanding of intermediary metabolism and digestive physiology and anatomy. The structure of the text is illustrated in Fig. P.1 where the 18 topics are grouped to show how they relate to each other.

We begin in Chapter 1 with a discussion of nutritional ecology, the study of how animals obtain and digest their food. This is an important part of animal nutrition, although it is not often included in nutrition texts. There are important differences in abilities to prehend and digest food, and in food preferences, and these determine the characteristics of feeding systems for different animals.

The second major topic (Chapters 2 and 3) is the definition, measurement and description of food nutritive value. Some understanding of food chemistry is needed to appreciate the discussion of the physico-chemical structure of food in Chapter 4, while knowledge of physico-chemical structure helps to explain the bases of food analysis and evaluation systems. It will be useful to revise Chapters 2–4 after an initial reading of these topics. Chapters 5 and 6 describe the nutritive value of concentrate foods. Chapters 2–6 thus form a group which deals with food composition, nutritive value and safety.

Digestion determines the nutrients which become available to the animal (Chapter 7). If 'intestinal' digestion occurs before microbial digestion, then the absorbed nutrients will reflect, fairly accurately, the composition of those constituents in the food. If microbial fermentation comes first, then the composition of the absorbed digestion products is usually quite different to the food. Of course, ruminants do have some advantages – they are more resistant to food-borne toxins than monogastric animals, and they have more flexibility in their requirements for some nutrients. The distinction between foregut fermentation and other types of digestive systems is important, and it informs the design of feeding systems and our expectations about animal product quality.

The functions, sources and metabolism of minerals, vitamins and water are discussed in Chapters 8–10. The metabolic roles of these nutrients, and the quantification of animals' requirements for them, are active fields of study. Examples of new information are the roles of vitamin D in controlling cell proliferation and differentiation and quantitative data about the water requirements and salinity tolerances of horses and deer.

Chapter 11 describes how the voluntary intake of foods is regulated. This is also an area where our understanding is rapidly improving, although it is far from complete. Some of the older ideas about voluntary intake control have been reassessed and have been found to be incomplete or erroneous. Control mediated via the

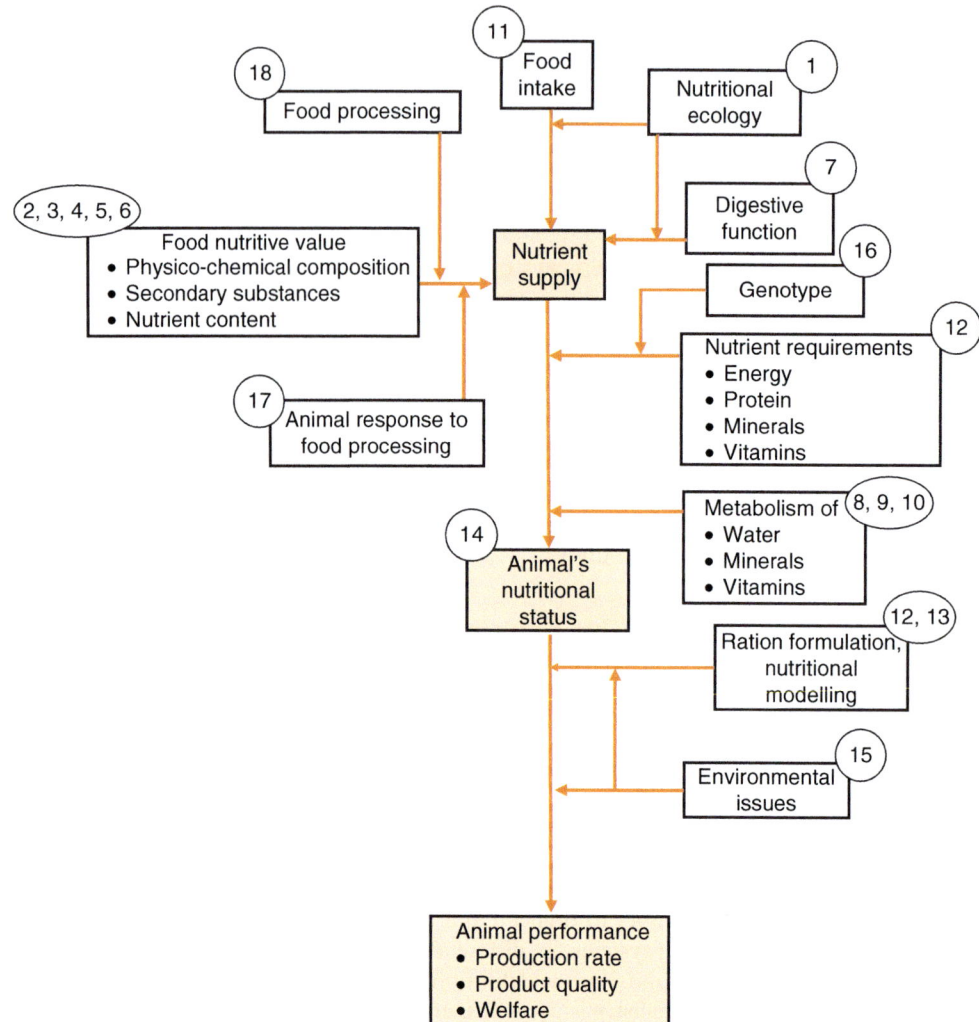

Fig. P.1. Nutritional factors which affect animal production and welfare (topic numbers refer to the chapters in this book).

neuro-hormonal systems is important in all animals, while digestive tract control is also important in ruminants, as the escape of undigested food from the rumen influences how much food these animals can eat.

Energy and protein requirements, and how these change with production type and rate, are discussed in Chapter 12; requirements for minerals and vitamins are considered in the preceding chapters. We consider energy and protein metabolism, and how ruminant and monogastric animals' requirements can be quantified. Ration formulation, described in Chapter 13, is the practical application of nutrient requirement data. Conventional methods optimize rations for a set of outcomes (e.g. least cost, or minimum nutrient excretion), while mathematical models of nutrient use describe the interplay of variables such as animal type and environmental factors, and can answer 'what if?' questions.

We may need to investigate situations where we think that our animal is incorrectly fed. The nature of the problem may be only apparent after careful examination. Objective procedures which can be used to investigate nutritional problems are described in Chapter 14.

The intensification of animal production has led to an increasing problem of nutrient wastage and environmental pollution, and Chapter 15 deals with current concerns about how feeding systems can affect the environment. Chapter 16 considers how the animal's genotype influences its nutrient requirements. With modern genetic technologies we can move past empirical observations that there are differences between genotypes, and we can now explain in detail how genotype influences an animal's ability to utilize nutrients. In the future we will be able to develop feeding systems tailored to individual animal's genotypes (this is already beginning to be done in human nutrition).

Food processing, the animal's response to this and the procedures needed for it are described in Chapters 17 and 18. This is a large topic and the whole subject could not be covered in this discussion. Milling, mixing and warehousing are central to the operation of all food processing facilities, and these topics and feed mill design are considered here.

This page intentionally left blank

Acknowledgements

Parts of this book were reviewed by Michael Evans (Applied Nutrition), Brenton Hosking (Better Blend Stockfeeds), Simone Hoskin (Massey University), André-Denis Wright (CSIRO), Rafat Al Jassim, John Gaughan, Patrick Moss, Clive Phillips and Dennis Poppi (The University of Queensland). I am grateful for their very helpful comments. Of course, any errors of omission or interpretation are mine.

I am indebted to my students: to my undergraduates who have shared in my experiments with nutrition syllabuses, and my postgraduates with whom I have shared the excitement of discovery and some of whose data are in this book.

It goes without saying (but I have great pleasure in saying it) that my wife and family have given me continued support (and forbearance) while I have been involved in this project. I give you my heartfelt thanks.

This page intentionally left blank

1 Nutritional Ecology

Nutrient Sources and Feeding Methods

Nutritionally, there are two fundamentally different types of organisms – the 'autotrophs', which are those that use CO_2 as their carbon source, and the 'heterotrophs', which are organisms that need carbon from preformed organic molecules. All animals are heterotrophs.

Food prehension and nutrient absorption

Autotrophs absorb their nutrients directly through their cell membranes. This applies regardless of whether the organism is unicellular or multicellular; for example, the higher plants absorb water and minerals directly from the soil through the walls of the root hair cells and absorb CO_2 directly through the membranes of their mesophyll cells.

Unicellular heterotrophs, e.g. some of the archaea and most bacteria, are small enough that they can feed themselves adequately by absorbing nutrients from the environment, provided that the environment is suitably nutrient-rich. These organisms obtain their nutrients through a combination of absorption of small, simple organic or inorganic molecules or ions (e.g. the absorption of hydrogen or formic acid by methanogens) and extracellular digestion of organic macromolecules followed by the absorption of the products of this digestion through the cell wall (e.g. the digestion of cellulose by fibrolytic bacteria and absorption of the resulting cellobiose and glucose).

Protozoa are larger than bacteria (actually about 1000 times larger, but still very small by human standards of measurement). They absorb small nutrient molecules through their cell membranes but also use phagocytosis as a way of prehending food. Phagocytosis occurs when the protozoan folds its cell membrane around the food particle and then takes it into the cell. The process (described by Alberts *et al.*, 2002) involves the 'recognition' of the substrate particle by the predator's cell membrane and the enclosing of the substrate by the cell membrane. Particles prehended in this way by rumen protozoa include starch granules, bacteria and other protozoa.

For the higher animals, nourishment involves the prehension of food, as well as its digestion within the animal's digestive tract. 'Prehension' is the act of taking food into the organism. For carnivores, prehension may be preceded by the act of catching prey animals and breaking the fairly tough carcass into pieces small enough to swallow. For herbivores, prehension is the act of collecting a mass of plant material and taking it into the mouth. This can be done by plucking individual fruits or leaves with the teeth and lips (e.g. horses and deer), wrapping the tongue around a clump of forage and jerking the head to break the leaves free (e.g. cows), plucking blades of grass by hand (gelada monkeys) or with a prehensile trunk (elephants) or standing on the hind legs and using the forelegs to pull twigs into position for browsing (the dik-dik). Digestion is preceded by the physical breaking up in the mouth (chewing) of food into pieces which are small enough to form a bolus, i.e. an amount of chewed food small enough to be swallowed. The breaking up of food is called 'comminution'. In non-ruminant animals this occurs during prehension and bolus formation; in ruminants it takes place during prehension and bolus formation and also during rumination.

We can see from this outline that the important events in the evolution of the higher animals and of modern digestive tracts and feeding behaviour were the development of teeth and jaws for prehension and comminution, a tubular and compartmented digestive tract to allow for the separation of the digestive process into sequential phases which may be accompanied by selective retention of digesta components, endogenous enzymes capable of digesting the eaten food and a symbiosis with microorganisms to supply those enzymes which are needed to digest plant cell walls.

Nutritional Classification

Animal species can be grouped into classes based on the main types of food they voluntarily eat. There are three major classes: carnivores, which eat meat (also called 'faunivores' or 'animal-eaters'); omnivores, which have a dietary range that includes foods of both vegetable and animal origins; and herbivores, which eat plants (Table 1.1). Carnivores also include insect-eaters ('insectivores') and fish-eaters ('piscivores'). Herbivores include animals which eat predominantly seeds ('granivores'), fruit ('frugivores') and plant leaves ('folivores' or 'graminivores'). There is considerable variation in the ways in which herbivores are classified. A very detailed classification is given by Damuth (1992). A herbivore classification system which is useful for describing ungulate feeding styles is that of Hofmann and Stewart (1972), who grouped these animals (including both ruminants and non-ruminants) into concentrate selectors (eaters of seeds, fruit, young foliage and similar sugar-, protein- and fat-rich foods), grazers (animals which eat mostly the leaves of grasses) and intermediate feeders (those which select their diet from among the whole range of plant foods according to availability). Another classification of herbivores is

Table 1.1. Nutritional classification of the Mammalia: characteristic foods and example species. (From Jones, 1998; Kruger *et al.*, 1999; Gursky, 2000; Watson and Owen-Smith, 2000; Hulbert and Andersen, 2001; Nugent *et al.*, 2001; Baldi *et al.*, 2004; Capitani *et al.*, 2004; Keiper and Johnson, 2004; Lentle *et al.*, 2004; Carvalho *et al.*, 2005; Hovens and Tungalaktuja, 2005; Evans *et al.*, 2006; Glen and Dickman, 2006; Munn and Dawson, 2006; Telfer and Bowman, 2006; Torstenson *et al.*, 2006.)

Nutritional class	Class	Example species	Characteristic foods
Carnivores			
Faunivores	Metatheria	Tasmanian devil (*Sarophilus laniarius*)	Small animals
		Spotted quoll (*Dasyurus maculata*)	Variety of small animals especially mammals
	Eutheria	Lion (*Panthera leo*)	Large animals
		Domestic cat (*Felis* spp.)	Animals and birds
		African wild dog (*Lycaon pictus*)	Animal carcasses (small and large ungulates)
		Domestic dog (*Canis lupus familiaris*)	Ungulate animals[a]
Insectivores	Metatheria	Echidna (*Tachyglossus aculeatus*)	Insects
	Eutheria	Shrews	Insects
		Spectral tarsier (*Tarsius spectrum*)	Insects
Omnivores			
	Metatheria	Bandicoot (*Isoodon obesulus peninsulae*)	Insects, plants, fruits, fungi
		Opossum (*Didelphis aurita*)	Insects, fruits
	Eutheria	Pig (*Sus scrofus*)	Small animals, seeds, plant roots
		Fox (*Vulpes vulpes*)	Small animals, insects, fruits
Herbivores			
Concentrate selectors	Metatheria	Brush-tailed possum (*Trichosurus vulpecular*)	Immature leaves, fruits, flowers, seeds
	Eutheria	Roe deer (*Capreolus capreolus*)	Immature leaves, fruits, flowers, seeds
		Eland (*Taurotragus oryx*)	Succulent, forb and woody species of browse
Mixed feeders	Metatheria	Parma wallaby (*Macropus parma*)	Grasses and browse
		Red kangaroo (*Macropus rufus*)	High-quality (low-fibre) forage
	Eutheria	Red deer (*Cervus elaphus*)	Plants, seeds, browse
		Goat (*Caprus hircus*)	Plants, seeds, browse
		Sheep (*Ovis ovis*)	Plants, seeds, forbs
Grazers	Metatheria	Common wombat (*Vombatus ursinus*)	Fibrous grasses
		Wallaroo (*Macropus robustus*)	Grasses
	Eutheria	Wapiti (*Cervus elaphus nelsoni*)	Predominantly grasses
		Horse (*Equus calabus*)	Grasses (leaves)
		Bison (*Bison bison*)	Grasses (leaves)
		Cow (*Bos taurus*)	Grasses (leaves)

[a]By analogy with the wolf (*Canus lupus*), from which the domestic dog is derived.

that used for African animals. This uses terms like 'browser' (which overlaps the mixed feeder and concentrate selector categories) and subdivides grazers into 'fresh grass' and 'dry grass' groups. We can also use the terms 'specialist' and 'generalist' to indicate if the dietary range is small or large. The spotted-tail quoll (*Dasyurus maculata*) eats only meat and has been called a 'hypercarnivore' or 'specialist carnivore'. Similarly, the koala (*Phascolarctos cinereus*) is a specialist herbivore and eats only the leaves of a few eucalypt species.

The Evolution of Feeding Behaviour

Primitive feeding behaviours

The most primitive organisms for which there is a fossil record were early archaea and eukaryotes (Schopf, 1993; Brocks *et al.*, 1999). These unicellular organisms first arose in the Precambrian aeon about 3500 million years ago (MYA). Figure 1.1 gives some information on geological ages.

Modern bacteria digest their food using enzymes which are either located in, or are secreted through,

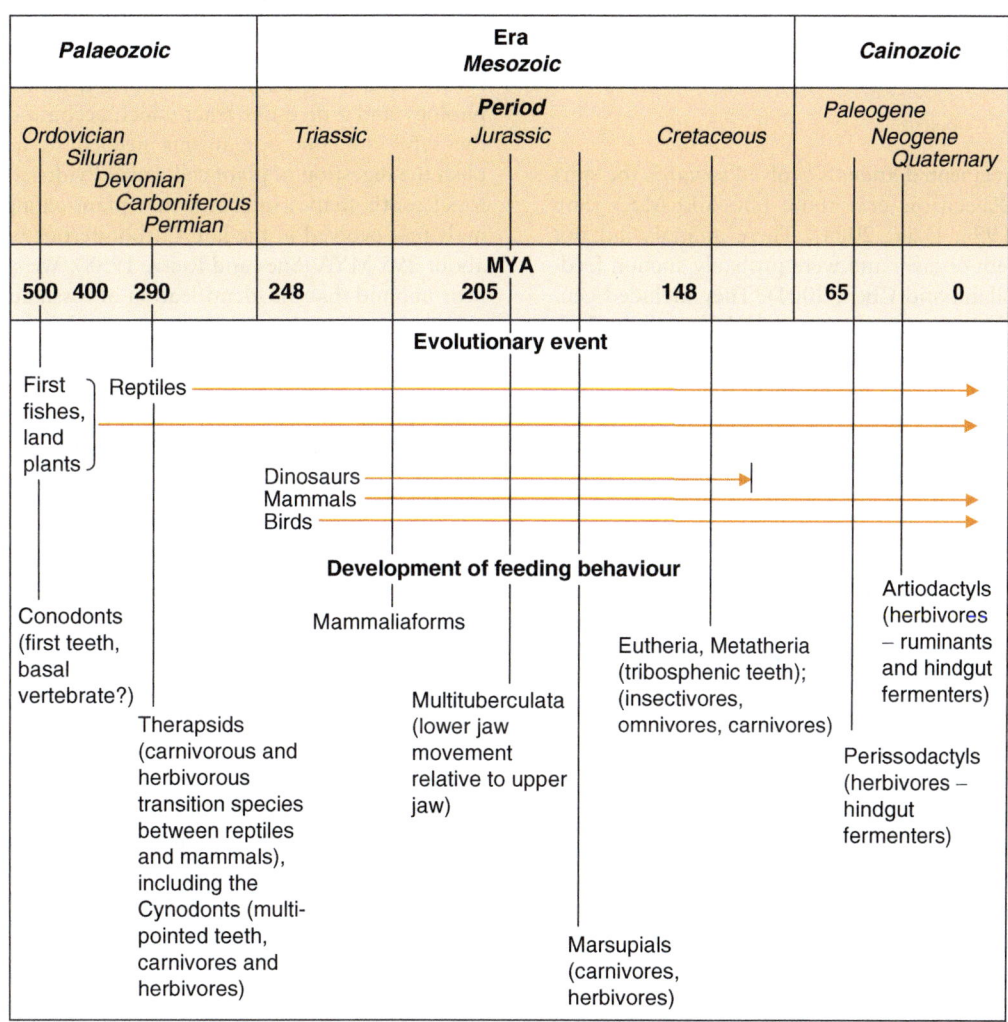

Fig. 1.1. Time line of the evolution of modern feeding styles. The Cainozoic era comprises three periods: the Paleogene (65–23.3 MYA) and Neogene (23.3–2.6 MYA), which contained the Palaeocene, Eocene, Oligocene, Miocene and Pliocene epochs (see Fig. 1.3), and the Quaternary (from 2.6 MYA), which comprised the Pleistocene and Holocene epochs. MYA = million years ago. (Dates are from ICS, 2007.)

their cell walls. The products of this enzyme action are then absorbed into the cell. A development of this primitive feeding behaviour is the inclusion of the DNA or organelles of the eaten organisms within the predator organism's cell, where the ingested structures continue to function and provide the predator with nutrients. Mitochondria and chloroplasts probably originated from the inclusion of symbiotic bacteria in the cells of higher animals and plants (Alberts *et al.*, 2002). More recent examples are the inclusion of bacterial genes coding for cellulolytic enzymes in protozoa (Ricard *et al.*, 2006) and the retention of plant chloroplasts in a specialized part of the mollusc digestive tract where they continue to photosynthesize for up to 4 months (Curtis *et al.*, 2006).

Evolution of teeth and jaws

Small, segmented animals evolved towards the start of the Palaeozoic era, about 600–500 MYA (Shu *et al.*, 1996; Dzik, 2004). These animals did not have teeth or jaws and were probably suction feeders (Holland and Chen, 2001). They included genera such as *Yunnanozoon* and *Haikouella*, which were probably the first vertebrates. Prehension may have been simply by pushing the open mouth against the food and 'forcing it into the oral cavity' (Mallat, 1996) or by sucking food into the mouth by contractions of the pharynx (Holland and Chen, 2001).

The first organisms to have operational teeth were the conodonts (the word means 'pointed tooth'). These date from about 525 MYA, i.e. the Cambrian period of the Palaeozoic era, and they became widely distributed during the Devonian period, about 400 MYA (Durand, 2005). Conodonts were generally small eel-like vertebrates about 4 cm long (although a giant conodont was 40 cm long) and had a notochord (Durand, 2005). They possessed small, cone-shaped teeth made of enamel and dentine, i.e. a similar chemical composition as modern teeth (Sansom *et al.*, 1994). Wear marks on conodont teeth indicate that the organisms which owned them used them to bite food (Purnell, 1995). However, conodonts were jawless, and biting was done by muscular contractions of the 'pharynx' (Donoghue and Purnell, 1999).

Jaws are assumed to have evolved from the anterior arches of the branchial basket (the cartilaginous arches which surround the gills). Mallat (1996) has proposed

that these structures first enlarged to allow the animal to close its jaws during the expiration of water through its gills, i.e. to improve the efficiency of gas exchange during breathing. Later, these structures evolved into jaws and were used for holding and chewing prey. Mallat (1996) proposed that these developments occurred about 450 MYA. There was a rapid diversification of jawed fish during the Devonian period.

Evolution of mammalian carnivory and herbivory

Carnivory and insectivory were the first feeding styles to evolve in land animals. As suggested by Sues and Reisz (1998) insectivory allowed animals indirect access to plant material as food. True herbivory requires specialized mouth and tooth morphology and a digestive tract which accommodates the symbiotic microorganisms needed to accomplish the digestion of plant cell walls. Evidence from fossil tooth shapes indicates that plant-eating animals first evolved in the late Carboniferous period, about 295 MYA (Sues and Reisz, 1998). We should bear in mind that the identification of fossil animals as herbivores, rather than the pre-evolving carnivores, is by comparing the shapes of fossil teeth with those of modern animals, some evidence from what appears to be fossilized stomach contents, and conjectures based on relationships between body size and the rate of digesta passage and energy requirements of modern animals. We cannot be sure that we have drawn the correct conclusions. Sues and Reisz (1998) recount an example: a reptilian fossil was thought to be a carnivore from its tooth shape but a later find of fossilized stomach contents indicated that it was probably a herbivore. Some modern herbivores have large canine teeth for display or fighting, e.g. the herbivorous gelada monkey (*Theropithecus gelada*; Jolly, 2001).

The Therapsids were a large group of animals intermediate between reptiles and mammals (Fig. 1.2). Therapsid fossils contain some intermediate anatomical forms showing a transition between reptilian jaws and inner ear structures and the mammalian forms of these organs (Durand, 2005). There were both carnivorous and herbivorous Therapsids. All had functional jaws and had developed tooth shapes which were appropriate to the tasks of holding prey and tearing the carcass into pieces, or of shearing and crushing plant material. In other words, 'heterodonty', or the differentiation of teeth

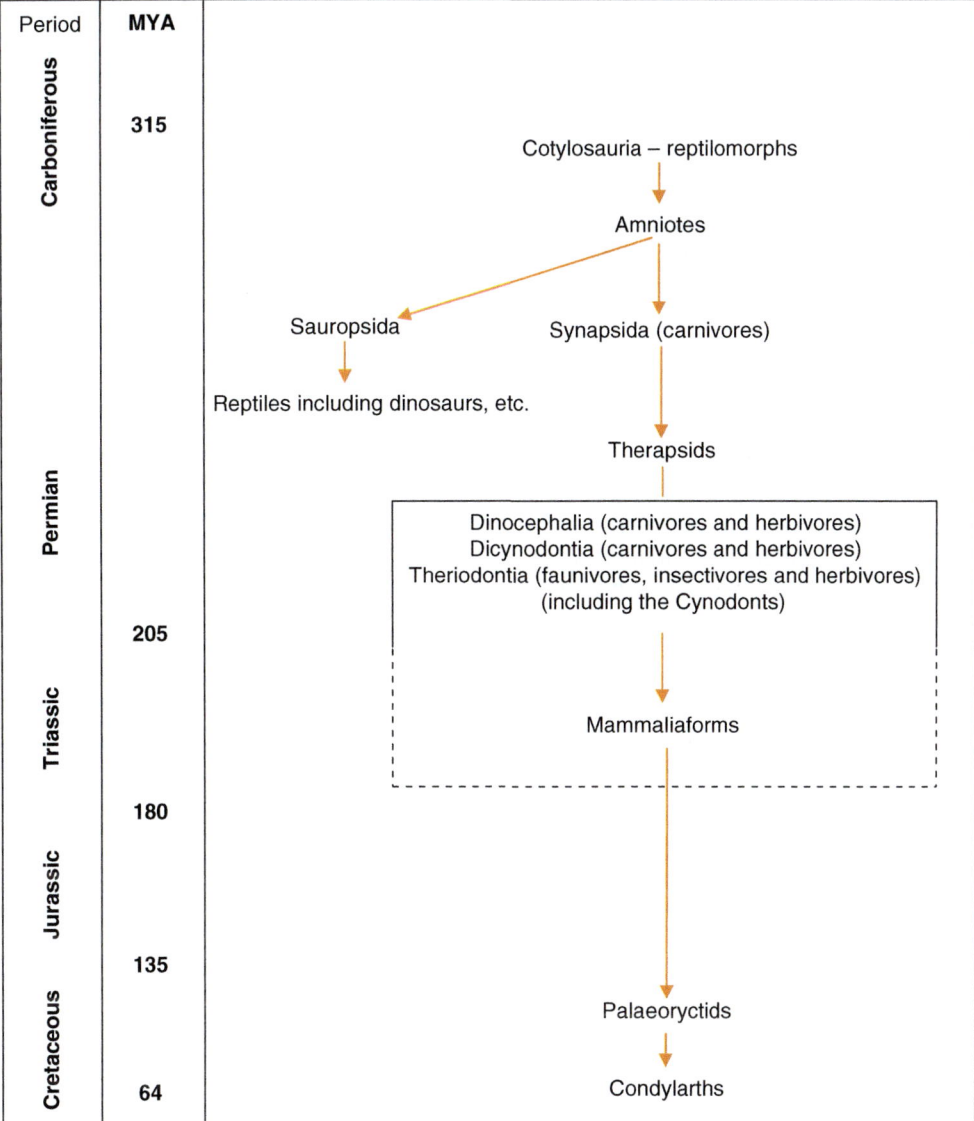

Fig. 1.2. Evolution of mammals and carnivorous and herbivorous feeding styles in the Palaeozoic and Mesozoic eras. MYA = million years ago. (From McKenna, 1969; Durand, 2005.)

into different types such as incisors and molars, had begun to develop at this time. The Cynodonts (one of the Therapsid groups) were the first animals to develop a secondary palate separating the respiratory and alimentary tracts. This allowed them to chew their food, a typically mammalian behaviour, rather than being forced to swallow it whole (Durand, 2005). The Therapsids gave rise to early mammals (Mammaliaforms) in the Triassic period

and these were the progenitors of the mammals which appeared in the Jurassic period.

The evolution of modern mammalian feeding

Mammaliaforms, the precursors of modern mammals, evolved in the early Jurassic period (Luo *et al.*, 2001). To digress: dinosaurs arose in the Triassic period and became extinct in the late Cretaceous/

Paleogene periods, so a discussion of their feeding habits is not relevant to the evolution of the feeding behaviour of modern animals. However, if you are interested in dinosaurs, read Farlow (1987) for some fascinating speculations about their diets, and the ways in which they may have digested them.

The Mesozoic era, and especially the Jurassic and Cretaceous periods, saw the appearance of several groups of mammals, including the Morganucodontidae, Docodonta, Eutriconodontia, Multituberculata, Monotremata, Symmetrodonta, Dryolestida, Metatheria and Eutheria (Meng *et al.*, 2006). Only the Monotremata, Metatheria and Eutheria have survived to the present. Mammals in the Mesozoic were small (2 g to 14 kg; Luo *et al.*, 2001; data cited by Meng *et al.*, 2006) and ate insects or fruits, or both, and there is evidence that some were carnivorous (Meng *et al.*, 2004).

The chewing teeth (molars and premolars) of Mesozoic mammals were of two main types: simple, shearing teeth, e.g. the three-pointed teeth of the Eutriconodonts and the Symmetrodonts; and the tribosphenic teeth of the Multituberculates and Therians. Further, all these animals had evolved some degree of differentiation between teeth at the front of the mouth which were possibly more used for prehension and those in the cheek (molariform teeth) which had evolved some degree of specialization for chewing. There were three important developments in eating and chewing mechanisms which first appeared during the Mesozoic era:

1. A molariform tooth structure which allowed simultaneous shearing and crushing movements. This 'tribosphenic' tooth pattern gave much greater efficiency in comminuting food; it is present in Eutherian and Metatherian fossils dating from 125 MYA (Meng *et al.*, 2006). There are less derived versions of this type of chewing tooth in earlier animals, such as the Multituberculates. We can assume, with reference to modern digestive physiology, that comminution of food allowed more complete (or faster) digestion and a faster rate of passage of undigested food through the digestive tract.

2. An angular process on the distal end of the lower jaw which in modern animals is the site for the attachment of the masseter muscle. This allows a more forceful biting action and is also present in the early Metatherian and Eutherian fossils (Meng *et al.*, 2006).

3. A capacity for the lower jaw to move in relation to the upper jaw as the animal chewed its food.

Although some dinosaurs were able to move their jaws relative to each other, this capacity first appeared in mammals in the Multituberculates (Meng *et al.*, 2006); the lower jaw moved backwards as the jaws came together during a bite. Modern herbivores also have this ability to move their lower jaw during chewing, although in the modern animal the jaw moves sideways.

The progenitors of several orders of modern mammals (including the Primates, Lagomorphs, Carnivores, Cetartiodactyls and Perissodactyls) arose about 100 MYA in the mid-Cretaceous period (Kumar and Hedges, 1998). These included the full range of modern feeding styles – carnivores, herbivores and omnivores. Most of the late Cretaceous/early Palaeocene mammals were small, rodent-like tree-dwellers. They were mostly insectivores although some were herbivorous. 'Modern' carnivores and herbivores, including the perissodactyls (e.g. horses, tapirs and rhinoceroses) and artiodactyls (e.g. cattle, deer, sheep, goats and pigs), arose during the Eocene epoch of the Cainozoic era. A detailed description of mammal evolution through the Cainozoic era is given by Agustí and Antón (2002).

Evolution of modern herbivory

Modern mammals have no endogenous enzymes which can digest the cellulose, hemicellulose or lignin of plant cell walls. This means that the feeding behaviour of grazing and intermediate-feeding herbivores must have co-evolved with a symbiosis with microorganisms. From a survey of methane production in 253 vertebrates, Hackstein and van Alen (1996) have postulated that gut-living methanogens and highly segmented gut structures like caeca and rumens occur together. This further suggests that the symbiosis between herbivores and microbes which we see in modern herbivores co-evolved with highly specialized, segmented digestive tracts. As noted previously, the earliest evidence of the evolution of herbivory was in the late Carboniferous period, about 295 MYA. Archaea, the modern methanogens (Bapteste *et al.*, 2005; Rother and Metcalf, 2005), and possibly early eukaryotes (Brocks *et al.*, 1999), had evolved many millions of years before that time. There is no direct fossil evidence about how the symbiosis between animals and microorganisms evolved. It could have developed simply through the ingestion of microorganisms with food. Sues and Reisz

(1998) have suggested that insectivores could have consumed 'ready-made' gut microorganisms inhabiting arthropods which had developed a herbivorous feeding habit in the early Carboniferous period.

The derivation of herbivorous and omnivorous ungulates through the Cainozoic era (65 MYA to the present day) is shown in Fig. 1.3. Judging from the evidence of fossil teeth, grazing as a feeding behaviour evolved from the early insectivores and concentrate selectors. Hofmann (1989) has suggested that ruminant digestion (i.e. foregut fermentation with the capacity for rumination) is the most highly evolved form of digestive physiology, and that grazing, rather than concentrate selection, is the most 'advanced' feeding style. Certainly, the enzyme systems needed to digest the proteins, lipids and non-fibrous carbohydrates in seeds, fruits and young shoots (the typical diet of a concentrate selector) do not require the evolution of a symbiosis

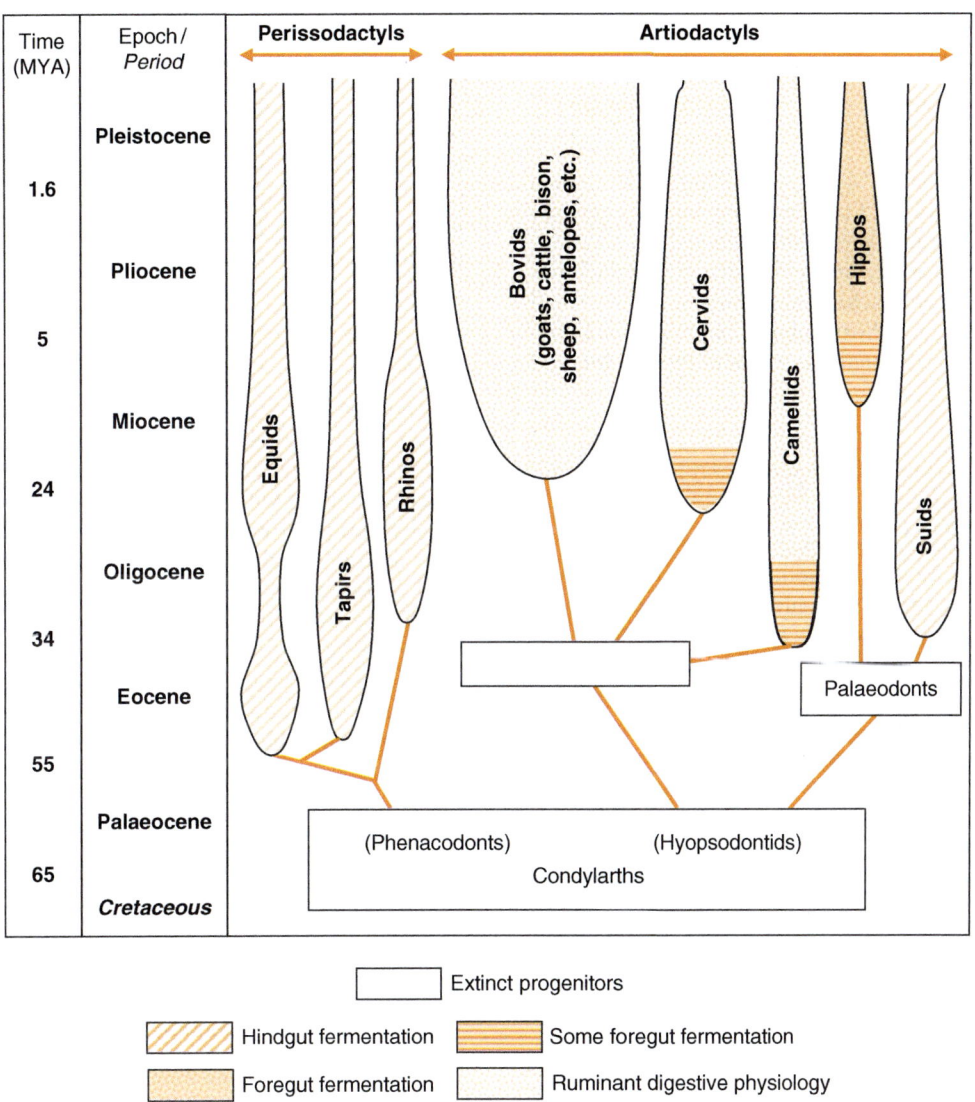

Fig. 1.3. Evolution of modern ungulate digestive strategies during the Cainozoic era. Several families of the Ruminantia are not shown. MYA = million years ago. (From McKenna, 1969; Janis, 1976; Potts and Behrensmeyer, 1992; Agustí and Antón, 2002.)

with microorganisms or of complex mechanisms to retain undigested plant fibre where it can be digested by microorganisms. Several other authorities (e.g. Janis, 1976; Demment and Van Soest, 1985) have suggested that it is erroneous to ascribe 'modernity' to foregut fermentation. We have no fossil evidence for ruminant physiology per se but if we assume that fossils with a ruminant leg structure infer a ruminant digestive physiology, then ruminants evolved some time during the end of the Miocene epoch (Agustí and Antón, 2002).

Feeding Styles and Digestive Physiologies of Modern Animals

Anatomical and physiological adaptations among modern carnivores

Carnivores have sharp, occluding teeth which allow them to hold captured prey and to chew the body tissues into pieces small enough to swallow. They have a short, simple digestive tract which has a glandular stomach; a short, large-diameter small intestine; and a short, tubular large intestine consisting of a very small caecum and short colon and rectum.

Herbivores are either foregut or hindgut fermenters

The foregut fermenters include the macropods (e.g. kangaroos, wallabies), the camelids (e.g. camels, llamas, alpacas), some suids (hippopotamus) and the ruminants (e.g. sheep, cattle, deer). All these animals have a large organ anterior to the gastric (or 'simple') stomach where microbial fermentation occurs. This allows them to benefit from the microbial digestion of forage cell walls before food passes into the stomach and lower digestive tract for further digestion. Diagrams of foregut fermenters' and ruminants' stomachs are in Hofmann (1973) and Langer (1988). Ruminants 'ruminate', i.e. they re-chew undigested forage so that they can increase the rate and extent of digestion in the rumen and promote the escape of undigested and possibly indigestible food pieces from the rumen.

Hindgut fermenters include the horse and koala as typical examples. Hindgut fermenters have a greatly enlarged caecum, or a greatly enlarged colon, or both (for diagrams see Langer, 1988). They subject their food to gastric and small intes-

tinal digestion before bacteria attack the plant cell wall and any remaining cell contents. This allows hindgut fermenters to obtain the benefit of any superior amino acid and/or fatty acid composition of the food, if the food has those qualities. Hindgut fermenters are less able to fully digest fibre (plant cell walls) and obtain the B vitamins and amino acids which result from the metabolism of food by microorganisms. Hindgut fermenters, like the non-ruminant foregut fermenters, have faster rates of passage of undigested food than ruminants. This can reduce the extent of food digestion, but because difficult-to-digest fibrous material is not retained in the digestive tract, a hindgut digestion system can allow these animals to eat larger amounts of less-digestible forages than ruminants. Both the horse (Drogoul *et al.*, 2000) and the koala (Hume, 1999) are able to selectively retain small particles and liquids while voiding larger, possibly indigestible plant fragments.

Mouth and tooth shapes are related to feeding style

A study of bears (Ursidae) by Sacco and van Valkenburgh (2004) summarizes many of the essential differences in mouth and teeth morphology of animals with different feeding styles. The herbivorous giant panda has large molars and can exert a large shearing action on its tough bamboo food. Carnivorous bears (e.g. the polar bear) have smaller molars, similar to other carnivorous species, but have relatively small carnassial teeth (these are enlarged molars or premolars which are adapted to cutting flesh and bone). The absence of large carnassials may be related to the relative size of their preferred prey. Omnivorous bears (black, sun and Andean bears) have an intermediate tooth morphology.

Three indexes of tooth and mouth shape are the hypsodontic index (HI), the relative muzzle width (RMW) and the incisor width ratio (IWR), as defined below and in Fig. 1.4.

$$\text{HI} = \text{(height of third molar)}/\text{(length of second molar)} \qquad (1.1)$$

$$\text{RMW} = \text{(palatal width)}/\text{(muzzle width)} \qquad (1.2)$$

$$\text{IWR} = \text{(width of central incisor)}/\text{(width of third incisor)} \qquad (1.3)$$

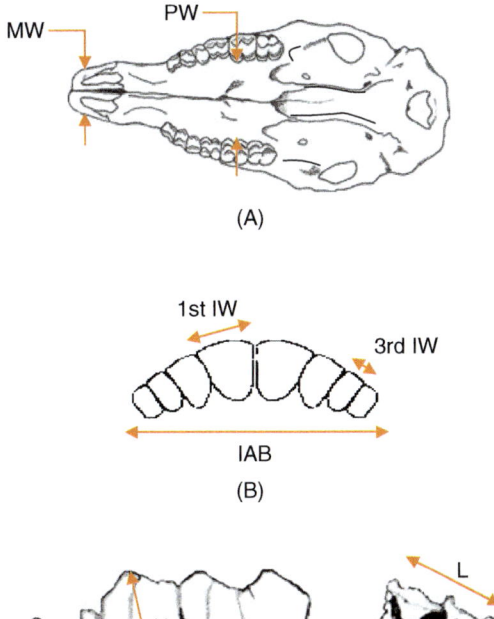

Fig. 1.4. Measurements of mouth and tooth morphology: (A) muzzle width (MW) and palatal width (PW) – the diagram shows the upper jaw; (B) incisor width (IW) and incisor arcade breadth (IAB) – lower jaw; (C) molar height (H) and length (L).

Within the herbivores, the shapes of the mouth and teeth are related to the type of food an animal eats, and the way in which it prehends it. Grazers have developed tall molars that can cope better with the wear resulting from macerating plant material which is fibrous, sometimes impregnated with silica and often contaminated with soil. Concentrate selectors have smaller muzzles and larger central incisors which allow closer cropping of the desired plant parts. Characteristic differences between grazing and concentrate-selecting herbivores are illustrated by data reported by Janis and Ehrhardt (1988). However, too broad an analysis of anatomical differences may confound the effects of phylogeny with those of feeding type. The Bovidae data of Janis and Ehrhardt (1988) give these average results:

1. Grazers have somewhat taller molar teeth; this shape gives them an HI of about 1.8, compared to 1.4 and 1.3 for intermediate feeders and concentrate selectors, respectively.
2. The incisor teeth of concentrate selectors are different in size, so their IWR is about 4.1, which is higher than intermediate feeders (2.7) and grazers (1.9).
3. Grazers have broader muzzles (RMW of 1.0) than the other types (1.4).

These characteristics are nicely illustrated by Janis and Ehrhardt (1988) and Van Soest *et al.* (1995), who showed that grazers (bovids and equids) cluster into groups with high HI and low RMW while concentrate selectors and intermediate feeders form a cluster with higher RMW and lower HI (Fig. 1.5). RMW is not constant over an animal's life. Younger deer have relatively narrower muzzles than older ones (Illius and Gordon, 1990; Dryden and Bisselling, 1999).

The shape and size of a grazing animal's mouth influence the amount of food it can prehend. The conceptual relationships between the incisor arcade breadth (IAB) and daily food (dry matter) intake are summarized in Fig. 1.6. If we accept that palatal width is an index of an animal's nutrient requirements because it varies isometrically with body size, and that the IAB can be an index of food intake (Fig. 1.6), then those animals that have smaller RMW are more likely to be undernourished if they are forced to eat fibrous, less-digestible plant material. More likely to be at risk are young herbivores and the concentrate selectors.

Concentrate selector and grazer herbivores have different stomachs and intestines

From dissections of a range of herbivores, Hofmann (1989) concluded that concentrate selectors (he used the roe deer as his example) and grazers (the African buffalo) differed in some systematic ways. These differences are summarized in Table 1.2. Concentrate selectors have a simpler and smaller rumen and larger liver, salivary glands and distal fermentation chamber than grazers.

Ecology, Diet and Digestive Physiology

Modern herbivorous animals have evolved four main ways of matching their energy (or nutrient) needs with the consumption of plant food:

1. Increasing body size, because energy requirements scale allometrically with body weight, while

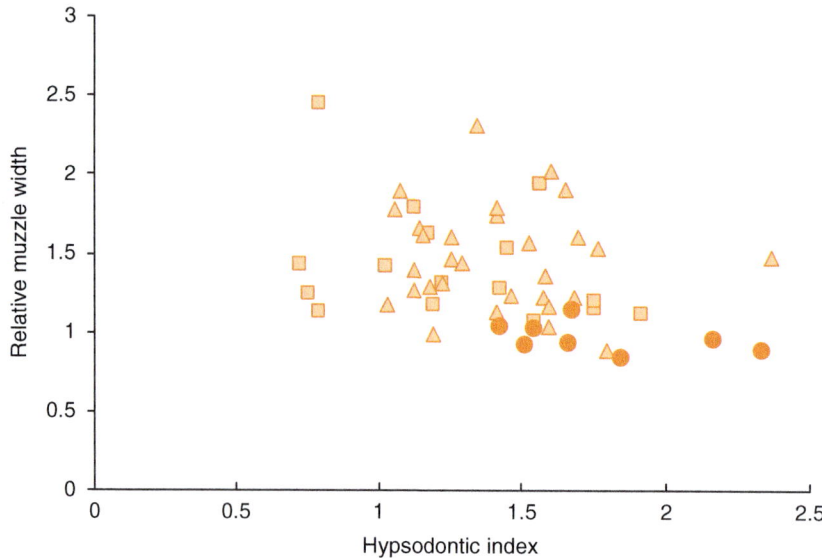

Fig. 1.5. Relationships between hypsodontic index and muzzle width ratio for the Bovidae (●grazers, ☐concentrate selectors, △intermediate feeders). (Calculated from the Bovidae data of Janis and Ehrhardt, 1988.)

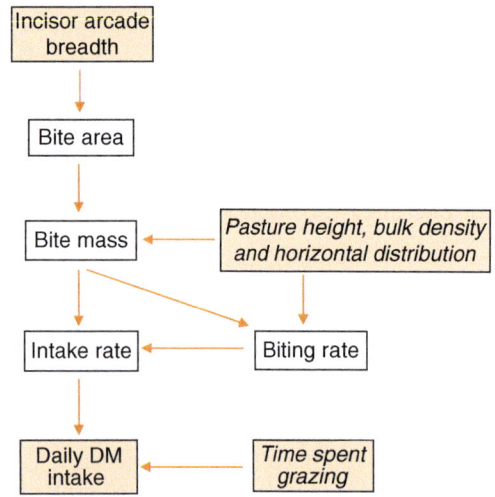

Fig. 1.6. Relationships between mouth size and pasture density, and food intake by grazing animals. (Based on Illius and Gordon, 1999.)

digestive tract capacity scales isometrically (Demment and Van Soest, 1985; Illius and Gordon, 1992) so that as the animal gets bigger its energy requirement decreases in relation to each unit of body weight while its digestive tract capacity increases at the same rate as its body size.

2. Being concentrate selectors, i.e. eaters of seeds, fruits and immature vegetative plant parts, which are rich in non-fibrous constituents that can be digested by mammalian enzymes.

3. Developing hindgut fermentation and not constricting the passage of undigested fibrous residues; this allows the easy elimination of undigested fibrous material and therefore a larger consumption of plant material.

4. Developing a ruminant digestive tract (foregut fermentation plus the capacity for rumination), because this allows the animal to maximize the extraction of digestible constituents from vegetative plant material.

Plant food is, by nature, fibrous. Plant 'fibre' is synonymous with plant cell walls. The large polysaccharides cellulose and hemicellulose, plus the indigestible substance lignin, are the main constituents of grass cell walls. The proportion of cell wall to cell contents increases as the plant matures. The proportion is greater in old leaf than young leaf, in stem than leaf and in vegetative material than seeds or fruits. As we have seen, neither cellulose nor hemicellulose can be digested by any endogenous mammalian enzyme. Animals whose diets are rich in plant material, especially the intermediate feeders and grazers which eat a leaf-rich diet rather than the seed- or fruit-rich diet eaten by concentrate selectors, need a symbiotic microbial population to assist in the digestion of this main food component. In turn, this creates a

Table 1.2. Anatomical differences between concentrate selecting and grazing herbivores. (From Janis and Erhardt, 1988; Hofmann, 1989; Snipes and Kriete, 1991.)

Characteristic	Concentrate selectors	Grazers
Stomach		
Rumen size, shape	Small, simple shape	Large, 2–3 blind sacs
Rumen suspension	No dorsal attachment, complete contraction of dorsal rumen	Dorsal attachment present, limited contraction of dorsal rumen
Rumen papillae	Dense and evenly distributed	Large ventrally, small dorsally
SEF[a]	Changes with seasons (20 in spring, 4.5 in winter)	1.5–3
Reticulum	Small	Large
Omasum	Small, few laminae	Larger, many laminae
Intestines		
Total length	12–15 times body length	25–30 times body length
Spiral colon length	24% of total intestine	18% of total intestine
Length ratio	Large intestine/small intestine = 7:15	Large intestine/small intestine = 15:7°
Volume ratio	Rumen/intestine = 10:1	Rumen/intestine = 30:1
Teeth and mouth morphology		
Incisor width ratio	About 3.2	About 1.9
Salivary glands	0.22% of body weight	0.06% of body weight
Liver		
Liver	Large	Small
Behavioural characteristics		
Feeding frequency	8–12 periods/day	5–6 periods/day
Rate of digesta passage	Fast rate	Slower rate

[a]SEF = surface enlargement factor (ratio of total surface area to area of basal membrane).

need for a capacious digestive tract, because the digestion of plant fibre is slow (about 2–10% per hour; Sniffen *et al.*, 1992), the digestibility of plant matter is not always high (from 45% to 75%), and so forage intakes are necessarily large.

Plasticity in Feeding Behaviour

Animals do not always eat what we expect them to. For example, the brush-tailed possum, which is considered to be a generalist folivore, 'combines eating of leaves with more readily digestible fruits and flowers, as well as fungi, invertebrates, birds eggs, and even meat' (Nugent *et al.*, 2001). Even when animals seem to confine themselves to their 'typical' diets, we find that they select within that dietary range. So we find that grazing cattle select green, young leaves or particular plant species according to their availability in different seasons (Lorimer, 1978). The bilby (*Macrotis lagotis*) is an opportunistic feeder which appears to select some dietary items almost randomly (Gibson, 2001) and Dasyurids maximize their energy gain from eating by choosing the largest cockroaches

that they can comfortably catch (Fisher and Dickman, 1993). A useful point has been made by Sprent and McArthur (2002) in relation to herbivore diets. They suggested that it is more sensible to compare diet selection, rather than diet composition. In their study of pademelons and wallabies they could not demonstrate differences in diet composition, breadth or overlap between these two sympatric ('living in the same place') species, but there were significant differences in the selection pressure exerted by these species, i.e. the proportions of grasses and forbs in the diets in comparison to the proportions of these plant types in the environment.

There is also physiological plasticity in food intake and digestion. Ruminants may adapt to seasonal changes in food supplies and quality (increasing fibrosity and decreasing digestibility as plants become more mature), or increased nutrient demand, by increasing the size of their rumens and altering rumen passage rates. Camels, sheep, goats and cattle increase the retention time of undigested food in the rumen when eating mature, fibrous plants (Lechner-Doll *et al.*, 1990; Aharoni *et al.*, 2004). Cows eat more food and retain digesta particles for a shorter time when lactating (Aharoni *et al.*, 2004). Some animals, e.g. male red deer (*Cervus elaphus*), have seasonal variations in food intake and digesta kinetics which are unrelated to food quality changes. Freudenberger *et al.* (1994) reported increased rumen volume and retention time in summer, which was not related to changes in food quality, and which they postulated would maintain fibre digestibility in a period of higher food consumption.

Although it seems that some species can expand their acceptable range of diet ingredients to quite a large extent, we can find numerous examples of animals being unable to tolerate unconventional ingredients, especially if they are used for long periods:

1. Lot-fed beef cattle fed grain-rich rations suffer from disorders such as acidosis caused by the excess production of volatile fatty acids accompanied by a reduced rumen pH and proliferation of lactate-producing bacteria (Martin *et al.*, 2006), and a range of disorders including rumenitis, rumen hyperkeratosis and liver abscess.
2. Dairy cattle must be given diets with sufficient (about 21%) digestible fibre, otherwise their milk fat content falls to unacceptably low levels (NRC, 2001) through the effect of low rumen pH on ruminal trans fatty acid production.
3. Horses fed grains that are rich in resistant starch may suffer from laminitis, which is associated with the proliferation of bacteria in the large intestine, eventually accompanied by damage to the laminae of the hoof (Milinovich *et al.*, 2006); on the other hand, they must be given diets which contain substantial amounts of cereal grains (e.g. up to 40% or 50%) or fats (up to about 8%) if they are to consume enough energy to support the activity levels required in racing and other equine sports.
4. Rusa deer will self-select a diet of up to 88% concentrate (cereal grain plus oilseed meal) but will suffer from diarrhoea if they are rapidly introduced to a concentrate-rich diet (Puttoo *et al.*, 1998).
5. Cats, which are carnivores, can be fed successfully on foods that contain large amounts of cereal grains and other ingredients derived from plants provided that taurine and other amino acids and retinol are included from a 'synthetic' source (Crissey *et al.*, 1997).

The Practical Uses of Nutritional Classification

Nutritional ecologists can use information on feeding styles to assess the carrying capacity of an ecosystem, e.g. to assess the amount of competition between kangaroos and cattle on Australian cattle properties (Wilson, 1991) or to develop suitable diets for captive wild animals. Animal production scientists can use nutritional classification to predict the performance of animals in novel feeding management, e.g. the lot-feeding of sheep (Bowen *et al.*, 2006); or starch-rich diets for dairy cattle (Knowlton, 2001); or the types of diet a newly domesticated species would perform best on such as farming systems for deer (Puttoo *et al.*, 1998; Tuckwell, 2003), reindeer (Eilertsen *et al.*, 2001), bison (Gegner, 2001), alpaca (McGregor, 2002) and crocodiles (Peucker *et al.*, 2005).

References

Agustí, J. and Antón, M. (2002) *Mammoths, Sabertooths and Hominids*. Columbia University Press, New York.

Aharoni, Y., Brosh, A., Orlov, A., Shargal, E. and Gutman, A. (2004) Measurements of energy balance of grazing beef cows on Mediterranean pasture, the effects of stocking rate and season – 1. Digesta kinetics, faecal output and digestible dry matter intake. *Livestock Production Science* 90, 89–100.

Alberts, B., Johnson, A., Lewis, J., Raff, M., Roberts, K. and Walter, P. (2002) *Molecular Biology of the Cell*, 4th edn. Garland Science, New York.

Baldi, R., Pelliza-Sbriller, A., Elston, D. and Albon, S. (2004) High potential for competition between guanacos and sheep in Patagonia. *Journal of Wildlife Management* 68, 924–938.

Bapteste, É., Brochier, C. and Boucher, Y. (2005) Higher-level classification of the Archaea: evolution of methanogenesis and methanogens. *Archaea* 1, 353–363.

Bowen, M.K., Ryan, M.P., Jordan, D.J., Beretta, V., Kirby, R.M., Stockman, C., McIntyre, B.L. and Rowe, J.B. (2006) Improving sheep feedlot management. *International Journal of Sheep and Wool Science* 54, 27–34.

Brocks, J.J., Logan, G.A., Buick, R. and Summons, R.E. (1999) Archean molecular fossils and the early rise of eukaryotes. *Science* 285, 1033–1036.

Capitani, C., Bertelli, I., Varuzza, P., Scandura, M. and Apollonio, M. (2004) A comparative analysis of wolf (*Canis lupus*) diet in three different Italian ecosystems. *Mammalian Biology* 69, 1–10.

Carvalho, F.M.V., Fernandez, F.A.S. and Nessimian, J.L. (2005) Food habits of sympatric opossums coexisting in small Atlantic forest fragments in Brazil. *Mammalian Biology* 70, 366–375.

Crissey, S.D., Swanson, J.A., Lintzenich, B.A., Brewer, B.A. and Slifka, K.A. (1997) Use of a raw meat-based diet or a dry kibble diet for sand cats (*Felis margarita*). *Journal of Animal Science* 75(8), 2154–2160.

Curtis, N.E., Massey, S.E. and Pierce, S.K. (2006) The symbiotic chloroplasts in the sacoglossan *Elysia clarki* are from several algal species. *Invertebrate Biology* 125, 336–345.

Damuth, J.D. (1992) Taxon-free characterization of animal communities. In: Behrensmeyer, A.K., Damuth, J.D., DiMichele, W.A., Potts, R., Sues, H.-D. and Wing, S.L. (eds) *Terrestrial Ecosystems Through Time*. Chicago University Press, Chicago, Illinois.

Demment, M.W. and Van Soest, P.J. (1985) A nutritional explanation for body-size patterns of ruminant and nonruminant herbivores. *American Naturalist* 125, 641–672.

Donoghue, P.C.J. and Purnell, M.A. (1999) Mammal-like occlusion in conodonts. *Paleobiology* 25, 58–74.

Drogoul, C., Poncet, C. and Tisserand, J.L. (2000) Feeding ground and pelleted hay rather than chopped hay to ponies. 1. Consequences for *in vivo* digestibility and rate of passage of digesta. *Animal Feed Science and Technology* 87, 117–130.

Dryden, G.McL. and Bisselling, I. (1999) Mouth structure and dentition of red (*Cervus elaphus*) and rusa (*Cervus timorensis*) deer, and implications for nutritional management. In: Stone, G., Forbes, T.D.A, Stuth, J.W. and Byers, F.M. (eds) *Nutritional Ecology of Herbivores*. April 11–16, 1999. San Antonio, Texas. Online poster. Available at: http://cnrit.tamu.edu/conf/isnh/post-online/post0005/

Durand, J.F. (2005) Major African contributions to Palaeozoic and Mesozoic vertebrate palaeontology. *Journal of African Earth Sciences* 43, 53–82.

Dzik, J. (2004) Anatomy and relationships of the early Cambrian worm *Myoscolex*. *Zoologica Scripta* 33, 57–69.

Eilertsen, S.M., Schelderup, I., Dryden, G.McL. and Mathiesen, S.D. (2001) High protein pastures in spring – effects on body composition in reindeer. *Rangifer* 21, 13–19.

Evans, M.C., Macgregor, C. and Jarman, P.J. (2006) Diet and feeding selectivity of common wombats. *Wildlife Research* 33, 321–330.

Farlow, J.O. (1987) Speculations about the diet and digestive physiology of herbivorous dinosaurs. *Paleobiology* 13, 60–72.

Fisher, D.O. and Dickman, C.R. (1993) Body-size prey size relationships in insectivorous marsupials – tests of 3 hypotheses. *Ecology* 74, 1871–1883.

Freudenberger, D.O., Toyakawa, K., Barry, T.N., Ball, A.J. and Suttie, J.M. (1994) Seasonality in digestion and rumen metabolism in red deer (*Cervus elaphus*) fed on a forage diet. *British Journal of Nutrition* 71, 489–499.

Gegner, L. (2001) *Bison Production. Livestock Production Guide*. ATTRA-National Sustainable Agriculture Information Service. Publ. No. IP151. Available at: http://attra.ncat.org/attra-pub/PDF/bison.pdf

Gibson, L.A. (2001) Seasonal changes in the diet, food availability and food preference of the greater bilby (*Macrotis lagotis*) in south-western Queensland. *Wildlife Research* 28, 121–134.

Glen, A.S. and Dickman, C.R. (2006) Diet of the spotted-tailed quoll (*Dasyurus maculatus*) in eastern Australia: effects of season, sex and size. *Journal of Zoology* 269, 241–248.

Gursky, S. (2000) Effect of seasonality on the behavior of an insectivorous primate. *Tarsius spectrum. International Journal of Primatology* 21, 477–495.

Hackstein, J.H.P. and van Alen, T.A. (1996) Fecal methanogens and vertebrate evolution. *Evolution* 50, 559–572.

Hofmann, R.R. (1973) *The Ruminant Stomach: Stomach Structure and Feeding Habits of East African Game Ruminants*. East African Literature Bureau, Nairobi, Kenya.

Hofmann, R.R. (1989) Evolutionary steps of ecophysiological adaptation and diversification of ruminants: a comparative view of their digestive system. *Oecologia* 78, 443–457.

Hofmann, R.R. and Stewart, D.R.M. (1972) Grazer or browser: a classification based on the stomach structure and feeding habits of East African ruminants. *Mammalia* 36, 226–240.

Holland, N.D. and Chen, J.Y. (2001) Origin and early evolution of the vertebrates: new insights from advances in molecular biology, anatomy, and palaeontology. *Bioessays* 23, 142–151.

Hovens, J.P.M. and Tungalaktuja, K. (2005) Seasonal fluctuations of the wolf diet in the Hustai National Park (Mongolia). *Mammalian Biology* 70, 210–217.

Hulbert, I.A.R. and Andersen, R. (2001) Food competition between a large ruminant and a small hindgut fermentor: the case of the roe deer and mountain hare. *Oecologia* 128, 499–508.

Hume, I.D. (1999) *Marsupial Nutrition*. Cambridge University Press, Cambridge.

Illius, A.W. and Gordon, I.J. (1990) Variation in foraging behaviour in red deer and the consequences for population demography. *Journal of Animal Ecology* 59, 89–101.

Illius, A.W. and Gordon, I.J. (1992) Modelling the nutritional ecology of ungulate herbivores: evolution of body size and competitive interactions. *Oecologia* 89, 428–434.

Illius, A.W. and Gordon, I.J. (1999) The physiological ecology of mammalian herbivory. In: Jung, H.-J.G. and Fahey, G.C. (eds) *Nutritional Ecology of Herbivores. Proceedings of the V International Symposium on the Nutrition of Herbivores*. American Society of Animal Science, Savoy, Illinois, pp. 71–96.

International Commission on Stratigraphy (ICS) (2007) *International Stratigraphic Chart*. Available at: http://www.stratigraphy.org/cheu.pdf

Janis, C. (1976) The evolutionary strategy of the equidae and the origins of rumen and cecal digestion. *Evolution* 30, 757–774.

Janis, C.M. and Ehrhardt, D. (1988) Correlation of relative muzzle width and relative incisor width with dietary preference in ungulates. *Zoological Journal of the Linnean Society* 92, 267–284.

Jolly, C.J. (2001) A proper study for mankind: analogies from the papionin monkeys and their implications for human evolution. *Yearbook of Physical Anthropology* 44, 177–204.

Jones, M.E. (1998) The function of vigilance in sympatric marsupial carnivores: the eastern quoll and the Tasmanian devil. *Animal Behaviour* 56, 1279–1284.

Keiper, P. and Johnson, C.N. (2004) Diet and habitat preference of the Cape York short-nosed bandicoot (*Isoodon obesulus peninsulae*) in north-east Queensland. *Wildlife Research* 31, 259–265.

Knowlton, K.F. (2001) High grain diets for dairy cattle. *Recent Advances in Animal Nutrition in Australia* 16, 19–28.

Kruger, S.C., Lawes, M.J. and Maddock, A.H. (1999) Diet choice and capture success of wild dog (*Lycaon pictus*) in Hluhluwe-Umfolozi Park, South Africa. *Journal of Zoology* 248, 543–551.

Kumar, S. and Hedges, S.B. (1998) A molecular timescale for vertebrate evolution. *Nature* 392, 917–920.

Langer, P. (1988) *The Mammalian Herbivore Stomach: Comparative Anatomy, Function and Evolution.* Gustav Fischer, Stuttgart, Germany.

Lechner-Doll, M., Rutagwenda, T., Schwartz, H.J., Schultka, W. and von Englehardt, W. (1990) Seasonal changes of ingesta mean retention time and forestomach fluid volume in indigenous camels, cattle, sheep and goats grazing a thornbush savannah pasture in Kenya. *Journal of Agricultural Science, Cambridge* 115, 409–420.

Lentle, R.G., Hume, I.D., Stafford, K.J., Kennedy, M., Springett, B.P., Browne, R. and Haslett, S. (2004) Temporal aspects of feeding events in tammar (*Macropus eugenii*) and parma (*Macropus parma*) wallabies. I. Food acquisition and oral processing. *Australian Journal of Zoology* 52, 81–95.

Lorimer, M. (1978) Forage selection studies. 1. The botanical composition of forage selected by sheep grazing *Astrebla* spp. pastures in north-west Queensland. *Tropical Grasslands* 12, 97–108.

Luo, Z.X., Crompton, A.W. and Sun, A.L. (2001) A new mammaliaform from the early Jurassic and evolution of mammalian characteristics. *Science* 292, 1535–1540.

McGregor, B.A. (2002) Comparative productivity and grazing behaviour of Huacaya alpacas and Peppin Merino sheep grazed on annual pastures. *Small Ruminant Research* 44, 219–232.

McKenna, M.G. (1969) The origin and early differentiation of therian mammals. *Annals of the New York Academy of Sciences* 167, 217–240.

Mallat, J. (1996) Ventilation and the origin of jawed vertebrates: a new mouth. *Zoological Journal of the Linnean Society* 117, 329–404.

Martin, C., Brossard, L. and Doreau, M. (2006) Mechanisms of appearance of ruminal acidosis and consequences on physiopathology and performances. *Productions Animales* 19, 93–107.

Meng, Q.J., Liu, J.Y., Varricchi, D.J., Huang, T. and Gan, C.L. (2004) Palaeontology: parental care in an ornithischian dinosaur. *Nature* 431, 145–146.

Meng, J., Hu, Y., Li, C. and Wang, Y. (2006) The mammal fauna in the early Cretaceous Jehol Biota: implications for diversity and biology of Mesozoic animals. *Geological Journal* 41, 439–463.

Milinovich, G.J., Trott, D.J., Burrell, P.C., van Eps, A.W., Thoefner, M.B., Blackall, L.L., Al Jassim, R.A.M., Morton, J.M. and Pollitt, C.C. (2006) Changes in equine hindgut bacterial populations during oligofructose-induced laminitis. *Environmental Microbiology* 8, 885–898.

Munn, A.J. and Dawson, T.J. (2006) Forage fibre digestion, rates of feed passage and gut fill in juvenile and adult red kangaroos *Macropus rufus* Desmarest: why body size matters. *Journal of Experimental Biology* 209, 1535–1547.

National Research Council (2001) *Nutrient Requirements of Dairy Cattle*, 7th rev. edn. National Academies Press, Washington, DC.

Nugent, G., Fraser, W. and Sweetapple, P. (2001) Top down or bottom up? Comparing the impacts of introduced arboreal possums and 'terrestrial' ruminants on native forests in New Zealand. *Biological Conservation* 99, 65–79.

Peucker, S.K.J., Davis, B.M. and van Barneveld, R.J. (2005) *Crocodile Farming Research: Hatching to Harvest*. Publ. No. 05/152. Rural Industries Research and Development Corporation, Canberra, Australia.

Potts, R. and Behrensmeyer, A.K. (1992) Late Cenozoic terrestrial ecosystems. In: Behrensmeyer, A.K., Damuth, J.D., DiMichele, W.A., Potts, R., Sues, H.-D. and Wing, S.L. (eds) *Terrestrial Ecosystems Through Time: Evolutionary Paleoecology of Terrestrial Plants and Animals*. University of Chicago Press, Chicago, Illinois, pp. 419–541.

Purnell, M.A. (1995) Microwear in conodont elements and macrophagy in the first vertebrates. *Nature* 374, 798–800.

Puttoo, K., Dryden, G.McL. and McCosker, J.E. (1998) Performance of weaned rusa (*Cervus timorensis*) deer given concentrates of varying protein content with sorghum hay. *Australian Journal of Experimental Agriculture* 38, 33–39.

Ricard, G. *et al.* (2006) Horizontal gene transfer from Bacteria to rumen Ciliates indicates adaptation to their anaerobic, carbohydrates-rich environment. *BMC Genomics* 7, 22.

Rother, M. and Metcalf, W.W. (2005) Genetic technologies for Archaea. *Current Opinion in Microbiology* 8, 745–751.

Sacco, T. and van Valkenburgh, B. (2004) Ecomorphological indicators of feeding behaviour in the bears (Carnivora: Ursidae). *Journal of Zoology* 263, 41–54.

Sansom, I.J., Smith, M.P., Armstrong, H.A. and Smith, M.M. (1994) Dentine in conodonts. *Nature* 368, 591.

Schopf, J.W. (1993) Microfossils of the early Archean Apex Chert: new evidence of the antiquity of life. *Science* 260, 640–646.

Shu, D.G., Morris, S.C. and Zhang, X.L. (1996) A Pikaia-like chordate from the lower Cambrian of China. *Nature* 384, 157–158.

Sniffen, C.J., O'Connor, J.D., Van Soest, P.J., Fox, D.G. and Russell, J.B. (1992) A net carbohydrate and protein system for evaluating cattle diets. II. Carbohydrate and protein availability. *Journal of Animal Science* 70, 3562–3577.

Snipes, R.L. and Kriete, A. (1991) Quantitative investigation of the area and volume in different compartments of the intestines of 18 mammalian species. *Z. Saugetierkunde* 56, 225–244.

Sprent, J.A. and McArthur, C. (2002) Diet and diet selection of two species in the macropodid browser-grazer continuum: do they eat what they 'should'? *Australian Journal of Zoology* 50, 183–192.

Sues, H.-D. and Reisz, R.R. (1998) Origins and early evolution of herbivory in tetrapods. *TREE* 13, 141–145.

Telfer, W.R. and Bowman, D.M.J.S. (2006) Diet of four rock-dwelling macropods in the Australian monsoon tropics. *Austral Ecology* 31, 817–827.

Torstenson, W.L.F., Mosley, J.C., Brewer, T.K., Tess, M.W. and Knight, J.E. (2006) Elk, mule deer, and cattle foraging relationships on foothill and mountain rangeland. *Rangeland Ecology and Management* 59, 80–87.

Tuckwell, C. (2003) *Deer Farming in Australia*. Rural Industries Research and Development Corporation, Canberra, Australia.

Van Soest, P.J., Dierenfeld, E.S. and Conklin, N.L. (1995) Digestive strategies and limitations of ruminants. In: Von Engelhardt, W. *et al.* (eds) *Ruminant Physiology: Digestion, Metabolism, Growth and Reproduction. Proceedings of the Eighth International Symposium on Ruminant Physiology*. Ferdinand Enke Verlag, Stuttgart, Germany, pp. 581–600.

Watson, L.H. and Owen-Smith, N. (2000) Diet composition and habitat selection of eland in semi-arid shrubland. *African Journal of Ecology* 38, 130–137.

Wilson, A.D. (1991) Forage utilization by sheep and kangaroos in a semi-arid woodland. *Rangeland Journal* 13, 81–90.

The Nutritive Value of Animal Foods: Introductory Concepts About Foods, Nutrients and Food Analysis

The Nutritive Value of Foods

Nutritive value is a concept which combines information on the content and availability of nutrients in a foodstuff, with considerations of the characteristic levels of intake, palatability, and the effects of the food on animal health and the quality of animal products (Fig. 2.1).

Most tables of food nutritive value deal only with nutrient content and nutrient availability. These are usually combined into the unifying concept of 'available nutrient content'. Examples are the available phosphorus and ileal-digestible lysine contents of pig diet ingredients (NRC, 1998), digestible energy contents of ingredients for pigs and horses (NRC, 1998, 2007) or ruminally degradable protein contents of ruminant diet ingredients (SCA, 1990; NRC, 2000, 2001).

Food Classification

'Food classification' is the grouping together of foods in classes, so that within a class we have foods which have similar nutritive values and can be used in similar ways. Foods which have similar origins have similar nutritive values. This extends beyond a simple consideration of their nutrient contents to the ways in which they affect animal production and the sort of use that is generally made of them. For example, soybean meal, cottonseed meal and sunflower meal are all by-products from the extraction of oil from oilseeds and are used to increase the protein content of diets; while wheat, maize, triticale and millet are all cereal grains and are rich sources of digestible energy. Not only do foods derived from similar sources have similar nutrient contents, but they also have similar effects on animal health and the quality of animal products, and are used in similar ways when we construct animal diets. Food classification will help predict the quality and potential use of a novel food (if we know what food class it belongs to), or to choose alternative foods for one which is no longer available.

There are several ways to classify animal foods. The International Network of Feed Information Centres (INFIC) identifies eight classes of foods (Harris et al., 1968, 1980):

1. Dry forages and roughages – subdivided into legume and non-legume hays, straws and forage crops either with or without the grain, and other materials with more than 18% crude fibre on a dry matter (DM) basis (this is equivalent to 22–25% acid detergent fibre (ADF), the total of cellulose plus most of the lignin).
2. Pasture, range plants and forages which have not been cut and cured – e.g. fresh pasture and forage crops, and fodder trees (however, this class does include 'stem-cured pasture plants', i.e. those which have become very mature and dry but which have not been cut for haymaking).
3. Silages – subdivided into maize (corn), grass and legume silages (but do not include ensiled cereal grains).
4. Energy foods (these foods contain less than 18% crude fibre and less than 20% protein) – subdivided into cereal grains, milled grain by-products, fruits, nuts and roots.
5. Protein supplements (these foods contain less than 18% crude fibre and 20% or more protein) – subdivided into foods of animal, bird, fish, milk or plant origin.
6. Mineral supplements.
7. Vitamin supplements.
8. Additives – subdivided into a group of non-nutritive substances used in foods, including antibiotics, colouring agents, anabolic agents, rumen modifiers, pellet binders and flavours.

This classification is used in the NRC feeding standards publications, and by some other author-

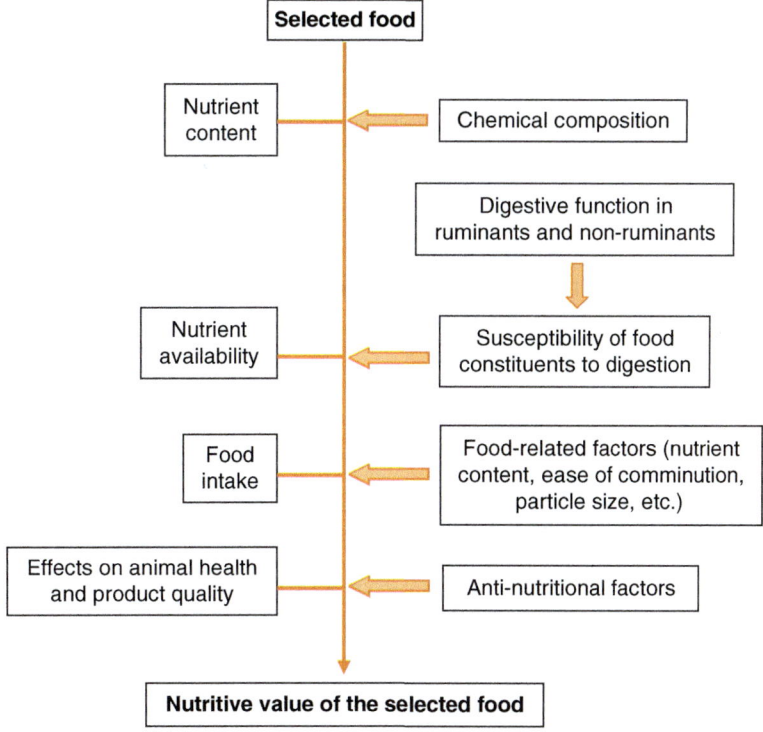

Fig. 2.1. Factors which influence the nutritive value of foods for animals.

ities (e.g. Kayongo-Male and Said, 1986; Ostrowski-Meissner, 1987, 1990). Several other systems have been developed, but no published classification system is completely satisfactory. The system described in Tables 2.1 and 2.2 has some additional features to that described by Harris *et al.* (1968).

The most important distinction is between concentrates and roughages, as the nutrient content of these foods is often very different. The fibre-based definition of roughage includes good-quality forages such as ryegrass (*Lolium* spp.) pasture or forage oats (*Avena sativa*), as well as low-quality foods such as cereal straws, mature pasture and sugarcane bagasse (Table 2.1). It is useful to distinguish between these. The terms 'roughages', 'forages' and 'dry foods' are sometimes erroneously used interchangeably. Roughages are fibrous foods; the term includes forages and also industrial by-products such as sugarcane bagasse or highly processed foods such as alkali-treated straws. Forages are foods which consist of the leaves and stems of plants. Dry or air-dry foods are those which have approximately 90% DM and which are chemically stable under most ambient temperatures and humidities. They include sun-

cured forages such as hays and straws, as well as the cereal grains and oilseed meals.

Energy and protein concentrates are large classes which are best divided into smaller subclasses, such as cereal grains and plant protein meals (Table 2.2). Mineral and vitamin sources are concentrated sources of these nutrients, but they are given separate classifications and are not included in the 'concentrate' food class. A final catch-all class for non-nutritive additives like colourants, flavourings, preservatives, pellet binders, enzymes and rumen modifiers is also needed.

Nutrients and Their Functions

Nutrients are food constituents which are released by digestion from their combination in food and absorbed from the digestive tract, or are the products of the metabolism of these constituents in the digestive tract. Nutrients are used for these three metabolic functions:

1. To provide the raw materials for the synthesis of body tissues and products;
2. To provide the energy which is needed for body tissue and animal product synthesis and for other

Table 2.1. A classification system for roughages.

Food class	Subclasses	Examples[a]
Green pasture and forage crops (living leguminous or graminaceous plants in a growth stage before senescence)		White clover, sainfoin, coastal Bermuda grass, prairie grass, oaten forage, triticale forage
Hays (usually sun-cured but sometimes artificially dried, air-dry forages)	Legume forage hays Graminaceous forage hays	Lucerne (alfalfa) hay, pea hay, soybean hay Timothy hay, ryegrass hay, barley forage hay
Low-quality air-dry forages	Dry, mature pasture	Mature black speargrass pasture, mature green panic pasture
	Cereal and other crop stubbles	Barley straw, wheaten straw, oaten straw, grain sorghum stubble
	Cereal crop stovers	Maize (corn) stem and leaves, sorghum stem and leaves
Silages (forage material which has been partly dried, then undergone anaerobic bacterial fermentation)	Cereal forage Pasture (grass) Legume	Immature or whole-crop barley forage Tropical and temperate grasses Lucerne (alfalfa) silage
Fodder trees		Sesbania, gliricidia, leucaena, mulga

[a]Botanical names: white clover, *Trifolium repens*; sainfoin, *Onobrychis viciifolia*; coastal Bermuda grass, *Cyanodon dactylon*; prairie grass, *Bromus willdenowii*; oats, *Avena sativa*; triticale, ×*Triticosecale rimpaui*; lucerne, *Medicago sativa*; pea, *Pisum* spp.; soybean, *Glycine soja*; timothy, *Phleum pratense*; ryegrass, *Lolium* spp.; black speargrass, *Heteropogon contortus*; green panic, *Panicum maximum* var. *trichoglume*; barley, *Hordeum vulgare*; wheat, *Triticum aestivum*; sorghum, *Sorghum bicolor*; maize, *Zea mays*; sesbania, *Sesbania sesban*; gliricidia, *Gliricidia sepium*; leucaena, *Leucaena leucocephala*; mulga, *Acacia aneura*.

purposes such as the basic life processes (e.g. breathing, blood circulation, digestion), movement and temperature regulation; and
3. To regulate and facilitate the metabolism of other nutrients.

The six major groups of nutrients, and their functions, are shown in Table 2.3. Energy is not included because, strictly, it is not a nutrient but the product of nutrient use. It is the chemical energy of a food which is released when amino acids, fatty acids and/or carbohydrates are oxidized in the body's cells, yielding ATP. When we talk about nutrients and the nutrient content of foods, we should refer to 'energy sources', but it has become a convenient convention in animal nutrition to refer to 'energy' as though it was a nutrient. To further complicate matters, instead of measuring directly the energy used (or usable) by animals cells, we have to take an indirect route in which the energy content of a food is evaluated by burning the food so that its chemical energy is converted to easily measured heat.

A complete list of the constituents of an animal's body is formidable. Clearly, an animal would be in an impossible position if it had to find, each day, a diet which contained all these substances in exactly the right amounts and proportions. Fortunately, this is not necessary, as animals can synthesize most of the enormous variety of substances which they need to build body tissue and sustain life.

However, they cannot make them all. This is easily demonstrated if animals are fed purified diets, and it quickly becomes obvious that there are some specific chemicals (i.e. nutrients) which are needed for normal life and health. The first recorded use of a purified diet was by M. Gannal, a Parisian glue-maker, in 1832. As described by Borek (1961) he fed his family on rendered animal connective tissue and bread. After several weeks they suffered from violent headaches and nausea due to deficiencies of essential amino acids, vitamins and other nutrients.

An 'essential nutrient' is a food constituent which is required for the proper functioning of the animal body and which cannot be made by the animal from other food constituents, or at least cannot be made in sufficient amounts to meet the animal's requirement. Note that in this context 'essential' means 'an essential component of the diet'. Some authorities use the

Table 2.2. A classification system for concentrate foods.

Food class	Subclasses	Examples[a,b]
Protein concentrates		
Animal protein meals (foods made from the rendered, i.e. fat-extracted, dried and ground residues of animal carcasses, or from dried blood or milk products)	Animal carcass by-products	Meat meals and meat and bone meals, blood meals
	Milk and milk by-products	Dried milk, casein, skimmed milk powder, buttermilk powder
	Fish meals	Foods made from whole fish or fish by-products
Plant protein meals	Oilseed meals (made from oil-rich seeds after they have been defatted, dried and ground)	Cottonseed meal, soybean meal, sunflower meal, canola (rapeseed) meal, palm kernel meal
	Grain legumes	Mung beans, cowpeas, lupins, navy beans
	Coconut meals	Copra meal, coconut meal
Energy concentrates		
Cereal grains		Wheat, barley, triticale, oats, rye, maize, sorghum, rice
Fats and oils		Vegetable oils, tallow, lard
Sugar-rich foods		Molasses, fruit pulp, biscuit waste, whey powder
Cereal by-products		Wheat bran, wheat pollard, wheat millrun, brewer's grains, rice bran

[a]Botanical names: cotton, *Gossypium hirsutum*; soybean, *Glycine soja*; sunflower, *Helianthus annuus*; canola, *Brassica napus*; oil palm, *Elaeis guineensis*; mung bean, *Vigna radiata*; cowpea, *Vigna* spp.; lupin, *Lupinus angustifolius*; navy bean, *Phaseolus vulgaris*; coconut, *Cocos nucifera*; wheat, *Triticum aestivum*; oats, *Avena sativa*; rye, *Secale cereale*; maize, *Zea mays*; sorghum, *Sorghum bicolor*; rice, *Oryza sativa*.
[b]For a glossary of by-product names and definitions of processing methods, see Harris *et al.* (1968, 1980).

term 'indispensable'. The list (Table 2.4) includes some ten amino acids (with the actual number varying with species and animal age), several fatty acids, all the minerals known to be required by animals and the vitamins (although most animals do not need a dietary source of ascorbic acid). Some tissues, such as the mammary gland, brain and the developing embryo, and biosynthetic pathways, require glucose (Battaglia and Meschia, 1988; McNay *et al.*, 2000; Père, 2003). Glucose can be made from other monosaccharides, the glucogenic amino acids, propionic acid and lactic acid, but not always in the required amounts.

Although the list of essential nutrients is large, conventional foods contain most of them in sufficient amounts. Sometimes, however, food may not have enough of particular essential nutrients. The nutrient which is in shortest supply and has the most serious detrimental effect is called the limiting nutrient. In practical terms, a 'limiting nutrient' is the one which is the most deficient in an animal's food and to which the animal will respond when

it is given as a supplement. The response can be improved health or productivity.

Which nutrients become limiting in practice will depend on the type and level of production normally expected from a particular sort of animal, and the nature of the animal's digestive tract. Monogastric animals have a larger list of potentially limiting nutrients than other animal types because of the synthesis of nutrients by the reticulo-ruminal and/or large intestinal microorganisms. For example, the growing pig produces up to 800 g of live weight daily and is fed a cereal grain-rich diet so that its energy intake will support this growth rate. However, cereal grains do not provide the pig with sufficient lysine, methionine, calcium, phosphorus, vitamin A and several other nutrients (Evans, 1985; NRC, 1998). In any given combination of food quality, animal sex, age and genotype, and environmental conditions, one of these essential nutrients will impose the greatest limitation to pig growth, and will be the limiting nutrient. Table 2.4 lists essential

Table 2.3. Nutrient groups and their functions.

Nutrient group	Usual form in foods	Function[a]	Example functions
Amino acids	Protein	1	Synthesis of lean tissue
		2	When surplus amino acids are deaminated and their C skeletons are oxidized, yielding ATP
		3	When used to synthesize enzymes or peptide hormones
Fatty acids	Fats	1	Synthesis of cell membranes
		2	When oxidized through the β oxidation reactions
		3	When used to synthesize steroid hormones
Carbohydrates	Plant fibre, starch, monosaccharides and disaccharides	1	Constituents of glycoprotein or glycolipid (e.g. glycosaminoglucans in cartilage, Hardingham and Fosang, 1992) although these precursors may be synthesized in the animal
		2	When oxidized through reactions such as the glycolytic pathway, yielding ATP
Minerals	In association with organic matter, e.g. bone	1	Mineralization of bone tissue
		3	When used as enzyme cofactors
Vitamins	As active vitamins, or as provitamins	3	When used as enzyme cofactors or in gene expression (e.g. the action of vitamin D in promoting the synthesis of Ca binding proteins, see Anderson and Toverud, 1994)
Water	Free water (cellular and adhering water), water of crystallization	1	As a major constituent of lean tissue, blood and milk
		3	As the medium in which the reactions of the body take place

[a]See text for definitions of the three functions of nutrients.

Table 2.4. Essential and potentially limiting nutrients in conventional animal foods.

Nutrient group	Individual nutrients
Amino acids	Lysine, methionine, cystine or cysteine Other essential amino acids are: arginine, histidine, isoleucine, leucine, phenylalanine, taurine (for cats), threonine, tryptophane, tyrosine and valine
Fatty acids	Linoleic acid The other essential fatty acids are the long-chain, ω3 and ω6 unsaturated fatty acids, including α-linolenic, arachidonic, eicosapentaenoic and docosahexaenoic acids
Minerals	Calcium, cobalt, copper, iodine, iron, magnesium, phosphorus, selenium, sodium Other minerals such as chlorine, chromium, manganese, molybdenum, potassium, silicon, sulfur and zinc are essential, but not usually limiting; arsenic, boron, bromine, fluorine, lithium, nickel, rubidium, tin and vanadium may be essential
Vitamins	Vitamin A (retinol), thiamine (vitamin Bl), cobalamin (vitamin B12), biotin, vitamin D (calciferol) Other vitamins include niacin, riboflavin (vitamin B2), pantothenic acid, pyridoxine (vitamin B6), folic acid, vitamin E (tocopherols) and vitamin K, but these are not likely to be limiting in practical diets. Vitamin C (ascorbic acid) may be needed by animals of some species and/or ages, e.g. the young pig; see Mahan *et al.* (2004). Additional dietary choline may be useful in the diets of newborn lambs and dairy calves (Al-Ali *et al.*, 1985; NRC, 2001) and lactating dairy cattle may respond to additional ruminally protected choline (Erdman and Sharma, 1991)

nutrients and indicates those which may limit the performance of animals given conventional diets.

The Units of Measurement Used to Describe Food Constituents and Nutrients

The contents of nutrients and constituents in foods can be expressed on either a whole-food basis (other terms for this are 'air-dry', 'as-received' or 'as-fed') or a DM basis. It is best to express nutrient contents on a DM basis as this reduces the errors which would otherwise occur when foods with different moisture contents are mixed to form a ration. Forages, in particular, can have widely different moisture contents (e.g. silage with 35% DM and hay containing 90% DM). As concentrates

have a rather uniform DM content (between 85% and 100%) it is not so important to use a DM basis for those foods which are typically fed to pigs, lot-fed beef cattle and household pets. The relationship between the DM and as-fed bases for expressing nutrient contents is illustrated in Fig. 2.2.

The concentrations in foods of most nutrients and food constituents are measured in g/kg or %, and expressed in reference to either the food DM or the total food. The basis of expression should be given, e.g. % or % DM. Energy is measured either MJ/kg or MJ/kg DM. The units used to measure the contents of nutrients and food constituents are given in Table 2.5.

Food energy contents may also be measured in calories. One calorie is the amount of heat energy needed to raise the temperature of 1 g of water by

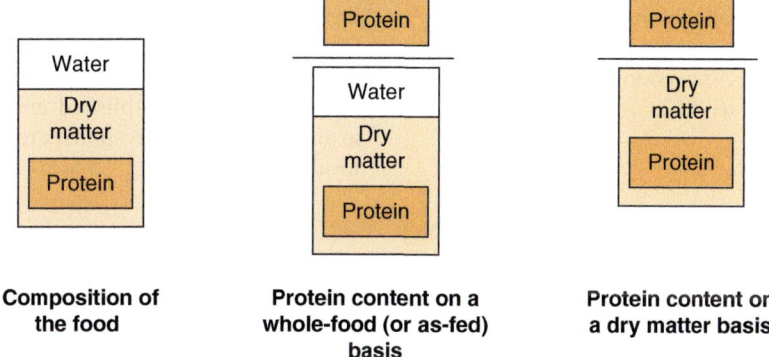

Fig. 2.2. The basis of expression: two ways in which the protein content of a food can be described.

Table 2.5. Units used to measure nutrients and food constituents.

Nutrient/constituent	Preferred unit[a]	Alternative unit
Dry matter (DM)	g/kg	%
Organic matter (OM)	g/kg DM	%
Energy	MJ/kg DM	Mcal or cal/kg DM
Acid detergent fibre (ADF)	g/kg DM	%
Protein	g/kg DM	%
Amino acids	g/kg DM	% (of DM or protein)
Fat	g/kg DM	%
Ash	g/kg DM	%
Minerals: major (e.g. Ca, P, Mg)	g/kg DM	%
trace (e.g. Cu, Zn, Co)	mg/kg DM	ppm
Vitamins: fat-soluble	mg or µg/kg DM	IU/kg
C and B group	mg/kg DM	

[a]10 g/kg = 1%; 1 mg/kg = 1 part per million (ppm); 4.184 J = 1 calorie (cal); 1,000,000 J = 1000 kJ = 1 MJ.

1°C (usually measured between 14.5°C and 15.5°C). Following the adoption of Système Internationale (SI, or metric) units throughout the biological sciences, a general unit for measuring energy, the joule (J), is now used by most animal nutritionists. One joule is the amount of energy expended when a mass of 1 kg is moved 1 m, accelerating at 1 m/s². There is no readily apparent relationship between this and the use of energy by an animal, but then there is only a tenuous connection between an animal's energy usage and the unit of heat (the calorie). The joule at least has the advantage of being part of an internationally adopted, systematic system of units.

Vitamin A, D and E activities are often quoted in International Units (IU). An IU is 'a unit of activity or potency of sera, hormones, vitamins, etc., defined individually for each substance in terms of the activity of a standard quantity or preparation' (Pearsall and Trumble, 2002). Using the IU as the unit of measurement allows the different forms of these vitamins and their provitamins to be equated. There are standard conversion factors, although these vary according to the animal species.

Food Analysis

Methods of food analysis have evolved gradually over the last 150 years, and this effort has been associated with some of the most luminous names in chemistry, including Davy, Boussingault and Liebig. The early development of food chemical analysis has been described by Tyler (1975) and Midkiff (1984). In 1805, Einhof extracted the 'fibrous' parts of foods (this is not synonymous with the modern definition of food 'fibre') by washing and rubbing the food to get a resistant residue. He may have considered (Einhof, 1806, cited in Tyler, 1975) that this was not a nutritious part of the food. Later workers, including Thaer, Davy, Boussingault and Wolff, specifically declared that this fibrous material was not digestible, although they apparently had no experimental evidence for this.

Gradually, more specific chemical methods were developed to isolate and measure those chemical fractions which were related to the food's nutritive value, including protein, fat and fibre. The Liebig/Dumas method for total N was developed in the 1830s and 1840s, and Kjeldahl's method was published in 1883. Henneberg and Stohmann developed their crude fibre analysis in 1859. In the half-century which followed this early work, there were substantial conceptual and practical advances (Armsby, 1902). By the beginning of the 20th century the foundations of food chemistry had been firmly established and there were data available on animal responses to different types of foods.

The proximate analysis

Following the work of Henneberg and Stohmann (1860, cited in Mertens, 2003), the state of food analysis was that summarized in Fig. 2.3. This scheme is called the 'proximate analysis', or the

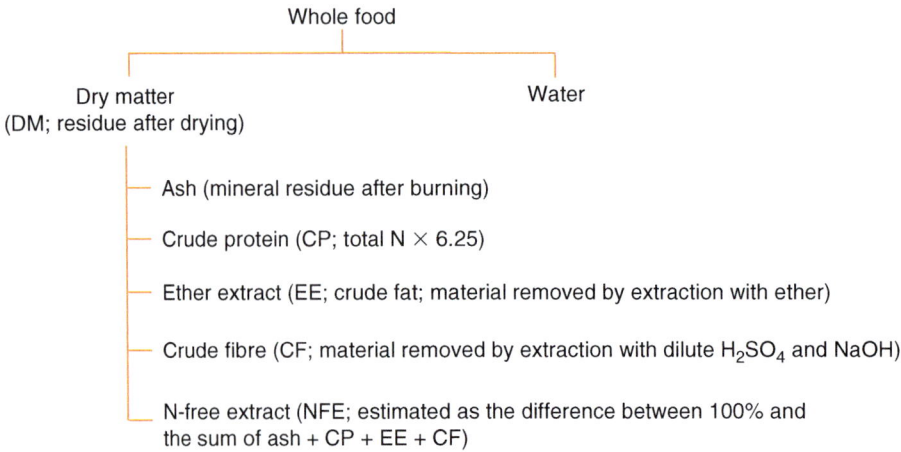

Fig. 2.3. Flowchart showing the components of the proximate analysis of foods.

'Weende analysis' after the research station at which Henneberg and Stohmann worked.

Elements of this analytical scheme are still used. We use food DM as a basis for expressing analytical data and in calculating food intake or formulating diets. The concept of multiplying total N by a factor to estimate the total protein content is still used because of its analytical simplicity and relative accuracy. The commonly used factor 6.25 derives from the fact that, in general, proteins contain 16% N, and 100/16 = 6.25. N in foods can be measured by the method of Kjeldahl (1883) in which protein-N is dissociated from its combination with other elements by digestion in concentrated H_2SO_4, followed by conversion to the hydroxide and subsequent distillation and titration. This is reasonably accurate and was, for many years, the standard method. However, the Kjeldahl method does not recover all the N of aromatic nitrogenous substances or of nitrates. The Dumas method, in which the sample is burnt and the N gas released is measured, was developed before the Kjeldahl method but has become popular only following the development of automated methods of carrying out the analysis. This method recovers all the sample N, and so may give slightly higher values than the Kjeldahl method depending on the sample analysed.

The other proximate analysis components are less used in modern food analysis. Ash is a mixture of food minerals. There is incomplete recovery of individual minerals as some, e.g. phosphorus, are volatile at the temperatures which are needed to burn completely the carbonaceous compounds. However, the food organic matter (OM) content (OM = DM – ash) is frequently used as a way of correcting data for mineral contamination, as will happen when measurements are made with grazing animals. Ether extraction (EE) is used to isolate lipids for more detailed fractionation into fatty acids and waxes, although the EE is still reported as a measure of total lipid. Note, though, that EE contains fat-soluble vitamins, waxes and pigments as well as fats. Nitrogen-free extract (NFE) was originally assumed to be mainly soluble carbohydrate and so was expected to be highly digestible. Unfortunately, the reagents used to measure crude fibre (CF) may remove up to 60% of the cellulose, about 80% of hemicellulose and a highly variable (10–95%) proportion of the lignin (Mertens, 2003). The NFE contains some of the plant cell wall material and can be less digestible than CF. Besides this, NFE contains all those chemical entities which are not measured by the other methods (some nitrogenous compounds and vitamins, in particular), as well as the errors inherent in the measurement of the other food constituents. NFE is now not used in food analysis. For the details of the proximate analytical methods, refer to AOAC International (2002).

Detergent-based analytical systems

Detergents were used by Van Soest and his colleagues (Van Soest, 1963; Van Soest and Wine, 1967) to remove plant cell content materials and leave the cell walls, or at least identifiable fractions of the cell wall. Acid detergent (cetyl trimethyl ammonium bromide in 0.5 M H_2SO_4), the earliest of the detergent methods, removes all plant cell constituents except cellulose and lignin, minerals and the cell wall-bound N. The neutral detergent reagent (sodium dodecyl sulfate in a pH 7, phosphate/EDTA/tetraborate buffer) removes only the cell contents. When the detergent analyses are combined with elements of the proximate analysis we have the scheme illustrated in Fig. 2.4. Although ash is shown in this diagram as a separate component, it is actually retained in the detergent fibre fractions and must be finally determined by ashing the residue. The original detergent methods have been modified to reduce interference from starch and protein. The sequence of analysis indicated in Fig. 2.4 is based on Van Soest and Robertson (1980). Details of the detergent methods are given in AOAC International (2002). These methods are widely used in animal nutrition.

Other analytical approaches

An alternative to the detergent methods for cell wall constituents is to measure the monosaccharides of the cell wall. These amounts can be summed to give estimates of the cell wall polysaccharides. In the Uppsala method (Theander et al., 1995) cellulose is taken to equal the total amount of glucose; hemicellulose is the sum of xylose, mannose and fucose; and pectin is the amount of uronic acids, galactose, arabinose and rhamnose residues.

It is difficult to measure lignin, not least because its chemical composition is not fixed. It is generally agreed that acid detergent lignin substantially underestimates the true lignin content (Jung et al., 1999; Fukushima and Hatfield, 2004; Jung and

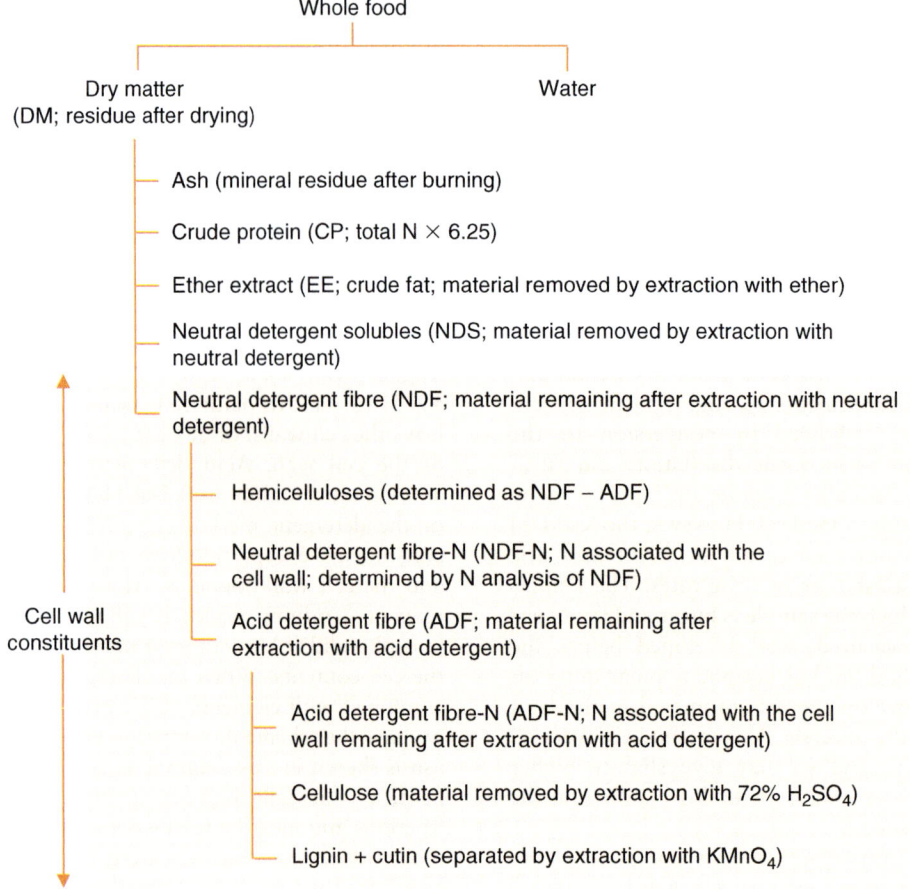

Fig. 2.4. Application of the detergent methods in the analysis of foods.

Lamb, 2004). Other lignin methods are described by Hatfield and Fukushima (2005), who suggest that Klason lignin, determined by sequential H_2SO_4 extraction, is a better measure of the plant lignin content.

Plant fibre and the plant cell wall

Plant cell walls, synonymous with plant fibre, consist of three main types of polysaccharides (cellulose, hemicellulose and pectin), lignin and some protein and waxes. The composition and molecular structure of these substances vary with plant age and species. A major problem in food analysis remains the definition and chemical isolation of plant fibre. A large part of the problem is that nutritionists have not yet decided on a suitable definition for fibre and until this is done, there can be no development of appropriate methods of fibre analysis. One defini-

tion, which goes beyond the limits of materials found only in the cell wall, is that of Pluske *et al.* (1997): 'nonstarch polysaccharides found in plant cell walls, resistant starch, and the α-galactosides (oligosaccharides) of the raffinose series . . . present mainly in legumes'. Giger-Reverdin (1995), Souffrant (2001), McLeary (2003) and Mertens (2003) have discussed the difficulties of definition and analysis. Most methods of cell wall fractionation rely on the gravimetric separation of residues which are insoluble in different reagents (Södekum, 1994). This introduces two problems: first, the fractions which are soluble in particular extractants may actually have different polysaccharide compositions if they come from different foods; and second, the accurate gravimetric analysis relies on 'clean' separations of the various food fractions. There is good evidence that different methods of measuring plant fibre give different results, and that the detergent methods do not give fractions of uniform composition. For example, John and Waghorn (1986) showed that different fibre analysis methods gave different rankings for the digestibility of pasture plants, Bailey and Ulyatt (1970) found that ADF isolated from clovers contained substantial amounts of pectin (the intercellular cement polysaccharide), and several workers (e.g. Bailey and Ulyatt, 1970; Morrison, 1980; McAllan and Griffith, 1984) have found that ADF, which is nominally cellulose plus lignin, contains monosaccharides which are expected to be constituents of hemicellulose.

A useful way of dividing up the various fibrous and non-fibrous carbohydrates in plant-based foods is to group them: first, according to whether they are cell wall constituents or not; second, whether they are insoluble or soluble; and third, whether they are structural or storage polysacchar-

ides, i.e. non-starch polysaccharides (NSPs) versus starch. This approach is summarized in Fig. 2.5.

Other Food Constituents

Our discussion of food analysis has, so far, been concerned with the quantitatively important constituents. However, we must also consider minerals, fatty acids, amino acids and vitamins. Techniques such as gas liquid chromatography and high-performance liquid chromatography are now used to separate and quantify fatty acids and amino acids. Atomic absorption spectrometry and inductively coupled plasma atomic emission spectrometry methods efficiently determine the concentrations of individual minerals. These methods are described in Linden (1996) and AOAC International (2002).

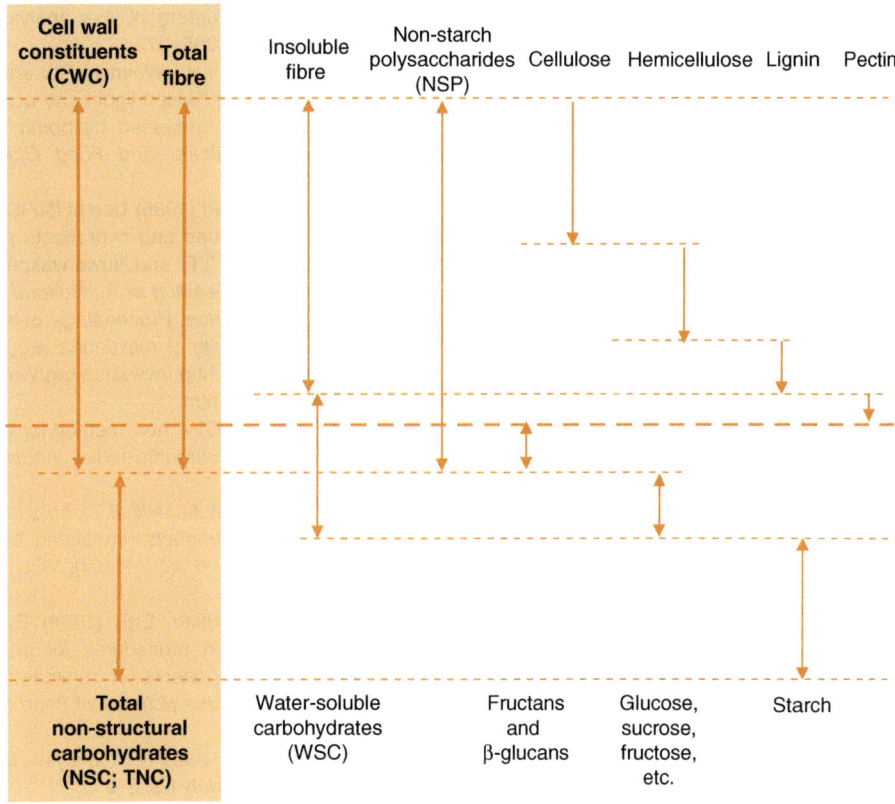

Fig. 2.5. Diagram showing plant carbohydrate groups and lignin, and the various ways in which these are grouped into chemically and/or nutritionally similar entities. The schematic is not to scale. 'Cell wall constituents' include components which are not retained in NDF; 'insoluble fibre' is approximately NDF; 'total fibre' and 'crude fibre' are not synonymous; NSC is not the total cell contents. (From Van Soest et al., 1980; Ciavarella et al., 2000; Bach-Knudsen, 2001; Frankow-Lindberg, 2001; Souffrant, 2001; Turner et al., 2002; Mertens, 2003.)

References

Al-Ali, S.J., Malouf, N.M. and Walker, D.M. (1985) Choline requirement of the preruminant lamb during the first two or three weeks of life. *Australian Journal of Agricultural Research* 36, 829–844.

Anderson, J.J.B. and Toverud, S.U. (1994) Diet and vitamin D: a review with an emphasis on human function. *Journal of Nutritional Biochemistry* 5, 58–65.

AOAC International (2002) *Official Methods of Analysis*, 17th edn. AOAC International, Arlington, Virginia.

Armsby, H.P. (1902) *Manual of Cattle Feeding*, 3rd edn. Wiley, New York.

Bach-Knudsen, K.E. (2001) The nutritional significance of dietary fibre analysis. *Animal Feed Science and Technology* 90, 3–20.

Bailey, R.W. and Ulyatt, M.J. (1970) Pasture quality and ruminant nutrition. II. Carbohydrate and lignin composition of detergent extracted residues from pasture grasses and legumes. *New Zealand Journal of Agricultural Research* 13, 591–604.

Battaglia, F.C. and Meschia, G. (1988) Fetal nutrition. *Annual Review of Nutrition* 8, 43–61.

Borek, E. (1961) *The Atoms Within Us*. Columbia University Press, New York, p. 98.

Ciavarella, T.A., Simpson, R.J., Dove, H., Leury, B.J. and Sims, I.M. (2000) Diurnal changes in the concentration of water-soluble carbohydrates in *Phalaris aquatica* L. pasture in spring, and the effect of short-term shading. *Australian Journal of Agricultural Research* 51, 749–756.

Einhof, E. (1806) Bemerkungen über die Nahrungsfähigkeit verschiedener vegetabilischer Produkte. *Annalen des Ackerbaues* 4, 627–659.

Erdman, R.A. and Sharma, B.K. (1991) Effect of dietary rumen-protected choline in lactating dairy cows. *Journal of Dairy Science* 74, 1641–1647.

Evans, M. (1985) *Nutrient Composition of Feedstuffs for Pigs and Poultry*. Queensland Department of Primary Industries, Brisbane, Australia.

Frankow-Lindberg, B.E. (2001) Adaptation to winter stress in nine white clover populations: changes in non-structural carbohydrates during exposure to simulated winter conditions and 'spring' regrowth potential. *Annals of Botany* 88, 745–751.

Fukushima, R.S. and Hatfield, R. (2004) Comparison of the acetyl bromide spectrophotometric method with other analytical lignin methods for determining lignin concentration in forage samples. *Journal of Agricultural and Food Chemistry* 52, 3713–3720.

Giger-Reverdin, S. (1995) Review of the main methods of cell wall estimation: interest and limits for ruminants. *Animal Feed Science and Technology* 55, 295–334.

Hardingham, T.E. and Fosang, A.J. (1992) Proteoglycans: many forms and many functions. *FASEB Journal* 6, 861–870.

Harris, L.E., Asplund, J.M. and Crampton, E.W. (1968) *An International Feed Nomenclature and Methods for Summarizing and Using Feed Data to Calculate Diets*. Bull. No. 479. Agricultural Experiment Station, Utah State University, Logan, Utah.

Harris, L.E., Jager, F., Leche, T.F., Mayr, H., Neese, U. and Kearl, L.C. (1980) *International Feed Descriptions, International Feed Names and Country Feed Names*. INFIC Publ. No. 5. International Feedstuffs Institute, Utah State University, Logan, Utah.

Hatfield, R. and Fukushima, R.S. (2005) Can lignin be accurately measured? *Crop Science* 45, 832–839.

Henneberg, W. and Stohmann, F. (1860) *Begründung einer Rationellen Futterung der Wiederkäuer*. Vol. 1. Schwetschtke u. Sohn, Braunschweig, Germany.

John, A. and Waghorn, G.C. (1986) Comparison of two methods for forage cellulose and hemicellulose determination and their use in digestibility measurements. *Proceedings of the Nutrition Society of New Zealand* 11, 109.

Jung, H.-J.G. and Lamb, J.F.S. (2004) Prediction of cell wall polysaccharide and lignin concentrations of alfalfa stems from detergent fiber analysis. *Biomass and Bioenergy* 27, 365–373.

Jung, H.-J.G., Varel, V.H., Weimer, P.J. and Ralph, J. (1999) Accuracy of Klason lignin and acid detergent lignin methods as assessed by bomb calorimetry. *Journal of Agricultural and Food Chemistry* 47, 2005–2008.

Kayongo-Male and Said (1986) Use of INFIC nomenclature on crop residues and byproducts produced in Kenya. In: Preston, T.R. and Nuwanyakpa, M.Y. (eds) *Towards Optimal Feeding of Agricultural Byproducts to Livestock in Africa*. Proceedings of a workshop held at the University of Alexandria, Egypt, October 1985. Available at: http://www.fao.org/Wairdocs/ILRI/x5487E/x5487e06.htm

Kjeldahl, J.G.C.T. (1883) A new method for the determination of nitrogen in organic bodies. *Analytical Chemistry* 22, 366.

Linden, G. (ed.) (1996) *Analytical Techniques for Foods and Agricultural Products* (translated by L. Dieter, English language ed. W.J. Hurst). VCH Publishers, New York.

McAllan, A.B. and Griffith, E.S. (1984) Evaluation of detergent extraction procedures for characterising carbohydrate components in ruminant feeds and digesta. *Journal of the Science of Food and Agriculture* 35, 869–877.

McLeary, B.V. (2003) Dietary fibre analysis. *Proceedings of the Nutrition Society* 62, 3–9.

McNay, E.C., Fries, T.M. and Gold, P.E. (2000) Decreases in rat extracellular hippocampal glucose concentration associated with cognitive demand during a spatial task. *Proceedings of the National Academy of Science of the USA* 97, 2881–2885.

Mahan, D.C., Ching, S. and Dabrowski, K. (2004) Developmental aspects and factors influencing the synthesis and status of ascorbic acid in the pig. *Annual Review of Nutrition* 24, 79–103.

Mertens, D.R. (2003) Challenges in measuring insoluble dietary fiber. *Journal of Animal Science* 81, 3233–3249.

Midkiff, V.C. (1984) A century of analytical excellence. The history of feed analysis, as chronicled in the development of AOAC official methods, 1884–1984. *Journal of the Association of Official Analytical Chemists* 67, 851–860.

Morrison, I.M. (1980) Hemicellulosic contamination of acid detergent residues and their replacement by cellulose residues in cell wall analysis. *Journal of the Science of Food and Agriculture* 31, 639–645.

National Research Council (NRC) (1998) *Nutrient Requirements of Swine*, 10th rev. edn. National Academy Press, Washington, DC.

National Research Council (NRC) (2000) *Nutrient Requirements of Beef Cattle*, 7th rev. edn. Update. National Academy Press, Washington, DC.

National Research Council (NRC) (2001) *Nutrient Requirements of Dairy Cattle*, 7th rev. edn. National Academy Press, Washington, DC.

National Research Council (NRC) (2007) *Nutrient Requirements of Horses*, 6th rev. edn. National Academy Press, Washington, DC.

Ostrowski-Meissner, H.T. (1987) *Australian Feed Composition Tables: Ruminants*. AFIC-CSIRO, Sydney, Australia.

Ostrowski-Meissner, H.T. (1990) *Australian Feed Composition Tables: Pigs and Poultry*, 2nd edn. AFIC-CSIRO, Sydney, Australia.

Pearsall, J. and Trumble, B. (2002) *The Oxford English Reference Dictionary*, 2nd edn. Oxford University Press, Oxford.

Père, M.-C. (2003) Materno-foetal exchanges and utilisation of nutrients by the foetus: comparison between species. *Reproduction, Nutrition and Development* 43, 1–15.

Pluske, J.R., Pethick, D.W., Durmic, Z., McDonald, D.E., Mullan, B.P. and Hempson, D.J. (1997) Diseases and conditions in pigs, horses and chickens arising from incomplete digestion and absorption of carbohydrates. *Recent Advances in Animal Nutrition in Australia* 14, 33–41.

Södekum, K.-H. (1994) Monosaccharide composition of cell-wall carbohydrates. Digestion and absorption. *Livestock Production Science* 39, 71–79.

Souffrant, W.B. (2001) Effect of dietary fibre on ileal digestibility and endogenous nitrogen losses in the pig. *Animal Feed Science and Technology* 90, 93–102.

Standing Committee on Agriculture (SCA) (1990) *Feeding Standards for Australian Livestock. Ruminants*. CSIRO, Melbourne, Australia.

Theander, O., Áman, P., Westerlund, E., Andersson, R. and Pettersson, D. (1995) Total dietary fiber determined as neutral sugar residues, uronic acid residues, and Klason lignin (the Uppsala method): collaborative study. *Journal of the Association of Official Analytical Chemists International* 78, 1030–1044.

Turner, L.B., Humphreys, M.O., Cairns, A.J. and Pollock, C.J. (2002) Carbon assimilation and partitioning into non-structural carbohydrate in contrasting varieties of *Lolium perenne. Journal of Plant Physiology* 159, 257–263.

Tyler, C. (1975) Albrecht Thaer's hay equivalents: fact or fiction? *Nutrition Abstracts and Reviews* 45, 1–11.

Van Soest, P.J. (1963) Use of detergents in the analysis of fibrous feeds. II. A rapid method for the determination of fiber and lignin. *Journal of the Association of Official Agricultural Chemists* 46, 825–829.

Van Soest, P.J. and Robertson, J.B. (1980) Systems of analysis for evaluating fibrous feeds. In: Pigden, W.J., Balch, C.C. and Graham, M. (eds) *Standardization of Analytical Methodology for Feeds*. International Development Research Centre, Ottawa, pp. 49–60.

Van Soest, P.J. and Wine, R.H. (1967) Use of detergents in the analysis of fibrous feeds. IV. Determination of plant cell wall constituents. *Journal of the Association of Official Agricultural Chemists* 50, 50–55.

Van Soest, P.J., Robertson, J.B. and Lewis, B.A. (1980) Methods for dietary fibre, neutral detergent fiber, and nonstarch polysaccharides in relation to animal nutrition. *Journal of Dairy Science* 74, 3583–3597.

3 Methods of Evaluating the Availability of Nutrients in Foods

Digestion Affects Nutrient Availability and the Types of Nutrients Absorbed

Food evaluation systems describe the nutritive value of foods by combining measurements of nutrient content with estimates of the availability of those nutrients. Available nutrient content is not an inherent, unchangeable, aspect of food quality because the interaction between the chemical composition of a food and the way that the animal digests it influences the nature and supply of nutrients. Foods are evaluated separately for carnivores, omnivores and herbivores; for monogastrics and ruminants; and for different species of herbivores (even within the foregut and hindgut fermenter groups).

Food Energy Which Is Available to Animals Can Be Partitioned into Several Different Fractions

The total energy content of most animal foods is close to 18.4 MJ/kg DM (Van Es, 1978); exceptions are those which are fat-rich (fat has 39.3 MJ/kg DM) or mineral-rich (ash has no energy). Therefore, if we want to describe the quality of a food as a source of energy to animals, there is little to be gained by simply quoting its total energy content as this will not distinguish it from other foods. A system of partitioning food energy is shown in Fig. 3.1. Depending on the level of availability we use, we can describe the available energy content of foods in three ways: by their digestible energy (DE), metabolizable energy (ME) or net energy (NE) contents.

Some authorities have felt that the users of energy feeding standards are more comfortable if these energy units are expressed as a 'food unit'. This idea may date from the use, in Europe in the 19th and 20th centuries, of straw, starch, barley and oats as comparative food standards before data on digestibility and metabolizability were available. Examples are the Dutch ruminant system (Van Es, 1978) where feed units for lactation and growth are

defined as the actual NE content (MJ/kg DM) divided by 6.9 (this value was chosen arbitrarily, but is similar to comparing the food to barley grain), and the French systems for horses and ruminants where the NE contents of foods are also expressed in relation to the NE content of barley grain (Jarrige, 1989; Martin-Rosset *et al.*, 1994).

Digestible Energy

'Digestion' is the series of chemical and physical processes which break down food constituents to molecules which can be absorbed through the membranes of the intestinal epithelial cells. 'Digestibility' is the extent to which a food or its constituents is digested. A 'digestibility coefficient' indicates the proportion of that food constituent which is digested and absorbed from the digestive tract when the food is eaten. Digestibility is greatly influenced by the physico-chemical composition of the food. For example, lignin limits the digestion of cell wall constituents in forages (Jung *et al.*, 1997), endosperm structure limits the digestion of starch in cereal grains (Ezeogu *et al.*, 2005; Taylor and Allen, 2005), and the digestibilities of carbohydrates and proteins may be restricted by plant tannins (Barry and Duncan, 1984; Kaitho *et al.*, 1998). Animal-related factors can also be important. There is a large difference between herbivores and non-herbivores in their abilities to digest plant fibre, because herbivores use a symbiotic relationship with microorganisms to accomplish this. The rate at which food passes through the digestive tract (the 'rate of passage') also influences the completeness of digestion (see further comments below).

We usually measure 'apparent' digestibility, i.e. we do not distinguish between the endogenous and exogenous (or dietary) constituents of the faeces. Some of the protein, fat and minerals excreted in the faeces are not undigested food constituents, but are derived from the animal's tissues (Mason, 1969; Dryden, 1982). Therefore, digestibility coefficients calculated from measures of total energy

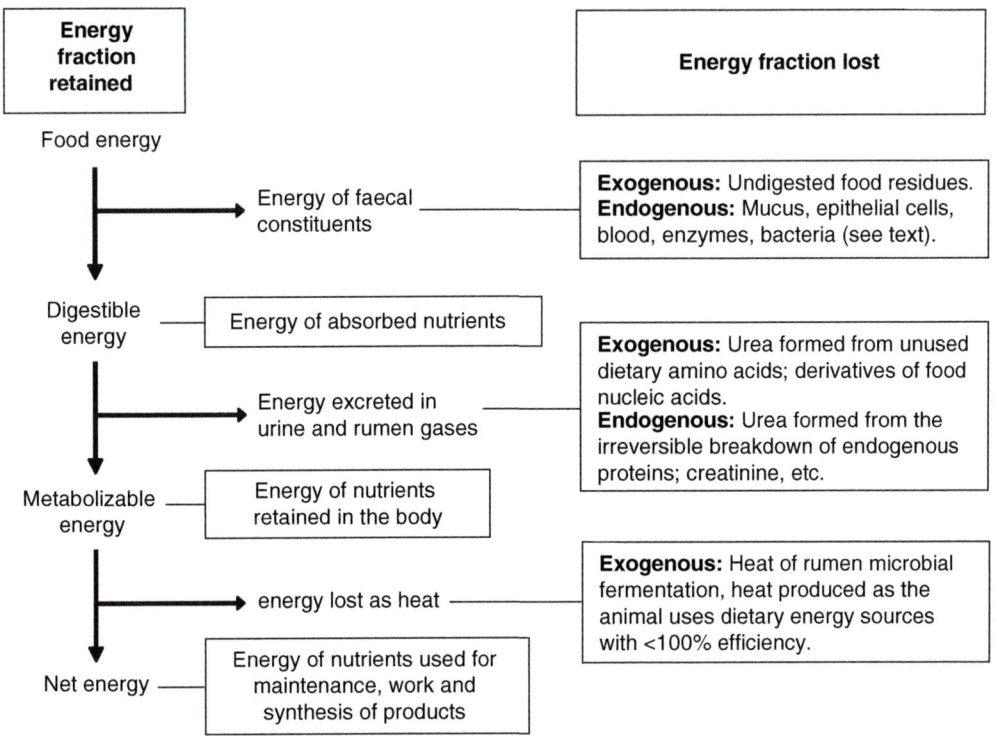

Fig. 3.1. Partitioning of food energy through digestion and metabolism.

excretion are actually lower than the true digestibility of food energy because the faeces contain energy from endogenous materials such as blood proteins, digestive enzymes, epithelial cells, mucus, bile salts and bile pigments. Bacterial cells voided in the faeces are partly 'endogenous' because these organisms in the large intestine are partly nourished by endogenous secretions (Dryden, 1982).

'True digestibilities' are values which are calculated from measurements of the amounts of undigested food residues excreted in faeces, uncontaminated with endogenous material. True digestibilities can be measured but the techniques are more difficult than the simple collection of the total faecal matter. Unless the contrary is stated, all published digestibility data are for apparent digestibilities.

Measurement of Digestibility and Digestible Energy Content *In Vivo*

DE content is easily measured, and indigestibility is the major source of energy loss when food is processed by monogastric animals. Food DE contents are calculated by measuring the apparent digestibility of food energy

(E_{dig}, expressed as either a decimal or a percentage) and applying this to the food total energy content:

$$E_{dig} \text{ (decimal)} = \frac{[(F_{food} \times E_{food}) - (F_{ref} \times E_{ref})] - (Fa_{out} \times E_{faeces})}{[(F_{food} \times F_{food}) - (F_{ref} \times E_{ref})]} \quad (3.1)$$

where
F_{food} = amount of food offered (g DM/day)
F_{ref} = amount of food refused (g DM/day)
E_{food} = total energy content of food (MJ/kg DM)
E_{ref} = total energy content of food refused (MJ/kg DM)
Fa_{out} = amount of faeces voided (g DM/day)
E_{faeces} = total energy content of faeces (MJ/kg DM)

Because neither the amount of food energy eaten nor the amount of food energy digested can be measured directly, we estimate these from the differences between food offered and refused, with appropriate allowances for the likely compositional differences between the offered food and the refusals (or 'orts'), and the difference between food energy eaten and faecal energy voided.

Example

A sheep is given 2 kg DM/day of grass hay which has 18.5 MJ total energy/kg DM. It leaves 0.5 kg DM/day of refusals. The refusals are rich in stalky material, which has a higher ash content than the whole food offered, and contain 17.5 MJ total energy/kg DM. The sheep voids 0.5 kg of faecal DM/day, containing 18 MJ total energy/kg DM. The digestibility of the energy in this hay is calculated as follows:

$$\begin{aligned} E_{dig} \text{(decimal)} &= \frac{[(F_{food} \times E_{food}) - (F_{ref} \times E_{ref})] - (Fa_{out} \times E_{faeces})}{[(F_{food} \times E_{food}) - (F_{ref} \times E_{ref})]} \\[6pt] &= \frac{[(2.0 \times 18.5) - (0.5 \times 17.5)] - (0.5 \times 18)}{[(2.0 \times 18.5) - (0.5 \times 17.5)]} \\[6pt] &= \frac{[37 - 8.75] - (9.0)}{[37 \times 8.75]} \\[6pt] &= \frac{19.25}{28.25} \\[6pt] &= 0.681 \text{(or 68.1\%)} \end{aligned}$$

Food DE content is the product of the food total energy content and its digestibility:

$$DE = E_{food} \times E_{dig} \qquad (3.2)$$

where
DE = digestible energy content (MJ/kg DM)
E_{dig} = energy digestibility (decimal)

Alternative Methods of Measuring Food Digestibility

The marker (or 'indicator') method

This approach is useful when we cannot make total faecal collections. Suitable markers are indigestible and non-absorbable, do not affect the digestion process and are otherwise harmless to the animal, are mixed evenly through the food, transit the digestive tract at the same rate as the rest of the food and are convenient to analyse. 'Internal' markers are natural food constituents. Examples are n-alkanes (Mayes et al., 1986; Dove and Mayes, 1991), lignin (Fahey and Jung, 1983; Clauss et al., 2004), indigestible cell wall (Lippke et al., 1986), ruminally undegraded dry matter (DM) (Ferret et al., 1999) and acid-insoluble ash (Van Keulen and Young, 1977). Internal markers are not completely reliable. If some of the marker is lost during passage through the tract the digestibility will be underestimated (Ferret et al., 1999; Hmeidan et al., 2000). Internal markers

should not be used unless they have been validated for the food under test.

Substances which are added to (and mixed evenly through) the food are called 'external' markers. Suitable external markers include chromium-mordanted plant fibre (Cr-NDF; Udén et al., 1980), lanthanum (Shaver et al., 1986) and ytterbium chloride (Teeter et al., 1984).

The ratios of food constituent to indigestible marker, in the food and the faeces, are used to calculate the digestibility of that constituent (Equation 3.3). The digestibility of any food constituent (protein in this example) is calculated thus:

$$Pr_{dig} = 1 - [(M_{food} \times Pr_{faeces}) / (M_{faeces} \times Pr_{food})] \quad (3.3)$$

where
Pr_{dig} = protein digestibility (decimal)
M_{food} = marker concentration in the food (mg/g DM)
M_{faeces} = marker concentration in the faeces (mg/g DM)
Pr_{food} = protein concentration in the food (g/kg DM)
Pr_{faeces} = protein concentration in the faeces (g/kg DM)

In vitro and in sacco methods

An in vivo digestibility experiment may not be appropriate if there is only a small amount of the test food available, if it is unpalatable or if it would cause digestive disorders when fed alone. Other approaches can be used. Several of these are reviewed by Mould (2003). In vitro and in sacco methods attempt to reproduce the chemistry of in vivo digestion in a laboratory setting. The in vitro method was invented by Tilley and Terry (1963) and originally involved incubating a small amount of milled food in rumen liquor for 48 h. The original method used strained rumen liquor (SRL) diluted in a phosphate buffer at pH 6.5 with added minerals. The residue from this fermentation was then suspended in acid pepsin for a further 48 h. The method has been widely modified, e.g. by the addition of an N source, and use of a neutral detergent extraction at the end of the rumen liquor incubation to get a measure approximating true rumen digestibility (Goering and Van Soest, 1970).

An alternative to the in vitro method is to suspend samples of milled food in the rumen of a ruminally fistulated animal. Samples of about 3 g of air-dry material are incubated in bags about $5 \times 10\,cm$, made of synthetic material (nylon or terylene). Incubation times may be 48 h to give data comparable to *in vitro* digestibilities, or up to 120 h to measure potential digestibility. *In sacco* digestibility values are influenced by the type of diet given to the host animal, the sample particle size, the amount of sample and the way that the bags are washed when they are taken out of the rumen (Mehrez and Ørskov, 1977; Ørskov *et al.*, 1980).

Gas production

CO_2 and methane are produced in stoichiometric amounts to the amount of organic matter fermented by rumen microorganisms. This implies that measurements of gas production can be used to estimate organic matter fermentation by rumen microorganisms. However, there were some significant plumbing problems to overcome before this method became practicable. These have now been successfully overcome, and gas production (i.e. the volume of gas produced per unit time) is now frequently used to investigate the nutritional properties of foods. The method is described by Theodorou *et al.* (1994).

Digestibility by difference

When we want to find digestibility coefficients for foods which cannot be fed alone we can measure the digestibilities of the constituents of a basal diet, and then of the same diet with the test food added. With the digestibility coefficients determined from the basal diet fed alone, we can calculate the amount of faeces which probably derived from the basal component of the basal + test ration, and then calculate the probable digestibility of the test food. Methods are described in detail by Schneider and Flatt (1975). This method has a major flaw, which is that foods interact with each other, especially during fermentation in the rumen. This effect, called 'associative digestibility', is caused by the extent or rate of digestion of one ingredient being either enhanced or inhibited by the presence of constituents in the other food. For example, soybean meal, which is rich in ruminally available N, will improve the digestion of a low-quality, low-N forage such as cereal straw. On the other hand, increased amounts of readily fermentable carbohydrates, such as we find in barley or wheat grains,

will reduce the extent of fibre digestion in low-quality forages. Economides (1998) has reported data which illustrate these points very well.

Potential Digestion and Rate of Digestion

So far, we have considered only the extent of digestion, i.e. how much of a particular food is digested, but an equally important concept especially for ruminant animals is the rate of digestion, or how fast digestion occurs. Digestion can be considered to be a first-order process, i.e. one in which the amount of digestible food constituent removed in each time period is a constant proportion of the amount present at the beginning of that period (Smith *et al.*, 1972; Waldo *et al.*, 1972; Mertens and Loften, 1980). Digestion rate can be measured by digesting food *in vitro* (e.g. using buffered rumen liquor) or in synthetic fibre bags suspended in the rumen, and recording the proportions of the food constituents which are digested over 72 or 120 h (Van Keuren and Heinemann, 1962; Smith *et al.*, 1972) or sometimes for longer periods. These digestion times are long enough to remove all, or nearly all, of the digestible material in a food. From this we can estimate the 'potential digestibility' of food constituents.

The disappearance of potentially digestible DM (PDDM) is determined by plotting the proportion of the sample which remains at each incubation time (Fig. 3.2A). If the assumption of first-order kinetics is correct, we will see an initial rapid loss of DM, with smaller losses as time progresses. The curve is extrapolated, if necessary, to find the proportion of the sample DM remaining when the plot plateaus. This plateau is the potentially indigestible DM (PIDM). Alternatively, we can assume that the proportion of DM remaining at the end of a long incubation is the PIDM. The PIDM is subtracted from the proportion of DM remaining at each incubation time to get the proportion of PDDM remaining at that time (Fig. 3.2B). The plot of PDDM remaining versus time is an exponentially declining curve. The natural logarithm versus time plot (Fig. 3.2C) gives an approximately straight line, the gradient of which is the fractional digestion rate of DM.

It is useful to separate the digestible component of foods into fast-digesting and slowly digesting fractions. We can treat the fast-digesting fraction as disappearing instantaneously, while the slowly

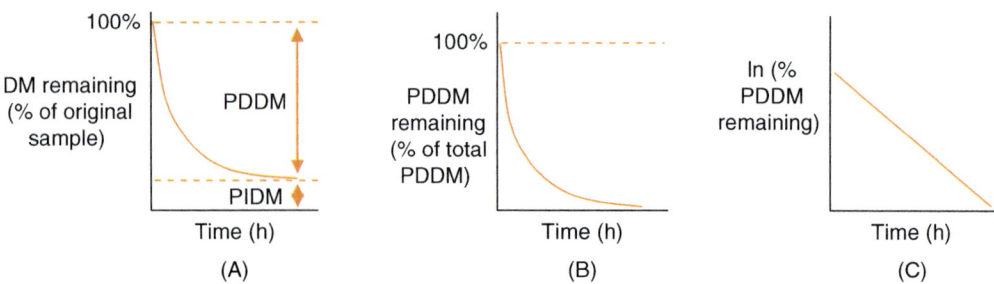

Fig. 3.2. Calculating the rate of food digestion: (A) plot of the proportion of sample DM remaining at each incubation time; (B) plot of the proportion of potentially digestible DM (PDDM = total DM – PIDM) remaining at each incubation time; (C) plot of the natural logarithm of PDDM remaining at each incubation time.

digesting fraction is digested at a rate which is largely determined by the chemistry of the food. If we know the amounts of each fraction, and the rate at which the slowly digesting fraction digests, then we can calculate the proportion of food which will be digested at any time (t h) after the start of digestion by the following relationship:

$$D_t = A + (B_0 - B_0 \times e^{-ct}) \qquad (3.4a)$$

Equation 3.4a simplifies to (Ørskov and McDonald, 1979):

$$D_t = A + B_0 (1 - e^{-ct}) \qquad (3.4b)$$

where

D_t = the proportion of the food which is digested after t h

A = the fast-digesting fraction of the food

B_0 = the proportion of the food which is the slowly digesting fraction at $t = 0$ h (the start of digestion)

= 1 – (A + proportion of potentially indigestible constituent)

$B_0 \times e^{-ct}$ = the proportion of slowly digesting food remaining after t h of digestion

c = the fractional digestion rate (/h) of the slowly digesting fraction

Effective Digestibility Is the Extent of Digestion Influenced by the Time Food Is Exposed to Digestion

'Effective digestibility' (or 'effective degradability' if we are talking about protein degradation in the rumen) is the degree to which food constituents are digested when this process is limited by the amount of time the food is exposed to digestive enzymes (Fig. 3.3). The concept was developed for the ruminal digestion of protein (Ørskov and McDonald, 1979; SCA, 1990).

The relationship between digestibility (D) and effective digestibility (eD) is modulated by the fractional digestion rate and the fractional passage rate, i.e. the rate at which food passes through the

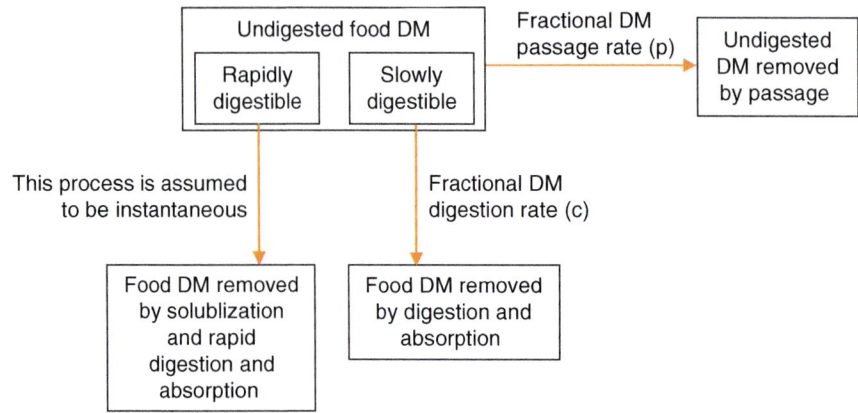

Fig. 3.3. Loss of dry matter (DM) from a digestive organ (e.g. the rumen) by digestion and passage.

organ where it is digested (e.g. the rumen). The effective digestibility (eD) is calculated thus (Ørskov and McDonald, 1979):

$$eD = A + B_0[c / (c + p)] \qquad (3.5)$$

where

eD = the proportion (decimal) of the food constituent which is digested in the digestive tract compartment of interest (often the rumen)

A = the fast-digesting fraction (decimal) of the food (measured as the intercept of the degradation curve at $t = 0$ h)

B_0 = the proportion of the food which constitutes the slowly digesting fraction at the start of digestion

c = the fractional digestion rate (/h) of the slowly digesting fraction

p = the fractional passage rate (i.e. the rate at which undigested food constituent moves through the digestive tract compartment) (/h)

This concept can be applied to the digestion of any food constituent (e.g. ruminal digestion of DM and starch; Ramírez et al., 2004; Jensen et al., 2005) in any section of the digestive tract.

Metabolizable Energy

The efficiency of the use of absorbed energy sources varies with the type of animal. Some 95% of the energy absorbed from the non-ruminant digestive tract (i.e. the DE) is presented to the body cells for use, with the remainder being excreted in the urine. Ruminants experience additional losses of energy during rumen fermentation. Some of the food energy which is used by the rumen microorganisms for metabolism and growth, and also the waste products resulting from microbial metabolism (especially methane, CO_2 and NH_3), are lost to the animal. These losses are conventionally charged against metabolizability, although a biologically correct scheme would charge rumen gas formation against digestibility (NRC, 1966). Efficiencies of conversion of DE to ME are (SCA, 1990; NRC, 1998):

Non-ruminants

$$M/D = 0.96 \, DE \qquad (3.6a)$$

Ruminants

$$M/D = 0.81 \, DE \qquad (3.6b)$$

where

M/D = metabolizable energy content (MJ/kg DM)

In ruminant nutrition, 'metabolizability' (q) means the ratio of ME/total food energy (ARC, 1980); equivalent to [(M/D)/18.4], assuming that the total energy content of foods is generally 18.4 MJ/kg DM.

Ruminant nutritionists use ME, rather than DE, to describe the available energy in foods because of the variations in efficiencies of utilization of ME which have been observed when ruminants use different diets for different purposes (Blaxter, 1962; ARC, 1980). These variations can lead to the overprediction of animal responses if DE is used, especially when fibre-rich foods are fed.

Net Energy

The final loss of energy occurs when ME is used by the body's cells. Heat is produced when cell constituents, or animal products, are synthesized. This is because the body's tissue- or product-building reactions are endergonic, i.e. they are chemical reactions in which there is more free energy in the products than in the reactants. These reactions are thermodynamically impossible unless they are coupled with exergonic reactions, which release more energy than is accumulated in the endergonic reaction (Conn and Stumpf, 1976). The energy difference is lost as heat. All of the heat produced by the animal is charged against nutrient use, i.e. the step between ME and NE, even though some is heat of microbial fermentation and is thus an energy loss which accrues during digestion.

Because ME is used by ruminants with varying efficiencies, each food has several possible NE contents. In practice, authorities report NE values for maintenance and growth, as these are substantially different. Values for lactation are similar to maintenance, so the unit $NE_{lactating \, cows}$ is used to combine both functions (NRC, 2001).

Glycaemic Index

The glycaemic index (GI) is a measure of the extent and rate of uptake into the blood of glucose from the digestion of carbohydrates (Jenkins et al., 1981). It is measured by giving a test meal and recording the area under the blood glucose concentration curve for 2 h after food consumption (Wolever et al., 2003), or can be predicted from the enzymic hydrolysis of foods (Englyst et al., 1999). GIs are increasingly used in the treatment of human diabetes and obesity (Chlup et al., 2004). Measured GIs have very large coefficients of variation of 20–60% (Wolever *et al.*, 2003), and it is best to class foods as low, medium and high GI, rather than to give a particular GI score. The concept is now being used in equine nutrition (Kronfeld *et al.*, 2004, 2005) where it may have application in the control of disorders of carbohydrate metabolism, including laminitis and myoglobinuria, and in the nutrition of performance horses. However, GI methodology, and its application to horses or other domestic animals, is still in its infancy.

Effective Neutral Detergent Fibre

Plant fibre, especially in forages, has several nutritional and physiological effects: these include maintenance of rumen wall health, milk fat content, and salivary flow and thus rumen buffering. From these considerations comes the concept of 'effective fibre' or 'effective NDF'. The theoretical basis for this was reviewed by Mertens (1997). Effective NDF (eNDF) is the amount of NDF in a ration which maintains milk fat content in lactating dairy cows (Pitt *et al.*, 1996). 'Physically effective NDF' (peNDF, Mertens, 1997) is the amount which maintains chewing activity and the presence of long particles in the rumen (the rumen 'mat' or 'raft'). Both the eNDF and peNDF contents of a food or diet must be determined by animal studies in which the amount of chewing (bites per gram of NDF), the levels of milk fat, or other physiologically relevant measurements, are determined experimentally. This is obviously laborious and expensive, and another approach is to estimate eNDF and peNDF contents by multiplying the total NDF content by an 'effectiveness factor'. Sniffens *et al.* (1992) have published a set of effectiveness values for concentrates and roughages, and Mertens (1997) has a table of physical effectiveness factors (pef) for hays, silages and some concentrates. Mertens' (1997) method for determining pef involves sieving the food sample and measuring the proportion which is retained on a 1.18 mm screen. This aperture size is consistent with measurements of the minimum particle size of forages which have a 100% probability of leaving the rumen through the reticulo-omasal orifice (Poppi *et al.*, 1985).

The Amino Acid Content of Foods for Monogastric Animals

The quality of the protein in a food for monogastric animals is described by the total protein content, to indicate the amount of total amino acids, and either the total or available contents of each of the essential amino acids. The list of amino acids so described may vary between data sets. Older data usually cite only lysine and methionine (or the total of methionine plus cystine). More recent data sets include valine, arginine and histidine, or perhaps all of the essential amino acids (e.g. NRC, 1998).

Availability of amino acids

The 'digestibility' of an amino acid is the proportion of that amino acid which can be released from its combination in protein by digestion and then absorbed from the digestive tract. Total collection and marker methods for measuring amino acid apparent digestibility are described by Jorgensen *et al.* (1984). The 'availability' or 'bioavailability' of an amino acid is the efficiency with which digested amino acid is retained in the body after it has been absorbed. Availability defined in this way includes the effects of not just digestibility and absorption, but also everything which influences the ability of the animal to use the amino acid for protein synthesis. This level of availability is sometimes called 'metabolic availability'. Ammerman *et al.* (1995) defined bioavailability as the 'degree to which an ingested nutrient in a particular source is absorbed in a form that can be utilized in metabolism by the animal'.

Values quoted for amino acid availabilities may vary because of the different ways in which this parameter is measured. Some of the more commonly used methods in pig nutrition are described below.

Ileal digestibility

This is the extent of digestion at the terminal ileum as the end point, rather than whole-tract digestibility measurements which use faeces as the end point. Ileal digestibilities are done by feeding rats a test diet which contains chromic oxide. After a suitable period (e.g. 7 days) the rats are slaughtered, and the ileal contents removed and analysed for each amino acid of interest. The digestibility of an amino acid is calculated in the same way as described for the marker digestibility method. Moughan *et al.* (1999) give a good description of the apparent ileal digestibility method. Either apparent (as described above) or true (where the data are corrected for the secretion of endogenous amino acid into the small intestine) ileal digestibilities can be measured.

Mobile nylon bag digestibility

Sauer *et al.* (1989) described a method in which 1 g samples of the test food are placed in 25×40 mm nylon bags which are given a pretreatment with pepsin, then inserted into the small intestine of the pig through a duodenal cannula. The amount of amino acid which disappears during passage through the intestines is called the 'available amino acid content'. This assay simply determines the proportion of the amino acid which is able to be digested and absorbed (or, more strictly, which is lost from the nylon bag during its transit through the small intestine). The advantage of this method over the standard *in vivo* technique is its rapidity and requirement for only a small sample size.

Fluorodinitrobenzene binding (for lysine only)

Lysine has two amino groups, one on each of the α and ε C atoms. When it binds to other amino acids in a polypeptide the peptide bonds involve the α amino group and the carboxyl group. This leaves the ε amino group free. If reactions such as the Maillard reaction occur, the ε amino group can be bound to non-protein cell constituents. This form of lysine is not easily released by digestive proteases, and so we refer to this lysine as 'unavailable'. The fluorodinitrobenzene (FDNB) binding method uses 1-fluoro-2,4-dinitrobenzene to bind to the free ε amino group. The amount of chemical so bound is determined by spectrophotometry and is an indication of the available lysine (Carpenter, 1960; Booth, 1971). This measures 'reactive' lysine, i.e. the lysine which is bound by peptide bonds to other amino acids but not to any other food constituent. The method gives results which are well correlated with slope-ratio assay values (Carpenter *et al.*, 1989). While the method is quick and cheap, it does not give a good prediction of metabolic availability for all foods. Moughan (2005) has discussed the use of FDNB and similar techniques combined with ileal digestibility measurements to predict the concentration of digestible reactive lysine in foods.

Slope-ratio assay

The method was devised by Batterham *et al.* (1979). The analysis involves feeding graded levels of the test ingredient in a standard food, and then repeating this with graded levels of a standard amino acid source. The animal's physiological response (often growth or some other production attribute, but also the accumulation of the test amino acid in the body) is measured, and the change in response between the test and standard ingredients is used to calculate the amino acid availability. For a careful description of a slope-ratio assay see Batterham *et al.* (1986). This analysis measures metabolic availability, rather than digestibility.

Indicator amino acid oxidation

This method of measuring the availability of essential amino acids relies on the fact that excess amino acid (i.e. that which cannot be used in the synthesis of protein because other required amino acids are not present in the required amounts) is oxidized with the release of CO_2. If we label the amino acid under study with ^{14}C we can measure the amount of ^{14}C label excreted in the expired air. This amount decreases as the amount of the test amino acid increases in the diet and more of the indicator amino acid is incorporated into body protein. The slope of the relationship between indicator amino acid oxidized and test amino acid included in the diet varies according to the availability of the test amino acid in the ingredient used to supply it. We can calculate the availability of the test amino acid in the food ingredient we are studying by finding the ratio of the oxidation line gradient of the test ingredient to that of the free amino acid (Fig. 3.4). The method was suggested by Black and Batterham (1987) and some preliminary studies in pigs were done by Ball *et al.* (1995). At this stage the method is under test but not yet properly validated.

Biological value of dietary protein

The biological value (BV) is an index of the similarity between the animal's amino acid requirements and the amino acid profile of the food protein. Animals have a very limited ability to store excess dietary protein, and any amino acids which are not used soon after having been absorbed are deaminated and their amino-N is excreted as urea. Accordingly, when the pool of absorbed dietary amino acids more closely matches those required, less of the absorbed amino acid is excreted. The BV (%) is the proportion of absorbed amino acid-N which is retained by the animal (Equation 3.7a).

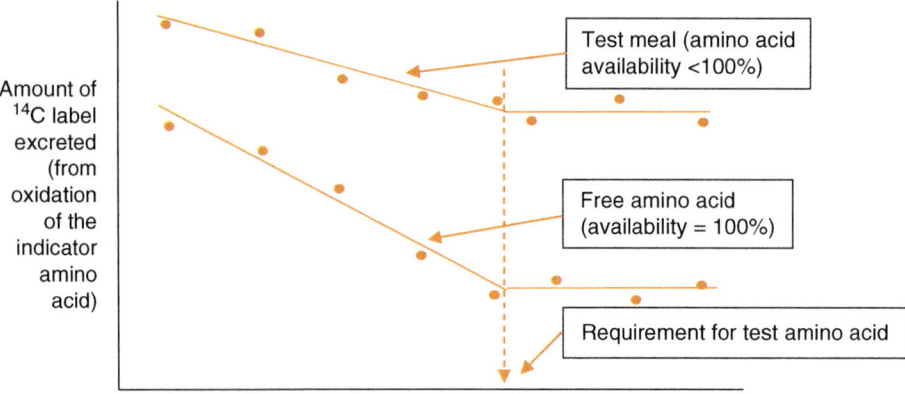

Fig. 3.4. Comparison of the regressions of ^{14}C label derived from the indicator amino acid against the amount of test amino acid in the diet, supplied from either the food ingredient under study or from free (i.e. chemically pure) amino acid.

$$BV = \frac{\text{Dietary amino acid absorbed and retained (g/day)}}{\text{Dietary amino acid absorbed (g/day)}} \times 100$$
(3.7a)

In terms of entities which can be measured in a feeding experiment, this becomes (Mitchell, 1924):

$$BV = \frac{NI - [(FN - MFN) + (UN - EUN)]}{[NI - (FN - MFN)]} \times 100$$
(3.7b)

where
NI = N intake (g/day)
FN = total faecal N excretion (g/day)
MFN = metabolic faecal N excretion (g/day)
UN = urine N excretion (g/day)
EUN = endogenous urinary N excretion (g/day)

The Availability of Protein in Ruminant Diets

Adequate nutrition of the rumen microorganisms is essential for proper rumen function. Ammonia is an essential nutrient for most bacteria (Wolin *et al.*, 1997). Ruminant foods must supply enough ruminally degradable protein (RDP) to maintain an acceptable rumen NH_3 content, or the animal will suffer the consequences of reduced food digestion and intake (Hunter and Siebert, 1985; Baumann *et al.*, 2004; Wickersham *et al.*, 2004). Food protein is digested by rumen microorganisms to varying extents depending on the chemical and physical composition of the food. Rumen degradability (dg) is thus an important characteristic of the dietary

protein in ruminant foods. It is measured by the *in vitro* or *in sacco* methods described for other food constituents.

In sacco residues are contaminated by microbial N and possibly by the influx of small food particles from the rumen liquor. Washing after removal from the rumen can affect the amount of different N fractions remaining in the residues (Silke and Udén, 2006). The dg measured in this way is an 'apparent' coefficient, which may underestimate the true degradability.

The extent to which a food protein is degraded in the rumen depends both on its inherent susceptibility to microbial digestion and the time that it remains in the rumen. Protein-rich foods may move through the rumen quite quickly – up to nearly 10% per hour at high levels of feeding (AFRC, 1992), so the proportion which will actually be degraded is less than the potential. The rate of passage of undigested food influences the effective degradability of protein (Edg). Edg is calculated as described previously (Equation 3.5).

Availability of Minerals

Mineral 'availability' has the same meaning as amino acid availability, and is measured by feeding trials, isotope dilution or increase in the amount of mineral contained in a particular metabolite or body compartment (see descriptions by Underwood and Suttle, 1999). Most nutrient content tables give values for the total amount of each individual mineral in the food, rather than the amounts of available minerals. This reflects a lack of information and also the multiplicity of factors which affect mineral absorption and retention. Mineral availability can be influenced by the animal's mineral status, age, type of diet, presence of other minerals or vitamins in the diet, and genotype. This makes it difficult to cite values for mineral availabilities. Nevertheless, some authorities (e.g. Jarrige, 1989; AFRC, 1991; NRC, 2001) apply estimated intestinal absorption coefficients to develop their mineral feeding standards.

The major constraint to the availability of phosphorus to monogastrics in plant-based diets is the proportion which is in the form of phytic acid. The availability to monogastrics of phytate-P varies, but it is generally low. The non-phytate phosphorus is an estimate of available phosphorus in these diet ingredients.

Mineral availability can be measured by comparing the change in retention accompanying a change in the amount of mineral digested (Underwood, 1981) when two levels of mineral are fed:

$$Min_{avail} = [(\Delta Min_{retained})/(\Delta Min_{consumed})] \times 100 \tag{3.8}$$

where

Min_{avail} = availability of the mineral (%)

$\Delta Min_{retained}$ = change in amount of mineral retained (mg/day)

$\Delta Min_{consumed}$ = change in amount of mineral consumed (mg/day)

Min_{avail} is similar to the regression coefficient obtained by plotting mineral retention against several levels of digestible mineral intake, i.e. an approach similar to the slope-ratio assay for amino acids.

References

Agricultural and Food Research Council (AFRC) (1991) A reappraisal of the calcium and phosphorus requirements of sheep and cattle. *Nutrition Abstracts and Reviews (Series B)* 61, 573–612.

Agricultural and Food Research Council (AFRC) (1992) AFRC Technical Committee on Responses to Nutrients. Report No. 9. Nutritive requirements of ruminant animals: protein. *Nutrition Abstracts and Reviews (Series B)* 62, 787–835.

Agricultural Research Council (ARC) (1980) *The Nutrient Requirements of Ruminant Livestock.* CAB, Farnham Royal, UK.

Ammerman, C.B., Baker, D.H. and Lewis, A.J. (eds) (1995) *Bioavailability of Nutrients for Animals: Amino Acids, Minerals, and Vitamins.* Academic Press, San Diego, California.

Ball, R.O., Batterham, E.S. and van Barneveldt, R.J. (1995) Lysine oxidation by growing pigs receiving diets containing free and protein-bound lysine. *Journal of Animal Science* 73, 785–792.

Barry, T.N. and Duncan, S.J. (1984) The role of condensed tannins in the nutritional value of *Lotus pedunculatus* for sheep. 1. Voluntary intake. *British Journal of Nutrition* 51, 485–491.

Batterham, E.S., Murison, R.D. and Lewis, C.E. (1979) Availability of lysine in protein concentrates as determined by the slope-ratio assay with growing pigs and rats and by chemical techniques. *British Journal of Nutrition* 41, 383–391.

Batterham, E.S., Lowe, R.F., Darnell, R.E. and Major, E.J. (1986) Availability of lysine in meat meal, meat and bone meal and blood meal as determined by the slope-ratio assay with growing pigs, rats and chicks and by chemical techniques. *British Journal of Nutrition* 55, 427–440.

Baumann, T.A., Lardy, G.P., Caton, J.S. and Anderson, V.L. (2004) Effect of energy source and ruminally degradable protein addition on performance of lactating beef cows and digestion characteristics of steers. *Journal of Animal Science* 82, 2667–2678.

Black, J.L. and Batterham, E.S. (1987) A proposed method for determining amino acid availability in pig diets. In: *Manipulating Pig Production.* Proceedings of the Inaugural Conference of the Australasian Pig Science Association. Australasian Pig Science Association, Werribee, Australia.

Blaxter, K.L. (1962) *The Energy Metabolism of Ruminants.* Hutchinson, London.

Booth, V.H. (1971) Problems in the determination of FDNB-available lysine. *Journal of the Science of Food and Agriculture* 22, 658–666.

Carpenter, K.J. (1960) The estimation of the available lysine in animal protein foods. *Biochemical Journal* 77, 604–610.

Carpenter, K.J., Steinke, F.H., Catignani, G.L., Swaisgood, H.E., Allred, M.C., MacDonald, J.L. and Schelstraete, M. (1989) The estimation of 'available lysine' in human foods by three chemical procedures. *Plant Foods for Human Nutrition* 39, 129–135.

Chlup, R., Bartek, J., Øezníèková, M., Zapletalová, J., Doubravová, B., Chlupová, L., Seèkaø, P., Dvoøáèková, S. and Šimánek, V. (2004) Determination of the glycaemic index of selected foods (white bread and cereal bars) in healthy persons. *Biomedical Papers* 148, 17–25.

Clauss, M., Schwarm, A., Ortmann, S., Alber, D., Flach, E.J., Kuhne, R., Hummel, J., Streich, W.J. and Hofer, H. (2004) Intake, ingesta retention, particle size distribution and digestibility in the Hippopotamidae. *Comparative Biochemistry and Physiology, Part A: Molecular and Integrative Physiology* 139, 449–459.

Conn, E.E. and Stumpf, P.K. (1976) *Outlines of Biochemistry,* 4th edn. Wiley, New York, pp. 150–152.

Dove, H. and Mayes, R.W. (1991) The use of plant wax alkanes as marker substances in studies of the nutrition of herbivores – a review. *Australian Journal of Agricultural Research* 42, 913–952.

Dryden, G.McL. (1982) Endogenous nitrogen in ruminant faeces. *Proceedings of the Nutrition Society of Australia* 7, 132–135.

Economides, S. (1998) The nutritive value of sunflower meal and its effect on replacing cereal straw in the diets of lactating goats and ewes. *Livestock Production Science* 55, 89–97.

Englyst, K.N., Englyst, H.N., Hudson, G.J., Cole, T.J. and Cummings, J.H. (1999) Rapidly available glucose in foods: an *in vitro* measurement that reflects the glycemic response. *American Journal of Clinical Nutrition* 69, 448–454.

Ezeogu, L.I., Duodu, K.G. and Taylor, J.R.N. (2005) Effects of endosperm texture and cooking conditions on the *in vitro* starch digestibility of sorghum and maize flours. *Journal of Cereal Science* 42, 33–44.

Fahey, G.C. and Jung, H.G. (1983) Lignin as a marker in digestion studies: a review. *Journal of Animal Science* 57, 220–225.

Ferret, A., Plaixats, J., Caja, G., Gasa, J. and Prió, P. (1999) Using markers to estimate apparent dry matter digestibility, faecal output and dry matter intake in dairy ewes fed Italian ryegrass hay or alfalfa hay. *Small Ruminant Research* 33, 145–152.

Goering, H.K. and Van Soest, P.J. (1970) *Forage Fiber Analyses (Apparatus, Reagents, Procedures, and Some Applications)*. Agriculture Handbook No. 379. Agricultural Research Service, USDA, Washington, DC.

Hmeidan, M.C., Dryden, G.McL. and McCosker, J.E. (2000) Predicting digestibility in rusa deer (*Cervus timorensis*) stags with internal and external markers. *Asian-Australasian Journal of Animal Sciences* 13B(Suppl.), 163.

Hunter, R.A. and Siebert, B.D. (1985) Utilization of low quality roughage by *Bos taurus* and *Bos indicus* cattle. 2. The effect of rumen-degradable nitrogen and sulphur on voluntary food intake and rumen characteristics. *British Journal of Nutrition* 53, 649–656.

Jarrige, R. (ed.) (1989) *Ruminant Nutrition. Recommended Allowances and Food Tables*. INRA/John Libbey Eurotext, Paris.

Jenkins, D.J., Wolever, T.M., Taylor, R.H., Barker, H., Fielden, H., Baldwin, J.M., Bowling, A.C., Newman, H.C., Jenkins, A.L. and Goff, D.V. (1981) Glycemic index of foods: a physiological basis for carbohydrate exchange. *American Journal of Clinical Nutrition* 34, 362–366.

Jensen, C., Weisbjerg, M.R., Nørgaard, P. and Hvelplund, T. (2005) Effect of maize silage maturity on site of starch and NDF digestion in lactating dairy cows. *Animal Feed Science and Technology* 118, 279–294.

Jorgensen, H., Sauer, W.C. and Thacker, P.A. (1984) Amino acid availabilities in soybean meal, sunflower meal, fish meal and meat and bone meal fed to growing pigs. *Journal of Animal Science* 58, 926–934.

Jung, H.G., Mertens, D.R. and Payne, A.J. (1997) Correlation of acid detergent lignin and klason lignin with digestibility of forage dry matter and neutral detergent fiber. *Journal of Dairy Science* 80, 1622–1628.

Kaitho, R.J., Nsahlai, I.V., Williams, B.A., Umunna, N.N., Tamminga, S. and Van Bruchem, J. (1998) Relationships between preference, rumen degradability, gas production and chemical composition of browses. *Agroforestry Systems* 39, 129–144.

Kronfeld, D., Rodiek, A. and Stull, C. (2004) Glycemic indices, glycemic loads, and glycemic dietetics. *Journal of Equine Veterinary Science* 24, 399–404.

Kronfeld, D.S., Treiber, K.H., Hess, T.M. and Boston, R.C. (2005) Insulin resistance in the horse: definition, detection, and dietetics. *Journal of Animal Science* 83(E. Suppl.), E22–E31.

Lippke, H., Ellis, W.C. and Jacobs, B.F. (1986) Recovery of indigestible fiber from feces of sheep and cattle on forage diets. *Journal of Dairy Science* 69, 403–412.

Martin-Rosset, W., Vermorel, M., Doreau, M., Tisserand, J.L. and Andrieu, J. (1994) The French horse feed evaluation systems and recommended allowances for energy and protein. *Livestock Production Science* 40, 37–56.

Mason, V.C. (1969) Some observations on the distribution and origin of nitrogen in sheep faeces. *Journal of Agricultural Science, Cambridge* 73, 99–111.

Mayes, R.W., Lamb, C.S. and Colgrove, P.M. (1986) The use of dosed and herbage *n*-alkanes as markers for the determination of herbage intake. *Journal of Agricultural Science* 107, 161–170.

Mehrez, A.Z. and Ørskov, E.R. (1977) A study of the artificial fibre bag technique for determining the digestibility of feeds in the rumen. *Journal of Agricultural Science, Cambridge* 88, 645–650.

Mertens, D.R. (1997) Creating a system for meeting the fiber requirements of dairy cows. *Journal of Dairy Science* 80, 1463–1481.

Mertens, D.R. and Loften, J.R. (1980) The effect of starch on forage fiber digestion kinetics *in vitro*. *Journal of Dairy Science* 63, 1437–1446.

Mitchell, H.H. (1924) Method of determining the biological value of protein. *Journal of Biological Chemistry* 58, 873–903.

Moughan, P.J. (2005) Absorption of chemically unmodified lysine from proteins in foods that have sustained damage during processing or storage. *Journal of AOAC International* 88, 949–954.

Moughan, P.J., Dong, G.Z., Pearson, G. and Wilkinson, B.H.P. (1999) Protein quality in blood meal. II. The effect of processing on *in vivo* nitrogen digestibility in rats, protein solubility and FDNB-available lysine. *Animal Feed Science and Technology* 79, 309–320.

Mould, F.L. (2003) Predicting feed quality – chemical analysis and *in vitro* evaluation. *Field Crops Research* 84, 31–44.

National Research Council (NRC) (1966) *Biological Energy Interrelationships and Glossary of Energy Terms*, rev. edn. National Academy of Sciences, Washington, DC.

National Research Council (NRC) (1998) *Nutrient Requirements of Swine*, 10th rev. edn. National Academy Press, Washington, DC.

National Research Council (NRC) (2001) *Nutrient Requirements of Dairy Cattle*, 7th rev. edn. National Academy Press, Washington, DC.

Ørskov, E.R. and McDonald, I. (1979) The estimation of protein degradability in the rumen from incubation measurements weighted according to rate of passage. *Journal of Agricultural Science, Cambridge* 92, 499–503.

Ørskov, E.R., Hovell, F.D.deB. and Mould, F. (1980) The use of the nylon bag technique for the evaluation of feedstuffs. *Tropical Animal Production* 5, 195–213.

Pitt, R.E., Van Kessel, J.S., Fox, D.G., Pell, A.N., Barry, M.C. and Van Soest, P.J. (1996) Prediction of ruminal volatile fatty acids and pH within the net carbohydrate and protein system. *Journal of Animal Science* 74, 226–244.

Poppi, D.P., Hendricksen, R.E. and Minson, D.J. (1985) The relative resistance to escape of leaf and stem particles from the rumen of cattle and sheep. *Journal of Agricultural Science, Cambridge* 105, 9–14.

Ramírez, R.G., Haenlein, G.F.W., García-Castillo, C.G. and Núñez-González, M.A. (2004) Protein, lignin and mineral contents and *in situ* dry matter digestibility of native Mexican grasses consumed by range goats. *Small Ruminant Research* 52, 261–269.

Sauer, W.C., den Hartog, L.A., Huisman, J., van Leeuwen, P. and de Lange, C.F.M. (1989) The evaluation of the mobile nylon bag technique for determining the apparent protein digestibility in a wide variety of feedstuffs for pigs. *Journal of Animal Science* 67, 432–440.

Schneider, B.H. and Flatt, W.P. (1975) *The Evaluation of Feeds Through Digestibility Experiments*. University of Georgia Press, Athens, Georgia, pp. 151–168.

Shaver, R.D., Nytes, A.J. and Satter, L.D. (1986) Influence of the amount of feed intake and forage physical form on digestion and passage of prebloom alfalfa hay in dairy cows. *Journal of Dairy Science* 69, 1545–1559.

Silke, D. and Udén, P. (2006) Losses of particulate N during filtration and handling of feed and rumen incubation residues. *Animal Feed Science and Technology* 125, 123–137.

Smith, L.W., Goering, H.K. and Gordon, C.H. (1972) Relationships of forage compositions with rates of cell wall digestion and indigestibility of cell walls. *Journal of Dairy Science* 55, 1140–1147.

Sniffens, C.J., O'Connor, J.J.D., Van Soest, P.J., Fox, D.G. and Russell, J.B. (1992) A net carbohydrate and protein system for evaluating cattle diets: II. Carbohydrate and protein availability. *Journal of Animal Science* 70, 3562–3577.

Standing Committee on Agriculture (SCA) (1990) *Feeding Standards for Australian Livestock. Ruminants*. CSIRO, Melbourne, Australia.

Taylor, C.C. and Allen, M.S. (2005) Corn grain endosperm type and brown midrib 3 corn silage: site of digestion and ruminal digestion kinetics in lactating cows. *Journal of Dairy Science* 88, 1413–1424.

Teeter, R., Owens, F. and Mader, T. (1984) Ytterbium chloride as a marker for particulate matter in the rumen. *Journal of Animal Science* 58, 465–473.

Theodorou, M.K., Williams, B.A., Dhanoa, M.S., McAllan, A.B. and France, J. (1994) A simple gas production method using a pressure transducer to determine the fermentation kinetics of ruminant feeds. *Animal Feed Science and Technology* 48, 185–197.

Tilley, J.M.A. and Terry, R.A. (1963) A two-stage technique for the *in vitro* digestion of forage crops. *Journal of the British Grasslands Society* 18, 104–111.

Udén, P., Colucci, P.E. and Van Soest, P.J. (1980) Investigation of chromium, cerium and cobalt as markers in digesta. Rate of passage studies. *Journal of the Science of Food and Agriculture* 31, 625–632.

Underwood, E.J. (1981) *The Mineral Nutrition of Livestock*, 2nd edn. CAB, Farnham Royal, UK.

Underwood, E.J. and Suttle, N.F. (1999) *The Mineral Nutrition of Livestock*, 3rd edn. CAB International, Wallingford, UK.

Van Es, A.J.H. (1978) Feed evaluation for ruminants. I. The systems in use from May 1977-onwards in the Netherlands. *Livestock Production Science* 5, 331–345.

Van Keulen, J. and Young, B.A. (1977) Evaluation of acid-insoluble ash as a natural marker in ruminant digestibility studies. *Journal of Animal Science* 44, 282–287.

Van Keuren, R.W. and Heinemann, W.W. (1962) Study of a nylon bag technique for *in vivo* estimation of forage digestibility. *Journal of Animal Science* 21, 340–345.

Waldo, D.R., Smith, L.W. and Cox, E.L. (1972) Model of cellulose disappearance from the rumen. *Journal of Dairy Science* 55, 125–129.

Wickersham, T.A., Cochran, R.C., Titgemeyer, E.C., Farmer, C.G., Klevesahl, E.A., Arroquy, J.I., Johnson, D.E. and Gnad, D.P. (2004) Effect of postruminal protein supply on the response to ruminal protein supplementation in beef steers fed a low-quality grass hay. *Animal Feed Science and Technology* 115, 19–36.

Wolever, T.M.S., Vorster, H.H., Bjorck, I., Brand-Miller, J., Brighenti, F., Mann, J.I., Ramdath, D.D., Granfeldt, Y., Holt, S., Perry, T.L., Venter, C. and Wu, X. (2003) Determination of the glycaemic index of foods: interlaboratory study. *European Journal of Clinical Nutrition* 57, 475–482.

Wolin, M.J., Miller, T.L. and Stewart, C.S. (1997) Microbe–microbe interactions. In: Hobson, P.N. and Stewart, C.S.(eds) *The Rumen Microbial Ecosystem.* Chapman & Hall, London, pp. 329–381.

4

Physico-chemical Composition and Digestibility of Forages and Cereal Grains

Plant Leaves Have Four Typical Tissues

The leaves and stems of monocotyledonous and dicotyledonous plants share many similarities. However, there are also typical differences, especially in the way that the mesophyll and the vascular tissues are arranged and in the polysaccharide composition of their cell walls (Carpita and Gibeaut, 1993; Carpita, 1996). Grasses are important components of the diets of domestic herbivores, so this discussion emphasizes grass anatomy and the digestibility of grass leaf tissues.

The quantitatively important tissues in plant leaves are:

- Epidermis – one-cell-thick layers of cells which line the upper and lower surfaces of the leaves, and which contain the stomata and other structures. The external walls of epidermal cells are covered with a cuticle, a waxy layer which helps to conserve water.
- Mesophyll – a tissue consisting of large, thin-walled cells which fill much of the space between the upper and lower epidermis and the vascular bundles.
- Schlerenchyma – ribs of heavily lignified cells which run parallel to the long axis of the leaf (in grasses) and which support the vascular bundles.
- Vascular bundles – tissues arranged along the long axis of the leaf blade (in grasses) or in a network (in dicotyledonous plants) and which contain the xylem vessels (heavily lignified) and phloem cells. The vascular bundles are surrounded by a one- or two-layered bundle sheath.

There Are Characteristic Differences in the Leaf Anatomies of C3 and C4 Species Grasses

Forage plants have two main ways of fixing the CO_2 needed in photosynthesis. This is done in the temperate, 'festucoid' or 'C3', grasses by the carboxylation of ribulose-1,5-diphosphate to form two molecules of 3-phosphoglyceric acid (these are 3-carbon molecules). Tropical, 'panicoid' or 'C4', grasses fix CO_2 in their mesophyll cells as oxaloacetate (a 4-carbon molecule) and then translocate the oxaloacetate into the bundle sheath cells, thus pumping carbon into these cells where (in C4 species) the photosynthetic enzymes are located. The C4 photosynthetic pathway enables tropical forages to continue high levels of carbohydrate synthesis at both high and low light intensities, and to use water and N very efficiently (Hatch, 1987). This influences their leaf anatomies, because C4 species have to efficiently transport oxaloacetate between the mesophyll and bundle sheath cells. These differences and the reasons for them are described in a classic paper by El-Sharkawy and Hesketh (1965).

We can thus group grasses into two types, based on their leaf anatomies (Brown and Hattersley, 1989); although there are also grasses, such as some of the *Panicum* genus, which have intermediate anatomies. Because both carbon fixation and photosynthesis occur in the mesophyll cells of the C3 photosynthetic species, these species have a large number of mesophyll cells between adjacent vascular bundles. In the C4 species, carbon fixation occurs in the mesophyll but carbohydrate synthesis in the bundle sheath, so these grasses have more vascular tissue and less mesophyll tissue than the C3 species (e.g. see Brown and Hattersley, 1989; Dengler *et al.*, 1994). In the C4 grasses the mesophyll cells are usually no more than one cell from the bundle sheath cells (Brown and Hattersley, 1989), and they surround the vascular bundles in a radial pattern (called the Kranz anatomy), although this is a less consistent feature of the panicoid species than the low intervascular mesophyll cell count. The anatomies of typical monocotyledon forage leaves are shown in Fig. 4.1.

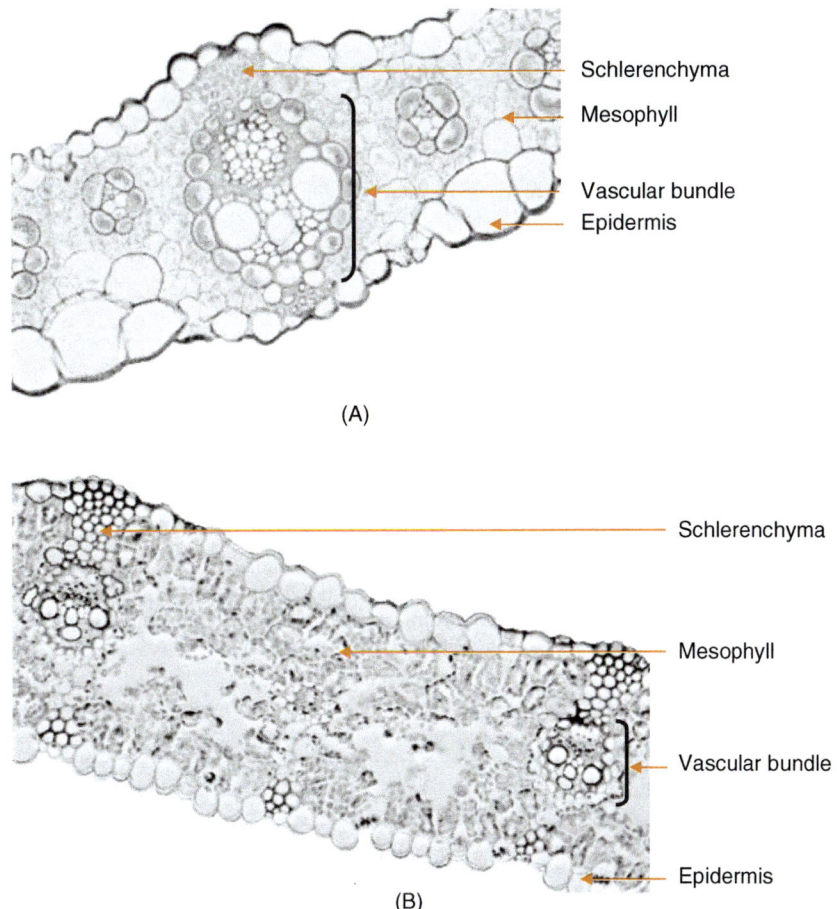

Schlerenchyma

Mesophyll

Vascular bundle

Epidermis

(A)

Schlerenchyma

Mesophyll

Vascular bundle

Epidermis

(B)

Fig. 4.1. Cross sections of typical C3 and C4 monocotyledonous plant leaves: (A) *Zea mays* (C4; panicoid; tropical); (B) *Bromus willdenowii* (C3; festucoid; temperate).

Whole-leaf Digestibility Is Influenced by Leaf Tissue Digestibility

The influence of plant anatomy on forage digestibility has been demonstrated in several studies. Several types of tissues resist digestion. Observations of leaf tissue digestion (Fig. 4.2) show that the mesophyll is rapidly and completely digested, while electron microscope studies have shown that the rigid and supportive cell walls of the vascular bundles (i.e. the inner bundle sheath cells and xylem vessels) which stain positive for lignin are totally undegraded by rumen microorganisms. Schlerenchyma (another lignified tissue) is only slightly degraded, and the parenchymal bundle sheaths and abaxial epidermis are also resistant. These and other observations (Akin, 1979; Harbers *et al.*, 1981) show that the susceptibility of plant tissues to degradability declines in the order shown in Fig. 4.3.

This order of susceptibility to microbial digestion (Fig. 4.3) is paralleled by the acid detergent fibre (ADF; lignocellulose) content of plant tissue. There is also a strong negative association between digestibility and the amounts of tissue which stain positively for lignin (Akin, 1988). Correlations between whole-leaf digestibility and the amounts of different leaf tissues show that plant digestibility is influenced by its tissue composition and the inherent resistance of each tissue to digestion (Table 4.1).

Grass Stems Contain Four Types of Tissues

The internodal stem has four tissues: epidermis, vascular tissue, parenchymal ground tissue (which is usually differentiated into cortex and pith) and

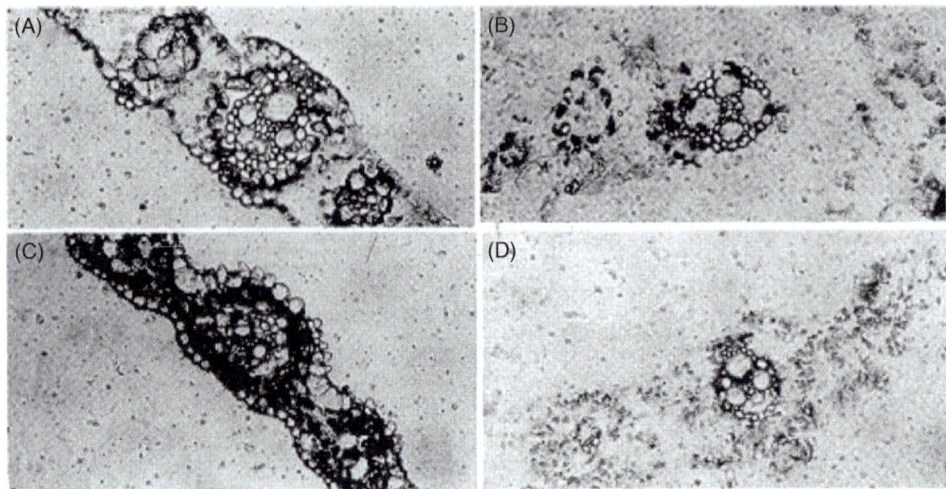

Fig. 4.2. The time course of digestion of *Panicum* leaves: (A) *Panicum virgatum* (C4 spp.) 6 h incubation in rumen liquor with loss of epidermis and mesophyll; (B) 48 h incubation with only vascular bundle (sheath and xylem) remaining; (C) *Panicum decipiens* (C3/4 spp.) 6 h incubation with no tissue loss; (D) 48 h incubation with some mesophyll, cuticle (non-cellular) and vascular bundle remaining. (Reprinted from Akin *et al.*, 1983. With permission.)

Most susceptible	>>>>	>>>>	Most resistant
Mesophyll		Epidermis	Schlerenchyma
Phloem		Bundle sheath	Lignified vascular tissue (xylem)

Fig. 4.3. Resistance of grass leaf tissues to microbial digestion.

Table 4.1. Proportions (%) of tissue types in *Panicum* spp. leaves, and correlations between tissue type and DM digestibility (DMD) and cell wall content (CWC) of the whole leaf. (From Wilson *et al.*, 1983.)

Leaf tissue	Proportions (%)[a]			Correlations[b]	
	C3	C3/4	C4	DMD	CWC
Mesophyll	66	48	43	0.62	−0.76
Bundle sheath	10	18	20	−0.67	0.69
Vascular tissue	3	6	8	−0.53	0.66
Epidermis	22	26	27	−0.23	0.44
Schlerenchyma	0.5	1.7	1.7	−0.29	0.51

[a]Proportions of tissue types in species having C3, C4 or intermediate C3/4 anatomies.
[b]Significant correlations between the proportion of each tissue type and either dry matter digestibility (DMD) or the content of cell wall constituents (CWC); pooled data over all species.

schlerenchyma (Brazle and Harbers, 1977; Juniper, 1979; Harbers *et al.*, 1982). The parenchyma provides a matrix for the vascular bundles and other structures. Where the vascular bundles are distributed in two concentric rings, the parenchyma is differentiated into the pith (or medulla), interfascicular parenchyma (or medullary rays) between the vascular bundles, and cortex. Pith cells are usually thin-walled and are often lignified (Harbers *et al.*, 1982). Their surfaces lining the pith cavity may be coated with cuticle (the medullary cuticle). Cells of the cortex may contain chloroplasts and silica (Juniper, 1979). Epidermal cells also frequently contain inclusions of silica (Harbers *et al.*, 1982) and are protected by a cuticle.

In the C3 cereal straws, the vascular bundles are distributed in two concentric bands around the periphery of the stem, while in C4 types (e.g. maize and sorghum) the bundles are distributed more evenly throughout the whole of the stem section (Juniper, 1979). The vascular bundles contain xylem and phloem, and may be surrounded by a schlerenchymal sheath. A ring of schlerenchyma is also found under the epidermis.

Grass Stem Tissues Have Variable Digestibilities

Microbial digestion of stems is initially fast while the susceptible tissue is attacked, but the rate subsequently falls. The digestion of wheat straw begins with the loss of the medullary cuticle and cracking through the parenchymal tissue to the schlerenchyma ring by 12 h, followed by a partial digestion of the longitudinal walls of the inner pith cells by 72 h (Harbers *et al.*, 1982). However, these stems maintain their general structure even after 72 h of microbial digestion, and the vascular tissue, schlerenchyma, outer cuticle and unexposed parenchymal cells remain undigested. Clover stems are less resistant to microbial digestion (Akin and Robinson, 1982) but the vascular tissue and interfascicular parenchyma are still identifiable after 24 h. The susceptibility of the stem tissues to bacterial attack is related to their lignification. Vascular (except the phloem) and schlerenchymal cells are extensively lignified, and lignin is present in the thin-walled parenchymal cells; these cells are not readily attacked by rumen bacteria (Akin *et al.*, 1995).

Lignification increases with plant age. The dry matter (DM) digestibility of the bottom portion of coastal Bermuda grass (*Cynadon dactylon*) is 15–20% units lower than the upper part of the plant (Akin *et al.*, 1977). Parenchymal cells in the upper stem are digested rapidly but the parenchymal tissue in the lower stem, or of vascular bundles or epidermis in either stem fraction, are much more resistant to digestion (Akin, 1988). These differences in digestibility are not due to changes in the proportions of tissue types, but to an increase in the amount of lignin in the parenchymal tissue of the lower stem.

The Amount and Digestibility of Cell Walls Determines Forage Digestibility

The cell contents of grasses are almost completely digestible (Van Soest, 1967; Minson, 1971), so the digestibility and intake of forages is limited mainly by the content and composition of their cell wall material. Figure 4.4 illustrates the relationship between forage cell composition and forage digestibility. As we have seen in Table 4.1 there are significant relationships between cell wall content and composition and digestibility. These also influence voluntary food intake (Ørskov *et al.*, 1988).

This discussion has concentrated on the disappearance of tissues from transverse sections, but vascular systems and fibres extend in a continuous three-dimensional network through the whole leaf or stem. The appearance of an area of undegraded xylem, say, in a cross section grossly understates the implications of that tissue's resistance to digestion, and especially its effects on voluntary intake. Plant fibres may be long: approximately 1 mm in reed canary grass and 3.5 mm in cereal straw (Rydholm, 1965; Finell and Nilsson, 2005). The presence of structures of this size which resist digestion will limit the rate of passage of food from the rumen (and thus food intake), as well as adversely affecting its digestibility.

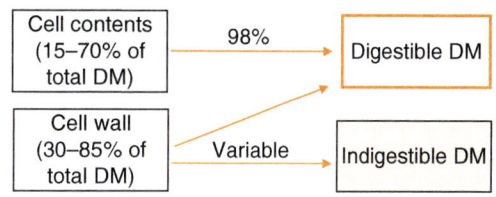

Fig. 4.4. Effect of cell composition on the digestibility of forage dry matter.

Chemistry of the Plant Cell Wall

The structure of a generalized mature plant cell wall is shown in Fig. 4.5. The primary wall is found in all cells; the secondary wall occurs in load-bearing or resistant tissues such as xylem vessels and fibres (Cosgrove, 2005; Minic and Jouanin, 2006).

The wall has four important constituents

The basic component is cellulose, a chemically resistant, inelastic, fibrous macromolecule made of glucose molecules joined in long, unbranched chains, which are further aggregated into microfibrils. The spaces between the microfibrils are filled with branched-chain matrix carbohydrates, the hemicelluloses and pectin. Further rigidity, especially in mature grasses, is given by the highly branched lignin macromolecule. Plant cell walls also contain a variety of enzymic and structural proteins. A pectin-rich middle lamella cements adjacent cell walls. In mature tissue it becomes infiltrated with cellulose and lignin.

There are large differences in the matrix carbohydrates between plant types, especially between dicotyledonous and non-grass monocotyledonous plants which have a Type I cell wall, and grasses which have a Type II wall. Type I walls (Carpita and Gibeaut, 1993) are rich in xyloglucans, pectin and glycoproteins; Type II walls (Carpita, 1996) are rich in glucuronoarabinoxylans. Typically grasses have very little pectin (e.g. 2.5% of DM in ryegrass; Ben-Ghedalia *et al.*, 1993), while pectin can be the quantitatively most important cell wall constituent in dicotyledonous plants (Table 4.2). The matrix carbohydrates also vary between tissues within the same plant, and with the stage of development. For example, in the developing maize coleoptile there is more glucomannan in the epidermal cell walls while the mesophyll walls contain more mixed-linkage β-glucans, and the epidermal cells have about 60% cellulose while the mesophyll cells contain about 30% (Carpita *et al.*, 2001).

Plant cell wall architecture

In the grass cell wall the cellulose microfibrils are tightly linked to mixed-linkage β-glucans (during active cell elongation), glucuronoarabinoxylans and glucomannans (Urbanowicz *et al.*, 2004). This structure is further embedded in a matrix of glucoronoarabinoxylans, glucomannans and pectins. Several schemes have been proposed for the 'linkages' between cellulose and the matrix wall components. These include (Cosgrove, 2005) H bonds, trapping of hemicellulose macromolecules between developing cellulose microfibrils, non-covalent and/or covalent bonding. These linkages, and covalent bonds

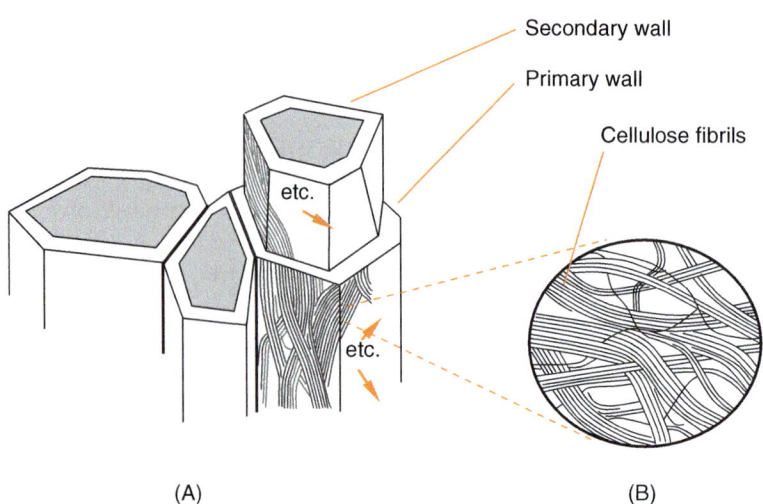

(A)   (B)

Fig. 4.5. Generalized structure of plant cells: (A) primary and secondary cell walls showing arrangements of cellulose microfibrils; (B) model of cellulose microfibril structure in the primary wall. (Drawn from data in Shafidazeh and McGinnis, 1971; Kögel-Knabner, 2002.)

Table 4.2. Cell wall composition of the leaf and stem of monocotyledonous and dicotyledonous plants.[a]

Plant (part)	Cellulose	Hemicelluloses	Pectins	Lignin	Source
Monocotyledonous plants					
Cocksfoot (whole)	31	34			Xu *et al.*, 2006
Ryegrass (whole)	28	41			Xu *et al.*, 2006
Barley (straw)	44	27		7	Carmona and Greenhalgh, 1972
Barley (straw)	42	28		7	Dryden and Leng, 1988
Oat (straw)	41	16		11	Saxena *et al.*, 1971
Rhodes grass (whole)	32	34		4	Granzin and Dryden, 2003
Chionochloa spp. (whole)	31	28		7	Connor *et al.*, 1970
Maize (whole)	28	15	8	15	Abedon *et al.*, 2006
Dicotyledonous plants					
Lucerne (stem)	41	17	21	22	Jung and Lamb, 2006
Lucerne (stem)	38	15	13	10	Jung and Lamb, 2004
Lucerne (xylem)	39	29	4	28	Grabber *et al.*, 2002
Lucerne (non-xylem)	30	30	25	15	Grabber *et al.*, 2002
Chickpea (straw)	30	20		10	Johnson and Pezo, 1975
Lotus pedunculatus (whole)	16	10	52	18	Barry *et al.*, 1986
Chicory (lamina)	21	57	67		Sun *et al.*, 2006

[a]The values in this table are not strictly comparable between plant types as they have been obtained with a variety of different extraction/fractionation methods.

between the hemicelluloses, pectins and lignin (Carpita, 1996), lock the wall components in place when cell growth has stopped.

Chemistry of the Cell Wall Carbohydrates

Cellulose

This is made by cellulose synthesis proteins in the plasma membrane. The cellulose macromolecule is a $(1\rightarrow4)$-β-D-glucan; an unbranched chain of glucose molecules linked by β1→4 glycosidic bonds (i.e. covalent bonds which link, with an intervening oxygen atom, the carbon atoms of adjacent glucose molecules at the 1 and 4 positions of the pyranose rings). Estimates of the number of glucoses in each chain vary from 10,000 to 25,000 (Brown *et al.*, 1996; O'Sullivan, 1997). Hydrogen bonds between the hydroxyl hydrogens and the oxygen atoms of the pyranose rings stabilize the structure (Nishiyama *et al.*, 2002; Fig. 4.6A). Cellulose macromolecules align parallel to each other to form a microfibril (Fig. 4.6B). These are aggregations of 30–36 cellulose macromolecules arranged in parallel (Vorwerk *et al.*, 2004), and are about 5 nm wide. In the plant secondary wall,

microfibrils further aggregate into macrofibrils (Delmer and Amor, 1995).

Cellulose I_β is the type which occurs in higher plants (Atalla and VanderHart, 1984). While most cellulose is crystalline, some (possibly 30% or more) of cellulose I_β is 'amorphous'. Some authors have suggested that amorphous cellulose is more easily digested. However, lack of crystallinity may simply indicate a particular structure of the microfibril surface and not a true lack of crystallinity (Shafizadeh and McGinnis, 1971; O'Sullivan, 1997). A detailed illustration of the surface appearance of crystalline cellulose is given by Baker *et al.* (2000).

A related macromolecule, a mixed-linkage $(1\rightarrow3,1\rightarrow4)$-β-D-glucan, is present in the walls of growing cells of grasses and cereals. It is found tightly bound to cellulose microfibrils in places where high mechanical loads are expected (Urbanowicz *et al.*, 2004).

Pectins

There are three main pectic substances (Vorwerk *et al.*, 2004). Homogalacturonan is an unbranched α-galacturonan of 25 or more galacturonic acid molecules linked by α1→4 glycosidic bonds, and with no

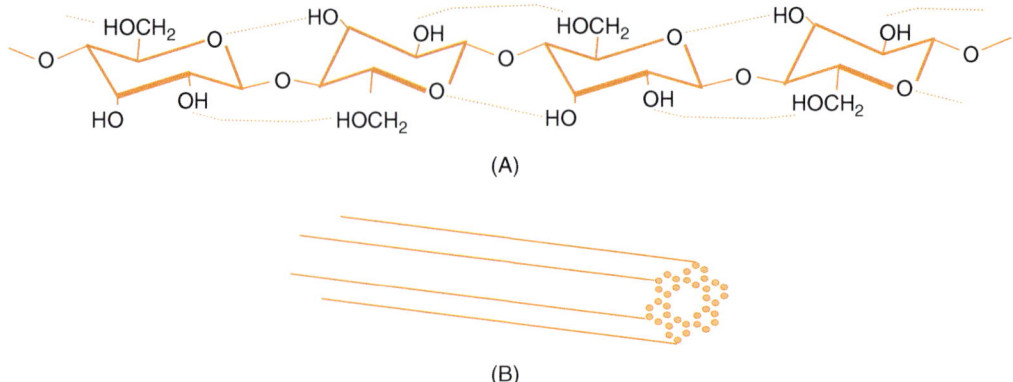

(A)

(B)

Fig. 4.6. The structure of cellulose: (A) part of a cellulose molecule showing the glucose molecules joined by β1→4 glycosidic bonds (⁻O⁻) and the structure stiffened by H bonding (····). (From Delmer and Amor, 1995.) (B) Structure of a cellulose microfibril. (From Varner and Lin, 1989; Carpita, 1996; Cosgrove, 2005.)

side chains, though bearing methyl and acetate groups (Pérez *et al.*, 2000). Rhamnogalacturonan I has a backbone of a disaccharide made from rhamnose and galacturonic acid molecules linked by α1→2 and α1→4 glycosidic bonds (Ridley *et al.*, 2001); there are about 1000 disaccharide units in each chain. Rhamnogalacturonan II has a galacturonan backbone with four types of highly branched side chains with both α and β glycosidic bonds and of varying monosaccharide composition including galacturonic acid, rhamnose, fucose, glucuronic acid, xylose, galactose and arabinose (Pérez *et al.*, 2000). These macromolecules are stabilized by Ca^{2+} ions and borate ester bonds (Vorwerk *et al.*, 2004).

The backbone of the pectin macromolecule consists of regions of unsubstituted (except for methyl groups) homogalacturonan, interspersed with regions of rhamanogalacturonan I bearing extensive side chains (Pérez *et al.*, 2000). An alternative structure with a rhamnogalacturonan I backbone and homogalacturonan side chains has been suggested by Vincken *et al.* (2003). The pectin macromolecules are present in all of the cell wall, but rhamnogalacturonan II is not found in the intercellular substance (Vorwerk *et al.*, 2004).

Hemicelluloses

These polysaccharides are synthesized (like the pectins) in the Golgi bodies, from where they migrate to fill the spaces between the cellulose microfibrils (Cosgrove, 2005). The hemicelluloses are polymers of a number of pentose and hexose sugars, including xylose, arabinose, mannose, galact-

ose and rhamnose, and glucuronic, galacturonic and O-methyl-galacturonic acids (Gaillard and Bailey, 1968). The important hemicelluloses are:

1. Glucuronoarabinoxylan has a backbone of about 200 β1→4 linked xylose molecules (i.e. a β1→4 xylan) to which are attached galacturonic acid and arabinose molecules (Fig. 4.7A). This polysaccharide is important in temperate and tropical cereals and grasses (Lawther *et al.*, 1995; Sun and Sun, 2002; Chaikumpollert *et al.*, 2004).
2. Xyloglucan has a (1→4)-β-D-glucan backbone with xylose branches; the substituent xylose may bear additional galactose and fucose molecules (Fig. 4.7B).
3. Arabinoxylan is a (1→4)-β-D-xylan with arabinose substituents on 70% of the xylose molecules (Minic and Jouanin, 2006).
4. Glucomannans are found in relatively small amounts in grass cell walls.

Lignin

Lignification (deposition of lignin throughout the cell wall) begins in the primary wall and middle lamella, and lignin infiltrates the whole wall structure as the cell becomes increasingly mature. Lignin is a highly branched polymer made by the condensing (joining together) of derivatives of cinnamic acid. There is a very accessible account of lignin formation and structure by Hatfield *et al.* (1999) and present understanding of the synthesis of lignin is well described by Grabber *et al.* (2004). Briefly, cinnamic acid is converted by a series of hydroxylation, methylation and reduction steps to *p*-coumaric, ferulic and sinapic acids, which are the precursors of the three lignin

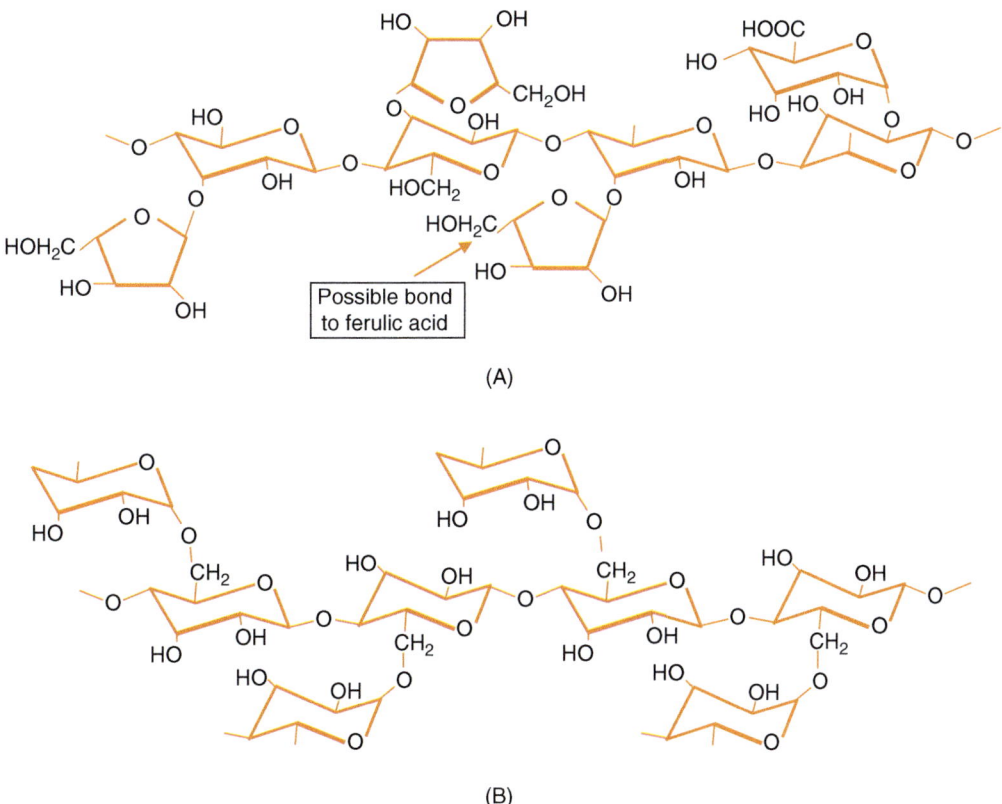

(A)

(B)

Fig. 4.7. Structure of two hemicellulose polysaccharides: (A) glucuronoarabinoxylan (from Carpita, 1996); (B) xyloglucan (from Pauly *et al.*, 1999).

monomers, i.e. *p*-coumaryl, coniferyl (called 'guaiacyl units') and synapyl ('syringyl units') alcohols. Grass lignin mainly consists of guaiacyl and syringyl units. Lignin can be divided into 'core' and 'alkali-soluble' (ferulic and *p*-coumaric acids) (Hatfield *et al.*, 1999). Lignin monomers become linked to each other by ester, ether and C–C bonds to form the polymerized, or core, lignin.

The structures of the basic chemical units of lignin, and the ways in which they may join together, are shown in Fig. 4.8; however, there is great variability. Durot *et al.* (2003) comment that 'there are as many structures of alkali-soluble lignins described in the literature as processes used to isolate them'. Removing plant cell wall structures intact so that they can be identified and measured is a problem which also applies to the investigation of cellulose and hemicellulose. Nevertheless, there is a consensus that there are several different types of lignin, and that the chemistry of lignin varies within the plant and between

plant species (Akin, 1982; Wallace *et al.*, 1991; Grabber *et al.*, 2004).

Lignin encrusts the cellulose microfibrils, rather than being chemically bound to them, but it is covalently bound to hemicelluloses and pectins by ferulic and *p*-coumaric acid bridges, which involve ester and ether linkages particularly to xylose and arabinose (Chesson *et al.*, 1983; Hatfield *et al.*, 1999; Xie *et al.*, 2000; Sun *et al.*, 2001).

Lignin negatively affects forage digestibility, but the mechanism is complex

Forage cell wall digestibility is negatively related to its lignin content (Harkin, 1973; Traxler *et al.*, 1998). However, a point of interest is that lignin concentration is related to the potential digestibility, i.e. the extent of digestion, rather than the digestion rate (Jung and Lamb, 2003). The role of lignin in constraining the digestion of plant material is complex. Zerbini and Thomas (2003) review this

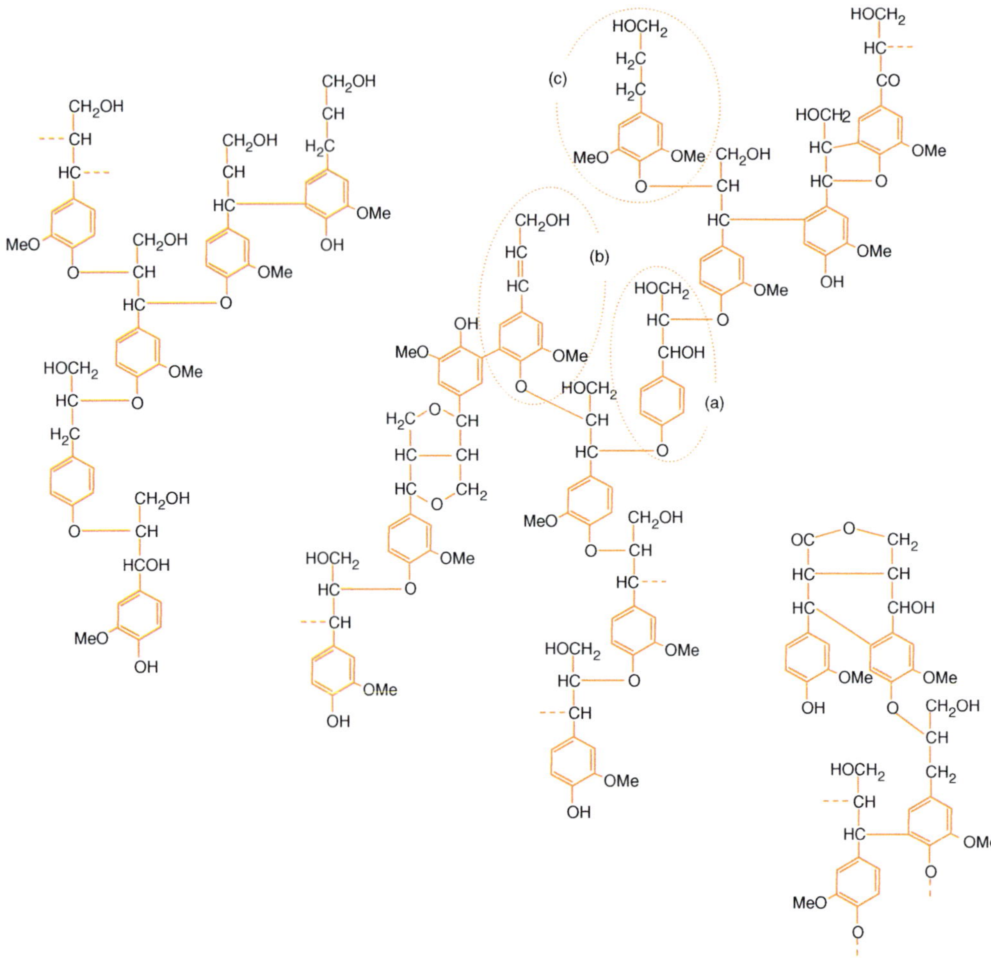

Fig. 4.8. Possible structures within the lignin macromolecule; incorporating derivatives of (A) p-coumaryl alcohol, (B) coniferyl alcohol and (C) synapyl alcohol. (Adapted and redrawn from Sarkanen and Ludwig, 1971.)

topic and the nutritional and non-nutritional ramifications of breeding low-lignin plants.

The degree of protection against digestion conferred by lignin is not linearly related to the lignin content of the DM, especially when data from different plant species are pooled. Differences in cellulose digestibility have been reported between forages with identical lignin contents (Johnston and Waite, 1965), and forages with the lowest lignin contents are not necessarily the most digestible (Kamstra et al., 1958). Temperate and tropical grass species may have similar digestible cell wall and digestible structural polysaccharide contents, even though the temperate species have less lignin.

McLeod and Minson (1974) showed that the ratio of indigestible polysaccharide to lignin in a selection of tropical grasses was almost constant between species, while the lignin to total polysaccharide ratio varied.

This observation is supported by the non-linear relationship between lignin content and the digestibility of neutral detergent fibre (NDF) (Traxler et al., 1998). If the relationship was linear, then we could argue for a simple, chemical protection of cell wall polysaccharides by lignin; a non-linear relationship implies a spatial as well as a chemical effect.

There is more evidence that this is the correct interpretation. When cell walls are digested lignin becomes concentrated at the digestion surface. Chesson (1984) exposed barley straw to microbial digestion and recorded 45% lignin at the digestion surface, but only 11% in the whole sample. He suggested that 'accumulation of phenolic material ultimately builds to a level where it effectively forms an inert layer protecting the underlying wall from further attack' (Chesson, 1988). The model of cell wall digestion in Fig. 4.9 illustrates

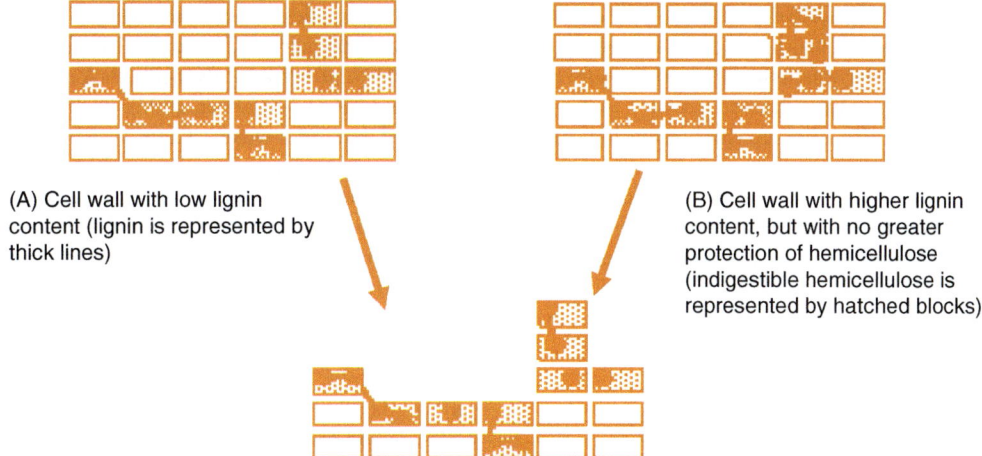

(A) Cell wall with low lignin content (lignin is represented by thick lines)

(B) Cell wall with higher lignin content, but with no greater protection of hemicellulose (indigestible hemicellulose is represented by hatched blocks)

(C) Cell wall after digestion by hemicellulases:
- Protection of hemicellulose depends on the spacial distribution of lignin, as well as the amount of lignin
- The lignin content of the undigested residue is much greater than that in the original cell wall

Fig. 4.9. A model of lignin protection. (Redrawn from the model of Chesson, 1993.)

this characteristic of the lignin effect on cell wall digestibility.

A major property of lignin is thus its ability to protect other cell wall constituents from digestion. There are several ways in which this may occur:

1. Lignin is inherently indigestible and dilutes the digestible components of the cell wall. The apparent digestibility of lignin has been reported to vary from –7.4% to 18.1%. Workers have consistently reported positive digestibility coefficients for the lignin in lucerne, negative coefficients for the lignin in sanfoin, and variably positive or negative values for the lignin in diploid and tetraploid ryegrasses. These coefficients (especially the negative ones) may be artefacts resulting from unsatisfactory techniques for lignin estimation or other analytical problems, e.g. the presence of Maillard substances from inappropriate drying (Van Soest, 1964). Alternatively, lignin might be chemically altered, or dissociated from the main part of the digesta, during passage through the ruminant digestive tract. Some 14–17% of lignin is soluble *in vitro* at pHs similar to those in the rumen (Van Soest, 1964). Solubilized lignin may leave the rumen more quickly than the remaining undigested cell wall constituents and then be re-precipitated in the more acid conditions of the abomasum. This might allow it to travel through the digestive tract at a different rate from the other feed constituents.

2. Tissues need not be lignified to have some protection against digestion. Digestion may be impeded by the presence of *p*-coumaric and ferulic acids ester-linked to hemicellulose (Jung, 1988; Akin, 2007), i.e. not the presence of lignin per se. In particular, *p*-coumaric acid greatly inhibits bacterial and protozoal metabolism and its presence reduces the digestion of mesophyll, a usually rapidly digested plant tissue (Akin, 1982).

3. Lignin may prevent the adhesion of cellulolytic bacteria. Cellulolytic bacteria adhere closely to the polysaccharides which they are digesting. Adhesion appears to be restricted at lignin-rich sites (Akin, 1979; Akin and Barton, 1983). Consistent with this, there is more bacterial adhesion to alkali-treated straw (Latham *et al.*, 1979), and this treatment removes some of the lignin.

4. Enzymes which digest cell wall polysaccharides may be inhibited by the bonding between lignin and hemicellulose. Lignin may protect cell wall carbohydrates by preventing bacterial enzymes from splitting those glycosidic bonds of hemicellulose which are closer than two or three monosaccharide units from an attachment between hemicellulose and lignin (Harkin, 1973). This argument was supported by the work of Fry (1984), who digested cell walls with a polysaccharase, and showed that ^{14}C labelled ferulic and *p*-coumaric acids were released from the cell wall

in association with disaccharide fragments. Further, most of the label, which was in the phenolic constituents of the cell walls, was associated with material which resisted the polysaccharase activity. In Fry's words, 'resistance [to the enzyme] may indicate polysaccharide cross-linking, leading to molecular domains from which hydrolytic enzymes are excluded'. Alternatively (or in addition) lignin protects other cell wall constituents against digestion because it prevents the swelling of the structural polysaccharides and their subsequent attack by enzymes (Millet *et al.*, 1976).

Cell Wall Protein

In a study of 14 grasses, Colburn and Evans (1967) found that 31% of the total plant protein was associated with the cell wall. In the non-grass cell wall structural proteins stiffen the wall, thus serving a similar role to that of lignin in grasses (Carpita, 1996; Cassab, 1998). There is little information on the digestibility of plant cell wall protein, but Hogan and Lindsay (1979) reported that about 50% of the cell wall protein of wheat straw was digested in the rumen and small intestine.

Cutin and Silica

Cutin and associated waxes and fats comprise the cuticle, which protects the outer surfaces of the epidermal cells from dessication and injury by pathogens. Cutin is a polymer of C16 and C18 dihydroxy and trihydroxy and epoxy fatty acids, linked mainly by ester bonds (Kögel-Knabner, 2002). In this matrix are deposited fats and waxes. Cutin resists digestion – fragments of cuticle can be recovered from the faeces and used to identify the plant species in herbivore diets.

Epidermal cells may be impregnated with silica, especially in rice (Vadiveloo, 1992) where it contributes to structural strength and disease resistance (Ma and Yamaji, 2006). Bae *et al.* (1997) have shown that silica impairs the attachment of fibrolytic bacteria to epidermal cells, thus reducing digestibility. Silica may reduce the digestibility of grasses (Van Soest and Jones, 1968), although Minson (1971) could not demonstrate this in *Panicum* spp.

The Physico-chemical Structure of Cereal Grains

A cereal grain has three major tissues: the embryo or 'germ', the endosperm and the 'seedcoat' (Fig. 4.10). These are described in detail by Fincher (1989) and Evers and Millar (2002).

The seedcoat is the fibrous and lignified outer covering of the seed which is derived from the ovary tissues. Its outer layers, the seed husk, are derived from the palea and lemma; they may be extended into an awn, e.g. in rice, barley, oats, triticale, rye

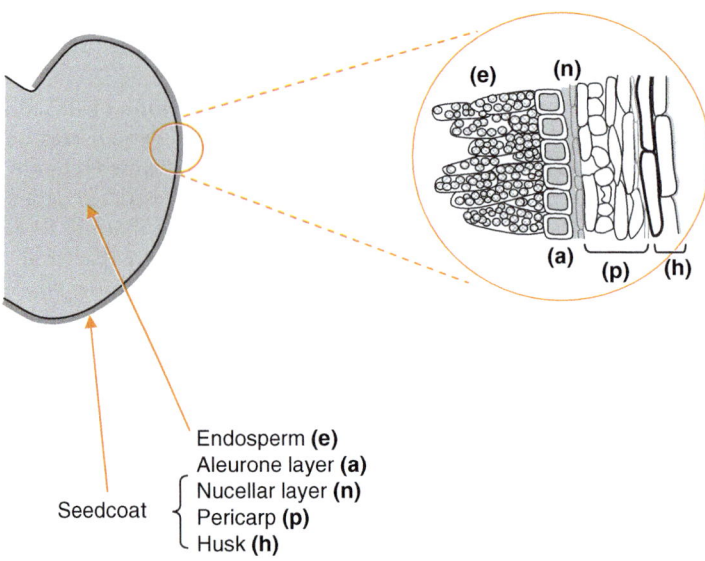

Endosperm **(e)**
Aleurone layer **(a)**
Seedcoat {
Nucellar layer **(n)**
Pericarp **(p)**
Husk **(h)**

Fig. 4.10. Anatomy of a cereal grain. (A generalized diagram adapted from Fincher, 1989; Evers and Millar, 2002; Jackowiak *et al.*, 2005.)

Chapter 4

and some wheat cultivars. The husks of rice, oats and barley are retained during harvesting, making them more fibrous than grains which are separated from the husk during threshing. The outer layer of the husk is covered with a waxy cuticle, and it may be heavily silicified (as in rice); some of the interior cell layers are thick-walled and lignified. The inner part of the seedcoat, the pericarp, includes the thick-walled hypodermis and epidermis cells, and the lignified tube cells and cross cells layers. The testa and the nucellar (or hyaline) layer lie between the tube cells and the aleurone layer. These are remnants of thin-walled cells which have been substantially crushed. Both of these layers have a cuticle.

The endosperm consists of the aleurone layer and the 'starchy' endosperm. The aleurone layer lies just underneath the nucellar layer and covers the starchy endosperm. It is one to three cells thick, depending on grain type (Olsen et al., 1999). The cells are thick-walled and heavily nucleated and are packed with protein and fat globules. The starchy endosperm consists of angular, tightly packed cells which are filled with starch granules (Fig. 4.11). In wheat and barley, the endosperm is uniform, but in maize and sorghum it is divided into the peripheral, corneous and floury endosperms (Kotarski et al., 1992; McAllister and Cheng, 1996). The starch granules are held within a protein matrix which consists of several proteins with different functions. One, 'puroindoline', is a small cysteine-rich protein which is found in wheat, rye, barley and oats grain (Morris, 2002). It occurs only in the endosperm (Wiley et al., 2007). It determines the firmness of attachment of starch granules to the matrix (more puroindoline is associated with a weaker attachment and 'softer' wheat).

Starch is the main constituent of the endosperm cells. It is a mixture of two polysaccharides: 20–30% amylose and 70–80% amylopectin (Rooney and Pflugfelder, 1986). Amylose (Svihus et al., 2005) is a straight-chain β-glucan of about 500–6000 glucose units linked by α1→4 glycosidic bonds. The molecule is wound into a helix, and lipid molecules can be found in the centre of that structure (Tester et al., 2004). Amylopectin is a highly branched β-glucan of up to 3 million glucose molecules. The structure consists of α1→4 linked chains of more than 10,000 glucose residues joined at intervals of about 20 glucoses by α1→6 glycosidic bonds (Kotarski et al., 1992; Tester et al., 2004). The α1→4 glucan chains of the amylopectin macromolecule are formed into double helices.

Endosperm starch is formed into granules which have alternating crystalline and amorphous regions. The amorphous layers are rich in amylopectin branch points while the crystalline layers are regions where the α1→4 glucan chains of the different amylopectin molecules have become aligned (Fig. 4.12).

Some grains, e.g. barley, oats, wheat and rye, have quantities of mixed-linkage (1→3, 1→4)-β-D-glucans and arabinoxylans in their endosperm and aleurone cell walls (Fincher and Stone, 1986; Lazaridou and Biliaderis, 2007). The mixed-linkage β-glucans are about 1000 glucose molecules with repeats of two or three 1→4 linked glucoses linked by 1→3 bonds. The arabinoxylans have a backbone of β1→4 xylan to which are attached arabinose molecules at the C-2 and C-3 atoms of xylose. Different grains and cultivars, grown under different conditions, have different arabinoxylan, β-glucan

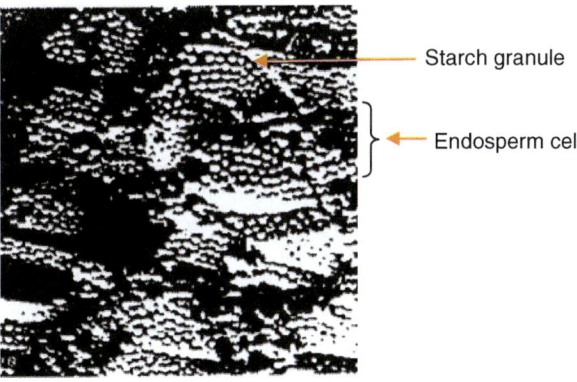

Fig. 4.11. Sorghum endosperm. (Electron micrograph courtesy of T.J. Kempton.)

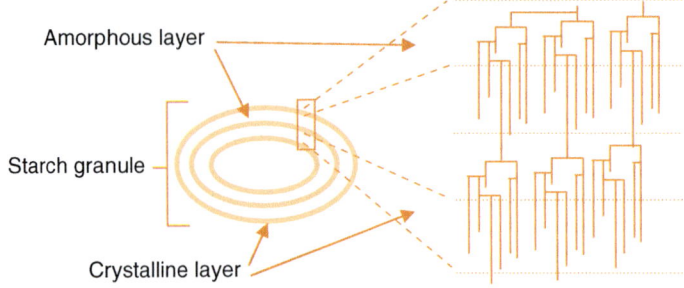

Amorphous layer

Starch granule

Crystalline layer

Fig. 4.12. Diagram of the alternating amorphous and crystalline layers of a starch granule. (Adapted from Jacobs and Delcour, 1998.)

and total pentosan + β-glucan (TPG) contents (Henry *et al.*, 1989). Rice has the lowest TPG level (about 1.3%), rye and oats have the highest (about 8%), and wheat, barley and triticale are intermediate (Henry, 1985). Waxy and high-amylose barleys have more protein and β-glucans (Holtekjølen *et al.*, 2006).

Digestion of Cereal Grains

The seedcoat provides the first level of protection against digestion of the grain. Its cuticle is not attacked by digestive enzymes and the fibrous, lignified cell layers probably have a nutritive value similar to that of a low-quality cereal straw (McAllister and Cheng, 1996). Second, the tight packing of endosperm cells prevents digestive enzymes from attacking starch in cells deep within the endosperm, as is indicated by the more extensive digestion of starch in small, rather than large, endosperm particles (McAllister *et al.*, 1993).

Several other characteristics of the endosperm inhibit digestion (for reviews see Svihus *et al.*, 2005; Rudi *et al.*, 2006). The large starch granules of barley are less well digested by α-amylase *in vitro* than the small granules (Stevnebø *et al.*, 2006) irrespective of their amylose content. Starches which are rich in amylose resist digestion in the small intestine (Gallant *et al.*, 1992); this is the 'resistant starch' (Brown, 1996). Sorghum and maize starches have appreciable resistance to digestion. Sniffen *et al.* (1992) suggest that 20–35% of maize starch and 20–40% of sorghum starch leaving the rumen resist digestion in the small intestine.

The proteins and phospholipids associated with the starch granule surface may impede the attachment between amylase and the starch, thus inhibiting starch digestion (Svihus *et al.*, 2005; Debet and

Gidley, 2006). McAllister *et al.* (1993) obtained greater digestion of barley starch (but not maize starch) if they treated the endosperm with a protease before using amylase. The reason for this result is not accurately known, but it may reflect the grain puroindoline content, as more puroindoline is associated with reduced digestibility of wheat (Swan *et al.*, 2006) and possibly barley (Beecher *et al.*, 2002).

Endosperm cell wall polysaccharides form gels in the small intestine. These gels inhibit the movement of digestive enzymes throughout the intestinal lumen and thus impede digestion of these grains and the absorption of the digestion products (Morel *et al.*, 2001; Pluske *et al.*, 2001).

References

Abedon, B.G., Hatfield, R.D. and Tracy, W.F. (2006) Cell wall composition in juvenile and adult leaves of maize (*Zea mays* L.). *Journal of Agricultural and Food Chemistry* 54, 3896–3900.

Akin, D.E. (1979) Microscopic evaluation of forage digestion by rumen micro-organisms – a review. *Journal of Animal Science* 48, 701–710.

Akin, D.E. (1982) Forage cell wall degradation and *p*-coumaric, ferulic and sinapic acids. *Agronomy Journal* 74, 424–428.

Akin, D.E. (1988) Biological structure of lignocellulose and its degradation in the rumen. *Animal Feed Science and Technology* 21, 295–310.

Akin, D.E. (2007) Grass lignocellulose. Strategies to overcome recalcitrance. *Applied Biochemistry and Biotechnology* 136–140, 3–15.

Akin, D.E. and Barton, F.E. (1983) Rumen microbial attachment and degradation of plant cell walls. *Federation Proceedings* 42, 114–121.

Akin, D.E. and Robinson, E.L. (1982) Structure of leaves and stems of arrowleaf and crimson clovers as related to *in vitro* digestibility. *Crop Science* 22, 24–29.

Akin, D.E., Robinson, E.L., Barton, F.E. II and Himmelsbach, D.S. (1977) Changes with maturity in anatomy, histolochemistry, chemistry and tissue digestibility of bermudagrass plant parts. *Journal of Agricultural and Food Chemistry* 25, 179–186.

Akin, D.E., Wilson, J.R. and Windham, W.R. (1983) Site and rate of tissue digestion in leaves of C3, C4 and C3/C4 intermediate Panicum species. *Crop Science* 23, 147–155.

Akin, D.E., Rigsby, L.L., Sethuraman, A., Morrison, W.H. III, Gamble, G.R. and Eriksson, K.-E.L. (1995) Alterations in structure, chemistry, and biodegradability of grass lignocellulose treated with the white rot fungi *Ceriporiopsis subvermispora* and *Cyathus stercoreus*. *Applied and Environmental Microbiology* 61, 1591–1598.

Atalla, R.H. and VanderHart, D.L. (1984) Native cellulose: a composite of two distinct crystalline forms. *Science* 223, 283–285.

Bae, H.D., McAllister, T.A., Kokko, E.G., Leggett, F.L., Yanke, L.J., Jakober, K.D., Ha, J.K., Shin, H.T. and Cheng, K.-J. (1997) Effect of silica on the colonization of rice straw by ruminal bacteria. *Animal Feed Science and Technology* 56, 165–181.

Baker, A.A., Helbert, W., Sugiyama, J. and Miles, M.J. (2000) New insight into cellulose structure by atomic force microscopy shows the I_α crystal phase at near-atomic resolution. *Biophysical Journal* 79, 1139–1145.

Barry, T.N., Manley, T.R. and Duncan, S.J. (1986) The role of condensed tannins in the nutritional value of *Lotus pedunculatus* for sheep. *British Journal of Nutrition* 55, 123–137.

Beecher, B., Bowman, J., Martin, J.M., Bettge, A.D., Morris, C.F., Blake, T.K. and Giroux, M.J. (2002) Hordoindolines are associated with a major endosperm-texture QTL in barley (*Hordeum vulgare*). *Genome/National Research Council Canada* 45, 584–591.

Ben-Ghedalia, D., Yosef, E. and Miron, J. (1993) Pectin fermentation and utilization by natural microflora during ryegrass ensilage. *Animal Feed Science and Technology* 41, 113–119.

Brazle, F.K. and Harbers, L.H. (1977) Digestion of alfalfa hay observed by scanning electron microscopy. *Journal of Animal Science* 46, 506–512.

Brown, I. (1996) Complex carbohydrates and resistant starch. *Nutrition Reviews* 54, S115–S119.

Brown, R.H. and Hattersley, P.W. (1989) Leaf anatomy of C3–C4 species as related to evolution of C4 photosynthesis. *Plant Physiology* 91, 1543–1550.

Brown, R.M., Saxena, I.M. and Kudlicka, S.M. (1996) Cellulose biosynthesis in higher plants. *Trends in Plant Science* 1, 149–156.

Carmona, J.F. and Greenhalgh, J.F.D. (1972) The digestibility and acceptability to sheep of chopped or milled barley straw soaked or sprayed with alkali. *Journal of Agricultural Science, Cambridge* 78, 477–485.

Carpita, N.C. (1996) Structure and biogenesis of the cell walls of grasses. *Annual Reviews of Plant Physiology and Plant Molecular Biology* 47, 445–476.

Carpita, N.C. and Gibeaut, D.M. (1993) Structural models of the primary cell walls in flowering plants: consistency of molecular structure with the physical properties of the walls during growth. *Plant Journal* 3, 1–30.

Carpita, N.C., Defernez, M., Findlay, K., Wells, B., Shoue, D.A., Catchpole, G., Wilson, R.H. and McCann, M.C. (2001) Cell wall architecture of the elongating maize coleoptile. *Plant Physiology* 127, 551–565.

Cassab, G.I. (1998) Plant cell wall proteins. *Annual Review of Plant Physiology and Plant Molecular Biology* 49, 281–309.

Chaikumpollert, O., Methacanon, P. and Suchiva, K. (2004) Structural elucidation of hemicelluloses from Vetiver grass. *Carbohydrate Polymers* 57, 191–196.

Chesson, A. (1984) Microbiology of the rumen in relation to the chemical or biological treatment of crops and by-products. In: *Improvements in the Nutritive Value of Crops and By-products by Chemical or Biological Treatments*. Ministry of Agriculture, Fisheries and Food, London, pp. 1–10.

Chesson, A. (1988) Lignin-polysaccharide complexes of the plant cell wall and their effect on microbial degradation in the rumen. *Animal Feed Science and Technology* 21, 219–228.

Chesson, A. (1993) Mechanistic models of forage cell wall degradation. In: Jung, H.-G., Buxton, D.R., Hatfield, R.D. and Ralph, J. (eds) *Forage Cell Wall Structure and Digestibility*. American Society of Agronomy, Madison, Wisconsin, pp. 347–376.

Chesson, A., Gordon, A.H. and Lomax, J.A. (1983) Substituent groups linked by alkali-labile bonds to arabinose and xylose residues of legume, grass and cereal straw cell walls and their fate during digestion by rumen microorganisms. *Journal of the Science of Food and Agriculture* 34, 1330–1340.

Colburn, M.W. and Evans, J.L. (1967) Chemical composition of the cell wall constituent and acid detergent fiber fractions of forages. *Journal of Dairy Science* 50, 1130–1135.

Connor, H.E., Bailey, R.W. and O'Connor, K.F. (1970) Chemical composition of New Zealand tall-tussocks (*Chionochloa*). *New Zealand Journal of Agricultural Research* 13, 534–554.

Cosgrove, D.J. (2005) Growth of the plant cell wall. Nature reviews. *Molecular Cell Biology* 6, 850–861.

Darvill, A., McNeill, M., Albersheim, P. and Delmer, D.P. (1980) The primary cell walls of flowering plants. In: Tolbert, N.E. (ed.) *The Biochemistry of Plants*, Vol. 1. Academic Press, New York.

Debet, M.R. and Gidley, M.J. (2006) Three classes of starch granule swelling: influence of surface proteins and lipids. *Carbohydrate Polymers* 64, 452–465.

Delmer, D.P. and Amor, Y. (1995) Cellulose biosynthesis. *The Plant Cell* 7, 987–1000.

Dengler, N.G., Dengler, R.E., Donnelly, P.M. and Hattersley, P.W. (1994) Quantitative leaf anatomy of C3 and C4 grasses (Poaceae): bundle sheath and mesophyll surface area relationships. *Annals of Botany* 73, 241–255.

Dryden, G.McL. and Leng, R.A. (1988) Effects of ammonia and sulphur dioxide gases on the composition and digestion of barley straw. *Animal Feed Science and Technology* 19, 121–133.

Durot, N., Gaudard, F. and Kurek, B. (2003) The unmasking of lignin structures in wheat straw by alkali. *Phytochemistry* 63, 617–623.

El-Sharkawy, M. and Hesketh, J. (1965) Photosynthesis among species in relation to characteristics of leaf anatomy and CO_2 diffusion resistances. *Crop Science* 5, 517–521.

Evers, T. and Millar, S. (2002) Cereal grain structure and development: some implications for quality. *Journal of Cereal Science* 36, 261–284.

Fincher, G.B. (1989) Molecular and cellular biology associated with endosperm mobilization in germinating cereal grains. *Annual Review of Plant Physiology and Plant Molecular Biology* 40, 305–346.

Fincher, G.B. and Stone, B.A. (1986) Cell walls and their components in cereal grain technology. *Advances in Cereal Science and Technology* 8, 207–295.

Finell, M. and Nilsson, C. (2005) Variations in ash content, pulp yield, and fibre properties of reed canarygrass. *Industrial Crops and Products* 22, 157–167.

Fry, S.C. (1984) Incorporation of [14C]cinnamate into hydrolase-resistant components of the primary cell wall of spinach. *Phytochemistry* 23, 59–64.

Gaillard, B.D.E. and Bailey, R.W. (1968) The distribution of galactose and mannose in the cell-wall polysaccharides of red clover (*Trifolium pretense*) leaves and stems. *Phytochemistry* 7, 2037–2044.

Gallant, D.J., Bouchet, B., Buléon, A. and Pérez, S. (1992) Physical characteristics of starch granules and susceptibility to enzymatic degradation. *European Journal of Clinical Nutrition* 46, 3–16.

Grabber, J.H., Panciera, M.T. and Hatfield, R.D. (2002) Chemical composition and enzymatic degradability of xylem and nonxylem walls isolated from alfalfa internodes. *Journal of Agricultural and Food Chemistry* 50, 2595–2600.

Grabber, J.H., Ralph, J., Lapierre, C. and Barrière, Y. (2004) Genetic and molecular basis of grass cell-wall degradability. I. Lignin-cell wall matrix interactions. *Comptes Rendus Biologies* 327, 455–465.

Granzin, B.C. and Dryden, G.M. (2003) Effects of alkalis, oxidants and urea on the nutritive value of rhodes grass (*Chloris gayana* cv. Callide). *Animal Feed Science and Technology* 103, 113–122.

Harbers, L.H., Brazle, F.K., Raiten, D.J. and Owensby, C.E. (1981) Microbial degradation of smooth brome and tall fescue observed by scanning electron microscopy. *Journal of Animal Science* 51, 439–446.

Harbers, L.H., Kreitner, G.L., Davis, G.V., Rasmussen, M.A. and Corah, L.R. (1982) Ruminal digestion of ammonium hydroxide-treated wheat straw observed by scanning electron microscopy. *Journal of Animal Science* 54, 1309–1319.

Harkin, J.M. (1973) Lignin. In: Butler, G.W. and Bailey, R.W. (eds) *Chemistry and Biochemistry of Herbage*, Vol. 1. Academic Press, London, pp. 323–368.

Hatch, M.D. (1987) C4 photosynthesis: a unique blend of modified biochemistry, anatomy and ultrastructure. *Biochimica et Biophysica Acta* 895, 81–106.

Hatfield, R.D., Ralph, J. and Grabber, J.H. (1999) Cell wall structural foundations: molecular basis for improving forage digestibilities. *Crop Science* 39, 27–37.

Henry, R.J. (1985) A comparison of the non-starch carbohydrates in cereal grains. *Journal of the Science of Food and Agriculture* 36, 1243–1253.

Henry, R.J., Martin, D.J. and Stewart, B.G. (1989) Cell-wall polysaccharides of rye-derived wheats: investigations of the biochemical causes of dough stickiness. *Food Chemistry* 34, 309–316.

Hogan, J.P. and Lindsay, J.R. (1979) The digestion of nitrogen associated with plant cell wall in the stomach and small intestine of the sheep. *Australian Journal of Agricultural Research* 31, 147–153.

Holtekjølen, A.K., Uhlen, A.K., Bråthen, E., Sahlstrøm, S. and Knutsen, S.H. (2006) Contents of starch and non-starch polysaccharides in barley varieties of different origin. *Food Chemistry* 94, 348–358.

Jackowiak, H., Packa, D., Wiwart, M. and Perkowski, J. (2005) Scanning electron microscopy of *Fusarium* damaged kernels of spring wheat. *International Journal of Food Microbiology* 98, 113–123.

Jacobs, H. and Delcour, J.A. (1998) Hydrothermal modifications of granular starch, with retention of the granular structure: a review. *Journal of Agricultural and Food Chemistry* 46, 2896–2905.

Johnson, W.L. and Pezo, D. (1975) Cell wall fractions and *in vitro* digestibility of Peruvian feedstuffs. *Journal of Animal Science* 41, 185–197.

Johnston, M.J. and Waite, R. (1965) Studies in the lignification of grasses. 1. Perennial rye-grass (S 24) and cocksfoot (S 37). *Journal of Agricultural Science, Cambridge* 64, 211–219.

Jung, H.-J.G. (1988) Inhibitory potential of phenolic-carbohydrate complexes released during ruminal fermentation. *Journal of Agricultural and Food Chemistry* 36, 782–788.

Jung, H.-J.G. and Lamb, J.F.S. (2003) Identification of lucerne stem cell wall traits related to *in vitro* neutral detergent fibre digestibility. *Animal Feed Science and Technology* 110, 17–29.

Jung, H.-J.G. and Lamb, J.F.S. (2004) Prediction of cell wall polysaccharide and lignin concentrations of alfalfa stems from detergent fiber analysis. *Biomass and Bioenergy* 27, 365–373.

Jung, H.G. and Lamb, J.F.S. (2006) Stem morphological and cell wall traits associated with divergent *in vitro* neutral detergent fiber digestibility in alfalfa clones. *Crop Science* 46, 2054–2061.

Juniper, B.E. (1979) The structure and chemistry of straw. *Agricultural Progress* 54, 18–27.

Kamstra, L.D., Moon, A.L. and Bentley, O.G. (1958) The effect of stage of maturity and lignification on the digestion of cellulose in forage plants by rumen microorganisms *in vitro*. *Journal of Animal Science* 17, 199–208.

Kögel-Knabner, I. (2002) The macromolecular organic composition of plant and microbial residues as inputs to soil organic matter. *Soil Biology and Biochemistry* 34, 139–162.

Kotarski, S.F., Waniska, R.D. and Thurn, K.K. (1992) Starch hydrolysis by the ruminal microflora. *Journal of Nutrition* 122, 178–190.

Latham, M.J., Hobbs, D.G. and Harris, P.J. (1979) Adhesion of rumen bacteria to alkali-treated plant stems. *Annales de Recherches Veterinaires* 10, 244–345.

Lawther, J.M., Sun, R.C. and Banks, W.B. (1995) Extraction, fractionation, and characterization of structural polysaccharides from wheat straw. *Journal of Agricultural and Food Chemistry* 43, 667–675.

Lazaridou, A. and Biliaderis, C.G. (2007) Molecular aspects of cereal β-glucan functionality: physical properties, technological applications and physiological effects. *Journal of Cereal Science* 46, 101–118.

Ma, J.F. and Yamaji, N. (2006) Silicon uptake and accumulation in higher plants. *Trends in Plant Science* 11, 392–397.

McAllister, T.A. and Cheng, K.-J. (1996) Microbial strategies in the ruminal digestion of cereal grains. *Animal Feed Science and Technology* 62, 29–36.

McAllister, T.A., Phillippe, R.C., Rode, L.M. and Cheng, K.J. (1993) Effect of the protein matrix on the digestion of cereal grains by ruminal microorganisms. *Journal of Animal Science* 71, 205–212.

McLeod, N.M. and Minson, D.J. (1974) Differences in carbohydrate factions between *Lolium perenne* and two tropical grasses of similar dry-matter digestibility. *Journal of Agricultural Science, Cambridge* 82, 449–454.

Millet, M.A., Baker, A.J. and Satter, L.D. (1976) Physical and chemical pretreatments for enhancing cellulose saccharification. *Biotechnology and Bioengineering Symposia* 6, 125–153.

Minic, Z. and Jouanin, L. (2006) Plant glycoside hydrolases involved in cell wall polysaccharide degradation. *Plant Physiology and Biochemistry* 44, 435–449.

Minson, D.J. (1971) Influence of lignin and silicon in the summative system for assessing the organic matter of digestibility of *Panicum*. *Australian Journal of Agricultural Research* 22, 589–598.

Morel, P.C.H., Padilla, R.M., Cottam, Y.H. and Coles, G.D. (2001) Influence of soluble and insoluble dietary fibre on growth, organ weights and blood lipid metabolites in weaner pigs. *Recent Advances in Animal Nutrition in Australia* 13, 22A.

Morris, C.F. (2002) Puroindolines: the molecular genetic basis of wheat grain hardness. *Plant Molecular Biology* 48, 633–647.

Nishiyama, Y., Langan, P. and Chanzy, H. (2002) Crystal structure and hydrogen-bonding system in cellulose Iβ from synchrotron X-ray and neutron fiber diffraction. *Journal of the American Chemistry Society* 124, 9074–9082.

Olsen, O.-A., Casper Linnestad, C. and Nichols, S.E. (1999) Developmental biology of the cereal endosperm. *Trends in Plant Science* 4, 253–257.

Ørskov, E.R., Reid, G.W., Kay, M. (1988) Prediction of intake by cattle from degradation characteristics of roughages. *Animal Production* 46, 29–34.

O'Sullivan, A.C. (1997) Cellulose: the structure slowly unravels. *Cellulose* 4, 173–207.

Pauly, M., Albersheim, P., Darvill, A. and York, W.S. (1999) Molecular domains of the cellulose/xyloglucan network in the cell walls of higher plants. *Plant Journal* 20, 629–639.

Pérez, S., Mazeau, K. and du Penhoat, C.H. (2000) The three-dimensional structures of the pectic polysaccharides. *Plant Physiology and Biochemistry* 38, 37–55.

Pluske, J.R., McDonald, D.E., Pethick, D.W., Mullan, B.P. and Hempson, D.J. (2001) Nutritional management of the gastrointestinal tract to reduce enteric diseases in pigs. *Recent Advances in Animal Nutrition in Australia* 16, 127–134.

Ridley, B.L., O'Neill, M.A. and Mohnen, D. (2001) Pectins: structure, biosynthesis, and oligogalacturonide-related signaling. *Phytochemistry* 57, 929–967.

Rondeau-Mouro, C., Bouchet, B., Pontoire, B., Robert, P., Mazoyer, J. and Buléon, A. (2003) Structural features and potential texturising properties of lemon and maize cellulose microfibrils. *Carbohydrate Polymers* 53, 241–252.

Rooney, L.W. and Pflugfelder, R.L. (1986) Factors affecting starch digestibility with special emphasis on sorghum and corn. *Journal of Animal Science* 63, 1607–1623.

Rudi, H., Uhlen, A.K., Harstad, O.M. and Munck, L. (2006) Genetic variability in cereal carbohydrate compositions and potentials for improving nutritional value. *Animal Feed Science and Technology* 130, 55–65.

Rydholm, S.A. (1965) *Pulping Processes*. Interscience, London.

Sarkanen, K.U. and Ludwig, G.H. (1971) (eds) *Lignins: Occurrence, Formation, Structure and Reactions*. Wiley, New York.

Saxena, S.K., Otterby, D.E., Donker, J.D. and Good, A.L. (1971) Effects of feeding alkali-treated oat straw supplemented with soyabean meal or non-protein nitrogen on growth of lambs and on certain blood and rumen liquor parameters. *Journal of Animal Science* 33, 485–490.

Shafizadeh, F. and McGinnis, G.D. (1971) Morphology and biogenesis of cellulose and plant cell-walls. *Advances in Carbohydrate Chemistry and Biochemistry* 26, 297–349.

Sniffen, C.J., O'Connor, J.D., Van Soest, P.J., Fox, D. G. and Russell, J.B. (1992) A net carbohydrate and protein system for evaluating cattle diets: II. Carbohydrate and protein availability. *Journal of Animal Science* 70, 3562–3577.

Stevnebø, A., Sahlstrøm, S. and Svihus, B.S. (2006) Starch structure and degree of starch hydrolysis of small and large starch granules from barley varieties with varying amylose content. *Animal Feed Science and Technology* 130, 23–38.

Sun, R.C. and Sun, X.F. (2002) Fractional and structural characterization of hemicelluloses isolated by alkali and alkaline peroxide from barley straw. *Carbohydrate Polymers* 49, 415–423.

Sun, R.-C., Sun, X.-F. and Zhang, S.-H. (2001) Quantitative determination of hydroxycinnamic acids in wheat, rice, rye, and barley straws, maize stems, oil palm frond fibre, and fast-growing poplar wood. *Journal of Agricultural and Food Chemistry* 49, 5122–5129.

Sun, X., Andrew, I.G., Joblin, K.N., Harris, P.J., McDonald, A. and Hoskin, S.O. (2006) Polysaccharide compositions of leaf walls of forage chicory (*Cichorium intybus* L.). *Plant Science* 170, 18–27.

Svihus, B., Uhlen, A.K. and Harstad, O.M. (2005) Effect of starch granule structure, associated components and processing on nutritive value of cereal starch. *Animal Feed Science and Technology* 122, 303–320.

Swan, C.G., Bowman, J.G.P., Martin, J.M. and Giroux, M.J. (2006) Increased puroindoline levels slow ruminal digestion of wheat (*Triticum aestivum* L.) starch by cattle. *Journal of Animal Science* 84, 641–650.

Tester, R.F., Karkalas, J. and Qi, X. (2004) Starch structure and digestibility. Enzyme-substrate relationship. *World's Poultry Science Journal* 60, 186–195.

Traxler, M.J., Fox, D.G., Van Soest, P.J., Pell, A.N., Lascano, C.E., Lanna, D.P.D., Moore, J.E., Lana, R. P., Vélez, M. and Flores, A. (1998) Predicting forage indigestible NDF from lignin concentration. *Journal of Animal Science* 76, 1469–1480.

Urbanowicz, B.R., Rayon, C. and Carpita, N.C. (2004) Topology of the maize mixed linkage (1→3),(1→4)-β-D-glucan synthase at the golgi membrane. *Plant Physiology* 134, 758–768.

Vadiveloo, J. (1992) Varietal differences in the chemical composition and *in vitro* digestibility of rice straw. *Journal of Agricultural Science, Cambridge* 119, 27–33.

Van Soest, P.J. (1964) Symposium on nutrition and forage and pastures: new chemical procedures for evaluating forages. *Journal of Animal Science* 23, 838–845.

Van Soest, P.J. (1967) Development of a comprehensive system of feed analyses and its application to forages. *Journal of Animal Science* 26, 119–128.

Van Soest, P.J. and Jones, L.H.P. (1968) Effect of silica in forages upon digestibility. *Journal of Dairy Science* 51, 1644–1648.

Varner, J.E. and Lin, L.-S. (1989) Plant cell wall architecture. *Cell* 56, 231–239.

Vincken, J.-P., Schols, H.A., Oomen, R.J.F.J., McCann, M.C., Ulvskov, P., Voragen, A.G.J. and Visser, R.G.F. (2003) If homogalacturonan were a side chain of rhamnogalacturonan I. Implications for cell wall architecture. *Plant Physiology* 132, 1781–1789.

Vorwerk, S., Somerville, S. and Somerville, C. (2004) The role of plant cell wall polysaccharide composition in disease resistance. *Trends in Plant Science* 9, 203–209.

Wallace, G., Chesson, A., Lomax, J.A. and Jarvis, M.C. (1991) Lignin-carbohydrate complexes in graminaceous cell walls in relation to digestibility. *Animal Feed Science and Technology* 32, 193–199.

Wiley, P.R., Tosi, P., Evrard, A., Lovegrove, A., Jones, H. D. and Shewry, P.R. (2007) Promoter analysis and immunolocalisation show that puroindoline genes are exclusively expressed in starchy endosperm cells of wheat grain. *Plant Molecular Biology* 64, 125–36.

Wilson, J.R., Brown, R.H. and Windham, W.R. (1983) Influence of leaf anatomy on the dry matter digestibility of C3, C4 and C3/C4 intermediate types of Panicum species. *Crop Science* 23, 141–146.

Xie, Y., Yasuda, S., Wu, H. and Liu, H. (2000) Analysis of the structure of lignin-carbohydrate complexes by the specific ^{13}C tracer method. *Journal of Wood Science* 46, 130–136.

Xu, F., Geng, Z.C., Sun, J.X., Liu, C.F., Ren, J.L., Sun, R.C., Fowler, P. and Baird, M.S. (2006) Fractional and structural characterization of hemicelluloses from perennial ryegrass (*Lolium perenne*) and cocksfoot grass (*Dactylis glomerata*). *Carbohydrate Research* 341, 2073–2082.

Zerbini, E. and Thomas, D. (2003) Opportunities for improvement of nutritive value in sorghum and pearl millet residues in South Asia through genetic enhancement. *Field Crops Research* 84, 3–15.

The Nutritive Value of Concentrate Foods

The Nutritive Value and Use of Concentrate Foods

Concentrates include nutrient-rich, low-fibre foods such as cereal grains, cereal grain by-products, molasses, grain legumes, oilseed meals and animal protein meals. There is relatively little data on food chemical composition published in peer-reviewed journals (see Udén *et al.*, 2005 for comment on this), so most of the information is from compilations of food nutritive value published by national or international authorities.

The nutrient content data discussed here are indicative only. They do not refer to any particular geographical region, agronomic condition or plant cultivar, or processing method. Regional and national food composition data have been published by Jarridge (1989) and Sauvant *et al.* (2004) (French); Jentsch *et al.* (2003) (German); Centraal Veevoederbureau (2004) (Dutch); MAFF (1992) (British); NRC (1982, 1989, 1998, 2000, 2001) (North American); Evans (1985) and Ostrowski-Meissner (1987, 1990) (Australian); and Göhl (1981) and Jarridge (1989) (tropical foods). FRG (undated) is an FAO multi-regional resource which has proximate analysis data on a wide variety of foods. Always use the most recent available information because modern plant cultivars can have different nutritional characteristics to older ones.

Cereal Grains

Maize and wheat are used in larger quantities than other cereal grains for animal feeding (Table 5.1). Smaller amounts of barley, sorghum and oats are used, although these amounts are substantially larger proportions of the total crop. Other grains such as rye and millet are also fed to animals. Sorghum and barley are often used in rations for pigs, poultry and ruminants. Maize (also called corn) is especially useful in pig and poultry diets because of its high available energy and linoleic acid

contents. Wheat is used in monogastric animal diets and is particularly useful where high energy contents are needed, e.g. pig creep and weaner diets. It is less suitable as a major constituent of ruminant rations because its high content of readily fermentable starch can lead to the rapid production of acids in the rumen, which can promote lactic acid production and predispose animals to grain engorgement poisoning. Triticale is a rye × wheat hybrid, but its nutritive value is more like wheat than rye (Metayer *et al.*, 1993). It is used in both monogastric and ruminant diets. Oats are often used in horse diets.

Nutrient Contents of Cereal Grains

Cereal grains have the highest usable energy contents of any conventional foods other than fats and oils. It is useful to divide cereal grains into low- and high-fibre types as there is a negative relationship between fibre (acid detergent fibre (ADF)) and usable energy contents (Fig. 5.1). Maize, wheat and rice have the lowest ADF contents (about 3% dry matter (DM) basis) and the highest metabolizable energy (ME) and digestible energy (DE) contents: on average 14 MJ ME for ruminants or 15.5 MJ ME/kg DM, and up to 17 MJ DE/kg for pigs (Table 5.2). Sorghum and barley have higher fibre contents (about 6% ADF, DM basis) and slightly lower usable energy contents: 13 MJ ME/kg DM for ruminants and 14.5 MJ ME/kg DM for pigs. Oats have about 15% ADF, and the lowest DE and ME values. Although triticale has a low ADF content (about 4.5% DM basis) and could be classified with maize and wheat on this basis, it has somewhat lower usable energy contents: about 13.5 MJ ME for ruminants and 15 MJ DE/kg DM for pigs (Evans, 1985).

There are positive relationships between grain protein and amino acid contents. The relationships ($P < 0.01$) below were published for Australian grains by Evans (1985). More prediction equations for the amino acids in cereal grains and other ingredients are

Table 5.1. Global use of cereal grains in animal feeding in 2002. (From FAOSTAT, 2004.)

Grain type	Human food Amount (1000 t)	Animal food Amount (1000 t)	Proportion of that grain used for animal food (%)	Proportion of all grains used for animal food (%)
Barley	7,207	89,625	92.6	13.4
Maize	110,051	400,169	78.4	59.7
Millet	18,288	2,021	9.9	0.3
Oats	3,174	18,725	85.5	2.8
Rice	355,653	7,075	1.9	1.1
Rye	6,299	10,453	62.4	1.6
Sorghum	24,046	24,319	50.3	3.6
Wheat	415,498	102,846	19.8	15.3
Other cereals	4,630	15,238	76.7	2.3
Total	944,846	670,471	41.5	100.0

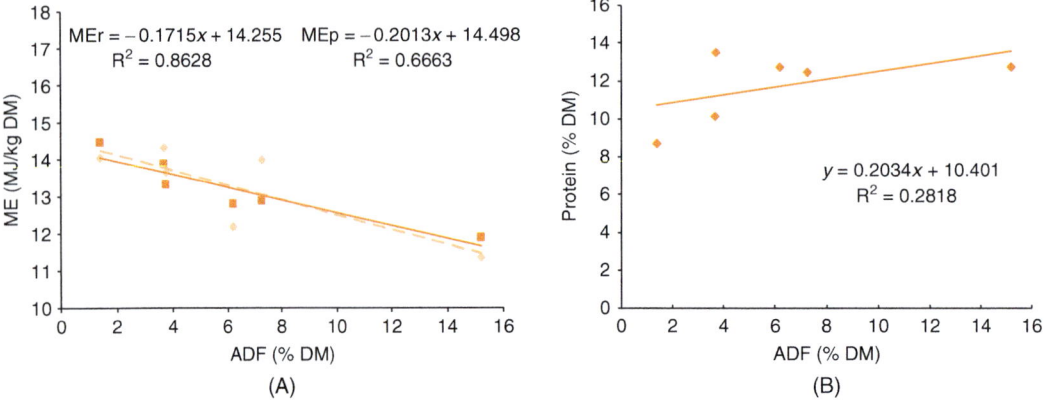

$MEr = -0.1715x + 14.255$ $R^2 = 0.8628$

$MEp = -0.2013x + 14.498$ $R^2 = 0.6663$

$y = 0.2034x + 10.401$ $R^2 = 0.2818$

(A) (B)

Fig. 5.1. Relationships between grain acid detergent fibre (ADF) content (% DM) and: (A) the contents of metabolizable energy for pigs (MEp) and ruminants (MEr) (MJ/kg DM); and (B) protein content (% DM). (Calculated from data in NRC, 1998, 2000; Jarridge, 1989.)

Table 5.2. Nutritive value of commonly used cereal grains. (From Taverner and Farrell, 1981; Evans, 1985; NRC, 1989, 1998, 2000; Jarridge, 1989.)

Constituent	Maize	Wheat	Sorghum	Barley	Oats	Rice
DE (pigs) (MJ/kg DM)[a]	16.4	15.6	15.9	14.4	13.3	17.3
DE (horses) (MJ/kg DM)[a]	16.1	16.2	14.9	15.1	13.3	15.9
ME (ruminants) (MJ/kg DM)[a]	13.9	13.3	12.9	12.8	11.9	14.4
Protein (% DM)	10.2	14.3	12.1	12.7	12.2	8.5
ADF (% DM)[b]	3.6	3.9	7.6	6.5	15.2	2.4
Linoleic acid (%)	1.91	0.94	1.12	0.88	1.51	0.24
Ca (% DM)	0.03	0.06	0.05	0.05	0.07	0.05
P (% DM)	0.32	0.35	0.34	0.36	0.39	0.26
Available P (% DM)[c]	0.07	0.18	0.08	0.11	0.09	0.05
Na (% DM)	0.02	0.11	0.02	0.03	0.04	0.07

[a]Values for grains which are milled and/or (for rice) polished.
[b]Acid detergent fibre (cellulose plus lignin, i.e. a fraction of the plant cell wall).
[c]Availability for pigs.

given by Taverner and Farrell (1981), Fuller *et al.* (1989) and NRC (1989). All cereal grains have much lower concentrations of lysine, cystine and methionine than is needed by most animals:

1. For barley:
   Lysine content (% DM) = 0.2286 + 0.0183 (protein content, % DM)
   Methionine + cystine content (% DM) = 0.0759 + 0.0251 (protein content, % DM)
2. For maize:
   Lysine content (% DM) = 0.0202 + 0.0250 (protein content, % DM)
   Methionine + cystine content (% DM) = 0.1860 + 0.0229 (protein content, % DM)
3. For sorghum:
   Lysine content (% DM) = 0.0944 + 0.0128 (protein content, % DM)
   Methionine + cystine content (% DM) = 0.0502 + 0.0252 (protein content, % DM)
4. For wheat:
   Lysine content (% DM) = 0.1320 + 0.0186 (protein content, % DM)
   Methionine + cystine content (% DM) = 0.0623 + 0.0299 (protein content, % DM)

Cereal grain protein is generally degradable in the rumen (Table 5.3). Only 20–30% of barley and wheat protein passes undigested into the small intestine. However, maize and sorghum grains are exceptions – their protein has greater natural protection and about 60% reaches the small intestine.

Some cereal grains are good sources of linoleic acid, one of the essential fatty acids. Maize in particular contains nearly 2% linoleic acid (Table 5.2). Cereal grains have very low Ca and Na contents (Table 5.2), and animal diets which contain large amounts of cereal grains must be supplemented with these minerals. Total P contents are fairly high, but most cereal grain P is present as phytin and only 15–50% of this P is available to pigs (Table 5.2).

Cereal grains are not good sources of vitamins. Compared to the requirements for pigs (NRC, 1998), they all have sufficient concentrations of biotin, choline, pyridoxine and thiamine, but not enough cobalamin, niacin or riboflavin. There is insufficient folic acid in sorghum and marginal amounts of pantothenic acid in maize and barley. Niacin in maize is incompletely available to monogastric animals, and the niacin in wheat and sorghum is probably also unavailable to pigs and rats (Luce *et al.*, 1967; Carpenter *et al.*, 1988). The B vitamins are located in the grain seedcoat or bran and so dehulled or polished grains are not well supplied with these nutrients. Even when the concentration of a particular vitamin is adequate, it is usual for feeders of pigs and other non-ruminants to use vitamin premixes which supply the total B group vitamin requirements. No cereal grain contains sufficient tocopherols, and they have inadequate concentrations of vitamin A (maize is the richest at about 200 IU/kg) or its carotenoid precursors.

Variability in the Nutrient Content of Cereal Grains

There is surprisingly little difference in the average nutrient contents of grains grown in different regions. On the other hand, there can be substantial variation between samples of grains, and regional, between-year and between-variety differences can

Table 5.3. Nutritive value (as-fed basis) of the protein in four cereal grains. (From Evans, 1985; NRC, 2000; Jarridge, 1989.)

Characteristic	Maize	Wheat	Sorghum	Barley
Protein content range (%)	7.8–11.3	10.0–18.9	7.0–15.6	9.0–14.0
Protein degradability (%)[a]	39–45	73–77	39–43	73–75
Lysine content (%)	0.28	0.43	0.34	0.21
Lysine availability (%)[b]	89	90	94	90
Cystine content (%)	0.22	0.26	0.16	0.22
Methionine content (%)	0.18	0.16	0.13	0.12
Tryptophane content (%)	0.09	0.13	0.10	0.12
Arginine content (%)	0.47	0.58	0.35	0.55

[a]Rumen degradability.
[b]Availability for pigs.

be expected (Taverner and Farrell, 1981; Fuller *et al.*, 1989; Mathison *et al.*, 1991; Metayer *et al.*, 1993; and reviews by Van Barneveld, 1999 and Kim *et al.*, 2005).

Several factors affect grain quality. Pinched grains, i.e. grains grown under low-moisture conditions, have a low bulk density (65 g/l or less). Bulk density, also called 'test weight' or 'standard weight', is associated positively with starch, DE and ME contents, and negatively with fibre and Ca contents. Protein content is high in crops fertilized with N, or where endosperm development is limited by water stress; conversely, it can be very low if the grain is plump and grown without N fertilizer. As an example, reported ranges in the protein contents of North American grains vary from 19% (maize) to 36% (wheat) of the mean values.

Variations in grain composition are influenced by the variety (Connor *et al.*, 1976; Sauvant *et al.*, 2004), locality, growing conditions, soil fertility and fertilizer usage (Taverner and Farrell, 1981; Fuller *et al.*, 1989; Hunt, 1996; Borrell *et al.*, 1999; Cooper *et al.*, 2001). The data presented by Coates *et al.* (1977) and Mathison *et al.* (1991) for Canadian grains, by Metayer *et al.* (1993) for French grains and by Evans (1985) for Australian grains give coefficients of variation (CVs) of 1.3–5.2% for ME, 3.3–16% for protein and 4.6–17.8% for fibre. The CVs associated with the values in Table 5.2 for DE and ME contents (calculated over all grain types and databases) for pigs and ruminants are only 9% and 7%, respectively. Although these between-grain differences in usable energy contents seem small, they are large enough to sometimes cause noticeable differences in the growth rates of grower pigs and lot-fed beef cattle (e.g. Owens *et al.*, 1997; Hongtrakul *et al.*, 1998).

It is unwise to use published nutrient content values as estimators of the actual nutrient content of a particular sample. The cost of having a sample analysed can be repaid many times over by savings in, especially, protein concentrates. For example, in a conventional 16% protein pig grower ration based on maize or sorghum and soybean meal, a 1% unit decrease in the protein content of the cereal grain component requires a 2.5% unit increase in the proportion of soybean meal.

Cereal grains are not digested very well if they are fed without processing. This is particularly so when whole grain is given to cattle. There may be up to 15% unit difference in digestibility between whole

and milled grain, e.g. 77% versus 98% for digestibility by cattle of the starch in whole versus steam-flaked maize (Lee *et al.*, 1982) and 52.9% versus 82.2% for the digestibility of the DM of whole versus cracked barley fed to cattle (Clarke *et al.*, 1996).

Cereal Grains and Animal Health

Most animals of all species find cereal grains palatable, although a few individuals may not. It is always prudent to introduce and manage cereal grain-rich diets carefully. Uncontrolled consumption increases the risk of lactic acid production from the bacterial fermentation of starch in the reticulo-rumen or large intestine (Enemark *et al.*, 2002; Schwartzkopf-Genswein *et al.*, 2003). When large amounts of starch are fermented the rumen pH falls to 5 or less. This leads to rumen stasis (lack of rumen movement), death of rumen protozoa and some bacteria, and absorption of acid into the blood. Blood bicarbonate is secreted into the rumen to neutralize rumen pH, and ultimately the animal may not be able to control its blood pH; this can be fatal (Owens *et al.*, 1998).

Some cereal grains contain substances which reduce nutrient availability. Tannins in 'bird-resistant' sorghum may reduce the digestibility and soluble protein content of those grains (Streeter *et al.*, 1990). Cereal grain non-starch polysaccharides (NSPs) include endosperm cell wall constituents such as cellulose, arabinoxylans and β glucans (Bach Knudsen and Hansen, 1991; Oscarsson *et al.*, 1998). Insoluble NSPs reduce pig growth and feed usage (Morel *et al.*, 2001); some soluble NSPs form gels which reduce nutrient absorption (Pluske *et al.*, 2001). These will reduce digestibility and nutrient absorption when included in monogastric animal diets. Amylose-rich starches, as are in some varieties of rice, maize, wheat, barley and other grains, are difficult to digest (see review by Svihus *et al.*, 2005). On the other hand, Spears and Fahey (2004) speculated that starch which is resistant to digestion by amylases (resistant starch) may help to maintain good digestive tract health in monogastrics by playing a role as 'potential proxy for dietary fibre'.

Cereal Grain By-products

Several by-products from wheat flour milling are available for animal feeding (Boucqué and Fiems,

1988). A mixture of large and small fibre-, fat- and protein-rich particles is produced during the first 'break' grindings of the wheat when the seedcoat and embryo are removed (Blasi *et al.*, 1998). The particular materials produced depend on the nature of these break grinds and of the separation procedures used. The bran (larger, more fibrous flakes) and pollard (smaller particles) fractions may be removed together. The resulting material may be used as is, called 'millrun', 'middlings' or 'mill feed', or may be sieved to give bran and pollard. These products can also contain amounts of ground screenings, i.e. seeds and chaff which may have contaminated the wheat sample. When it is produced separately, pollard consists of small particles of seedcoat and is richer in the aleurone and nucellar layers than bran. Pollard may contain amounts of adhering endosperm and germ (embryo).

The minerals, protein and fibre of grains are not evenly distributed throughout the seed and so the by-products of grain milling are richer in those constituents which are located in the seedcoat. Bran and millrun have twice the fat content of the whole grain, one and a half times the protein content, more lysine and arginine and most of the grain Ca, P and Na (Table 5.4). Grain by-products are good sources of amino acids and are reasonably rich in the B vitamins. However, they also have most of the grain fibre and consequently lower ME and DE contents than the whole grain. Indicative ME and DE values of by-products for ruminants and pigs are in Table 5.4. Takagi *et al.* (2003) have

reported a DE content of wheat bran for horses of 12.7 MJ/kg DM (compare this with the DE content for the whole wheat grain in Table 5.2). Pollard has a slightly higher protein content (16%) than bran or millrun, a lower fibre content and higher availability of nutrients. Pollard is very palatable and can be used in the diets of most animals.

Rice bran is widely used as a protein and energy source. It is rich in oil – 13–26% of the DM (Palipane and Swarnasiri, 1985; Soren *et al.*, 2003). The oil is rich in oleic and linoleic fatty acids, but is prone to oxidation and thus rancidity from endogenous lipoxygenase enzymes (Warren and Farrell, 1990). It can have widely varying ADF (10.5–15.5%), protein (10–18%) and Ca (0.02–0.09%) contents (DM bases). Indicative nutrient contents for rice bran are in Table 5.4.

Grain milling by-products are often used when a palatable, less energy-rich (higher fibre content) or lower glycaemic index (GI) alternative to cereal grains is required or acceptable. However, pigs, especially younger animals, can experience reduced food intake and digestibility, and poorer growth rates and feed/gain ratios, if these by-products are used injudiciously (Le Goff *et al.*, 2002; Wang *et al.*, 2002; Soren *et al.*, 2003). Diets for sows and finisher pigs generally contain no more than 30% bran or millrun, and about 15% in diets for grower pigs.

Brewers' grains are a fibre- and protein-rich residue of malted (i.e. sprouted) barley after the partly digested endosperm has been extracted with water (the extract is used to make the wort for beer

Table 5.4. Nutritive value of the by-products of barley, wheat and rice processing. (From Kornegay, 1973; Evans, 1985; Warren and Farrell, 1990; NRC, 1998, 2000; Pereira *et al.*, 1998.)

Constituent	Rice bran	Wheat bran	Wheat millrun	Wheat pollard	Brewers' grains
DE (pigs) (MJ/kg DM)	14.4	11.6	13.6	13.7	9.2
ME (pigs) (MJ/kg DM)	13.3	10.7	14.2	13.4	8.9
DE (horses) (MJ/kg DM)	12.1	13.3	14.5	–	–
ME (ruminants) (MJ/kg DM)	10.6	10.6	10.6	13.3	10.0
Protein (% DM)	14.5	17.3	17.8	17.3	29.1
Protein degradability (%)[a]	75.0	78.0	77.5	–	47.5
Lysine (% DM)	0.63	0.67	0.68	0.79	1.06
Acid detergent fibre (% DM)	17.7	14.1	10.6	9.6	25.6
Linoleic acid (% DM)	4.58	1.94	2.00	2.21	3.32
Ca (% DM)	0.09	0.16	0.15	0.13	0.30
P (% DM)	1.70	1.27	1.01	0.84	0.66
Available P (% DM)[b]	0.45	0.38	0.36	0.25	0.18
Na (% DM)	0.13	0.05	0.17	0.02	0.22

[a]Rumen degradability.
[b]Availability for pigs.

production). The product may be sold wet, or dried to about 10% moisture. Brewers' grains (Table 5.4) are useful as a ruminant food. Dairy farmers frequently use it as an energy source in lactating cow diets. It provides some protein, but although it has between 16% and 31% ADF (DM basis), it may not be a good source of digestible fibre (Younker *et al.*, 1998). Brewers' grains are not well digested by pigs and inclusion in diets at more than 20% is likely to reduce food intake, growth rates and feed/gain ratios (Quemere *et al.*, 1983; Chawla and Sikka, 1985). However, they are useful sources of lysine, methionine and cystine.

Foods Produced from Sugar Refining – Molasses, Sugarbeet Pulp and Sugarcane Bagasse

Molasses is a by-product of sugar refining, and also from citrus and wood processing. It is a black, viscous liquid which contains about 76% DM and has a bulk density of 1.5 kg/l. Different types of molasses vary in composition, especially in the DM content (Perez, 1995). Commercially, the consistency of molasses is described by its Brix value (degrees Brix). This is similar to a measure of specific gravity, but there is no linear relationship between Brix and either specific gravity or sucrose content. Molasses should have at least 71°Brix.

Sugars make up 70–80% of molasses DM. Sucrose is up to 40% of the DM in cane molasses and 66% in beet molasses. There are very small amounts (about 1% of the DM) of glucose and fructose in beet molasses but more (about 10% each) in cane molasses (Steg and van der Meer, 1985; OECD, 2002). The organic matter is 80–

95% digestible in both cane and beet molasses (Steg and van der Meer, 1985; Perez, 1995; OECD, 2002). Molasses contains 5% (cane) to 15% (beet) protein (DM basis). Steg and van der Meer (1985) reported that 27% of the total N in beet molasses was present as betaine-N (equivalent to 3–4% of the DM) and 33% was amino acid-N. For cane molasses these values were 1% and 23%, respectively. Neither type of molasses is a good source of protein or essential amino acids (lysine 0.04%, methionine + cystine 0.1% DM in beet molasses; OECD, 2002). Molasses is particularly rich in K, Ca and S (Table 5.5). Cane molasses has about 4 mg Co, 9 mg Cu, 16 mg Zn and 200 mg Fe/kg DM (Stewart, 1976).

Most animals find molasses very palatable and it is often used to mask the taste of unpalatable ingredients such as urea, meat meals and straws. It is viscous and helps to bind small particles such as mineral and vitamin sources into the mass of mixed food. Molasses is widely used as an animal food ingredient and has been used successfuly in diets for rabbits (Le Thu Ha *et al.*, 1996), cattle (Garrett *et al.*, 1989; Granzin and Dryden, 2005), goats (Amin and Adam, 1993) and pigs (Perez, 1995). However, there are possible disadvantages in using molasses at more than about 20% of the diet. Ruminants given molasses-based diets grow more slowly than when grain is used as the major energy source, even when the protein and mineral deficiencies of molasses are rectified (Tomkins *et al.*, 2004). This poorer performance is partly due to the high mineral content which can lead to diarrhoea, and partly to the predominance of acetic and butyric acids in the end products of rumen fermentation (Araba *et al.*, 2002). Although sucrose yields glucose when it is digested in the monogastric

Table 5.5. Nutritive value of the by-products of sugar processing. (From Stewart, 1976; NRC, 1982, 2000; Evans, 1985; Steg and van der Meer, 1985; OECD, 2002.)

| Constituent | Molasses | | Sugarbeet pulp | Sugarcane bagasse |
	Cane	Beet		
DE (pigs) (MJ/kg DM)	14.1	13.5	13.5	–
ME (ruminants) (MJ/kg DM)	11.5	12.8	11.9	7.1
Protein (% DM)	4.9–5.9	6.6–14.8	6.6–9.7	1.6–2.9
ADF (% DM)	0.1	0–0.4	19–27	60
Ca (% DM)	1.1–1.4	0.1–0.5	0.6–1.1	0.9
P (% DM)	0.06–0.11	0.02–0.06	0.1–0.2	0.3
Na (% DM)	0.04–0.22	0.6–1.9	0.1–0.5	0.2
K (% DM)	3.8–5.7	3.2–6.1	0.2–1.6	0.5
S (% DM)	0.5–1.1	0.2–0.7	0.22	0.1

intestine, rumen fermentation yields predominantly ketogenic volatile fatty acids and so ruminants can become glucose deficient when they are given a molasses-rich ration. Molasses is not suitable for pre-ruminant calves, kids or lambs because they do not have an endogenous sucrase (Kreikemeier et al., 1990). Molasses will increase the GI (Brand-Miller et al., 2003) of horse foods. Kronfeld et al. (2004) have suggested that low-GI foods may protect horses against osteochondrosis dissecans. Horses are also susceptible to what appear to be food-related temperament problems. Relevant to this is the observation of Hanstock et al. (2004) that rats given a molasses-rich diet showed symptoms of anxiety and aggression.

Sugarbeet pulp is the dried residue remaining after the extraction of sugar from sugarbeet. It has from 4% to 10% sugars, 7% to 10% protein (DM bases) and useful concentrations (for pigs) of leucine (0.36–0.6% DM), valine (0.36–0.57% DM), histidine (0.19–0.29% DM) and arginine (0.24–0.41% DM) (OECD, 2002).

Bagasse is the fibrous residue remaining after sugarcane stalks have been crushed to remove the sugar-rich juice. It is rich in cell wall material, heavily lignified (about 20–25% in the DM; Zandersons et al., 1999), not very digestible, and has low concentrations of nutrients. It has a limited role in animal feeding but can be improved by chemical treatment. For example, Reddy et al. (1989) and Tudor and Inkerman (1989) increased the DM digestibility of bagasse from 4% to 81% and 30% to 55%, after treatment with H_2O_2 and NaOH solutions, respectively.

Grain Legumes

Grain legumes (also called 'pulses') typically have (Hoover and Ratnayake, 2002) approximately 25% protein, 20–37% starch (of which 22–30% is amylose) and about 10% fibre (ADF) (Evans, 1985). Compared with animal requirements, most grain legumes are rich in lysine (availabilities are between 85% and 95%) but have low levels of available P, cystine, methionine and tryptophane (Table 5.6).

Many grain legumes have poisonous secondary substances. These can often be detoxified by heating, but the problem is more usually managed by restricting the inclusion levels in pig and other monogastric animal diets (Evans, 1985). Heat-treated navy beans should be restricted to 25% because of low levels of cystine and methionine rather than any practical toxicity problem, although raw beans are not recommended for use in pig diets. Do not feed uncooked cowpeas to pigs. They should not be included in weaner pig diets, and no more than 15% put in diets for older stock. An upper limit of 20–40% is recommended for peas because of the presence of a trypsin inhibitor and a hemagglutinin.

Oilseeds and Oilseed Meals

Oilseed meals (also called 'oilseed cakes') are the most commonly used plant protein concentrates. They are products made from 'the residues left after the extraction of oil from the seeds of annual crops' (UNECE, 1995). Soybean meal is the pre-eminent oilseed-derived protein food. Consumption by animals was

Table 5.6. Nutrient content of grain legumes. (From Evans, 1985; NRC, 1982, 1998, 2000; Freer and Dove, 1984; Jarridge, 1989; Yu et al., 2002.)

Constituent	Cowpea	Lupins (sweet)	Mung bean	Navy bean	Pea
DE (pigs) (MJ/kg DM)	15.0	16.0	17.2	15.8	16.0
ME (ruminant) (MJ/kg DM)	10.7	14.2	10.1	–	13.0
Protein (% DM)	23.4	35.9	26.6	26.8	24.8
Protein degradability (%)[a]	–	62–95	–	–	79–90
Lysine (% DM)	1.58	1.57	1.91	1.90	1.64
Lysine availability (%)[b]	94	84	91	83	94
ADF (% DM)	38.6	18.0	–	6.3	8.2
Ca (% DM)	0.30	0.22	0.24	0.21	0.16
P (% DM)	0.47	0.46	0.63	0.48	0.40
Available P (% DM)[b]	0.14	0.10	0.19	0.14	0.09
Na (% DM)	0.33	0.04	–	0.06	0.11

[a]Rumen degradability.
[b]Availability for pigs.

118 million t in 2000/2002 (USDA, 2002, cited by Bajjalieh, 2004). Rapeseed (21 million t), cottonseed (11 million t) and sunflower (10 million t) meals are used in relatively large amounts, followed by groundnut, palm kernel, copra and coconut, and linseed (made from flax, *Linum usitatissimum*) meals (between 1 and 6 million t annually). Meals made from sesame (*Sesamum orientale*), safflower (*Carthamus tinctorius*) and mustard (*Brassica juncea*) are used in lesser quantities.

Oilseed meals are the residues which remain after the oil has been extracted from the raw oilseed, either by mechanical pressing (giving an expeller-extracted meal) or by dissolving it with hexane (solvent-extracted meals). 'Pre-press solvent extraction' is where some of the oil is first removed by mechanical extraction. There can be quite large differences between expeller and solvent meals in the rumen degradability of their protein and the digestibility of their amino acids. Exposing soybean meals to temperatures typical of those used in extrusion (110–150°C) has been reported to reduce the availability of lysine to pigs by up to 20% units (Faldet *et al.*, 1992; Marty and Chavez, 1995), the whole-tract digestibility of lysine in mink by 9% (Ljøkjel *et al.*, 2000) and the rumen degradability of soybean meal protein in cattle by up to 58% units, respectively (Plegge *et al.*, 1985; Faldet *et al.*, 1992; Ljøkjel *et al.*, 2000). There are different effects of heating on the rumen degradabilities of the individual amino acids with methionine being more adversely affected than lysine or the total protein (Ljøkjel *et al.*, 2000).

If the seedcoat is removed during processing the resulting meal is called 'dehulled' or 'decorticated'. The protein contents of oilseed meals vary inversely with the amount of seedcoat remaining in the meal after processing. Low-protein oilseed meals (especially sunflower and safflower meals), in which a proportion of the hull remains, are also known as 'lo-pro' or 'undecorticated' meals. 'Lo-pro' sunflower meal contains about 85% of the protein in the dehulled variety (SCA, 1987).

Oilseed meal nutritive values are described in Table 5.7. They have more lysine than the cereal grains – mostly between 1% and 2%, while rapeseed and soybean meals have between 2% and 3% on a DM basis. Linseed, safflower and coconut meals have 1.3%, 1.3% and 0.6% lysine (DM basis), respectively (Evans, 1985). The higher lysine contents of the oilseed meals compared to cereal grains reflect the higher protein content of these feeds, and also indicate differences in the amino acid compositions of cereal grain and oilseed proteins. The cereal grains have about 0.3% lysine and 12% protein, giving a lysine/protein ratio of 1:40. This ratio is half of that found in the oilseed meals (1:20). Soybean meal has substantially higher essential amino acid contents than are needed by pigs (NRC, 1998), and cottonseed, linseed, groundnut and rapeseed meals also have excellent amino acid profiles. Coconut meal has marginal to deficient concentrations of all essential amino acids except arginine. While most oilseed meals supply useful amounts of lysine in non-ruminant diets and some bypass protein in

Table 5.7. Nutritive value of solvent-extracted oilseed meals. (From Rundgren, 1983; Evans, 1985; NRC, 1989, 2000; SCA, 1990; Faldet *et al.*, 1992; Chiou *et al.*, 1999; O'Mara *et al.*, 1999; Woods *et al.*, 1999; Carvalho *et al.*, 2005.)

Constituent	Rapeseed/ canola meal	Cottonseed meal	Palm kernel meal	Soybean meal	Sunflower meal
DE (pigs) (MJ/kg DM)	13.4	13.0	–	16.2	13.1
ME (ruminants) (MJ/kg DM)	11.3	11.2	10.0–12.1	13.1	9.8
Protein (% DM)	39.9	46.4	15.8–20.4	50.8	41.4
Protein degradability (%)[a]	68–74	46–90	38	52–80	78–80
Lysine (% DM)	2.31	1.92	–	3.21	1.26
Lysine availability (%)[b]	83	78	–	95	72
ADF (% DM)	18.5	16.6	24.1–57	5.8	18.7
Ca (% DM)	0.75	0.22	0.31	0.34	0.39
P (% DM)	1.19	1.20	0.67	0.72	0.96
Available P (% DM)[b]	0.24	0.20	–	0.19	0.18
Na (% DM)	0.05	0.05	–	0.22	0.03

[a]Rumen degradability.
[b]Availability for pigs.

ruminant diets, the availability of lysine and the degradability of protein vary considerably between meals. Rumen degradabilities are 38% for palm kernel meal and between 35% and 77% in soybean meal depending on the processing method. Lysine availabilitites vary between 70% and 95% in different types of meal.

Oilseed meals have DE contents (for pigs) and ME contents (for ruminants) which are similar to those of the cereal grains, i.e. about 13–16 MJ DE/kg DM and 10–13 MJ ME/kg DM. This allows us to replace cereal grains with oilseed meals when we formulate rations, and to increase the protein content without greatly reducing the ration DE or ME content. The processing method influences the energy content. The greater residual oil levels in expeller-extracted meals are associated with higher ME values.

Oilseed meals are only modest sources of minerals. Their Ca contents (between 0.15% and 0.3% in the DM of most oilseed meals) are not adequate for highly productive animals. They contain between 0.7% and 1.5% P (DM basis), which is sufficient for ruminants, but only 1–20% of the P in oilseed meals is available to pigs because most of the P occurs as phytin. The P available to pigs in these feeds is about 0.2%, which is marginal for highly productive animals. The Na content of oilseed meals is 0.02–0.04%, similar to that in the cereal grains and insufficient for most animals.

Groundnut, rapeseed, safflower and sunflower meals are adequate to good sources of the B group vitamins. Most have substantially more choline than pigs need. However, coconut meal lacks sufficient pantothenic acid and thiamine for growing pigs (NRC, 1998). Oilseed meals have no vitamin B12. Vitamin E contents are deficient to marginal and they have very low concentrations of vitamin A.

Oilseeds often contain anti-nutritive substances, such as oxalic acid, phytin, substances with oestrogenic or goitrogenic activity and various alkaloids (Vaughan, 1970). As a general rule, these poisons are more toxic to non-ruminant animals. Microbial digestion of feed in the rumen may detoxify some secondary substances and allow others to react with feed constituents or rumen bacteria and so prevent them from being absorbed from the lower digestive tract. Some oilseed toxins cannot be completely removed or detoxified (usually by heating) and there are recommended upper levels for the inclusion of most oilseed meals in non-ruminant diets.

Nutritional Characteristics of Individual Oilseed Meals

Soybean meal

This is one of the most widely used oilseed meals, because of its palatability and nutrient content. Raw soybeans contain several toxins but these are inactivated by adequate heat treatment. There is generally no limit to the inclusion of soybean meal in any animal diet, except that it may be prudent to restrict its use for young pigs and pre-ruminant calves to avoid a type of food allergy (Lalles, 1993; Bailey *et al.*, 2001). Meals made from genetically engineered soybeans are not suitable for feeding animals which are intended for organic food markets. Soybean meal quality is influenced by the growing conditions, and meals produced from soybeans grown in different regions may have different protein and amino acid contents. Grieshop and Fahey (2001) reported variations of up to 3% in total protein and 11% in total essential amino acids.

Rapeseed (canola) meal

Some rapeseed meals contain levels of secondary substances which adversely effect pig performance. These include erucic acid, glucosinolates, tannins and phytic acid. Erucic acid is a fatty acid which increases the requirements for Se and other antioxidants and may cause heart disorders and growth retardation (Griffiths *et al.*, 1998). Glucosinolates are glycosides which can be hydrolysed to produce goitrogens (Griffiths *et al.*, 1998), and which are unpalatable and can reduce feed intake (Rundgren, 1983). Rapeseed cultivars with low glucosinolate and erucic acid contents are called 'canola' or '00' cultivars. Canola has 2% or less of erucic acid in the oil and less than 30 μM/g glucosinolates in the oil-free meal (Dupont *et al.*, 1989; Murdock *et al.*, 1992). Rapeseed meal with high glucosinolate levels can cause reproductive problems in pigs (Opalka, 1996), and they should be restricted to about 10% in grower, and 3% in sow, diets (SCA, 1987). Canola or 00 rapeseed meals do not have these effects and can be used at higher levels, e.g. 20–30% (King *et al.*, 2001).

Sunflower meal

Although sunflower seeds contain about 3% phenols, mostly chlorogenic acid, there have been no reports of any adverse effect of feeding sunflower meal to any animal. Indeed, sunflower seed phenols can stabilize sunflower oil by acting as antioxidants (De Leonardis *et al.*, 2003). Raw sunflower seeds may be given to horses. Ground, unprocessed seeds can be used in pig diets where their high-fibre content can be used to dilute the ration energy content. Sunflower hulls can also be used as a fibre source in pig diets but they may be unpalatable. Due to its low available lysine and high-fibre contents, it is recommended that no more than 10% sunflower meal be included in adult pig diets and none in creep or weaner diets (Evans, 1985), unless diets are formulated using available amino acid values.

Cottonseed meal

The use of cottonseed meal for poultry, pigs and horses is limited by the presence of gossypol (about 0.05% in solvent-extracted meal). If eaten in large enough quantities, this meal can reduce fertility and may cause death (Panigrahi and Hammonds, 1990; Chenoweth *et al.*, 2000). Cottonseed meal should be limited to no more than 5–10% in pig and horse diets and should not be used in pig creep or weaner diets (Evans, 1985; SCA, 1987). While whole cottonseed (also called 'white' or 'fuzzy' cottonseed) is a valuable ruminant food, it can have high levels of gossypol and the reported effects of feeding it for 60–200 days include delayed puberty in bulls (Chase *et al.*, 1994) and impaired liver and kidney function (Colin-Negrete *et al.*, 1996).

Groundnut meal

Groundnut meal can be fed to pigs of all ages where it gives results which are comparable to soybean meal (Tartrakoon *et al.*, 1999; Shelton *et al.*, 2001). Groundnut meal is a good substrate for mould growth. Care should be taken with moulds in groundnuts as the fungus *Aspergillus flavus* produces aflatoxins which are very poisonous and carcinogenic (Forsyth, 1991). At levels which do not cause these serious effects, aflatoxins can cause loss of appetite and reduced growth rate. Vitamin A has a protective effect against aflatoxin B1 and so the presence of this toxin will increase the animal's vitamin A requirement (Bhattacharya *et al.*, 1989).

Palm kernel meal

Palm kernel meal is one of the less palatable plant protein concentrates (Moss and Givens, 1994) and should be introduced gradually into a diet. It has a relatively saturated fat (Norulaini *et al.*, 2004). The cell walls of this meal are rich in mannose (Dusterhoft *et al.*, 1992). Animal responses to palm kernel meal are not always good, and adverse effects on pig carcasses have been reported when large amounts (40%) were included (Rhule, 1996). Palm kernel meal should be limited to about 10% of pig diets.

Coconut meal (also called 'copra meal' or 'coconut oil meal')

There is some confusion in the literature about coconut products: strictly speaking, copra is the unextracted, dried flesh of the coconut, while coconut meal is the residue which remains after the extraction of coconut oil. Both names can be used to refer to the extracted product. Copra is an expensive commodity, and so coconut meal is generally the material used in animal feeding. Coconut meal has a modest protein content (21–25%) which lacks useful amounts of lysine (Thorne *et al.*, 1990), and it is rich in fibre (about 49% NDF). Its available energy varies with the residual oil content which can vary from 1% to 20% (FAO, undated). Indicative values are 16.5 MJ DE/kg DM for pigs when incorporated in rations at 20% (O'Doherty and McKeon, 2000) and 9 MJ ME/kg DM for ruminants (Krishnamoorthy *et al.*, 1995). The residual oil in coconut meal is rich in lauric and myristic fatty acids. When coconut meal is fed to pigs, dairy cows and goats these fatty acids can increase the amounts of these fatty acids in pig carcass fat and the hardness of butterfat. These changes in fatty acid composition can also alter the flavour of milk and pig meat (Thorne *et al.*, 1992; Guarte *et al.*, 1996; FAO, undated). The food intake of pigs declines with increasing coconut meal, possibly because of increasing fibre levels but not due to changes in amino acid contents (Thorne *et al.*, 1992; O'Doherty and McKeon, 2000). Coconut meal should be gradually introduced, at levels up to 20% in pig and 60% in goat diets (O'Doherty and McKeon, 2000; Aregheore, 2006).

Full-fat oilseed meals

These are prepared by milling the unextracted seed. Full-fat soybean meal contains about 39% protein, with an energy content similar to maize and wheat grains. It can be used as described for the extracted meal, but it is prudent to check that the meal has been sufficiently heated (urease and nitrogen solubility index tests will show if the meal has been correctly processed; Balloun *et al.*, 1953; SCA, 1987; Araba and Dale, 1990). Full-fat soybean meal is suitable for creep and weaner diets, and to increase the energy content of breeding sow diets. Its use by ruminants reduces methane production (Moss and Givens, 1994). Full-fat rapeseed meals can have as much as 17 MJ ME/kg DM for pigs (Rundgren, 1983) compared to 10.5–12.5 MJ ME/kg DM in the extracted meal. Genetic improvements in sunflowers have increased the oil content of the seed. Qiao and Thacker (2004) have reported a DE content of 16 MJ/kg DM which they suggest is a better indication of the available energy content of modern sunflower cultivars. The DE content for pigs of full-fat palm kernel meal is approximately 19.5 MJ/kg DM (Agunbiade *et al.*, 1999).

Full-fat oilseed meals are made from ground and heated, but otherwise unprocessed, oilseeds. They must be heated to inactivate their endogenous poisons. They are a rich energy source but there is a possibility of mould growth and the associated risk of mould toxins, and the oil can become rancid through contact with air.

Animal Protein Meals

The animal protein meals include meat meals made from poultry, cattle and sheep carcasses; blood meal; fish meals; and milk by-products. Meals made from animal carcasses usually include bone, but are usually called 'meat meals' nevertheless, instead of the more correct 'meat and bone meal'.

Some jurisdictions limit the use of animal protein meals in animal foods because they might be a vector for the transmission of 'mad cow disease' (bovine spongiform encephalopathy (BSE)). The disease is fatal to the infected animal, and people who eat meat from infected animals may possibly contract the fatal disease variant Creutzfeld Jacob Disease (vCJD). Cases of vCJD have been reported in Europe since the 1980s (Sneden, 2001; FAO, 2002). For reviews of the causes and effects of bovine and human transmissible spongiform encephalopathies read Dealler and Lacey (1991) or WHO (2002).

The European Community (Anonymous, 2000) adopted recommendations that ruminant-derived meat meals should not be fed to ruminants. Fish meal was banned in the UK from 2001 and in some other jurisdictions because of the risk of adulteration with meat meal (WHO, 2002). Regulations about the use of animal protein meals are continually changing, and vary widely between different jurisdictions.

Table 5.8 illustrates the nutrient content of some typical animal protein meals. They are good sources of essential amino acids, and particularly of available lysine, provided that the meals have been properly prepared from good-quality raw materials. Meat meals are good sources of Ca and P. Monogastric animals do not find blood, meat and fish meals very palatable.

Table 5.8. Nutrient contents of meat and fish meals.[a] (From Evans, 1985; NRC, 1989.)

Constituent	Meat meal (55% protein)	Meat meal (45% protein)	Blood meal (spray dried)	Fish meal (65% protein)
DE (pigs) (MJ/kg DM)	12.6	12.6	15.8	14.7
DE (horses) (MJ/kg DM)	–	–	–	12.6
ME (ruminants) (MJ/kg DM)	9.7	7.6	11.6	10.9
Protein (% DM)	58.0	49.6	92.4	70.0
Protein degradability (%)[b]	44	–	25	40
Lysine (% DM)	3.3	2.5	7.9	5.6
Lysine availability (%)[c]	89	72	94	94
ADF (% DM)	5.9	1.7	2.4	–
Linoleic acid (% DM)	0.85	0.43	0.18	0.23
Ca (% DM)	9.05	11.83	0.44	4.55
P (% DM)	4.55	5.69	0.31	2.90
Available P (% DM)[c]	4.34	5.69	0.27	2.90
Na (% DM)	0.77	0.68	0.56	0.79

[a]These foodstuffs are prohibited in ruminant foods in some jurisdictions.
[b]Rumen degradability.
[c]Availability for pigs, but these values vary according to the degree of heat damage.

Nutritional Characteristics of Individual Animal Protein Meals

Meat meals

Meals made from lean tissue are good protein and amino acid sources, and Ca and P both occur in readily available forms. Poor-quality meat meals include amounts of trimmings, tendons, hide, etc. These tissues are rich in connective tissue proteins which have a poor essential amino acid composition. Large amounts of bone in the meal are associated with low-protein contents and high Ca and P levels. The disadvantage of low-protein content is obvious. Excess Ca may interfere with Zn absorption and cause parakeratosis in pigs (Lewis *et al.*, 1956). Affected animals grow slowly and develop a type of dermatitis. Meat meals should not be included at levels which will raise the total diet Ca to more than 12 g/kg DM (see SCA, 1987), especially where diet ingredients have high phytate levels (Oberleas *et al.*, 1962; Fordyce *et al.*, 1987), unless a Zn supplement is used. Meat meals have 9–12% fat, with up to 20% in meals made from poultry tissue.

Meat meals should be autoclaved at 133°C for 20 min at 300 kPa (FAO, undated). While this will reduce the risk of the meal carrying pathogens and stabilize the raw material by drying it to a water content of about 5–8%, it reduces amino acid availability and protein degradability in the rumen.

Fish meals

These are generally superior to any grade of meat meal, but their quality is influenced by the type of raw material (fish species and whether the whole fish or trimmings are used) and the temperature reached during drying (Rand *et al.*, 1959). Over-heating the meal reduces both amino acid availability and protein degradability in the rumen. Fish meals (recommended limit of 7% in pig rations; Evans, 1985) contain about 10% oil, which can produce off-tastes in pork.

Blood meal

This has a high protein content of good amino acid composition and adequate (for pigs) concentrations of Cu, Fe, K, Na and Se. Amino acid availabilities reflect the amount of heat used in processing. Less heating gives higher amino acid availabilities

(Waibel *et al.*, 1977). The recommended inclusion limit in pig diets is 5% (M'ncene *et al.*, 1999).

Milk by-products

Buttermilk, skimmed milk powders and casein are protein-rich meals with an excellent amino acid profile (FAO, undated). They are palatable, easily digested by immature animals such as calves, and are often included in milk replacer mixtures where they provide essential amino acids, digestible protein and readily available minerals. Casein is included in milk replacers for very young ruminants where, provided that it has not been heat-damaged, it facilitates the formation of a protein/fat clot in the abomasum. This is considered to ensure the complete digestion of milk solids (see discussion by Longenbach and Heinrichs, 1998). Milk by-products are generally expensive and are reserved for applications where these characteristics are required and cannot be provided from cheaper ingredients.

References

Agunbiade, J.A., Wiseman, J. and Cole, D.J.A. (1999) Energy and nutrient use of palm kernels, palm kernel meal and palm kernel oil in diets for growing pigs. *Animal Feed Science and Technology* 80, 165–181.

Amin, A.E. and Adam, S.E. (1993) Determination of the toxicity of molasses in Nubian goats. *Veterinary and Human Toxicology* 35, 213–216.

Anonymous (2000) *Council Decision of 4 December 2000 Concerning Certain Protection Measures with Regard to Transmissible Spongiform Encephalopathies and the Feeding of Animal Protein (2000/766/EC). Official Journal of the European Communities*, L 306/ 32–33. Available at: http://europa.eu.int/eur-lex/pri/en/oj/dat/2000/ l_306/l_30620001207en 00320033.pdf

Araba, A., Byers, F.M. and Guessous, F. (2002) Patterns of rumen fermentation in bulls fed barley/molasses diets. *Animal Feed Science and Technology* 97, 53–64.

Araba, M. and Dale, N.M. (1990) Evaluation of protein solubility as an indicator of over processing soybean meal. *Poultry Science* 69, 76–83.

Aregheore, E.M. (2006) Utilization of concentrate supplements containing varying levels of copra cake (*Cocos nucifera*) by growing goats fed a basal diet of napier grass (*Pennisetum purpureum*). *Small Ruminant Research* 64, 87–93.

Bach Knudsen, K.E. and Hansen, I. (1991) Gastrointestinal implications in pigs of wheat and oat fractions. 1. Digestibility and bulking properties of polysaccharides and other major constituents. *British Journal of Nutrition* 65, 217–232.

Bailey, M., Plunkett, F.J., Rothkotter, H.J., Vega Lopez, M.A., Haverson, K. and Stokes, C.R. (2001) Regulation of mucosal immune responses in effector sites. *Proceedings of the Nutrition Society* 60, 427–435.

Bajjalieh, N. (2004) Proteins from oilseeds. In: *Protein Sources for the Animal Feed Industry*. Expert Consultation and Workshop, Bangkok, 29 April–3 May, 2002. FAO-UN, Rome, Italy, pp. 141–159.

Balloun, S.L., Johnson, E.L. and Arnold, L.K. (1953) Laboratory estimation of the nutritive value of soybean oil meal. *Poultry Science* 32, 517–527.

Bhattacharya, R.K., Prabhu, A.L. and Aboobaker, V.S. (1989) *In vivo* effect of dietary factors on the molecular action of aflatoxin B1: role of vitamin A on the catalytic activity of liver fractions. *Cancer Letters* 44, 83–88.

Blasi, D.A., Kuhl, G.L., Drouillard, J.S., Reed, C.L., Trigo-Stockli, D.M., Behnke, K.C. and Fairchild, F.J. (1998) *Wheat Middlings Composition, Feeding Value and Storage Guidelines. Contribution No. 99-35-E from Kansas Agricultural Experiment Station.* Available at: http://www.oznet.ksu.edu/library/lvstk2/mf2353.pdf

Borrell, A.K., Garside, A.L., Fukai, S. and Reid, D.J. (1999) Grain quality of flooded rice is affected by season, nitrogen rate, and plant type. *Australian Journal of Agricultural Research* 50, 1399–1408.

Boucqué, Ch.V. and Fiems, L.O. (1988) Vegetable by-products of agro-industrial origin. *Livestock Production Science* 19, 97–135.

Brand-Miller, J., Wolever, T.M.S., Foster-Powell, K. and Colagiuri, S. (2003) *The New Glucose Revolution: The Authoritative Guide to the Glycemic Index*. Marlowe and Company, New York.

Carpenter, K.J., Schelstraete, M., Vilicich, V.C. and Wall, J.S. (1988) Immature corn as a source of niacin for rats. *Journal of Nutrition* 118, 165–169.

Carvalho, L.P.F., Melo, D.S.P., Pereira, C.R.M., Rodrigues, M.A.M., Cabrita, A.R.J. and Fonseca, A.J.M. (2005) Chemical composition, *in vivo* digestibility, N degradability and enzymatic intestinal digestibility of five protein supplements. *Animal Feed Science and Technology* 119, 171–178.

Centraal Veevoederbureau (2004) *Tabellenboek Veevoeding 2004–2005*. Centraal Veevoederbureau, Lelystad, The Netherlands.

Chase, C.C. Jr, Bastidas, P., Ruttle, J.L., Long, C.R. and Randel, R.D. (1994) Growth and reproductive development in Brahman bulls fed diets containing gossypol. *Journal of Animal Science* 72, 445–452.

Chawla, J.S. and Sikka, S.S. (1985) Growth performance and carcass quality of large white Yorkshire pigs fed graded levels of brewers' spent grains (BSG). *Indian Journal of Animal Nutrition* 2, 19–22.

Chenoweth, P.J., Chase, C.C. Jr, Risco, C.A. and Larsen, R.E. (2000) Characterization of gossypol-induced sperm abnormalities in bulls. *Theriogenology* 53, 1193–1203.

Chiou, P.W.-S., Yu, B. and Wu, S.-S. (1999) Protein subfractions and amino acid profiles of rumen-undegradable protein in dairy cows from soybean, cottonseed and fish meals. *Animal Feed Science and Technology* 78, 65–80.

Clarke, L.C., Cummins, L.J., Flinn, P.C., Arnold, D.M. and Heazlewood, P.G. (1996) A comparison of triticale and bandicoot oats and the influence of cracking barley grain on digestibility in cattle and sheep. *Animal Production in Australia* 21, 286–289.

Coates, B.J., Slinger, S.L., Summers, J.D. and Bayley, H.S. (1977) Metabolisable energy values and chemical and physical characteristics of wheat and barley. *Canadian Journal of Animal Science* 57, 195–207.

Colin-Negrete, J., Kiesling, H.E., Ross, T.T. and Smith, J.F. (1996) Effect of whole cottonseed on serum constituents, fragility of erythrocyte cells, and reproduction of growing Holstein heifers. *Journal of Dairy Science* 79, 2016–2023.

Connor, J.K., Neill, A.R. and Barram, K.M. (1976) The metabolizable energy content for the chicken of maize and sorghum grain hybrids grown at several geographical regions. *Australian Journal of Experimental Agriculture and Animal Husbandry* 16, 699–703.

Cooper, M., Woodruff, D.R., Phillips, I.G., Basford, K.E. and Gilmour, A.R. (2001) Genotype-by-management interactions for grain yield and grain protein concentration of wheat. *Field Crops Research* 69, 47–67.

Dealler, S. and Lacey, R. (1991) Beef and bovine spongiform encephalopathy: the risk persists. *Nutrition and Health* 7, 117–133.

De Leonardis, A., Macciola, V. and DiRocco, A. (2003) Oxidative stabilization of cold-pressed sunflower oil using phenolic compounds of the same seeds. *Journal of the Science of Food and Agriculture* 83, 523–528.

Dupont, J., White, P.J., Johnston, K.M., Heggtveit, H.A., Mcdonald, B.E., Grundy, S.M. and Bonanome, A. (1989) Food safety and health effects of canola oil. *Journal of the American College of Nutrition* 8, 360–375.

Dusterhoft, E.M., Posthumus, M.A. and Voragen, A.G.J. (1992) Nonstarch polysaccharides from sunflower (*Helianthus annuus*) meal and palm-kernel (*Elaeis guineensis*) meal investigation of the structure of major polysaccharides. *Journal of the Science of Food and Agriculture* 59, 151–160.

Enemark, J.M.D., Jørgensen, R.J. and Enemark, P.S. (2002) Rumen acidosis with special emphasis on diagnostic aspects of subclinical rumen acidosis: a review. *Veterinarija ir Zootechnika, T* 20(42), 16–29.

Evans, M. (1985) *Nutrient Composition of Feedstuffs for Pigs and Poultry*. QDPI, Brisbane, Australia.

Faldet, M.A., Son, Y.S. and Satter, L.D. (1992) Chemical, *in vitro*, and *in vivo* evaluation of soybeans heat-treated by various processing methods. *Journal of Dairy Science* 75, 789–795.

FAOSTAT (2004) *Food Balance Sheets (updated February, 2004)*. Available at: http://faostat.fao.org/ faostat/ form?collection=FBS&Domain=FBS&servlet= 1&hasbulk=0&version=ext& language=EN

Feed Resources Group (FRG) (undated) *Animal Feed Resources Information System*. Available at: http://www.fao.org/WAICENT/FAOINFO/AGRICULT/aga/ agap/FRG/afris/default.htm

Food and Agriculture Organization (FAO) (undated) *Animal Feed Resources Information System*. Available at: http://www.fao.org/ag/aga/agap/frg/afris/index_ en.htm

Food and Agriculture Organization (FAO) (2002) *BSE as a National and Trans-Boundary Food Safety Emergency. In: FAO/WHO Global Forum of Food Safety Regulators, Marrakesh, Morocco, 28–30 January 2002*. Available at: http://www.fao.org/DOCREP/ MEETING/004/ Y2038E.HTM

Fordyce, E.J., Forbes, R.M., Robbins, K.R. and Erdman, J.W. (1987) Phytate × calcium zinc molar ratios – are they predictive of zinc bioavailability. *Journal of Food Science* 52, 440–444.

Forsyth, D.M. (1991) Mycotoxins in swine feeds. In: Miller, E.R., Ullrey, D.E. and Lewis, A.J. (eds) *Swine Nutrition*. Butterworth-Heinemann, Stoneham, Massachusetts, pp. 425–437.

Freer, M. and Dove, H. (1984) Rumen degradation of protein in sunflower meal, rapeseed meal and lupin seed placed in nylon bags. *Animal Feed Science and Technology* 11, 87–101.

Fuller, M.F., Cadenhead, A., Brown, D.S., Brewer, A.C., Carver, M. and Robinson, R. (1989) Varietal differences in the nutritive value of cereal grains for pigs. *Journal of Agricultural Science, Cambridge* 113, 149–163.

Garrett, J.E., Guessous, F. and Eddebbarh, A. (1989) Utilization of sugar beet molasses and monensin for finishing dairy bullocks. *Animal Feed Science and Technology* 25, 11–21.

Göhl, B. (1981) *Tropical Feeds: Feed Information Summaries and Nutritive Value*, new edn. Food and Agriculture Organization of the United Nations, Rome, Italy.

Granzin, B.C. and Dryden, G.McL. (2005) Monensin supplementation of lactating cows fed tropical grasses and cane molasses or grain. *Animal Feed Science and Technology* 120, 1–16.

Grieshop, C.M. and Fahey, G.C. Jr (2001) Comparison of quality characteristics of soybeans from Brazil, China, and the United States. *Journal of Agricultural and Food Chemistry* 49, 2669–2673.

Griffiths, D.W., Birch, A.N.E. and Hillman, J.R. (1998) Antinutritional compounds in the *Brassicaceae*: Analysis, biosynthesis, chemistry and dietary effects. *Journal of Horticultural Science and Biotechnology* 73, 1–18.

Guarte, R.C., Muhlbauer, W. and Kellert, M. (1996) Drying characteristics of copra and quality of copra and coconut oil. *Postharvest Biology and Technology* 9, 361–372.

Hanstock, T.L., Clayton, E.H., Li, K.M. and Mallet, P.E. (2004) Anxiety and aggression associated with the fermentation of carbohydrates in the hindgut of rats. *Physiology and Behavior* 82, 357–368.

Hongtrakul, K., Goodband, R.D., Behnke, K.C., Nelssen, J.L., Tokach, M.D., Bergström, J.R., Nessmith, W.B. Jr and Kim, I.H. (1998) The effects of extrusion processing of carbohydrate sources on weanling pig performance. *Journal of Animal Science* 76, 3034–3042.

Hoover, R. and Ratnayake, W.S. (2002) Starch characteristics of black bean, chick pea, lentil, navy bean and pinto bean cultivars grown in Canada. *Food Chemistry* 78, 489–498.

Hunt, C.W. (1996) Factors affecting the feeding quality of barley for ruminants. *Animal Feed Science and Technology* 62, 37–48.

Jarridge, R. (ed.) (1989) *Ruminant Nutrition Recommended Allowances and Feed Tables*. INRA, John Libbey Eurotext, Paris, France.

Jentsch, W., Chudy, A. and Beyer, M. (2003) *Rostock Feed Evaluation System*. Plexus Verlag, Miltenberg, Germany.

Kim, J.C., Simmins, P.H., Mullan, B.P. and Pluske, J.R. (2005) The digestible energy value of wheat for pigs, with special reference to the post-weaned animal [Review]. *Animal Feed Science and Technology* 122, 257–287.

King, R.H., Eason, P.E., Kerton, D.K. and Dunshea, F.R. (2001) Evaluation of solvent-extracted canola meal for growing pigs and lactating sows. *Australian Journal of Agricultural Research* 52, 1033–1041.

Kornegay, E.T. (1973) Digestible and metabolizable energy and protein utilization values of brewers dried by-products for swine. *Journal of Animal Science* 37, 479–483.

Kreikemeier, K.K., Harmon, D.L., Peters, J.P., Gross, K.L., Armendariz, C.K. and Krehbiel, C.R. (1990) Influence of dietary forage and feed intake on carbohydrase activities and small intestinal morphology of calves. *Journal of Animal Science* 68, 2916–2929.

Krishnamoorthy, U., Solled, H., Steingass, H. and Menke, K.H. (1995) Energy and protein evaluation of tropical feedstuffs for whole tract and ruminal digestion by chemical analyses and rumen inoculum studies *in vitro*. *Animal Feed Science and Technology* 52, 177–188.

Kronfeld, D., Rodiek, A. and Stull, C. (2004) Glycemic indices, glycemic loads, and glycemic dietetics. *Journal of Equine Veterinary Science* 24, 399–404.

Lalles, J.P. (1993) Nutritional and antinutritional aspects of soyabean and field pea proteins used in veal calf production: a review. *Livestock Production Science* 34, 181–202.

Lee, R.W., Galyean, M.L. and Lofgreen, G.P. (1982) Effects of mixing whole shelled and steam flaked corn in finishing diets on feedlot performance and site and extent of digestion in beef steers. *Journal of Animal Science* 55, 475–485.

Le Goff, G., Van Milgen, J. and Noblet, J. (2002) Effets du niveau et de l'origine botanique des parois vegetales sur l'utilisation digestive de l'aliment et le transit des digesta chez le porc en croissance et la truie adulte. *34emes Journees de la Recherche Porcine, sous l'egide de l'Association Franccaise de Zootechnie*, Paris, France, 5–7 fevrier 2002, pp. 75–80.

Le Thu Ha, Nguyen Quang Suc, Dinh Van Binh, Le Thi Bien and Preston, T.R. (1996) Replacing concentrates with molasses blocks and protein-rich tree leaves for reproduction and growth of rabbits. *Livestock Research for Rural Development 8(3)*. Available at: http://www.cipav.org.co/lrrd/lrrd8/3/ha83.htm

Lewis, P.K., Hoekstra, W.G., Grummer, R.H. and Phillips, P.H. (1956) The effect of certain nutritional factors including calcium, phosphorus and zinc on parakeratosis in swine. *Journal of Animal Science* 15, 741–751.

Ljøkjel, K., Harstad, O.M. and Skrede, A. (2000) Effect of heat treatment of soybean meal and fish meal on amino acid digestibility in mink and dairy cows. *Animal Feed Science and Technology* 84, 83–95.

Longenbach, J.I. and Heinrichs, A.J. (1998) A review of the importance and physiological role of curd formation in the abomasum of young calves. *Animal Feed Science and Technology* 73, 85–97.

Luce, W.G., Peo, E.R. Jr and Hudman, D.B. (1967) Availability of niacin in corn and milo for swine. *Journal of Animal Science* 26, 76–84.

Marty, B.J. and Chavez, E.R. (1995) Ileal digestibilities and urinary losses of amino acids in pigs fed heat processed soybean products. *Livestock Production Science* 43, 37–48.

Mathison, G.W., Hironaka, R., Kerrigan, B.K., Vlach, I., Milligan, L.P. and Weisenburger, R.D. (1991) Rate of starch degradation, apparent digestibility and rate and efficiency of steer gain as influenced by barley grain volume-weight and processing method. *Canadian Journal of Animal Science* 71, 867–878.

Metayer, J.P., Grosjean, F. and Castaing, J. (1993) Study of variability in French cereals. *Animal Feed Science and Technology* 43, 87–108.

Ministry of Agriculture, Fisheries and Food (MAFF) (1992) *Feed Composition. UK Tables of Feed Composition and Nutritive Value for Ruminants*, 2nd edn. Chalcombe Publications, Canterbury, UK.

M'ncene, W.B., Tuitoek, J.K. and Muiruri, H.K. (1999) Nitrogen utilization and performance of pigs given diets containing a dried or undried fermented blood/molasses mixture. *Animal Feed Science and Technology* 78, 239–247.

Morel, P.C.H., Padilla, R.M., Cottam, Y.H. and Coles, G.D. (2001) Influence of soluble and insoluble dietary fibre on growth, organ weights and blood lipid metabolites in weaner pigs. *Recent Advances in Animal Nutrition in Australia* 13, 22A.

Moss, A.R. and Givens, D.I. (1994) The chemical composition, digestibility, metabolisable energy content and nitrogen degradability of some protein concentrates. *Animal Feed Science and Technology* 47, 335–351.

Murdock, L., Herbek, J. and Riggins, S.K. (1992) *Canola Production and Management*. Available at: www.ca.uky.edu/agc/pubs/id/id114/id114.htm

National Research Council (NRC) (1982) *United States-Canadian Tables of Feed Composition: Nutritional Data for United States and Canadian Feeds*, 3rd rev. edn. National Academy Press, Washington, DC.

National Research Council (NRC) (1989) *Nutrient Requirements of Horses*, 6th rev. edn. National Academy Press, Washington, DC.

National Research Council (NRC) (1998) *Nutrient Requirements of Swine*, 10th rev. edn. National Academy Press, Washington, DC.

National Research Council (NRC) (2000) *Nutrient Requirements of Beef Cattle*, 7th rev. edn. Update. National Academy Press, Washington, DC.

National Research Council (NRC) (2001) *Nutrient Requirements of Dairy Cattle*, 7th rev. edn. National Academy Press, Washington, DC.

Norulaini, N.A.N., Zaidul, I.S.M., Anuar, O. and Omar, A.K.M. (2004) Supercritical enhancement for separation of lauric acid and oleic acid in palm kernel oil (PKO). *Separation and Purification Technology* 39, 133–138.

Oberleas, D., Muhrer, M.E. and Dell, B.L. (1962) Effects of phytic acid on zinc availability and parakeratosis in swine. *Journal of Animal Science* 21, 57–61.

O'Doherty, J.V. and McKeon, M.P. (2000) The use of expeller copra meal in grower and finisher pig diets. *Livestock Production Science* 67, 55–65.

OECD (2002) *Consensus Document on Compositional Considerations for New Varieties of Sugar Beet: Key Food and Feed Nutrients and Antinutrients*. Series on the Safety of Novel Foods and Feeds, No. 3. Organisation for Economic Co-operation and Development, Paris, France.

O'Mara, F.P., Mulligan, F.J., Cronin, E.J., Rath, M. and Caffrey, P.J. (1999) The nutritive value of palm kernel meal measured *in vivo* and using rumen fluid and enzymatic techniques. *Livestock Production Science* 60, 305–316.

Opalka, M. (1996) Effect of feeding based on diets containing rapeseed meal on reproductive processes of sows and gilts. *Acta Academiae Agriculturae ac Technicae Olstenensis, Zootechnica* 45, 209–220.

Oscarsson, M., Andersson, R., Åman, P., Olofsson, S. and Jonsson, A. (1998) Effects of cultivar, nitrogen fertilization rate and environment on yield and grain quality of barley. *Journal of the Science of Food and Agriculture* 78, 359–366.

Ostrowski-Meissner, H.T. (ed.) (1987) *Australian Feed Composition Tables: Ruminants*. AFIC-CSIRO, Sydney, Australia.

Ostrowski-Meissner, H.T. (ed.) (1990) *Australian Feed Composition Tables: Pigs and Poultry*, 2nd edn. AFIC-CSIRO, Sydney, Australia.

Owens, F.N., Secrist, D.S., Hill, W.J. and Gill, D.R. (1997) The effect of grain source and grain processing on performance of feedlot cattle: a review. *Journal of Animal Science* 75, 868–879.

Owens, F.N., Secrist, D.S., Hill, W.J. and Gill, D.R. (1998) Acidosis in cattle: a review. *Journal of Animal Science* 76, 275–286.

Palipane, K.B. and Swarnasiri, C.D.P. (1985) Composition of raw and parboiled rice bran from common Sri Lankan varieties and from different types of rice mills. *Journal of Agricultural and Food Chemistry* 33, 732–734.

Panigrahi, S. and Hammonds, T.W. (1990) Effects of dietary cottonseed meal and iron-treated cottonseed meal in different laying hen genotypes. *British Poultry Science* 31, 107–120.

Pereira, J.C., Carro, M.D., Gonzalez, J., Alvir, M.R. and Rodriguez, C.A. (1998) Rumen degradability and intestinal digestibility of brewers' grains as affected by origin and heat treatment and of barley rootlets. *Animal Feed Science and Technology* 74, 107–121.

Perez, R. (1995) Molasses. *Appendix. Extracts from the FAO Tropical Feeds Database. First FAO Electronic Conference on Tropical Feeds and Feeding Systems.* Available at: http://www.fao.org/waicent/faoinfo/agricult/AGA/AGAP/FRG/

Plegge, S.D., Berger, L.L. and Fahey, G.C. (1985) Effect of roasting temperature on the proportion of soyabean meal nitrogen escaping degradation in the rumen. *Journal of Animal Science* 61, 1211–1218.

Pluske, J.R., McDonald, D.E., Pethick, D.W., Mullan, B.P. and Hempson, D.J. (2001) Nutritional management of the gastrointestinal tract to reduce enteric diseases in pigs. *Recent Advances in Animal Nutrition in Australia* 16, 127–134.

Qiao, S.Y. and Thacker, P.A. (2004) Digestible energy content of traditional and non-traditional feeds for swine determined using the mobile nylon bag technique. *Journal of Animal and Veterinary Advances* 3, 371–377.

Quemere, P., Fourdrinier, R., Lefranc, A. and Willequet, F. (1983) Utilisation de la dreche de brasserie deshydratee par le porc en croissance-finition. *15emes Journees de la Recherche Porcine en France 1983*, pp. 325–334.

Rand, N.T., Collins, V.K., Vamer, D.S. and Mosser, J.D. (1959) Biological evaluation of the factors affecting the protein quality of fish meals. *Poultry Science* 39, 45–53.

Reddy, D.V., Mehra, U.R. and Singh, U.B. (1989) Effect of hydrogen peroxide treatment on the utilization of lignocellulosic residues by rumen micro-organisms. *Biological Wastes* 28, 133–141.

Rhule, S.W.A. (1996) Growth rate and carcass characteristics of pigs fed on diets containing palm kernel cake. *Animal Feed Science and Technology* 61, 167–172.

Rundgren, M. (1983) Low-glucosinolate rapeseed products for pigs – a review. *Animal Feed Science and Technology* 9, 239–262.

Sauvant, D., Perez, J.-M. and Tran, G. (2004) *Tables of Composition and Nutritional Value of Feed Materials*, 2nd rev. and corrected edn. Wageningen Academic Publishers and INRA, Wageningen, The Netherlands.

Schwartzkopf-Genswein, K.S., Beauchemin, K.A., McAllister, T.A., Gibb, D.J., Streeter, M. and Kennedy, A.D. (2003) Effect of feed delivery fluctuations and feeding time on ruminal acidosis, growth performance, and feeding behavior of feedlot cattle. *Journal of Animal Science* 82, 3357–3365.

Shelton, J.L., Hemann, M.D., Strode, R.M., Brashear, G.L., Ellis, M., McKeith, F.K., Bidner, T.D. and Southern, L.L. (2001) Effect of different protein sources on growth and carcass traits in growing–finishing pigs. *Journal of Animal Science* 79, 2428–2435.

Sneden, C. (2001) Testing times for BSE. *Nature, London* 409, 658–659.

Soren, N.M., Bhar, R., Chhabra, A.K. and Mandal, A.B. (2003) Performance of crossbred gilts fed on diets with higher levels of fat and fibre through addition of rice bran. *Asian-Australasian Journal of Animal Sciences* 16, 1650–1655.

Spears, J.K. and Fahey, G.C. Jr (2004) Resistant starch as related to companion animal nutrition. *Journal of the AOAC International* 87, 787–791.

Standing Committee on Agriculture (SCA) (1987) *Feeding Standards for Australian Livestock. Pigs.* CSIRO, Melbourne, Australia.

Standing Committee on Agriculture (SCA) (1990) *Feeding Standards for Australian Livestock. Ruminants.* CSIRO, Melbourne, Australia.

Steg, A. and van der Meer, J.M. (1985) Differences in chemical composition and digestibility of beet and cane molasses. *Animal Feed Science and Technology* 13, 83–91.

Stewart, G.A. (1976) Mineral nutrients in Australian molasses for fattening cattle. *Proceedings of the Australian Society of Animal Production* XI, 373–376.

Streeter, M.N., Wagner, D.G., Hibberd, C.A., Mitchell, E. D. Jr and Oltjen, J.W. (1990) Effect of variety of sorghum grain on digestion and availability of dry matter and starch *in vitro*. *Animal Feed Science and Technology* 29, 279–287.

Svihus, B., Uhlen, A.K. and Harstad, O.M. (2005) Effect of starch granule structure, associated components and processing on nutritive value of cereal starch: a review. *Animal Feed Science and Technology* 122, 303–320.

Takagi, H., Hashimoto, Y., Yonemochi, C., Ishibashi, T., Asai, Y. and Watanabe, R. (2003) Digestibility of cereals, oil meals, brans and hays in thoroughbreds. *Journal of Equine Science* 14, 119–124.

Tartrakoon, W., Thinggaard, G., Chakeredza, S., Vearasilp, T. and Meulen, U. (1999) An evaluation of feedstuffs quality for pigs in Thailand 2. Ileal and faecal digestibilities of crude protein and amino acids in soybean, peanut and sesame meals. *Thai Journal of Agricultural Science* 32, 453–464.

Taverner, M.R. and Farrell, D.J. (1981) Availability to pigs of amino acids in cereal grains. 4. Factors influencing the availability of amino acids and energy in grains. *British Journal of Nutrition* 46, 181–192.

Thorne, P.J., Wiseman, J. and Cole, D.J.A. (1990) Copra meal. In: Thacker, P.A. and Kirkwood, R.N. (eds) *Non-Traditional Feed Sources for Use in Swine Production*. Butterworth, London, pp. 127–134.

Thorne, P.J., Wiseman, J., Cole, D.J.A. and Machin, D.H. (1992) Effects of level of inclusion of copra meal in balanced diets supplemented with synthetic amino acids on growth and fat deposition and composition in growing pigs fed ad libitum at a constant temperature of 25°C. *Animal Feed Science and Technology* 40, 31–40.

Tomkins, N.W., Fenwicke, C.T. and Hunter, R.A. (2004) High molasses diets for feedlot cattle. *Animal Production in Australia* 25, 184–187.

Tudor, G.D. and Inkerman, P.A. (1989) Alkali treated bagasse – potential as feed for ruminants. *Recent Advances in Animal Nutrition in Australia* 10, 53–66.

Udén, P., Robinson, P.H. and Wiseman, J. (2005) Editorial. Use of detergent system terminology and criteria for submission of manuscripts on new, or revised, analytical methods as well as descriptive information on feed analysis and/or variability. *Animal Feed Science and Technology* 118, 181–186.

United Nations Economic Commission for Europe (UNECE) (1995) *Handbook on Economic Statistics. Agricultural Statistics. 6. Oil-bearing Crops and Derived Products*. Available at: http://www.unece.org/stats/econ/iwg.agri/handbook.oil.html

USDA (2002) *Counsellor and Attaché Reports Official Statistics, USDA Estimates*, FAS, April 2002. Available at: http://www.fas.usda.gov/oilseeds_arc.asp

Van Barneveld, R.J. (1999) Chemical and physical characteristics of grains related to variability in energy and amino acid availability in pigs: a review. *Australian Journal of Agricultural Research* 50, 667–687.

Vaughan, J.G. (1970) *The Structure and Utilization of Oil Seeds*. Chapman & Hall, London.

Waibel, P.E., Cuperlovic, M., Hurrell, R.F. and Carpenter, K.J. (1977) Processing damage to lysine and other amino acids in the manufacture of blood meal. *Journal of Agricultural and Food Chemistry* 25, 171–175.

Wang, J.F., Jensen, B.B., Jorgensen, H., Li, D.F. and Lindberg, J.E. (2002) Ileal and total tract digestibility, and protein and fat balance in pigs fed rice with addition of potato starch, sugar beet pulp or wheat bran. *Animal Feed Science and Technology* 102, 125–136.

Warren, B.E. and Farrell, D.J. (1990) The nutritive value of full-fat and defatted Australian rice bran. I. Chemical composition. *Animal Feed Science and Technology* 27, 219–228.

WHO (2002) *Fact Sheet No. 113. Bovine Spongiform Encephalopathy*. Available at: http://www.who.int/mediacentre/factsheets/fs113/en/index.html

Woods, V.B., Moloney, A.P., Mulligan, F.J., Kenny, M.J. and O'Mara, F.P. (1999) The effect of animal species (cattle or sheep) and level of intake by cattle on *in vivo* digestibility of concentrate ingredients. *Animal Feed Science and Technology* 80, 135–150.

Younker, R.S., Winland, S.D., Firkins, J.L. and Hull, B.L. (1998) Effects of replacing forage fiber or nonfiber carbohydrates with dried brewers grains. *Journal of Dairy Science* 81, 2645–2656.

Yu, P., Goelema, J.O., Leury, B.J., Tamminga, S. and Egan, A.R. (2002) An analysis of the nutritive value of heat processed legume seeds for animal production using the DVE/OEB model: a review. *Animal Feed Science and Technology* 99, 141–176.

Zandersons, J., Gravitis, J., Kokorevics, A., Zhurinsh, A., Bikovens, O., Tardenaka, A. and Spince, B. (1999) Studies of the Brazilian sugarcane bagasse carbonization process and products properties. *Biomass and Bioenergy* 17, 209–219.

6 Secondary Substances in Concentrates and Roughages

Introduction

Secondary substances (also called 'plant secondary metabolites' or 'anti-nutritional factors') are compounds made by plants as defences against microbial, insect and animal predation. When present in foods they have an adverse effect on animal health and performance. Grain legumes, oilseeds and oilseed meals are the commonly used foods which are likely to contain poisonous secondary substances (Liener, 1975; Evans, 1980; Van der Poel *et al.*, 1993; Muzquiz *et al.*, 2004).

The plant toxins in oilseeds (Liener, 1975) include proteins (e.g. protease inhibitors and lectins), glycosides (e.g. goitrogens, cyanogens, saponins and oestrogens), phenols (e.g. gossypol and tannins), as well as antivitamins which may increase the requirements for vitamins A, D, E and B_{12}. Non-protein amino acids, cyanogens, protease inhibitors, saponins and lectins occur commonly in legume seeds (Wink and Mohamed, 2003). Saponins are widely distributed throughout all plant genera (Vincken *et al.*, 2007). Phenols occur commonly and in a wide variety (e.g. gossypol, isoflavones, catechins, proanthocyanins, alkylresorcinols and tannins; Hengtrakul *et al.*, 1991; Kozubek, 1999; Matthäus and Angelini, 2005; Naczk and Shahidi, 2006). Phytin is found throughout the cereal grains and oilseeds (Maga, 1982). Moulds and bacteria growing on the food also elaborate poisons. Examples are the aflatoxins synthesized by the mould *Aspergillus* spp., and toxins produced by *Salmonella* bacteria.

The effects of secondary substances on animals range from acute or chronic poisoning to changes in animal product composition, and sometimes to favourable effects on nutrient supply or metabolism. In many cases, adverse effects can be limited to acceptable levels by restricting the amount of the problematic ingredient in an animal's diet (Table 6.1). In some cases there are specific antagonists to the toxin or methods of treating the raw food (often by heating it) to produce an acceptably safe product. However, there are some examples where there is no satisfactory treatment and the food should not be fed to specific classes of animals.

Protease Inhibitors

There are two forms of protease inhibitors in legume seeds: the Kunitz and Bowman-Birk types. The Bowman-Birk inhibitors are small, dimeric molecules which appear to be progressively inactivated as the seed matures. Kunitz (1945) isolated a protein from defatted soybean flakes which inhibited the action of the protein-digesting enzyme, trypsin. The Kunitz trypsin inhibitor (KTI) is a globular, monomeric protein with 181 amino acid residues and is the protease inhibitor which has the greatest practical effect on animals. KTI is a chemical which is elaborated near a wound in plant tissue and potentiates a sequence of events which leads to the elaboration of antibiotics at the wound (Park *et al.*, 2001). The chemistry of soybean KTI and the nature of its complex with porcine trypsin have been described by Song and Suh (1998). Soybeans are heated (by steam, or roasting at 100°C) so as to destroy these substances. A discussion of the susceptibility of soybean KTI to heat is given by Osman *et al.* (2002). Soybean KTI reduces the ability of the animal's trypsin to digest dietary protein. The animal's response to KTI ingestion is to synthesize more enzymes, but this is not effective and just leads to increased endogenous N excretion and pancreatic hypertrophy (Liener and Kakade, 1980).

A Kunitz-type protease inhibitor in leucaena has been described by Sattar *et al.* (2004). The protein blocks enzymes involved in blood clotting and fibrinolysis, has anti-inflammatory effects and decreases bradykinin release.

Lectins

Lectins are proteins or glycoproteins (Liener, 1997). They are widely distributed in plants, especially

Table 6.1. Processing methods and recommended maximum inclusion levels in the diet of pigs for potentially toxic legume seeds. (From Liener, 1975; Evans, 1980; SCA, 1987; Love *et al.*, 1990; Morrison *et al.*, 2007.)

Food	Major toxin(s)	Example of detoxification method	Maximum inclusion level for pigs
Legume seeds and meals			
Chickpea	Protease inhibitors	Heat	30%
Cowpea	Protease inhibitors	Heat	15% for raw pea; nil in diets of young pigs
Faba bean	Protease inhibitors, tannins	Heat	10–20%
Field pea	Protease inhibitors	Heat	40%
Mung bean	Protease inhibitors	Heat	Raw beans may be used at 15%
Navy bean	Lectins, protease inhibitors	Heat	Nil for raw beans; 25% for heat-treated meal
Pigeonpea	Protease inhibitors, tannins	Heat	10–15%
Soybean	Antigens, antivitamins, goitrogens, isoflavones, lectins, protease inhibitors, tannins, phytic acid	Heat	No limit for heat-treated meal
Oilseed meals			
Cottonseed	Gossypol	Add Fe^{2+} iron to equal the gossypol content	3–5% for sows; 10–15% for growers
Linseed	Cyanogen	Heat	5%
Rapeseed	Glucosinolates, erucic acid	Hot water extraction; micronization	0–3% for breeding stock; 5% for young pigs; 10–15% for growers/finishers
Cereal grains and other energy sources			
Barley	Phytin, non-starch polysaccharides	Addition of phytase	
Cassava	Cyanogen	Water extraction and air-drying	
Rye	Tannins, phytin		20% for adult pigs, nil for young pigs
Triticale	Tannins, phytin		40–50% for mature pigs, nil for young pigs

legumes, where they may constitute up to 10% of the protein. Lectins may have a role in packing storage proteins into cell vacuoles (Einhoff *et al.*, 1986), but they are certainly elaborated by plants as defence proteins (Santimone *et al.*, 2004). They impede digestion by binding to the intestinal mucosal cells (Sauvion *et al.*, 2004) or by acting as α-amylase inhibitors (Moreno and Chrispeels, 1989; Rougé *et al.*, 1993). They can be absorbed from the intestine and damage cells within the body (Wang *et al.*, 1998; Sauvion *et al.*, 2004).

Allergens

Glycinin and β-conglycinin (the major storage proteins) are important allergens in soybean meals. KTIs have also been implicated as allergens in soy-

beans and other foodstuffs (Moroz and Yang, 1980; Seppälä *et al.*, 2001). Other enzyme inhibitors and storage proteins (e.g. glutenin and gliadin) are also allergens (Breiteneder and Radauer, 2004).

Allergens in soybean meal are especially dangerous when fed to pigs and dairy calves. Bailey *et al.* (2001) have proposed that when pigs are weaned early (at 3–5 weeks) they lose the protection of maternal IgA previously available to them via the milk before they have established proper immune function. If these pigs are then fed soybean meal they rapidly develop antibodies to soy protein. Allergic responses to food proteins include damage to the intestinal epithelium (vascular congestion, bleeding, oedema and deepening of the crypts of the intestinal epithelium, increased digesta transit rate, diarrhoea), respiratory distress (coughing, wheezing, cyanosis) and anaphylactic shock (Dreau and Lalles, 1999; Helm *et al.*, 2002).

Isoflavones

Isoflavones are polyphenolic compounds which can have an oestrogenic effect on animals. Some isoflavones, especially genistein and daidzein which are found in soybeans and formononetin and coumestans in clover, can have serious adverse effects on animal reproduction (Cox and Braden, 1974). Equol is an oestrogenic metabolite of formononetin metabolism by intestinal bacteria (Batterham *et al.*, 1971; Erkoç *et al.*, 2005). High levels of equol in plasma are responsible for the infertility in sheep grazing legume pastures (Shutt, 1976; and review by Adams, 1995) and infertility in cattle, guinea pigs, rabbits and mice, as well as endometrial hyperplasia in sheep and guinea pigs. The potential of a food to interfere with animal reproduction depends on the concentration of the isoflavones and also on the activity of intestinal microorganisms against these substances. Studies in monkeys (Rafii *et al.*, 2004) and rats (Bowey *et al.*, 2003) demonstrate that there are substantial between-individual differences in the abilities of their intestinal bacteria to metabolize isoflavones to equol.

Gossypol

Gossypol is the most important toxin in raw cottonseed and the commercial meal. It is a phenolic substance which the seed elaborates as a protection against insects. It has serious adverse effects on animals, especially non-ruminants, because it binds to Fe and appears to promote lipid peroxidation (Velasquez-Pereira *et al.*, 1998). The latter effect can be reversed by vitamin E supplementation, in some cases (Velasquez-Pereira *et al.*, 1999). Gossypol is metabolized in the liver, and the half-life of free gossypol in the pig is about 80 h, but gossypol can be detected for at least 20 days after administration (Abou-Donia and Dieckert, 1975).

Gossypol causes impaired growth, infertility or premature birth (Love *et al.*, 1990), and sometimes death (Velasquez-Pereira *et al.*, 1999). Other effects include hypertrophy of the heart and liver and pulmonary oedema (Burlatschenko, 2003). The lowered haematocrit found in intoxicated pigs may be the result of gossypol binding to Fe in the liver with the complex then being excreted in bile (Skutches *et al.*, 1974). The increased osmotic fragility of

erythrocytes reported by several workers may be related to its apparent effect on lipid oxidation.

Many of the effects are related to reproduction. For example, when pigs were fed a diet containing 0.07% free gossypol, delayed oestrus was observed in sows, decreased plasma testosterone and libido in boars and the mortality rate increased (Ling-yun *et al.*, 1984). Degenerative changes have been reported in the prostate gland of guinea pigs (Wong and Tam, 1989). Bulls may experience an increase in the number of abnormal sperm, and possible degeneration of the testicles (Chenoweth *et al.*, 2000; Hassan *et al.*, 2004). Many reproductive effects remain for several weeks after cottonseed is removed from the diet. However, 200 days of feeding a gossypol-free diet to bulls returned sperm production to control levels (Hassan *et al.*, 2004). It seems that cows are much less susceptible to gossypol-related infertility than males.

Whole (white or fuzzy) cottonseed (the raw by-product obtained from cotton ginning) is used as a fat- and protein-rich food for ruminants. This is possible, even though its gossypol content may be high, because of the relatively greater resistance of ruminants to gossypol poisoning, and also because cattle tend to self-limit their intake of cottonseed (Prewett and Dryden, 1996).

The concentration of active gossypol in the extracted meal is reduced, but not eliminated, by heating during oil extraction (Batterham, 1989). Gossypol toxicity can be reduced or avoided by adding Fe^{2+} (e.g. as the sulfate) at a rate similar to the expected concentration of free gossypol (about 0.06% in cottonseed meal), by using meal made from glandless cottonseed, which has a low gossypol content, or by restricting the cottonseed meal content of the animal's diet.

Phytin

Phytic acid (phytin, phytate, *myo*-inositol(1,2,3, 4,5,6) *hexakis*-dihydrogen phosphate) is a phosphorus-rich substance which is widely distributed in plant organs and particularly in seeds (Raboy, 2001) where it is about 75% of total phosphorus. Phytate-P is essentially unavailable to pigs (Dilger and Adeola, 2006) and this leads to very low availability of the total P in seeds. For example, availabilities for pigs are 13–50% in cereal grains, with wheat and triticale having considerably higher values than other grains, 1–21% in oilseed meals, and 23–31% in legume seeds and meals (NRC, 1998).

Phytate binds other minerals, including Ca, Fe, Cu, Mn and Zn, and prevents the animal from absorbing these and other nutrients (House, 1999). Adding phytase to grain-based pig diets will improve the availability of P, and also Zn, Fe, Ca, Mg and Cu (Simons *et al.*, 1990; Revy *et al.*, 2004; Shelton *et al.*, 2005). On the other hand, there is no convincing evidence that addition of phytase improves the availability of P in equine diets (e.g. Hainze *et al.*, 2004; Van Doorn *et al.*, 2004).

We can use two approaches to improve the performance (e.g. growth, bone strength, phosphorus retention) of monogastric animals fed diets rich in phytic acid. The first is to add a microbial phytase enzyme. Addition of phytase to pig diets raises the availability of P by 20–30% units (Table 6.2) and also releases other minerals from phytate complexes. The second approach is to breed low-phytate grains, as has been done with barley and maize (Spencer *et al.*, 2000; Veum *et al.*, 2001, 2002). The availability of P in low-phytate maize is about five times that in normal maize, and that in barley about twice the level in normal barley (Spencer *et al.*, 2000; Veum *et al.*, 2001, 2002).

The addition of phytase or the use of low-phytate grains also improves the utilization of energy and amino acids. The reasons for this are not clear. Selle *et al.* (2000) have suggested that '[i]t may be that the release of protein from protein-phytate complexes occurring naturally in feed ingredients, the prevention of formation of binary and ternary protein-phytate complexes within the gut, the alleviation of the negative impact of phytate on digestive enzymes and the reduction in endogenous amino acid losses are all contributing factors'.

Erucic Acid

Erucic acid is a long-chain mono-unsaturated fatty acid (22:1), which interferes with the metabolism of essential unsaturated fatty acids and increases the likelihood of tissue fat oxidation. Erucic acid occurs in rapeseed and the seeds of other brassicas. Diets rich in erucic acid may cause reduced growth and heart disease (Nera *et al.*, 1971).

Glucosinolates

Glucosinolates are present in rapeseed meals. The major one is progoitrin, which, when hydrolysed by the rapeseed enzyme myrosinase, yields isothiocyanates and oxazolidinethiones, which inhibit the uptake of iodine by the thyroid gland into thyroxine (Liener, 1975). These goitrogens cause poor growth, reduced food intake and nitrogen retention, enlarged thyroid gland and fatty infiltration in the liver, impaired fertility and taints in meat (Oliver *et al.*, 1971; Evans, 1980).

Cyclopropenoid Fatty Acids

Malvalic and sterculic acids, which occur in cottonseed meal (Evans, 1980), have a range of toxic effects. They impair reproduction (Tumbelaka *et al.*, 1994), liver function and growth (Andrianaivo-Rafehivola *et al.*, 1995). Sterculic acid inhibits the Δ9 desaturase enzyme which is responsible for synthesis of oleic acid (Jeffcoat and Pollard, 1977) and inhibits the ability of the ovine corpus luteum to synthesize steroid hormones (Slayden and Stormshak, 1990), along with other adverse effects on lipid metabolism.

Table 6.2. Effects of low-phytate grains and addition of phytase on the availability of phosphorus in cereal grains to pigs (values are for the whole diet).

Food	Total P (%)	P availability (%)	Reference
Barley	0.57	52.6[a]	Veum *et al.*, 2002
Low-phytate barley	0.50	60.0[a]	Veum *et al.*, 2002
Maize	0.54	55.5[a]	Veum *et al.*, 2001
Low-phytate maize	0.45	66.6[a]	Veum *et al.*, 2001
Maize	0.30	12.5	Spencer *et al.*, 2000
Low-phytate maize	0.35	45.7	Spencer *et al.*, 2000
Maize + soybean meal, no phytase	–	53.8	Revy *et al.*, 2004
Maize + soybean meal, + phytase	–	73.3	Revy *et al.*, 2004

[a]Assumes that phytate-P is unavailable and non-phytate-P is completely available.

Cyanogens

Linseed contains a cyanogenetic glucoside, linamarin, which releases hydrogen cyanide on hydrolysis by the enzyme linamarase (Liener, 1975). Heat inactivates the enzyme. The cyanogens linamarin and lotaustralin are found in cassava (Evans, 1980), a tuber which is harvested from the cassava or manioc plant and which is widely used in Asia, Africa and South America as a foodstuff for humans and animals. Cassava can be poisonous unless the glucosides are removed by soaking in water, followed by drying.

All forage sorghums and related species, such as Johnson grass and Sudan grass, contain the cyanogenic glucoside, dhurrin (Wheeler, 1994). The cyanogenic potential of the forage is usually from 100 to 1000 mg hydrogen cyanide equivalent/kg DM, although potentials of 2300 and 2450 mg/kg DM have been reported in Zulu and Silk cultivars (Wheeler *et al.*, 1990). The risk of cyanide poisoning is greatest when the plant is young (this includes young regrowth), and may be increased by N fertilization. It is positively correlated with the leaf/stem ratio and leaf width (Mulcahy *et al.*, 1992). Interestingly, these authors found that sheep and cattle selected against forages with wide leaves. The plant's cyanide potential may be increased by moisture stress, phosphate fertilizer, changes in light intensity and changes in temperature, although the evidence for this is equivocal (Wheeler *et al.*, 1990). The cyanide potential of conserved forage is reduced with longer ensiling periods, or by sundrying (Fleischer and Tackie, 1996).

In its acute form, cyanide poisoning results in death. Subacute poisoning may cause fetal deformities (neural tube defects, hydropericardium and crooked tail), enlargement of the thyroid gland (thiocyanate, a detoxification product of cyanide, is goitrogenic), reduced weight gain and anaemia (Seaman *et al.*, 1981; Frakes *et al.*, 1985; Soto-Blanco *et al.*, 2001). In horses, sheep and cattle grazing the leaves and stems of forage and grain sorghums, incoordination of the hind legs, urinary incontinence and haematuria, serous nasal discharge, erratic and small increase in body temperature, depression and reduced appetite have been reported (Varshney *et al.*, 1996). Cyanide is detoxified in the rumen to thiocyanate. This uses S, and so reduces the amount of S available for microbial protein synthesis, and also animal performance and appetite (Wheeler, 1994). Providing S as a sulfurized salt block can reduce the incidence of cyanide poisoning.

Non-starch Polysaccharides

Non-starch polysaccharides (NSPs) are polysaccharides typical of those found in plant cell walls (e.g. cellulose, other β-glucans, pectins, xylans and arabinoxylans), which are present in cereal grains (Black, 2000). NSPs are either water-soluble or insoluble; the soluble fraction is largely mixed β-glucans originating in the endosperm (Theander and Aman, 1979). Barley grain has about 15% of β-glucans, while other cereal grains (especially wheat and rye) have between 5% and 10% of NSPs (Pluske *et al.*, 2001). NSPs can adversely influence the performance and health of non-ruminant animals. They bind to digestive enzymes and form gels which increase the viscosity of the intestinal contents (Edwards *et al.*, 1988) by combining with mucus. These effects interfere with the digestion of starch and other grain endosperm constituents. Because they are not readily digested in the small intestine, the soluble NSPs provide a growth medium for bacterial pathogens in the large intestine and there is indirect evidence that this can promote diseases like swine dysentery (Pluske *et al.*, 2001).

Tannins

Tannins are a group of polyphenolic compounds found widely in plants. They are found in pasture legumes (e.g. lucerne, lespedeza, desmodium), fodder trees (e.g. *Acacia aneura* and *Gliricidia* spp.), cereal grains and oilseeds (Table 6.1).

The biochemistry and effects of tannins have been reviewed by Barry and McNabb (1999; and also see Brooker and Acamovic, 2006). The hydrolysable tannins have a glucose core, and when they are hydrolysed, they yield amounts of gallic or similar acids, and a large molecular weight hepatotoxic substance. Thus, hydrolysable tannins can be poisonous if they are eaten in large amounts and the products of digestion are absorbed (Lowry *et al.*, 1996). The major class of plant tannins is the condensed tannins (also called proanthocyanidins). These are polymeric flavonoids in which flavandiol units are linked mainly through the 4, 6 and 8 C positions to form complexes of about 10 units, with a molecular weight of 2000–4000 (Barry and McNabb, 1999).

Although condensed tannins are not toxic per se, as they are not absorbed unless the intestinal wall is damaged, they generally cause more problems in animal nutrition than the hydrolysable tannins as they are able to (reversibly) bind to cell protein and thus make it unavailable, especially in the rumen. This may be a disadvantage if a large proportion of the protein is bound, as in some fodder trees where animal performance is reduced due to the low availability of the protein. For example, mulga leaf protein is largely unavailable to the animal unless the diet is supplemented with polyethylene glycol, which binds, instead of the plant protein, to tannins (Pritchard et al., 1988). Tannins may also reduce forage palatability and food intake if they are present at very high levels, but the relationship between tannin content and food intake is not clear.

One of the more interesting nutritional characteristics of condensed tannins is that at certain levels (probably up to 9% of the dry matter (DM); Barry and McNabb, 1999) they can have beneficial effects. For example, tannin binding reduces the likelihood of rumen bloat, can increase the amount of dietary protein flowing to the duodenum (Lowry et al., 1996; Barry and McNabb, 1999) and may reduce methane emissions (Puchala et al., 2005).

Tannins have anthelmintic activity in vitro, e.g. they impair the development of Trichostrongylus colubriformis eggs and the motility of various nematode larvae (Molan et al., 1999, 2000). The improved performance of animals grazing on some tannin-containing pastures (Ramirez-Restrepo et al., 2004) may be a combination of the effects of a better supply of nutrients, especially essential amino acids, to the small intestine, and the possibly protective effect of tannins against intestinal helminths.

Mimosine

Mimosine is an amino acid which comprises 4–5.5% of the dry weight of leucaena (Leucaena leucocephala) (Bray et al., 1988). Rumen bacteria metabolize mimosine to the substance DHP (3-OH, 4-(1H)-pyridone). This reduces the direct effects of mimosine but results in secondary toxicity from the DHP, preventing the synthesis of thyroxine. This causes the thyroid gland to enlarge and the animal shows the typical signs of goitre. General symptoms of mimosine and/or DHP poisoning in ruminants include hair loss, listlessness, profuse salivation, ulceration of the mouth and poor

growth (Jones, 1979; Ram et al., 1994). Pigs suffer from inappetence and muscular uncoordination (Zakayo and Krebs, 1997). Animals may die unless the intake of leucaena is restricted. These effects were prevented by introducing into the rumens of unadapted animals in northern Australia a bacterium isolated from goats grazing leucaena in Indonesia (Jones and Lowry, 1984). This organism destroys the DHP produced by other rumen bacteria and inoculated ruminants show no symptoms of either mimosine or DHP toxicity and are able to produce at substantially higher levels than unsupplemented animals.

Nitrate/nitrite

Under certain conditions grasses may take up large amounts of nitrogen. This is converted into nitrate (NO_3^-) prior to the synthesis of amino acids. If the plant cannot convert the nitrate-N into amino acid N because of excessive N fertilization, water stress, lack of sunlight, frost, etc. (Nicholls and Miles, 1980; Vermunt and Visser, 1987), large amounts of nitrate can be accumulated. Nitrate poisoning has been reported from many pasture grasses and forage crops including Lolium spp., Festuca arundinacae, Dactylis glomerata, Sorghum spp., Echinochloa polystachya, Pennisetum purpureum and Dactyloctenium radulans (Murphy and Smith, 1967; Nicholls and Miles, 1980; Carrigan and Gardner, 1982; Vermunt and Visser, 1987; McKenzie et al., 2004; da Silva et al., 2006).

Subacute nitrate poisoning has been associated with abortion and retained placenta, sudden death in cattle, impairment of thyroid function, interference with vitamin A metabolism, anaemia, severe gastritis in monogastric animals, and reduced growth and production (Bruningfann and Kaneene, 1993; ElBahri et al., 1997).

Nitrites in water and food, or produced by the reduction of nitrates in the rumen, may be absorbed into the blood and combine with haemoglobin, forming methaemoglobin. This cannot carry oxygen, and so nitrite poisoning can result in death. Nitrites will produce methaemoglobinaemia in cattle, sheep, swine, dogs, guinea pigs and rats (Bruningfann and Kaneene, 1993).

Oxalate

Some plants have high levels of oxalic acid, which binds much of their calcium. This may induce a calcium deficiency in animals eating these plants, as oxalate is not affected by intestinal digestion and oxalate-bound calcium cannot be absorbed from the small intestine. Horses are susceptible to oxalate poisoning and suffer from the disorder of osteodystrophia fibrosa (ODF). The condition is manifested by lameness, weight loss and swelling of the mandibles and maxillae. The incidence and nature of this condition has been reviewed by McKenzie (1985). Oxalate poisoning has also been reported in goats and sheep (McKenzie *et al.*, 1988; Peet *et al.*, 1990). Affected goats had excessive salivation, abdominal distension, apparent constipation, continual bleating and ataxia; sheep showed stiff-legged gait, staggering, then collapsed and died.

Pasture grasses which contain high concentrations of oxalic acid, and which cause ODF, include buffel grass (*Cenchrus cillaris*; 3% oxalic acid) and setaria (especially the Kazungula cultivar – Nandi contains least, and Narok is intermediate), kikuyu (*Pennisetum clandestinum*), green panic (*Panicum maximum*), para grass (*Brachiaria mutica*), pangola grass (*Digitaria eriantha* subsp. *pentzii*) and purple pigeon grass (*Setaria incrassata*) (Jones and Ford, 1972; McKenzie, 1988). Silcock and Smith (1983) recorded 3.5% oxalate in the Australian native grass *Paspalidium constrictum* and 10.3% in *Portulaca* spp. Oxalic acid contents are highest in young forage (e.g. lush kikuyu and setaria regrowth) and after a period of wet weather. The level of oxalic acid in grasses increases during the night and decreases during the day. Consequently, if suspect pastures must be grazed, it is best to do this in the late afternoon. Animals, especially horses, which graze these pastures, should be given a calcium supplement.

References

Abou-Donia, M. and Dieckert, J.W. (1975) Metabolic fate of gossypol: the metabolism of [14C] gossypol in swine. *Toxicology and Applied Pharmacology* 31, 32–46.

Adams, N.R. (1995) Detection of the effects of phytoestrogens on sheep and cattle. *Journal of Animal Science* 73, 1509–1515.

Andrianaivo-Rafehivola, A.A., Siess, M.H. and Gaydou, E.M. (1995) Modifications of hepatic drug metabolizing enzyme activities in rats fed baobab seed oil containing cyclopropenoid fatty acids. *Food and Chemical Toxicology* 33, 377–382.

Bailey, M., Plunkett, F.J., Rothkotter, H.J., Vega Lopez, M.A., Haverson, K. and Stokes, C.R. (2001) Regula-

tion of mucosal immune responses in effector sites. *Proceedings of the Nutrition Society* 60, 427–435.

Barry, T.N. and McNabb, W.C. (1999) Review article. The implications of condensed tannins on the nutritive value of temperate forages fed to ruminants. *British Journal of Nutrition* 81, 263–272.

Batterham, E.S. (1989) Advances in the use of cottonseed meal in diets for growing pigs. *Recent Advances in Animal Nutrition in Australia* 10, 164–171.

Batterham, T.J., Shutt, D.A., Hart, N.K., Braden, A.W.H. and Tweeddale, H.J. (1971) Metabolism of intraruminally administered [4–14C]formononetin and [4–14C]biochanin A in sheep. *Australian Journal of Agricultural Research* 22, 131–138.

Black, J.L. (2000) Bioavailability: the energy component of a ration for monogastric animals. In: Moughan, P. J., Verstegen, M.W.A. and Visser-Reyneveld, M.I. (eds) *Feed Evaluation: Principles and Practice.* Wageningen Pers, Wageningen, The Netherlands, pp. 133–152.

Bowey, E., Adlercreutz, H. and Rowland, I. (2003) Metabolism of isoflavones and lignans by the gut microflora: a study in germ-free and human flora associated rats. *Food and Chemical Toxicology* 41, 631–636.

Bray, R.A., Cooksley, D.G., Hall, T.J. and Ratcliff, D. (1988) Performance of fourteen Leucaena lines at five sites in Queensland. *Australian Journal of Experimental Agriculture* 28, 69–76.

Breiteneder, H. and Radauer, C. (2004) A classification of plant food allergens. *Journal of Allergy and Clinical Immunology* 113, 821–830.

Brooker, J.D. and Acamovic, T. (2006) Phytochemicals in livestock production systems. *Animal Feed Science and Technology* 121, 1–4.

Bruningfann, C.S. and Kaneene, J.B. (1993) The effects of nitrate, nitrite, and n-nitroso compounds on animal health. *Veterinary and Human Toxicology* 35, 237–253.

Burlatschenko, S. (2003) Suspected gossypol toxicosis in a sow herd. *Journal of Swine Health and Production* 11, 137–139.

Carrigan, M.J. and Gardner, I.A. (1982) Nitrate poisoning in cattle fed Sudax (sorghum sp hybrid) hay. *Australian Veterinary Journal* 59, 155–157.

Chenoweth, P.J., Chase, C.C., Jr, Risco, C.A. and Larsen, R.E. (2000) Characterization of gossypol-induced sperm abnormalities in bulls. *Theriogenology* 53, 1193–1203.

Cox, R.I. and Braden, A.W. (1974) The metabolism and physiological effects of phyto-oestrogens in livestock. *Proceedings of the Australian Society of Animal Production* 10, 122–129.

da Silva, D.M., Riet-Correa, F., Medeiros, R.M.T. and de Oliveira, O.F. (2006) Toxic plants for livestock in the western and eastern Serido, state of Rio Grande do Norte, in the Brazilian semiarid. *Pesquisa Veterinaria Brasileira* 26, 223–236.

Dilger, R.N. and Adeola, O. (2006) Estimation of true phosphorus digestibility and endogenous phosphorus loss in growing pigs fed conventional and low-phytate soybean meals. *Journal of Animal Science* 84, 627–634.

Dreau, D. and Lalles, J.-P. (1999) Contribution to the study of gut hypersensitivity reactions to soybean proteins in preruminant calves and early-weaned piglets. *Livestock Production Science* 60, 209–218.

Edwards, C.A., Johnson, I.I. and Read, N.W. (1988) Do viscous polysaccharides slow absorption by inhibiting diffusion or convection? *European Journal of Clinical Nutrition* 42, 307–312.

Einhoff, W., Fleischmann, G., Freier, T., Kummer, H. and Rüdiger, H. (1986) Interactions between lectins and other components of leguminous protein bodies. *Biological Chemistry Hoppe-Seyler* 367, 15–25.

ElBahri, L., Belguith, J. and Blouin, A. (1997) Toxicology of nitrates and nitrites in livestock. *Compendium on Continuing Education for the Practicing Veterinarian* 19, 643–649.

Erkoç, F., Yilmazer, M. and Erkoç, S. (2005) Theoretical investigations of the equol molecule: semi-empirical and density functional theory calculations. *Journal of Molecular Structure: THEOCHEM* 713, 37–42.

Evans, M. (1985) *Nutrient Composition of Feeds for Pigs and Poultry*. Queensland Department of Primary Industries, Brisbane, Australia.

Fleischer, J.E. and Tackie, A.M. (1996) The possibility of making silage from 'wild sorghum' (*Sorghum arundinaceum*). *Bulletin of Animal Health and Production in Africa* 44, 45–50.

Frakes, R.A., Sharma, R.P. and Willhite, C.C. (1985) Developmental toxicity of the cyanogenic glycoside linamarin in the golden hamster. *Teratology* 31, 241–246.

Hainze, M.T.M., Muntifering, R.B., Wood, C.W., McCall, C.A. and Wood, B.H. (2004) Faecal phosphorus excretion from horses fed typical diets with and without added phytase. *Animal Feed Science and Technology* 117, 265–279.

Hassan, M.E., Smith, G.W., Ott, R.S., Faulkner, D.B., Firkins, L.D., Ehrhart, E.J., and Schaeffer, D.J. (2004) Reversibility of the reproductive toxicity of gossypol in peripubertal bulls. *Theriogenology* 61, 1171–1179.

Helm, R.M., Furuta, G.T., Stanley, J.S., Ye, J., Cockrell, G., Connaughton, C., Simpson, P., Bannon, G.A. and Burks, A.W. (2002) A neonatal swine model for peanut allergy. *Journal of Allergy and Clinical Immunology* 109, 136–142.

Hengtrakul, P., Lorenz, K. and Mathias, M. (1991) Alkylresorcinol homologs in cereal grains. *Journal of Food Composition and Analysis* 4, 52–57.

House, W.A. (1999) Trace element bioavailability as exemplified by iron and zinc. *Field Crops Research* 60, 115–141.

Jeffcoat, R. and Pollard, M.R. (1977) Studies on the inhibition of the desaturases by cyclopropenoid fatty acids. *Lipids* 12, 480–485.

Jones, R.J. (1979) Value of *Leucaena leucocephala* as a feed for ruminants in the tropics. *World Animal Review* 31, 13–23.

Jones, R.J. and Ford, C.W. (1972) The soluble oxalate content of some tropical pasture grasses grown in southeast Queensland. *Tropical Grasslands* 6, 201–204.

Jones, R.J. and Lowry, J.B. (1984) Australian goats detoxify the goitrogen 3-hydroxy-4(1H) pyridone (DHP) after rumen infusion from an Indonesian goat. *Experientia* 40, 1435–1436.

Kozubek, A. (1999) Resorcinolic lipids, the natural non-isoprenoid phenolic amphiphiles and their biological activity. *Chemical Reviews* 99, 1–25.

Kunitz, M. (1945) Crystallization of a trypsin inhibitor from soybean. *Science* 101, 668–669.

Liener, I.E. (1975) Endogenous toxic factors in oilseed residues. In: *Animal Feeds '74*. Tropical Products Institute, London, pp. 179–188.

Liener, I.E. (1997) Plant lectins: properties, nutritional significance, and function. *Antinutrients and Phytochemicals in Food. ACS Symposium Series* 662, 31–43.

Liener, I.E. and Kakade, M.L. (1980) Protease inhibitors. In: *Toxic Constituents of Plant Food Stuffs*, 2nd edn. Academic Press, New York, pp. 7–71.

Ling-yun, Q., Shu-haun, Z., Gia-xiang, W., He-Ping, Z., Xi-yun, L., Qiang, G., Zhi-kui, S., Te-fu, N., Ke-xian, Z., Sheng-rong, L. and Xin-min, H. (1984) Effect of cottonseed meal on reproduction performance in boars and sows. *Acta Veterinaria et Zootechnica Sinica* 15, 157–162.

Love, R.J., Peacock, A.J. and Evans, G. (1990) Premature farrowings caused by feeding cottonseed meal. *Australian Veterinary Journal* 67, 223–226.

Lowry, J.B., McSweeney, C.S. and Palmer, B. (1996) Changing perceptions of the effect of plant phenolics on nutrient supply in the ruminant. *Australian Journal of Agricultural Research* 47, 829–842.

McKenzie, R.A. (1985) Plant toxicology. *Proceedings of the Australia–U.S.A. Poisonous Plants Symposium*. Brisbane, Australia, May 14–18, 1984, pp. 150–154.

McKenzie, R.A. (1988) Purple pigeon grass (*Setaria incrassata*): a potential cause of nutritional secondary hyperparathyroidism of grazing horses. *Australian Veterinary Journal* 65, 329–330.

McKenzie, R.A., Bell, A.M., Storie, G.J., Keenan, F.J., Cornack, K.M. and Grant, S.G. (1988) Acute oxalate poisoning of sheep by buffel grass (*Cenchrus ciliaris*). *Australian Veterinary Journal* 65, 26.

McKenzie, R.A., Rayner, A.C., Thompson, G.K. and Burren, B. (2004) Nitrate–nitrite toxicity in cattle and sheep grazing *Dactyloctenium radulans* (button grass) in stockyards. *Australian Veterinary Journal* 82, 630–634.

Maga, J.A. (1982) Phytate: its chemistry, occurrence, food interactions, nutritional significance, and methods of analysis. *Journal of Agricultural and Food Chemistry* 30, 1–9.

Matthäus, B. and Angelini, L.G. (2005) Anti-nutritive constituents in oilseed crops from Italy. *Industrial Crops and Products* 21, 89–99.

Molan, A.L., Hoskin, S.O., Barry, T.N. and McNabb, W.C. (2000) Effect of condensed tannins extracted from four forages on the viability of the larvae of deer lungworms and gastrointestinal nematodes. *Veterinary Record* 147(2), 44–48.

Molan, A.L., Waghorn, G.C. and McNabb, W.C. (1999) Condensed tannins and gastro-intestinal parasites in sheep. *Proceedings of the New Zealand Grassland Association* 61, 57–61.

Moreno, J. and Chrispeels, M.J. (1989) A lectin gene encodes the alpha-amylase inhibitor of the common bean. *Proceedings of the National Academy of Sciences of the USA* 86, 7885–7889.

Moroz, L.A. and Yang, W.H. (1980) Kunitz soybean trypsin inhibitor: a specific allergen in food anaphylaxis. *New England Journal of Medicine* 302, 1126–1128.

Morrison, S.C., Savage, G.P., Morton, J.D. and Russell, A.C. (2007) Identification and stability of trypsin inhibitor isoforms in pea (*Pisum sativum* L.) cultivars grown in New Zealand. *Food Chemistry* 100, 1–7.

Mulcahy, C., Hedges, D.A., Rapp, G.G. and Wheeler, J.L. (1992) Correlations among potential selection criteria for improving the feeding value of forage sorghums. *Tropical Grasslands* 26, 7–11.

Murphy, L.S. and Smith, G.E. (1967) Nitrate accumulations in forage crops. *Agronomy Journal* 59, 171–174.

Muzquiz, M., Hill, G.D., Cuadrado, C., Pedrosa, M.M. and Burbano, C. (eds) (2004) Recent advances of research in antinutritional factors in legume seeds and oilseeds. *Proceedings of the Fourth International Workshop on Antinutritional Factors in Legume Seeds and Oilseeds, Toledo, Spain, 8–10 March 2004*. Wageningen Academic Publishers, Wageningen, The Netherlands.

Naczk, M. and Shahidi, F. (2006) Phenolics in cereals, fruits and vegetables: occurrence, extraction and analysis. *Journal of Pharmaceutical and Biomedical Analysis* 41, 1523–1542.

National Research Council (NRC) (1998) *Nutrient Requirements of Swine*, 10th rev. edn. National Academies Press, Washington, DC.

Nera, E.A., Beare-Rogers, J.L. and Heggtveit, H.A. (1971) Cardiotoxicity of rapeseed oil. *American Journal of Pathology* 62, 34a.

Nicholls, T.J. and Miles, E.J. (1980) Nitrate–nitrite poisoning of cattle on ryegrass pasture. *Australian Veterinary Journal* 56, 95–96.

Oliver, S.L., McDonald, B.E. and Opuszyfiska, T. (1971) Weight gain, protein utilization, and liver histochemistry of rats fed low- and high-thioglucoside-content rapeseed meals. *Canadian Journal of Physiology and Pharmacology* 49, 448.

Osman, M.A., Reid, P.M. and Weber, C.W. (2002) Thermal inactivation of tepary bean (*Phaseolus acutifolius*), soybean and lima bean protease inhibitors: effect of acidic and basic pH. *Food Chemistry* 78, 419–423.

Park, D.-S., Graham, M.Y. and Graham, T.L. (2001) Identification of soybean elicitation competency factor, CF-1, as the soybean Kunitz trypsin inhibitor. *Physiological and Molecular Plant Pathology* 59, 265–273.

Peet, R.L., Dickson, J. and Hare, M. (1990) Kikuyu poisoning in goats and sheep. *Australian Veterinary Journal* 67, 229–230.

Pluske, J.R., McDonald, D.E., Pethick, D.W., Mullan, B.P. and Hempson, D.J. (2001) Nutritional management of the gastrointestinal tract to reduce enteric diseases in pigs. *Recent Advances in Animal Nutrition in Australia* 13, 127–134.

Prewett, S.A. and Dryden, G.McL. (1996) Whole cottonseed supplements for cattle grazing black speargrass. *Animal Production in Australia* 21, 464.

Pritchard, D.A., Stocks, D.C., O'Sullivan, B.M., Martin, P.R., Hurwood, I.S. and O'Rourke, P.K. (1988) The effect of polyethylene glycol (PEG) on wool growth and liveweight of sheep consuming a mulga (*Acacia aneura*) diet. *Proceedings of the Australian Society of Animal Production* 17, 290–293.

Puchala, R., Min, B.R., Goetsch, A.L. and Sahlu, T. (2005) The effect of a condensed tannin containing forage on methane emission by goats. *Journal of Animal Science* 83, 182–186.

Raboy, V. (2001) Seeds for a better future: 'low phytate' grains help to overcome malnutrition and reduce pollution. *Trends in Plant Science* 6, 458–462.

Rafii, F., Hotchkiss, C., Heinze, T.M. and Park, M. (2004) Metabolism of daidzein by intestinal bacteria from rhesus monkeys (*Macaca mulatta*). *Comparative Medicine* 54, 165–169.

Ram, J.J., Atreja, P.P., Chopra, R.C. and Chhabra, A. (1994) Mimosine degradation in calves fed a sole diet of *Leucaena leucocephala* in India. *Tropical Animal Health and Production* 26, 199–206.

Ramirez-Restrepo, C.A., Barry, T.N., Lopez Villalobos, N., Kemp, P.D. and McNabb, W.C. (2004) Use of *Lotus corniculatus* containing condensed tannins to increase lamb and wool production under commercial dryland farming conditions without the use of anthelmintics. *Animal Feed Science and Technology* 117, 85–105.

Revy, P.S., Jondreville, C., Dourmad, J.Y. and Nys, Y. (2004) Effect of zinc supplemented as either an organic or an inorganic source and of microbial phytase on zinc and other minerals utilisation by weanling pigs. *Animal Feed Science and Technology* 116, 93–112.

Rougé, P., Barre, A., Causse, H., Chatelain, C. and Porth, G. (1993) Arcelin and α-amylase inhibitor from the seeds of common bean (*Phaseolus vulgaris* L.) are truncated lectins. *Biochemical Systematics and Ecology* 21(6/7), 695–703.

Santimone, M., Koukiekolo, R., Moreau, Y., Le Berre, V., Rougé, P., Marchis-Mouren, G. and Desseaux, V. (2004) Porcine pancreatic α-amylase inhibition by the kidney bean (*Phaseolus vulgaris*) inhibitor (α-AI1) and structural changes in the α-amylase inhibitor complex. *Biochimica et Biophysica Acta – Proteins and Proteomics* 1696, 181–190.

Sattar, R., Ali, S.A., Kamal, M., Khan, A.A. and Abbasi, A. (2004) Molecular mechanism of enzyme inhibition: prediction of the three-dimensional structure of the dimeric trypsin inhibitor from *Leucaena leucocephala* by homology modeling. *Biochemical and Biophysical Research Communications* 314, 755–765.

Sauvion, N., Nardon, C., Febvay, G., Gatehouse, A.M.R. and Rahbé, Y. (2004) Binding of the insecticidal lectin Concanavalin A in pea aphid, *Acyrthosiphon pisum* (Harris) and induced effects on the structure of midgut epithelial cells. *Journal of Insect Physiology* 50, 1137–1150.

Seaman, J.T., Smeal, M.G. and Wright, J.C. (1981) The possible association of a sorghum (*Sorghum sudanese*) hybrid as a cause of developmental defects in calves. *Australian Veterinary Journal* 57, 351–352.

Selle, P.H., Ravindran, V., Caldwell, R.A. and Bryden, W.L. (2000) Phytate and phytase: consequences for protein utilization. *Nutrition Research Reviews* 3, 255–278.

Seppälä, U., Majamaa, H., Turjanmaa, K., Helin, J., Reunala, T., Kalkkinen, N. and Palosuo, T. (2001) Identification of four novel potato (*Solanum tuberosum*) allergens belonging to the family of soybean trypsin inhibitors. *Allergy* 56, 619–626.

Shelton, J.L., Le Mieux, F.M., Southern, L.L. and Bidner, T.D. (2005) Effect of microbial phytase addition with or without the trace mineral premix in nursery, growing, and finishing pig diets. *Journal of Animal Science* 83, 376–385.

Shutt, D.A. (1976) The effects of plant oestrogens on animal reproduction. *Endeavour* 35, 110–113.

Silcock, R.G. and Smith, F.T. (1983) Soluble oxalates in summer pastures on a mulga soil. *Tropical Grasslands* 17, 179–181.

Simons, P.C., Versteegh, H.A., Jongbloed, A.W., Kemme, P.A., Slump, P., Bos, K.D., Wolters, M.G., Beudeker, R.F. and Verschoor, G.J. (1990) Improvement of phosphorus availability by microbial phytase in broilers and pigs. *British Journal of Nutrition* 64, 525–540.

Skutches, C.L., Herman, D.L. and Smith, F.H. (1974) Effect of dietary free gossypol on blood components and tissue iron in swine and rats. *Journal of Nutrition* 104, 415–422.

Slayden, O. and Stormshak, F. (1990) *In vivo* and *in vitro* effects of a cyclopropenoid fatty acid on ovine corpus luteum function. *Endocrinology* 127, 3166–3171.

Song, H.K. and Suh, S.W. (1998) Kunitz-type soybean trypsin inhibitor revisited: refined structure of its complex with porcine trypsin reveals an insight into the interaction between a homologous inhibitor from *Erythrina caffra* and tissue-type plasminogen activator. *Journal of Molecular Biology* 275, 347–363.

Soto-Blanco, B., Gorniak, S.L. and Kmura, E.T. (2001) Physiopathalogical effects of the administration of chronic cyanide to growing goats – a model for ingestion of cyanogenic plants. *Veterinary Research Communications* 25, 279–389.

Spencer, J.D., Allee, G.L. and Sauber, T.E. (2000) Phosphorus bioavailability and digestibility of normal and genetically modified low-phytate corn for pigs. *Journal of Animal Science* 78, 675–681.

Standing Committee on Agriculture (SCA) (1987) *Feeding Standards for Australian Livestock. No. 2 Pigs.* CSIRO, Melbourne, Australia.

Theander, O. and Aman, P. (1979) The chemistry, morphology and analysis of dietary fiber components. In: Inglett, G.E. and Falkehag, S.I. (eds) *Dietary Fibers: Chemistry and Nutrition.* Academic Press, New York, pp. 215–244.

Tumbelaka, L.I., Slayden, O. and Stormshak, F. (1994) Action of a cyclopropenoid fatty acid on the corpus luteum of pregnant and nonpregnant ewes. *Biology of Reproduction* 50, 253–257.

Van der Poel, A.F.B., Huisman, J. and Saini, H.S. (eds) (1993) *Recent Advances of Research in Antinutritional Factors in Legume Seeds: Analytical Methods, Animal Nutrition, Feed (Bio)technology, Plant Breeding.* Proceedings of the Second International Workshop on Antinutritional Factors (ANFs) in Legume Seeds, Wageningen, The Netherlands, 1–3 December 1993. Wageningen Pers, Wageningen, The Netherlands.

Van Doorn, D.A., Everts, H., Wouterse, H. and Beynen, A.C. (2004) The apparent digestibility of phytate phosphorus and the influence of supplemental phytase in horses. *Journal of Animal Science* 82, 1756–1763.

Varshney, J.P., Gupta, A.K. and Yadav, M.P. (1996) Occurrence of ataxia-cystitis syndrome in horses fed on sorghum vulgare in India. *Indian Veterinary Journal* 73, 985–986.

Velasquez-Pereira, J., Prichard, D., McDowell, L.R., Chenoweth, P.J., Risco, C.A., Staples, C.R., Martin, F.G., Calhoun, M.C., Rojas, L.X., Williams, S.N. and Wilkinson, N.S. (1998) Long-term effects of gossypol and vitamin E in the diets of dairy bulls. *Journal of Dairy Science* 81, 2475–2484.

Velasquez-Pereira, J., Risco, C.A., McDowell, L.R., Staples, C.R., Prichard, D., Chenoweth, P.J., Martin, F.G., Williams, S.N., Rojas, L.X., Calhoun, M.C. and Wilkinson, N.S. (1999) Long-term effects of feeding gossypol and vitamin E to dairy calves. *Journal of Dairy Science* 82, 1240–1251.

Vermunt, J. and Visser, R. (1987) Nitrate toxicity in cattle. *New Zealand Veterinary Journal* 35, 136–137.

Veum, T.L., Ledoux, D.R., Raboy, V. and Ertl, D.S. (2001) Low-phytic acid corn improves nutrient utilization for growing pigs. *Journal of Animal Science* 79, 2873–2880.

Veum, T.L., Ledoux, D.R., Bollinger, D.W., Raboy, V. and Cook, A. (2002) Low-phytic acid barley improves calcium and phosphorus utilization and growth performance in growing pigs. *Journal of Animal Science* 80, 2663–2670.

Vincken, J.P., Heng, L., de Groot, A. and Gruppen, H. (2007) Saponins, classification and occurrence in the plant kingdom. *Phytochemistry* 68, 275–297.

Wang, Q., Yu, L.-G., Campbell, B.J., Milton, J.D. and Rhodes, J.M. (1998) Identification of intact peanut lectin in peripheral venous blood. *Lancet* 352, 1831–1832.

Wheeler, J.L. (1994) International workshop on cassava safety, Ibadan, Nigeria, March 1–4, 1994. *Acta Horticulturae* 375, 251–259.

Wheeler, J.L., Mulcahy, C., Walcott, J.J. and Rapp, G.G. (1990) Factors affecting the hydrogen cyanide potential of forage sorghum. *Australian Journal of Agricultural Research* 41, 1093–1100.

Wink, M. and Mohamed, G.I.A. (2003) Evolution of chemical defense traits in the *Leguminosae*: mapping of distribution patterns of secondary metabolites on a molecular phylogeny inferred from nucleotide sequences of the *rbc*L gene. *Biochemical Systematics and Ecology* 31, 897–917.

Wong, Y.C. and Tam, C.C. (1989) Structural changes of the guinea pig prostatic epithelial cells after gossypol treatment. *Acta Anatomica* 134, 18–25.

Zakayo, G. and Krebs, G.L. (1997) *Leucaena leucocephala* as a protein supplement for growing pigs. *Recent Advances in Animal Nutrition in Australia* 14, 244.

7 Digestion and the Supply of Nutrients

Comparative Digestive Function: All Digestive Tracts Share Some Important Functions

A digestive tract is a tube that runs from a food-gathering organ to an orifice through which waste is voided. The 'tube' is not simply a conduit, but a place where digesta is held, exposed to various physical and chemical processes (chewing, mixing, sieving, chemical buffering and enzymic hydrolysis), the end products of digestion are absorbed and water economy is regulated. Digestive tracts are also modified so that parts of them are good places for microbial growth, there are regions which are used for waste excretion, and there are mechanisms for the recycling of certain body chemicals.

In addition to the tract itself, ancillary organs are important to digestion. In the mouth, these include the lips, teeth and tongue, which are used to prehend food, reduce its particle size and form it into a bolus. The salivary glands, which are also located in the buccal cavity, secrete water and mucus which lubricate the bolus. Saliva contains digestive enzymes: many animals produce a salivary amylase and lipase or esterase (Altmann and Dittmer, 1968) although there are species differences; e.g. amylase is absent from sheep and goat saliva, and there is no esterase in red kangaroo saliva (Beal, 1998). The bicarbonate and phosphate ions in saliva, especially in that of ruminants, constitute a chemical buffering system. The pancreas secretes into the duodenum enzymes, which are responsible for most of the digestion of proteins and starches and for the digestion of fats and nucleic acids. Fat digestion is aided by bile salts which are synthesized in the liver and (in most animals, except for example, horses, deer and rats; Wilkie, 1926; Altmann and Dittmer, 1968; Hudson, undated) stored in the gall bladder prior to release into the duodenum.

Digestive Tract Function Involves the Sequence of Mammalian and Microbial Digestion and Control of Digesta Flow

Functional diagrams (in contrast to anatomical diagrams) of the digestive tract (Fig. 7.1) show the locations of the different types of digestion (animal or microbial) and their sequence, the location of ancillary organs, where the mixing of digesta occurs and where the flow of digesta is controlled. 'Digesta' and 'ingesta' are names given to the partially digested food which travels through the digestive tract.

Carnivores and Omnivores Have Simple Digestive Systems

Carnivores and omnivores are monogastric (or 'single-stomached') animals, i.e. they do not have the stomach compartments anterior to the gastric stomach which are found in some herbivores, for example. A carnivore's diet (e.g. that of the cat or the quoll) has about 15–20% protein, 5–30% fat and 9–15% bone. These diets reflect the body compositions of their typical prey. Beef cattle, sheep and pigs (ARC, 1980; Kempster *et al.*, 1986; Dolezal *et al.*, 1993) and the blowfly and mealworm (Toshio and Fraenkel, 1966; Collatz and Hoeger, 1980) have 7–25% protein, 3–20% fat and 5–15% carbohydrate, with 2–4% cuticle in the blowfly. All of these food constituents (except the insect cuticle) are potentially digestible by mammalian digestive enzymes. Omnivores, such as the pig or the fox, eat plant material which contains starch, simple sugars and oils. Again, much of this plant material can be digested by mammalian enzymes.

Monogastric animals have a glandular stomach and a short small intestine. Digestion of proteins and some fats begins in the stomach, with most digestion occurring in the small intestine. They

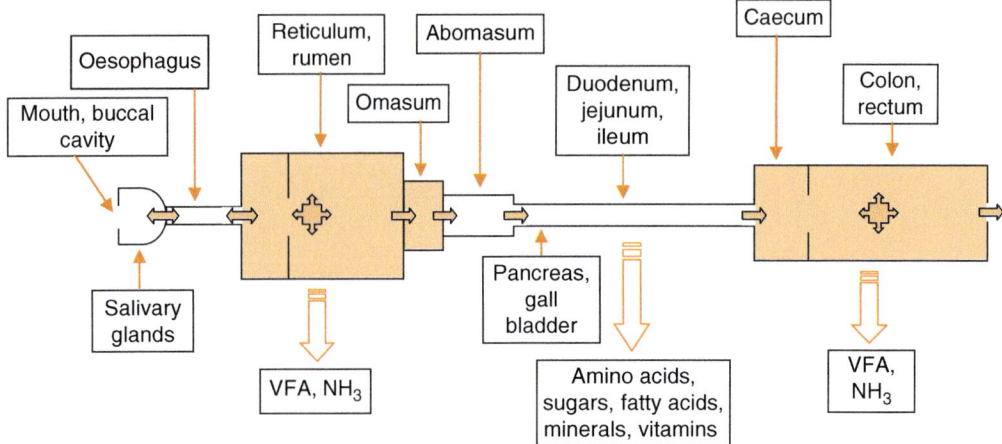

(A) Ruminantia (ruminant, foregut fermenter; cow)

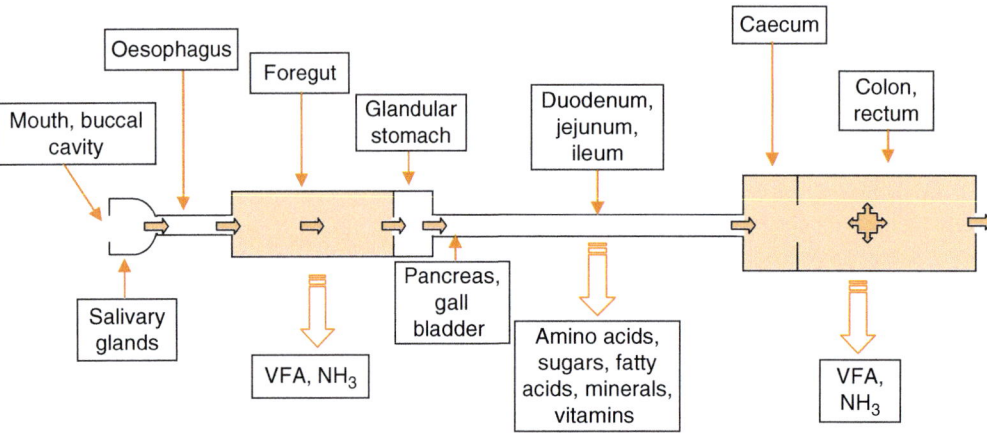

(B) Macropodidae (foregut fermenter; kangaroo)

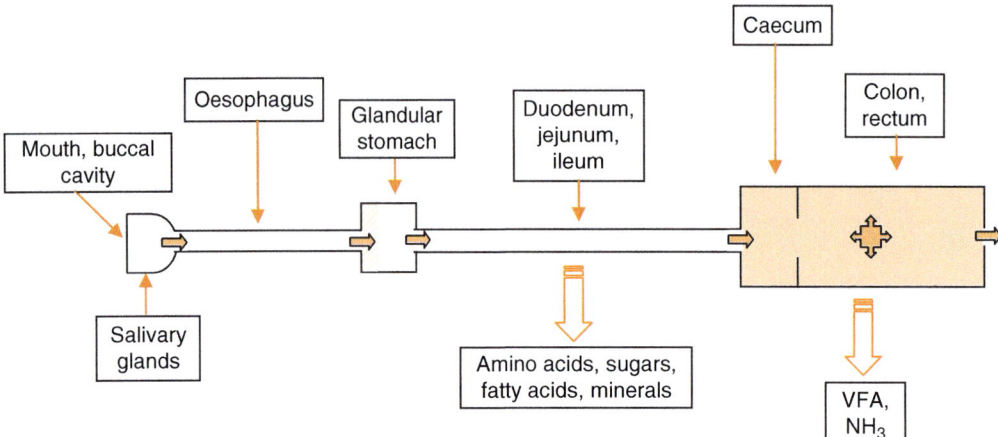

(C) Hippomorpha (non-ruminant herbivore, hindgut fermenter; horse)

Fig. 7.1. Functional diagrams of animal digestive systems (continued on next page).

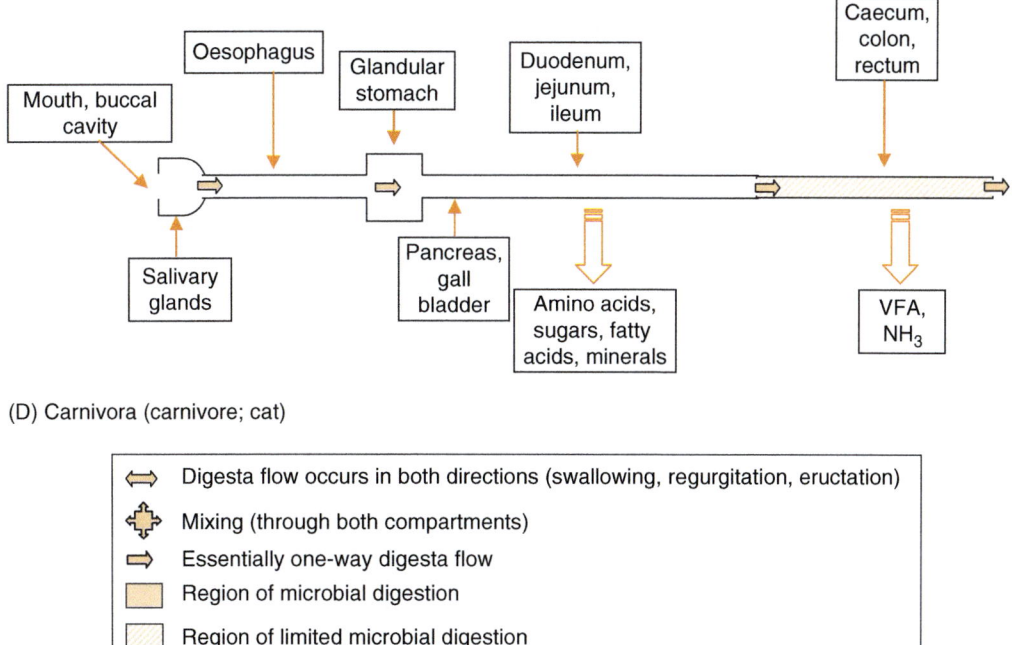

(D) Carnivora (carnivore; cat)

⟺	Digesta flow occurs in both directions (swallowing, regurgitation, eructation)
⟴	Mixing (through both compartments)
⟹	Essentially one-way digesta flow
▨	Region of microbial digestion
▨	Region of limited microbial digestion
▢	Region of mammalian digestion

Fig. 7.1. *Continued.*

have a population of bacteria in the large intestine, and in some species, e.g. the pig, these can have digestive significance. The microorganisms in the pig's ileum and large intestine easily degrade β-glucans and arabinoxylans from oat grain (Knudsen *et al.*, 1993; Johansen *et al.*, 1997). The contribution of microbial end products (mainly the volatile fatty acids (VFA) – acetate, propionate and butyrate) to the energy substrates absorbed from the gastrointestinal tracts of non-ruminants has not been extensively examined. However, the data reviewed by Bugaut (1987) suggest that 7–40% of these animals' basal energy requirements may be met from VFA produced by microbial metabolism, mainly in the large intestine.

Herbivores Have More Complex Digestive Systems

Plant fibre is between 50% and 85% of the dry matter (DM) in herbivore foods. There are no mammalian enzymes which can digest the plant cell wall carbohydrates, so all herbivores have developed a symbiotic relationship with microbes (bacteria, protozoa and fungi) which have the required enzymes. These organisms produce, as end products of their metabolism of glucose and the other monosaccharides released by digestion from plant fibre, VFA that are used as energy sources by the host animal. Ruminants regulate the flow of digesta from the rumen, so that undigested forage particles are retained for longer exposure to microbial enzymes.

Herbivores must optimize the flow through their digestive tracts of forage which is usually not completely digested by microbial enzymes. Grazers are particularly disadvantaged, because their diets typically include hard-to-digest plant fibre. Some of the grazer species have evolved a digestive tract which gives little impediment to the flow of undigested forage particles. An example is the horse (see Fig. 7.1C). The equine digestive tract has no compartment analogous to the rumen to retain particles of long, undigested forage so that these can be more completely digested. Ruminants (e.g. cattle, sheep, goats, deer) and tylopods (e.g. camels, llamas) have evolved rumination behaviour (Harfoot, 1978). This allows them to reduce the size of long forage particles so that these can escape from the fore-stomach into the hindgut. These animals comminute (physically break up by chewing) their forage as they eat it, but also regurgitate the longer particles in the rumen digesta and re-chew it. Ruminants have also evolved an eructation behaviour, which allows them to remove the large volumes of gases (methane and CO_2) produced during pregastric microbial digestion.

Caecotrophy and Coprophagy Are Techniques Used by Some Plant-eaters to Harvest Microbial Nutrients

Colonic separation movements of two types exist in some hindgut fermenters. These are described by Pei *et al.* (2001). The 'mucus-trap' type retains bacteria in the caecum. Bacteria passing from the caecum into the proximal colon are trapped in mucus, with very few food particles, and this mixture of mucus and bacteria is returned to the caecum by antiperistalsis. The 'wash-back' type returns bacteria and also solutes and small food particles. This mixture is moved in a stream of water from the proximal colon to the caecum by antiperistalsis, assisted by the net secretion of water into the distal colon and net absorption from the caecum. The koala has a wash-back separation mechanism (Cork and Warner, 1983) which allows it to retain the fluids and fine particles of its digesta while it voids the more difficult-to-digest fibrous particles.

Several small hindgut fermenting herbivores (e.g. rabbits and hares, rats and mice, voles and the ring-tailed possum; Hornicke, 1981; Chilcott and Hume, 1985; Cranford and Johnson, 1989; Pei *et al.*, 2001) harvest the nutrients produced by hindgut microbial metabolism through 'caecotrophy' or 'coprophagy' (the latter term also refers to the eating of faeces for non-nutritional reasons; Hornicke, 1981). Caecotrophy is the eating of caecal digesta which is voided during rest. This material is not the same as normal faeces. It is rich in B vitamins, VFA, minerals and microbial protein which would otherwise be voided, and may improve the availability of vitamin A (Hornicke, 1981; Belenguer *et al.*, 2005; Kertin *et al.*, 2005).

Herbivores Use Several Different Digestive Strategies

The main differences between the different types of herbivores are: (i) the extent to which they control diet quality (and thus nutrient yield) by diet selection; (ii) the location of their region(s) of microbial digestion, especially whether or not there is a pregastric region of microbial digestion; (iii) the extent to which the animal can regulate the passage of undigested food through its digestive tract; (iv) behaviour which promotes increased comminution of plant particles, i.e. chewing during eating versus chewing during eating and rumination; (v) whether the animal has any behaviours that allow it to harvest nutrients which are produced by microbial metabolism in the large intestine; and (vi) the capacity in some herbivores (e.g. the deer) to detoxify plant poisons such as tannins (Table 7.1).

Nutrient Supply in Herbivores Depends on the Sequence of Mammalian and Microbial Digestion

One of the most important factors which determines the types of nutrients absorbed from a digestive tract is the sequence of microbial and mammalian digestion. It is important for the best use of the results of microbial metabolism that this occurs before the digesta flows to a region of mammalian digestion, because this allows the microbial cells to be digested. Thus, ruminants, and to a lesser extent macropods, make better use of bacterial metabolism than horses and some other hindgut fermenters. The rate of digesta passage is also influenced by digestive tract anatomy, and in turn this influences the extent of digestion and the nature of digestion end products.

The influence of digestive tract anatomy and function is illustrated in Table 7.2, which shows the digestion processes and outcomes when a horse and a cow are fed practical rations that are based on grass hay and various concentrates.

There are some other advantages and disadvantages for both foregut and hindgut fermenters which are not made explicit in Table 7.2. The ability of the ruminant to regulate the passage of undigested forage particles allows more complete forage digestion, but the intake of low-quality forages is reduced, and the animal is at risk of rumen impaction. In comparison, the relative inability of the horse to regulate digesta passage allows that animal to maintain its intake of hard-to-digest forage, although this is at the expense of a reduced extent of digestion of all forages.

In hindgut fermenters such as the horse, mammalian digestion occurs first, followed by microbial metabolism in the large intestine. When mammalian digestion occurs first, the nutrient content of the digestible part of the food is reflected fully in the nutrients absorbed into the blood. The chemistries of food amino acids and fatty acids are unchanged by

Table 7.1. Diet selection and digestion strategies of herbivores.

(Sub)Order Family (example)	Herbivore type[a]	Location of microorganisms[b]	Regulation of forward digesta flow[c]	Other behaviours
Ruminantia				
Bovidae (cow)	Grazer (savannah)	Rumen and hindgut	From the rumen and abomasum	Rumination, eructation
Bovidae (goat)	Mixed feeder	Rumen and hindgut	From the rumen and abomasum	Rumination, eructation
Cervidae (red deer)	Mixed feeder (forest fringe)	Rumen and hindgut	From the rumen and abomasum	Rumination, eructation
Marsupialia				
Macropodidae (kangaroo)	Grazer	Foregut and hindgut	From the glandular stomach	
Lagomorpha				
Leporidae (hare)	Mixed feeder	Hindgut	From the glandular stomach and in the large intestine	Caecotrophy
Hippomorpha				
Equidae (horse)	Grazer	Hindgut[d]	From the glandular stomach and in the large intestine	
Suiformes				
Suidae (pig)	Omnivore	Hindgut	From the glandular stomach	

[a]These classifications are based on Hoffman (1985) and Janis and Erhardt (1988): grazers' diets are more than 90% grass, mixed feeders' diets are between 10% and 90% grass.
[b]Microorganisms are found in all parts of the digestive tract; the compartments mentioned here are the major fermentation sites.
[c]Flow from the mouth to the anus; regulation includes both the prevention of backwards flow and the retention of digesta components in a digestive compartment (e.g. large food particles in the rumen or small particles in the large intestine).
[d]Bacteria occur in the equine stomach (Al Jassim *et al.*, 2005) but their digestive significance is not known.

digestion. In foregut fermenters such as the cow, microbial digestion and metabolism occur first. This allows ammonia-N produced from the hydrolysis of non-protein N substances such as urea to be incorporated into bacterial protein, which is subsequently digested in the small intestine, and the harvesting of vitamins synthesized by the rumen bacteria. But it also ensures that most food energy precursors are absorbed as VFA with reduced efficiencies of utilization of metabolizable energy (ME), saturation (hydrogenation) of dietary fatty acids and destruction of food amino acids by bacterial metabolism, and susceptibility to rumen disorders, e.g. lactic acidosis.

Digestion in the Stomach and Small Intestines: Mammalian Digestion Occurs in Stages

Protein and nucleic acid digestion

Proteins consist of several hundred amino acid molecules joined by peptide bonds. This primary structure is coiled and folded, giving the secondary and tertiary structures of the protein. Many proteins are complexes of several monomers. These molecules are too large to be absorbed through the walls of the small intestinal epithelial cells and so they must be reduced in size. This is done by hydrolysing the peptide bonds and breaking the polypeptide chain into smaller fragments. Food and microbial nucleic acids are digested in the small intestine to their component nucleotides by the combined activity of pancreatic nucleases and phosphoesterases (Cosgrove, 1998).

Protein hydrolysis is accomplished by a group of protease enzymes (Corring, 1982). Proteases are secreted as inactive proenzymes (also called 'zymogens'). Clearly, it would be dangerous for cells to synthesize digestive proteases in an active form because these could begin to digest the cells in which they are synthesized. Proenzymes have the active site obscured by a polypeptide 'tail' which prevents a substrate from binding to the enzyme. The tail is removed when the proenzyme is activated. An

Table 7.2. Nutrients obtained by horses and cattle eating the same diet ingredients.

Ingredient	Horse	Cow
	Outcome of the digestive process[a]	
Barley grain	• Starch is digested in the small intestine → glucose	• Most starch is digested in the rumen → VFA (relatively rich in Pr)
Cottonseed meal	• Protein is digested in the stomach and small intestine → amino acids • Gossypol is absorbed into the body: binds Fe and promotes lipid peroxidation	• Most protein is digested in rumen → NH_3 → bacterial protein, later digested in the small intestine → amino acids • Gossypol complexes with soluble rumen proteins: action against body cells protected
Synthetic lysine	• Is absorbed in the small intestine → contributes to the body lysine pool	• Is deaminated in the rumen → contributes to the rumen NH_3 pool
Molasses	• Makes the ration more palatable • Is digested in the small intestine → glucose + sucrose	• Makes the ration more palatable • Is digested in the rumen → VFA (relatively rich in Bu)
Canola oil	• Is digested in small intestine → various glycerides + FA	• May inhibit rumen function • FA are saturated by rumen bacteria • Is digested in the small intestine → various glycerides + FA
Grass hay	• Fibre is digested in the large intestine → VFA • Is needed for digestive tract health	• Fibre is digested in the rumen and large intestine → VFA (Ac, Pr, Bu) • Is needed for rumen health

[a]VFA = volatile fatty acids; Ac = acetic acid; Pr = propionic acid; Bu = butyric acid; FA = fatty acids.

example of protease activation is given by Nielsen and Foltmann (1995). Pepsinogens, secreted from the chief cells in the gastric glands, are activated by exposure to low pH (caused by the secretion of HCl from the parietal cells of the gastric glands) and by autocatalysis by previously activated pepsin molecules (Sjaastad *et al.*, 2003). The pancreatic proteases trypsin, chymotrypsins, elastase and carboxypeptidases are also synthesized as proenzymes. Trypsinogen is activated by enterokinase (a brush border enzyme; Hinsberger and Sandhu, 2004) and by autocatalysis. Trypsin then activates the other pancreatic proenzymes. The intestinal epithelial peptidases do not need to be activated.

Proteases act either in the middle of the substrate protein molecule ('endopeptidases') or at the ends ('exopeptidases'). The pepsins, trypsins, chymotrypsins and elastase are endopeptidases, and their action releases polypeptide fragments. These fragments are then digested by the exopeptidases (carboxypeptidases and the epithelial peptidases) which release individual amino acids. Peptidases of the intestinal epithelium are located in the 'brush border'. The luminal surface of the epithelial cells of

the small intestinal villi (i.e. the enterocytes) is raised into about 3000–6000 microvilli (Sjaastad *et al.*, 2003) called the brush border (see Maroux *et al.*, 1988 and Steffanson *et al.*, 2004 for detailed descriptions of the brush border).

Protein digestion in monogastric animals begins in the stomach where the pepsins (five have been identified; Narita *et al.*, 2002) preferentially attack peptide bonds which involve aromatic amino acids (Corring, 1982). The products of these enzymes are oligopeptides (often 2–6 amino acids in length), and di- and tripeptides. These are then hydrolysed by the carboxypeptidases and the brush border peptidases, releasing the constituent amino acids (Fig. 7.2), although some di- and tripeptides are absorbed as such.

Unless ruminants are fed a diet rich in ruminally undegradable (or 'bypass') protein, most of the protein which enters the abomasum is bacterial. Ruminants have evolved a means of digesting this protein through an adaptation of lysozyme c, an enzyme which is usually synthesized by animals as a means of combatting infection by bacteria (Dobson *et al.*, 1984). Gram positive rumen bacteria are

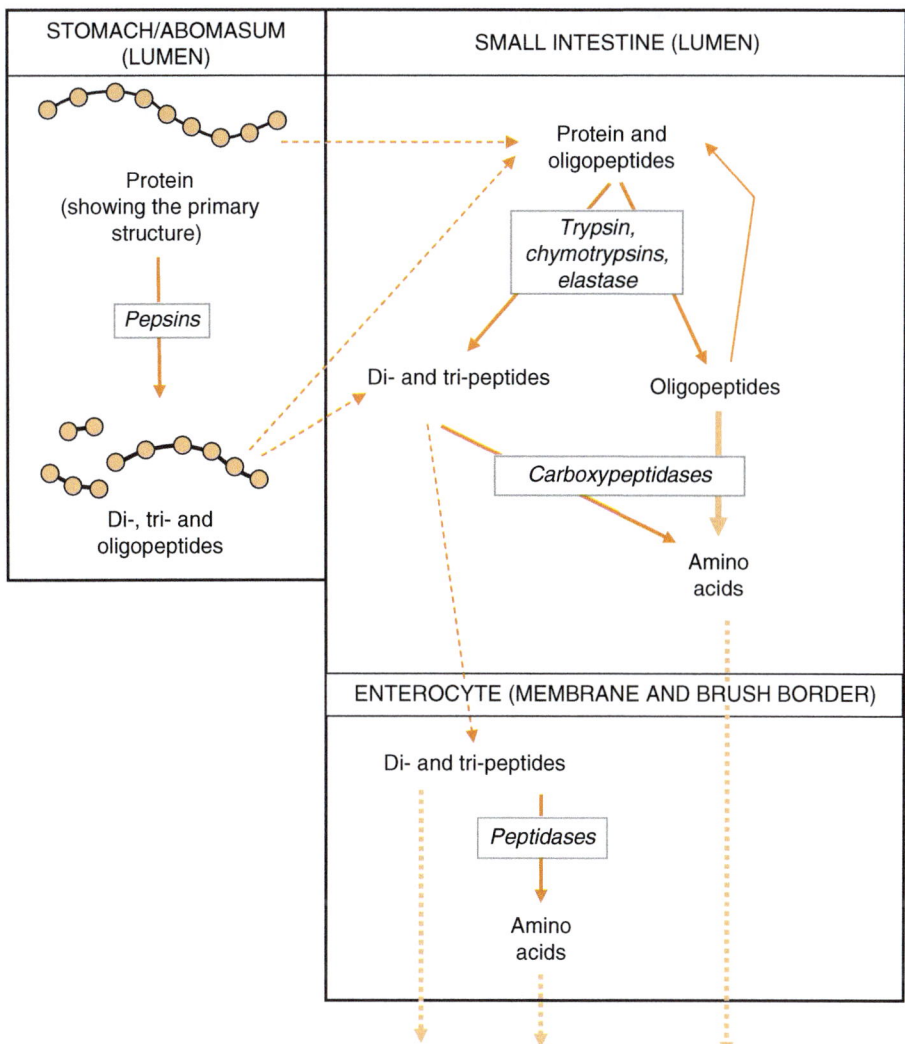

Fig. 7.2. The sequence of protein digestion in the stomach and small intestine (→ chemical changes (digestion) as a result of digestive enzyme action; --▶ flow of substrate; ··▶ absorption of digestion products).

digested by this enzyme, but Gram negative bacteria are not, and there is some evidence of rumen bacteria developing resistance to bovine gastric lysozyme (Domínguez-Bello *et al.*, 2004); the possible consequences of this are thought-provoking!

Polysaccharide and disaccharide digestion

Amylose and amylopectin (the polysaccharides of starch) are digested by salivary and intestinal amylases. Some animals (e.g. dog, rabbit, rat and pig; Altmann and Dittmer, 1968) have a salivary α-amylase, which has a similar action to the pancreatic

enzyme. Salivary α-amylase is not active at the low pH of the stomach, so its digestive role is not clear, although it will continue to act in the stomach until the gastric juice penetrates into the interior of the swallowed bolus.

In non-ruminants, most starch digestion occurs in the small intestine. Pancreatic α-amylase is an endoglucosidase, i.e. it breaks α1-4 glucosidic bonds in the interior of both the amylase and amylopectin molecules. Maltose and isomaltose are produced when both amylose and amylopectin are digested, but as α-amylase cannot break α1-6 glucosidic bonds, we are left with a small oligosaccharide of up to seven glucose units (Richardson and Gorton, 2003) when the enzyme has digested all the susceptible bonds in amylopectin. This oligosaccharide is called a 'limit dextrin'. Thus, the products of α-amylase digestion of starch are maltose, isomaltose and the limit dextrin (Gray, 1992; Harmon *et al.*, 2004). The final hydrolyses release glucose; this is done by a series of enterocyte brush border enzymes which hydrolyse α1-4 and α1-6 glucosidic bonds

(Fig. 7.3; Gray, 1992). Starch digestion in ruminants is similar to that described above, but there are two important differences: ruminants do not have a salivary amylase or an intestinal sucrase (Siddons, 1968; Kreikemeier *et al.*, 1990). There are other subtle differences between ruminants and non-ruminants. Starch digestion in the ruminant's small intestine is reviewed by Harmon (1993) and Harmon *et al.* (2004).

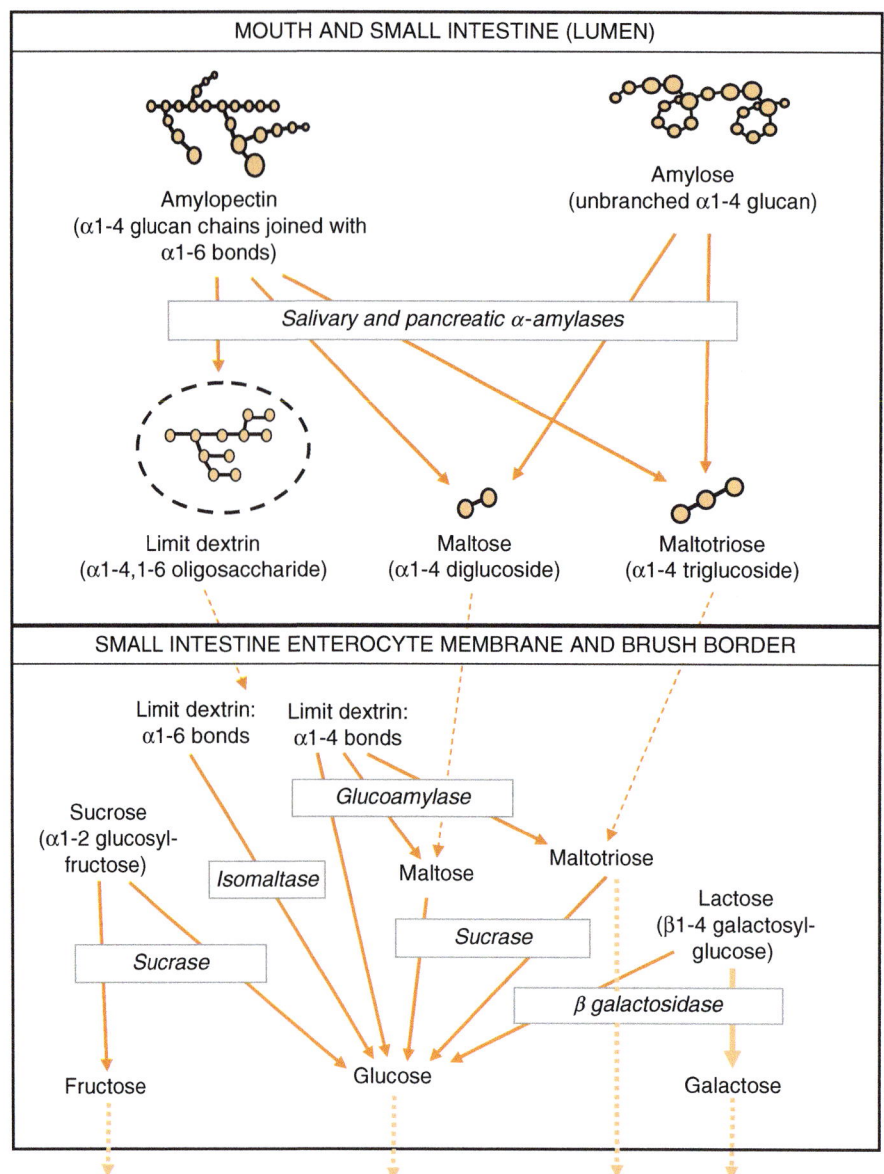

Fig. 7.3. The sequence of starch and disaccharide digestion in the stomach and small intestine (→ chemical changes (digestion) as a result of digestive enzyme action; --→ flow of substrate; ···▶ absorption of digestion products).

Lipid digestion

Lipids in food include triacylglycerols (especially in plant oils and animal fat), phospholipids (components of animal cell membranes) and galactoglycerides (an important lipid of grass). A triacylglycerol (also called a 'triglyceride') is a small macromolecule in which a molecule of glycerol is esterified to a fatty acid at each of the three glycerol carbons; thus, we have a 1,2,3-triacylglycerol. Phospholipids consist of glycerol esterified to one or two fatty acids, with a phosphate group bonded at the third glycerol carbon, while galactoglycerides are diacylglycerols with a galactose (sugar) molecule bonded at the third glycerol carbon.

Up to 10–30% of dietary fat may be digested in the stomach (Carey et al., 1983; Miled et al., 2000) by enzymes which are secreted from cells in the tongue, pharynx or stomach (Hayes et al., 1994; Embleton and Pouton, 1997). The pre-ruminant calf has a salivary esterase which is particularly active against milk fat, which contains the shorter-chain fatty acids (Sun et al., 2002). Pre-intestinal lipases preferentially hydrolyse ester bonds at the outer glycerol carbons, i.e. the 1 and 3 positions (Embleton and Pouton, 1997). Dietary fats undigested by salivary or gastric lipases may be digested by pancreatic lipase. Lipase action in the duodenum is assisted by the emulsification of dietary fats by bile salts (Na^+ and K^+ taurocholates and glycocholates), bile lipids and the products of pre-intestinal lipase action (Embleton and Pouton, 1997). The emulsified fats have a greater surface area for lipase action. Bile salts also remove the liberated fatty acids from the region of lipase action. This action also tends to remove the enzyme molecule from the surface of the lipid droplet. However, a protein, colipase, also secreted from the pancreas, prevents this by binding to both the lipase molecule and the substrate lipid. The pancreas elaborates other lipid-digesting enzymes, including pancreatic cholesterol esterase, which hydrolyses cholesterol (Jamry et al., 1995), and phospholipase A2 which assists in the digestion of phospholipids by hydrolysing the ester bond at the second carbon position (Dennis, 1994).

Digestion occurs in stages, as first one of the exterior fatty acids (attached at either the first or third glycerol carbon) is removed, then the other (Fig. 7.4). The most common end product of lipase action is a 2-monoacylglycerol. The digestion products are absorbed as micelles, which are mixtures of lipid surrounded by a bile salt/phospholipid membrane. Embleton and Pouton (1997) and Miled et al. (2000) have reviewed lipase action and the absorption of lipase products.

The absorption of the products of lipid digestion is more complicated than the absorption of amino acids and sugars, because there is some reassembly of lipid molecules in the epithelial cell. The mechanisms of lipid absorption are described by Thomson et al. (1993).

Mammalian Digestive Enzymes Cannot Digest All the Components of Animal Foods

The tissues of the higher animals can be digested by animal enzymes, although in practice this does not happen completely because the food passes through the tract too quickly for pieces of bone and connective tissue to be completely digested. In plant foods, most starch, protein and lipid can be digested by the animal's gastric and intestinal enzymes. On the other hand, there are some common food constituents which higher animals cannot digest. An example from animal-derived foods is the chitin in the exoskeletons of insects and marine organisms. This is probably little digested by higher animals, although chitinases have been identified in the digestive tracts of several vertebrates (Suzuki et al., 2002).

It is in the realm of the plant-food-eaters that the limitations of mammalian digestive enzymes have their biggest impact. Cellulose and hemicellulose make up about 80% of plant DM and are rich in β-glycosidic bonds, which cannot be digested by mammalian enzymes (although invertebrate animals may have an endogenous cellulase; e.g. the redclaw crayfish, Crawford et al., 2004; also see Watanabe and Tokuda, 2001). In the mammals, cellulose and hemicellulose are degraded by microbial β-glycosidases elaborated in the rumen and large intestine, and this symbiosis between mammals and microorganisms is very important for the digestion of foods of plant origin. Microorganisms live and are active (in a digestive sense) in most animal digestive tracts (Fonty and Gouet, 1989). Mammals have active microbial populations in the large intestine, and ruminants and other foregut fermenters such as the macropods have large populations in the rumen or foregut. The bacterial species in the ruminant large intestine are similar to those in

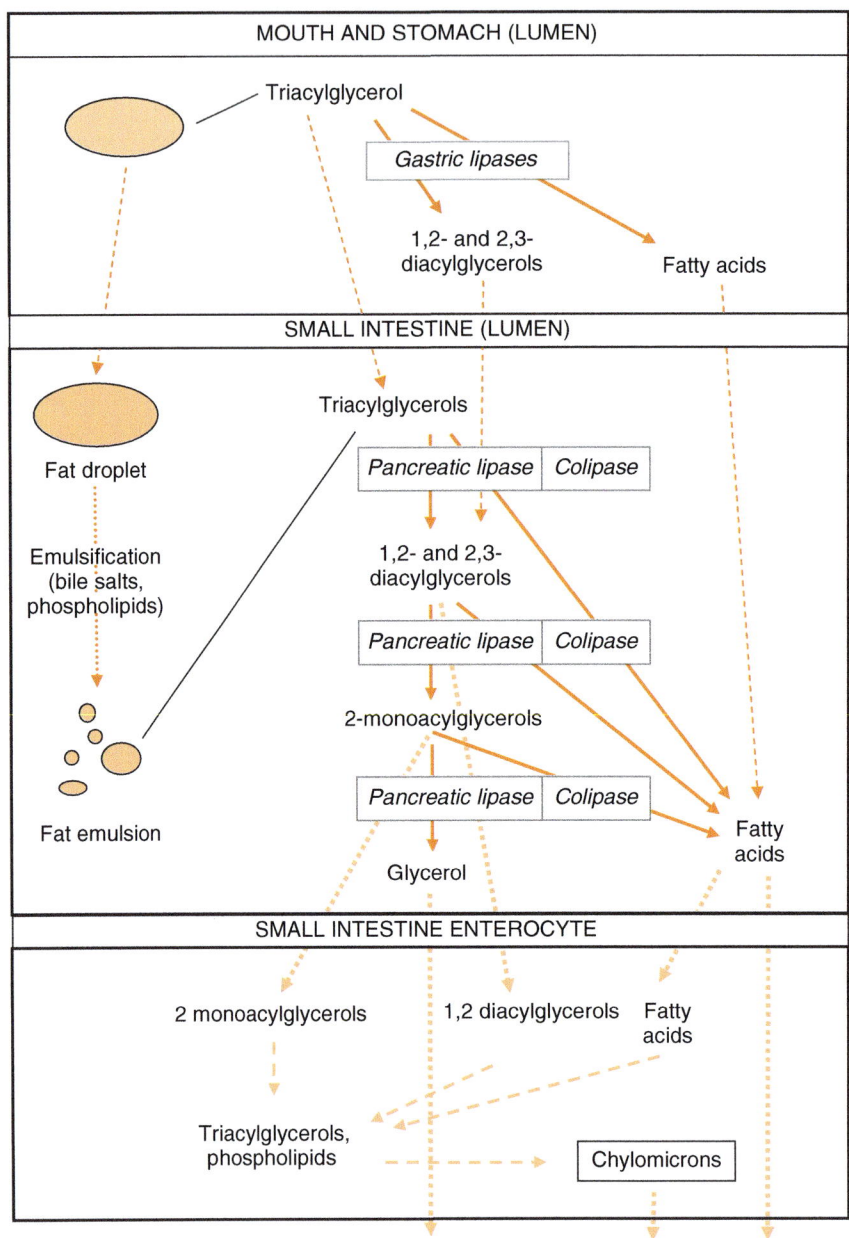

Fig. 7.4. The sequence of lipid digestion in the stomach and small intestine (→ chemical changes (digestion) as a result of digestive enzyme action; ⇢ physical and chemical changes associated with chylomicron synthesis; ⋯▸ physical changes associated with fat droplet emulsification; ⇢ flow of substrate; ⋯▸ absorption of digestion products).

the rumen, but there are only about a tenth of the numbers, and there are fewer cellulolytic species (Mould *et al.*, 2005). In the following discussion we will use the rumen as our model for a discussion of microbial digestion and its contribution to the animal's nutrient supply.

The Ecology of the Rumen Microorganisms

Studies of digestion in the ruminant show that some 60% of the total digestible organic matter of roughages is digested in the rumen, in comparison with intestinal and caecal digestion. With low-quality roughages the proportion may be even higher: values of up to 80% have been reported for ground wheaten straw treated with NaOH (Demeyer, 1981). Rumen microbial function is a major determinant of the digestion of both roughages and concentrates by ruminants. One of the most interesting aspects of rumen function is the interrelationships between the various rumen microorganisms, and between these and the animal's food. In many cases, effects of food type on animal health and the efficiency of animal performance (e.g. growth and lactation) can be traced back to the effects of these interactions on rumen function. Rumen ecology is a large topic and one which is continually advancing, especially now that genetic techniques enable better identification of the rumen microorganisms and their enzymology (Hobson and Stewart, 1997; Russell and Rychlik, 2001; Mould *et al.*, 2005).

The Rumen Works Like a Fermentation Vat

The conditions of temperature, pH, provision of substrate, removal of waste products, etc., which make the rumen a suitable microbial ecosystem, have been discussed by Hungate (1966) and the authors referred to above. In summary, these are the key features of the rumen as an environment for its organisms.

pH

The rumen pH varies between 5.5 and 7.5 in roughage-fed animals. This relative constancy is maintained by the rumen's buffering capacity: there are both phosphate and bicarbonate buffering systems. Phosphate buffers appear to be most effective at high rumen pH (6–8) while acceptable conditions at lower pH (5–6) are maintained by the bicarbonate buffer (Emmanuel *et al.*, 1969, 1970) and absorption of VFA and by the dilution of the rumen liquor by drinking water and saliva.

Anaerobiosis

The rumen liquor is effectively oxygen-free; those small amounts of oxygen which enter the rumen rapidly disappear.

Temperature

The rumen is maintained at approximately 39°C, although this varies temporarily when water is consumed.

Removal of end products

Bacteria are poisoned by the accumulation of their metabolic end products. These are continually removed from the rumen by passage to the lower gut, by absorption through the rumen wall and by eructation of methane and CO_2.

Supply of substrate

Grazing ruminants generally eat periodically (typically in two or three episodes daily), but the food consumed is digested fairly slowly (e.g. fibre is digested in the rumen at about 3–10% per hour; Sniffen *et al.*, 1992) and so there is almost always a supply of fermentable substrate for microbial metabolism. The situation can be different with animals which are fed at defined times (e.g. dairy cattle which are given a concentrate supplement while they are being milked). In these cases, the supply of fermentable carbohydrate and other nutrients (e.g. ruminally available N) can vary substantially during the day.

Mixing

Rumen movements continually cycle digesta around the rumen; this promotes the reattachment of bacteria to undigested substrate, and the passage of indigestible material to the hindgut by presenting particles of appropriate size and density to the reticulo-omasal orifice.

Comminution

We might expect that the loss of DM during microbial digestion would reduce the size of food particles, but this is not so (Van Soest, 1975). Ruminants need to chew their food to physically reduce the size of large food particles, especially forages, so that difficult-to-digest food can pass into the abomasum and lower gut. Food is initially broken up by chewing during prehension and bolus formation (Dryden *et al.*, 1995) but this is not completely effective in reducing particle size. After some time in the rumen, the large food particles are brought back into the mouth (by regurgitation) for further comminution.

There Are Three Main Types of Rumen Microorganisms

Bacteria and archaea

There are some 10^9–10^{11} cells/g in the rumen liquor of animals eating highly digestible roughages. Most of the rumen bacterial species are also found in the ruminant large intestine. Although *Fibrobacter succinogenes*, *Ruminococcus albus* and *Ruminococcus flavefaciens* are found in the digestive tracts of many species (Hungate, 1984), there are differences in the species of bacteria found in ruminants and other herbivores. Daly *et al.* (2001), for example, noted that 89% of the species they found in the equine large intestine were not found elsewhere.

Archaea are organisms which are similar to bacteria but which have been recently classed in a separate domain. In this discussion, we will consider the bacteria and the archaea together.

The species and numbers of rumen bacteria have been described by Hungate (1966), Hungate *et al.* (1964), Demeyer (1981) and Mould *et al.* (2005). Some of the more important bacterial species are listed in Table 7.3. Over 200 bacterial species have been identified in the rumen (McAllister and Cheng, 1996) but we have probably identified only 10–20% of the species which live in the ruminant digestive tract. In particular, a recent discovery is a group of acetogenic bacteria which compete with methanogens for H and produce acetic acid from sugars and lactic acid (Morvan *et al.*, 1994; Rieu-Lesme *et al.*, 1996). The (lack of) depth of our understanding of digestive tract microbiology is described by Daly *et al.* (2001) and McSweeney *et al.* (2005).

The majority of the rumen bacteria are in one of the three populations. Most are associated with food particles, either firmly attached or loosely attached (Michalet-Doreau *et al.*, 2001; Trabalza-Marinucci *et al.*, 2006). These attached bacteria are grouped in microcolonies (Cheng *et al.*, 1984; Fig. 7.5), which are groups of symbiotic (or consortial) bacteria growing near each other apparently to facilitate the transfer of nutrients between their cells. These microcolonies are also called a 'biofilm'. The attached bacteria may be in 'better' nutritional condition than the unattached bacteria – they have higher polysaccharidase enzyme activities (Michalet-Doreau *et al.*, 2001) and higher ratios of total cell protein to nucleic acids (see references in Trabalza-Marinucci *et al.*, 2006). The remaining 20–40% is found free in the rumen liquor (the liquid-associated bacteria). This includes those organisms which have become detached from solid substrates (Faichney *et al.*, 1997), and also those which are always found in this ecological niche. The latter group may detoxify nitrates and cyanogenetic glycosides. Other bacteria (about 1%; Cheng and Costerton, 1984) are found attached to the rumen epithelium where they are involved in the hydrolysis of blood urea as it diffuses back into the rumen and in scavenging oxygen (McCowan *et al.*, 1980; Stewart *et al.*, 1988) and others are attached to protozoa and fungal sporangia (Miron *et al.*, 2001). As many as one-third of the methanogenic organisms are found attached to protozoa (Regensbogenova *et al.*, 2004). The *Methanobrevibacter* genus is an important member of the rumen archaea.

Protozoa

There are some 10^5–10^6 protozoal cells/ml of rumen liquor in animals eating good-quality roughages. More than 100 rumen protozoa species have been identified (McAllister and Cheng, 1996), but no more than 15–20 species are present in any given animal. The important genera are *Dasytricha*, *Isotricha*, *Diplodinium*, *Entodinium*, *Polyplastron*, *Eudiplodinium*, *Epidinium* and *Ophryscolex*. Many of the common rumen protozoa are found in other herbivores although there are some which are found only in particular hosts (Hungate, 1966; Ogimoto and Imai, 1981).

Fungi

Orpin (1977) demonstrated the presence of a phycomycetous fungus in the rumen and Bauchop (1979) demonstrated the digestive role of these organisms. Fungi comprise some 8% (Orpin, 1984) to 20% (Rezaeian *et al.*, 2004) of the rumen microbial mass, with zoospore densities of 10^3–10^6/g (Obispo and Dehority, 1992). Important fungal genera are *Neocallimastix*, *Piromyces*, *Sphaeromyces* and *Orpinomyces*.

Bacteriophages and yeasts

Bacteriophages (viruses) occur in the rumen in large numbers, at about 10^9 particles/g (Klieve and Swain, 1993). Their morphology has been described by Ritchie *et al.* (1970, cited by Ogimoto and Imai,

Table 7.3. Important species of rumen bacteria. (From Hungate, 1966; Stewart *et al.*, 1988, 1997; Mould *et al.*, 2005.)

Species	Description	Energy sources[a]	Fermentation products[b]						
			Fo	Ac	Pr	Bu	iBu	La	Su
Principal fibre digesters			+	+		+			
Butyrivibrio fibrisolvens	Gram +ve[c] rods	Xylans, pectin (β glucans, starch)							
Fibrobacter succinogenes	Gram –ve[c] rods	β glucans, glucose (pectin, starch)	+	+					+
Ruminococcus albus	Gram –ve single or paired cocci	β glucans, cellobiose, xylans	+	+					
Ruminococcus flavefaciens	Catalase –ve streptococci	β glucans (xylans)	+	+					+
Principal starch and soluble sugar digesters									
Eubacterium ruminantium	Gram +ve rods	Glucose (xylan, pectin)	+			+		+	
								+	
Lactobacillus spp.	Gram +ve rods	Glucose							
Megasphaera elsdenii	Gram –ve large cocci	Lactate, glucose (glycerol)		+	+	+			
Prevotella amylophilus		Starch, monosaccharides	+	+	+		+		+
Prevotella ruminicola	Gram –ve rods	Glucose and other monosaccharides (xylans, pectin, starch)	+	+			+		+
Ruminobacter amylophilus	Gram –ve rods	Starch (xylose)	+	+					+
Selenomonas lactilytica		Lactate, sugars		+	+				+
Selenomonas ruminantium	Gram –ve rods	Glucose, xylose and other monosaccharides (starch, sucrose)		+	+			+	
Streptococcus bovis	Gram +ve cocci	Starch, glucose (xylans, pectin)						+	
Succinomonas amylolytica	Gram –ve rods	Starch, maltose, fructose							+

[a]Alternative substrates for some strains are shown in brackets.
[b]Fo = formic acid; Ac = acetic acid; Pr = propionic acid; Bu = *n*-butyric acid; iBu = iso-butyric acid; La = lactic acid; Su = succinic acid.
[c]+ve = positive; –ve = negative.

1981). Several yeasts are found in the rumen; *Candida* and *Trichosporon* are the most frequently reported (Ogimoto and Imai, 1981). The role of these organisms in rumen metabolism is unclear, although the yeasts may play an important role in removing oxygen which enters during eating or diffuses through the rumen wall (Newbold *et al.*, 1996). Bacteriophages may lyse cellulolytic bacteria and thus impact on cellulolysis within the rumen (Klieve *et al.*, 2004).

Ruminant Neonates Rapidly Develop a Mixed Ruminal Microbiota

Ruminants are born with a sterile digestive tract (Salmon, 1999), and one of the important proc-esses in achieving adult digestive function is the gaining of an appropriate set of digestive tract microorganisms. Lambs can obtain acetogenic and methanogenic organisms within a day of birth (Morvan *et al.*, 1994; Skillman *et al.*, 2004). Inoculation of newborn herbivores with bacteria and protozoa is usually by direct contact between mother and offspring, although young animals may develop a normal population without direct contact (Eadie, 1962; Dehority, 2003). Protozoa and fungi begin to establish in the developing rumen by about 2 weeks (see data reviewed by Stewart *et al.*, 1988) but the protozoa seem to need the presence of bacteria before they can become properly established.

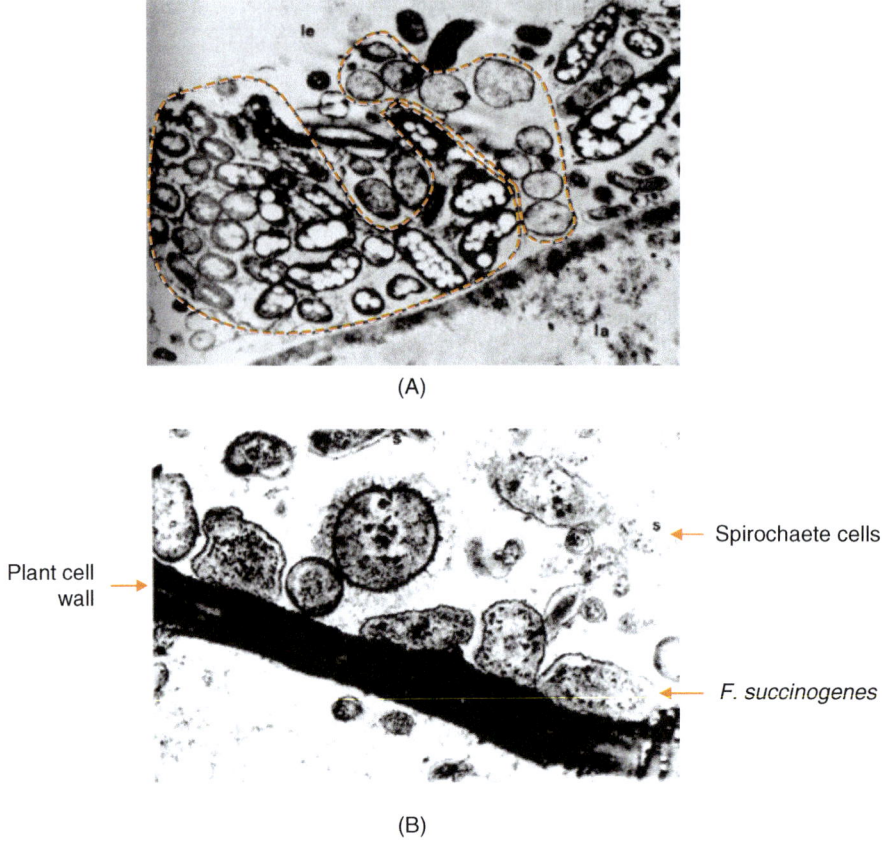

(A)

Plant cell wall →

s ← Spirochaete cells

← *F. succinogenes*

(B)

Fig. 7.5. Examples of bacterial microcolonies (consortia): (A) transmission electron micrograph of a lucerne leaf colonized by mixed rumen bacteria; microcolonies are shown by dotted lines. (B) Transmission electron micrograph of a coastal Bermuda grass leaf incubated with a mixed rumen bacteria; cells of *F. succinogenes* have colonized the plant cell wall fragment, and spirochaete cells (*Borrelia* sp.) can be seen adjacent. (Reprinted from Cheng *et al.*, 1984. Elsevier. With Permission.)

Several Environmental Factors, Other than the Diet, Influence the Rumen Microbial Population

Geographical differences

An example of geographical differences in bacteria is the differences in the ability of ruminants to destroy mimosine, the toxic amino acid of the fodder tree *Leucaena leucocephala*. Bacteria which have this capacity are present in Japanese and Indonesian goats but were not found in Australian ruminants prior to their deliberate introduction to northern Australia (Jones and Lowry, 1984). Certain species of protozoa are present in some countries, but not in others, e.g. *Isotricha* and *Polyplastron* (Clarke, 1964; Hungate *et al.*, 1964; Ogimoto and Imai, 1981).

Seasonal and diurnal changes

Regular changes in the numbers of both bacteria and protozoa in the rumen of grazing sheep are correlated with the seasonal changes in the availability of green pasture (Moir, 1951). The diurnal variation in the numbers of rumen bacteria is related to changes in the amount of ruminal digesta and available carbohydrate. The concentration (number/ml) of rumen bacteria is greatest just before feeding and is reduced by the diluting effects of feed and water consumption and saliva flow (Bryant and Robinson, 1961), and then increased by bacterial growth in response to the increased supply of feed. This diurnal variation may depend on the nature of the diet. Longer periods without food lead to large reductions in the number of rumen bacteria, but numbers rapidly increase when the animal is fed. Diurnal variations in protozoa (number/ml) have been recorded in sheep fed once daily (Purser and Moir, 1959; Petkov, 1976).

Each Microorganism Has Its Preferred Carbohydrate Substrate and Changes in the Diet Induce Changes in the Rumen Microbial Population

Bacteria and archaea

Bacteria and protozoa contribute about equally to the rumen microbial biomass, but because the bacteria are metabolically more diverse they may be responsible for most of the feed digestion in the rumen (Cheng *et al.*, 1991). Bacterial activities against some substrates are summarized in Table 7.3. These data were obtained by incubations in pure cultures, but the metabolism of microorganisms in the rumen is not particularly well modelled by pure-culture methods.

Bacteria are either 'specialists' or 'generalists' in their use of energy substrates (Prins, 1988). For example, *Ruminobacter amylophilus* (a specialist starch-digesting species) and *Streptococcus bovis* predominate when diets are rich in concentrates. *F. succinogenes* (a fibrolytic species) and *M. ruminantium* (which produces methane) predominate when low-quality forages are fed. However, most rumen bacteria are 'generalists', such as *Prevotella ruminicola*, *S. bovis* and *Butyrivibrio fibrisolvens*, which degrade a wide range of polysaccharides, or *Selenomonas ruminantium*, which utilizes sugars and a variety of bacterial end products (Mould *et al.*, 2005).

Starch-rich foods, like cereal grains and oilseed meals, are relatively easily digested and when used in animal rations they encourage the rapid growth of bacterial populations. The principal starch-digesting bacteria are *S. bovis*, *R. amylophilus*, *P. ruminicola*, *B. fibrisolvens*, *Succinimonas amylolytica* and *S. ruminantium* (Cotta, 1988; McAllister and Cheng, 1996; Table 7.3).

Forages have high concentrations of structural polysaccharides and low levels of soluble carbohydrates and protein. Under these conditions, the rate-limiting step in the intake and digestion of DM (by the animal) is the activity of the rumen fibre digesters. These organisms degrade the feed to VFAs used in mammalian energy metabolism and supply the non-cellulolytic organisms with substrates for their growth. In contrast, digestion of soluble carbohydrate, especially starch, is rate-limited by the protozoa which ingest starch granules and hold them unavailable to bacterial digestion.

The physical form of forage may influence the numbers and types of bacteria. Milling of hay causes the loss of cellulolytic organisms and the proliferation of those producing propionate and lactate, and an increase in the number of bacteria present in the rumen liquor of cattle fed the ground material (Thorley *et al.*, 1968). These effects are reflected in reductions in cell wall digestibility and fermentation rate (Udén, 1988) and there may be a change in the proportions of the different VFA (Beever *et al.*, 1981; Hu *et al.*, 2005).

Roughage-rich diets induce a rumen microbial population in which fibre digesters and *Methanobrevibacter* predominate (Wright *et al.*, 2004). Not all roughages have the same effect, though, on the bacterial population. Low-quality roughages (i.e. those with high cell wall contents) are often lignin-rich, deficient in protein and minerals, and not good substrates for bacterial growth. On the other hand, forages such as the legumes (which are rich in pectin), or immature temperate grasses and beet pulp (which have high soluble sugar contents), are associated with large population sizes and species which are more typical of those found when concentrates are fed (Wolstrup *et al.*, 1974). Feeding increased amounts of starch-rich concentrate causes species such as *Succinivibrio* and lactate-using bacteria to proliferate. These effects are illustrated by data from Latham *et al.* (1971; Table 7.4) and Mackie *et al.* (1978; Fig. 7.6). In the latter study, ciliate protozoa increased from 1.9 × 10^5/g to 6.4 × 10^5/g as the diet concentrate content increased from 10% to 71%.

Latham *et al.* (1971) published their results prior to several recent changes in bacterial taxonomy and nomenclature. These included dividing from the *Bacteroides* genus, the genera *Fibrobacter*, *Prevotella* and *Ruminobacter*, and renaming *Peptostreptococcus elsdenii* as *Megasphaera elsdenii*. Consequently, some of the bacteria cannot be identified precisely. Nevertheless, *Butyrivibrio* spp. declined from 38% to 10% of the bacteria identified and *Selenomonas* and *Megasphaera* spp. increased from 5% to 12% and 0% to 8%, respectively.

Protozoa

Almost all of the rumen ciliate protozoa are found associated with food particles in the

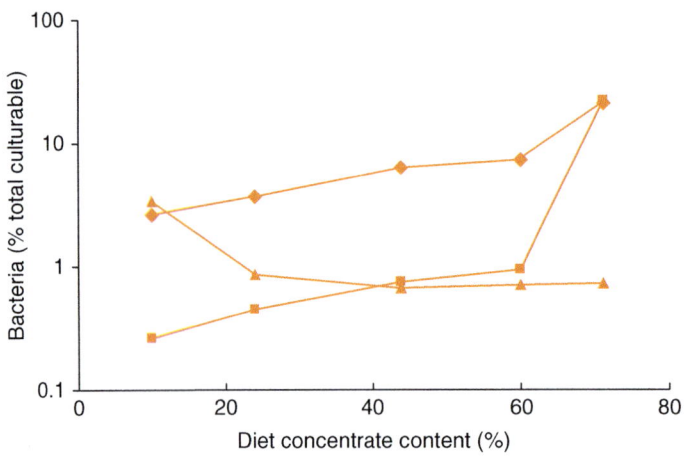

Fig. 7.6. Changes in the proportions of amylolytic (starch-digesting, ♦–♦), lactate-utilizing (■–■) and fibrolytic (cell wall-digesting, ▲–▲) bacteria. (Drawn from data in Mackie *et al.*, 1978.)

Table 7.4. Effect of diet on the types of bacteria (mean $\log_{10}$ counts) in the rumens of cattle fed hay, or hay plus flaked maize. (From Latham *et al.*, 1971.)

Isolated bacterial types	Hay (100%)	Hay (80%) plus maize (20%)
Fibrolytic bacteria	7.17	3.74
Lactobacillus spp.	7.78	8.68
Streptococcus spp.	6.06	7.10
Lactate-utilizing bacteria	7.63	7.91
Tributyrin-hydrolysing bacteria	7.43	7.04

rumen digesta (Orpin, 1984). They rapidly attach to newly eaten feed particles, and if they are small enough, they ingest them. They use starch and cellulose as energy sources, and bacteria and insoluble protein as sources of N (Coleman, 1986; Jouany, 1996).

Diets containing concentrates favour the proliferation of protozoa because they provide a ready source of digestible energy and the protozoa are able to ingest and store starch granules (Mendoza *et al.*, 1993; McAllister and Cheng, 1996). This gives them a competitive advantage over most bacteria. Provided that starch-rich ingredients are introduced slowly, protozoa will proliferate and will be the rate-limiters in the digestion of feed in the rumen (Schwartz and Gilchrist, 1975; Mendoza *et al.*, 1993). On the other hand, if starch-rich ingredients are rapidly introduced to a ruminant's diet the rumen pH can quickly fall because concentrate-rich diets are easily digested and the VFA concentration rises rapidly. Different protozoa species have different pH tolerances, but protozoa are completely eliminated at pH 5.5 or less (Purser and Moir,

1959; Eadie, 1962; Christiansen *et al.*, 1964; Vance *et al.*, 1972).

There is good evidence that rumen protozoa have a negative effect on the protein economy of the host animal. Defaunated sheep and cattle have improved live-weight gains and wool growth (Bird and Leng, 1984; Ivan *et al.*, 1992; Bird *et al.*, 1994; Santra and Karim, 2000). Hristov *et al.* (2001) reported lower rumen pH and ammonia concentrations and higher concentrations of soluble protein in steers with 42% fewer protozoa, although in this study the results were confounded with differences in barley grain intake. However, Santra and Karim (2002) obtained similar results with defaunated lambs fed similar diets. Rumen ammonia levels are reduced and the flow of non-ammonia N into the duodenum is increased when animals are defaunated (Bird *et al.*, 1979; Santra and Karim, 2002). Fibre digestion is consistently reduced in defaunated animals (Williams and Withers, 1991; Santra and Karim, 2000, 2002). These observations are largely supported by a quantitative meta-analysis by Eugène *et al.* (2004).

Fungi

Under normal conditions, the fungi colonize highly lignified fibrous material (Akin and Rigsby, 1987), and penetrate and weaken lignified tissue and cuticle (Bauchop, 1979; Akin *et al.*, 1988). In some circumstances, they are as effective as bacteria in digesting plant DM (Akin *et al.*, 1988). The array of carbohydrate substrates summarized by Theodorou *et al.* (1988) suggests that they are well equipped to digest plant cell walls, although there are substantial differences between species in their ability to degrade difficult-to-digest plant material (Nielsen *et al.*, 2002). *Neocallimastix patriciarum* cannot attack lignin (McSweeney *et al.*, 1994). McAllister and Cheng (1996) noted that at least four fungal species can digest starch. Windham and Akin (1984) commented that the contribution of fungi to the digestion of forages was not then quantified. The situation has not changed noticeably since then, although there have been reports of improved food digestion and nutrient yield when animals were inoculated with *Orpinomyces* (Lee *et al.*, 2000) or *Piromyces* (Paul *et al.*, 2004).

Other Nutrient Requirements of Rumen Microorganisms

The VFA butyrate, iso-butyrate, valerate and iso-valerate are required by many bacteria (Bryant and Robinson, 1962; Andries *et al.*, 1987). These branched-chain VFAs are incorporated intact into bacterial cells, where they are the precursors of long-chain fatty acids or of the branched-chain amino acids valine, leucine and isoleucine.

Many bacteria require ammonia as an essential nutrient, while most can use it for protein synthesis (Bryant and Robinson, 1962; Stewart *et al.*, 1997; Wolin *et al.*, 1997). The principal fibrolytic bacteria utilize ammonia as their main N source, but some may also use amino acids.

The rumen bacteria need several B group vitamins, including biotin, para-amino-benzoic acid, folic acid, vitamin B_6, pantothenic acid and thiamin. The thiamin requirement may be not less than $0.6 \mu g/ml$ of rumen liquor, which is equivalent to about $3 \mu g/day$ for a sheep. Niacin has been widely shown to be required by bacteria and may also be essential for protozoa. However, many bacteria synthesize net amounts of B vitamins, which may be used by the protozoa, and by the animal host (e.g. the net synthesis of thiamin; Breves *et al.*, 1980).

Protozoa and the cellulolytic bacteria may depend on synthesis of vitamin B12 by non-cellulolytic bacteria. Responses of mixed rumen organisms to Co supplementation may be due to a stimulation of protozoal metabolism. The ciliate protozoa require vitamin B12, but this requirement is less if Co is provided in the presence of bacteria.

There Is Extensive Interaction Between Rumen Microorganisms

There are three types of beneficial interaction between the rumen microorganisms. First, in some cases a substrate can be digested by bacteria only after it has been physically or chemically altered by the actions of other bacteria, protozoa or fungi. For example, amylolytic (starch-digesting) bacteria cannot attack the starch in cereal grain endosperm cells until fibrolytic bacteria have digested the β-glucan polysaccharides of the endosperm cell walls, or (in the case of maize grain) until fungi have digested the protein network in which that grain's starch granules are enmeshed (McAllister and Cheng, 1996).

Second, an organism may produce an end product which is used as an energy source by another species. McAllister and Cheng (1996) put it like this: 'Primary colonizing bacteria produce end products which attract secondary colonizers to the site of digestion to form adherent multispecies microbial biofilms.' Thus, bacteria form consortia because of the nutrient-rich environment which some species create for other species. For example, Table 7.3 shows that several bacteria produce succinic acid, but this acid is not found in the rumen liquor because it is immediately used as a substrate by other bacteria.

The third form of symbiosis is where one species synthesizes a non-energy nutrient which is needed by another species. CO_2 or bicarbonate is required by the rumen bacteria, especially the fibrolytic species (such as *F. succinogenes*, *R. flavefaciens* and *B. fibrisolvens*; Dehority, 1971) that require large amounts for the synthesis of succinate, which, for these organisms, is the sink for electrons released during glycolysis. *Selenomonas* spp. use succinate as an energy substrate (Strobel and Russell, 1991; Evans and Martin, 1997) and in turn produce some of the CO_2 required by the fibrolytic bacteria (e.g. Fig. 7.7). Similarly, lactate is a major end product of *Lactobacillus*, and if it accumulates in the rumen it can cause serious effects on both the animal (acidosis) and on other rumen microorganisms (reduced pH causing death of protozoa and very low levels of cellulolysis). Some species of bacteria (e. g. *M. elsdenii*) use lactate as an energy source. This sharing of substrate and end product is facilitated by the arrangement of bacteria in consortia, as we have seen previously. Ciliate protozoa secrete H_2 from hydrogenosomes (an organelle which is equivalent to a mitochondrion, but produces H_2 gas instead

of water), and protozoa and fibrolytic bacteria are H sources for the methanogens. Methanogenic archaea adhere externally to protozoa (Krumholz *et al.*, 1983) and are found within protozoal cells (Finlay *et al.*, 1994).

Not all the interactions between bacteria are benign – there is fierce competition for substrate in the rumen as the microbial population expands

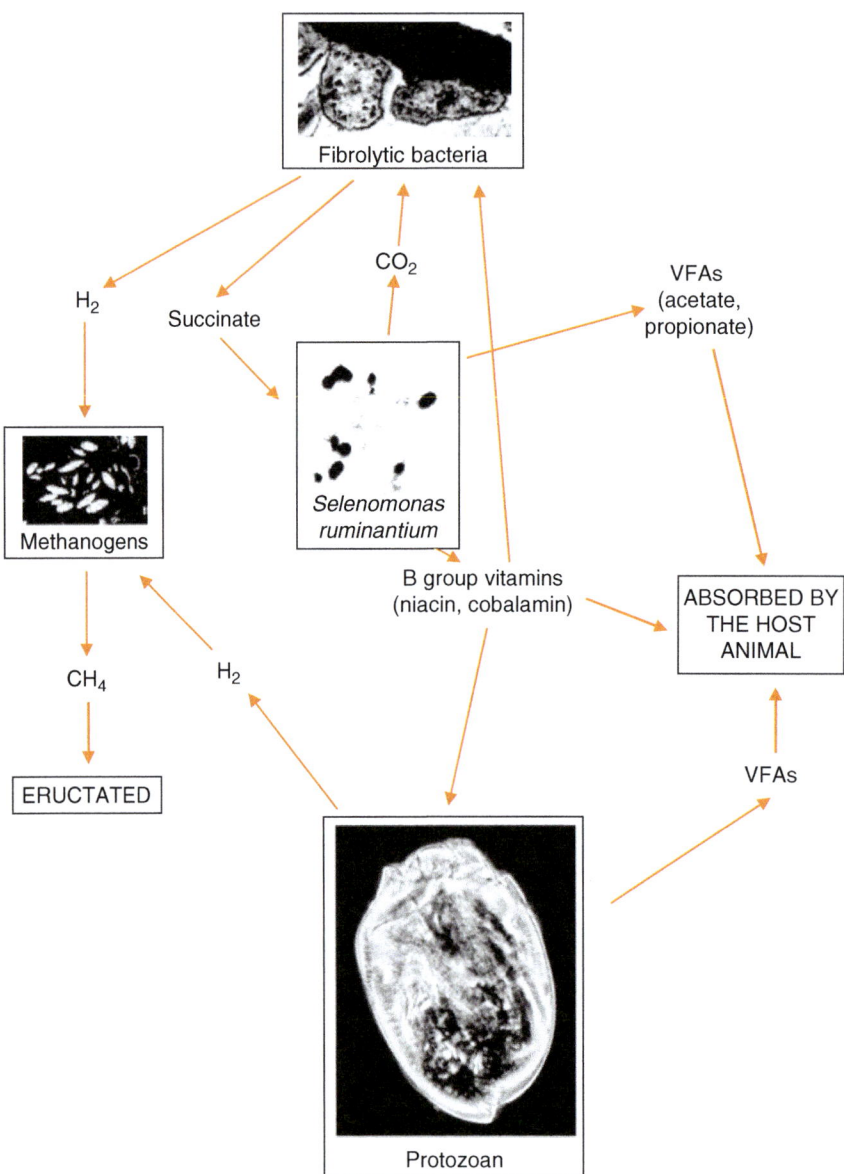

Fig. 7.7. Suggested interrelationships between some bacteria and protozoa. The micrographs are not to the same scale. (Electron micrographs courtesy of R. Al Jassim.)

until it is limited by the available nutrient supply. Anaerobic lactate-producing and fibre-digesting Gram positive bacteria produce bacteriocins (proteins and peptides which have a bacteriocidal action) to remove other bacterial species which would compete for the available substrate (Tagg *et al.*, 1976; Kalmokoff and Teather, 1997; Cookson *et al.*, 2004). Protozoa engulf bacteria and smaller protozoa and these serve as sources of nutrients (Coleman and Sandford, 1979; Coleman and Hall, 1984).

Plant Cell Wall Digestion Is Mainly Accomplished by Three Bacterial Species

R. albus, *R. flavefaciens* and *F. succinogenes* are the most important fibre digesters (i.e. 'fibrolytic' species; Table 7.5). Cellulose and hemicellulose are the principal substrates for *Ruminococcus* although a few strains utilize glucose as a minor energy source. *F. succinogenes* is very actively fibrolytic. It digests the resistant cellulose of matured forages more completely than other species and colonizes straw and other substrates which other fibrolytic organisms are unable to attack. Although *F. succinogenes* digests hemicellulose, it appears to be unable to utilize the digestion product, xylose (Chesson and Forsberg, 1997). *R. flavefaciens* may possess an enzyme system which allows it to ferment the less-accessible carbohydrate in heavily lignified material. *B. fibrisolvens* ferments a wide variety of carbohydrates including hemicellulose, although most strains of this organism are not fibrolytic.

The fibre digesters adhere to the plant cell wall in several different ways. The fibrolytic bacteria adhere intimately to plant cell walls (and *F. succinogenes* forms pits in the cell wall; Cheng *et al.*, 1984) and their cellulases appear to remain membrane-bound

(Chesson and Forsberg, 1997). The fibrolytic bacteria adhere to forage particles using finger-like projections ('fimbriae') from their cell membranes, specific molecules which attach to the substrate, and binding domains which are part of their glycocalyx (a mucopolysaccharide-rich coating on the cell membrane), or are associated with the digestive enzymes. These mechanisms are reviewed by Miron *et al.* (2001).

Cellulose digestion and VFA production are enhanced when protozoa are added to a defaunated rumen. This could be due to either provision of nutrients for the cellulolytic bacteria or cellulolysis by protozoa. Protozoa engulf plant particles, and these are digested within the cell. It was thought that this digestion of fibre by protozoa was by cellulases which originated from ingested bacteria. However, it is now clear that protozoa have their own endogenous cellulases, as earlier suggested by Orpin (1984). Genes for endogenous cellulase expression have been demonstrated in *Eudiplodinium maggii*, *Polyplastron multivesiculatum* and *Epidinium caudatum* (Devillard *et al.*, 1999; Takenaka *et al.*, 1999; Béra-Maillet *et al.*, 2005).

The role of phycomycetous fungi in the digestion of forage, especially the fibrous tissue, has been demonstrated (Bauchop, 1979). Fungi colonize the vascular tissue of grass and legume leaves and stem, and even more lignified material such as wheat straw stem, and sisal. Fungi isolated from the sheep rumen are active against cell wall constituents including acid detergent fibre. The amount of feed which is digested by fungi under normal conditions is not yet known, although they have the enzyme systems needed to degrade cellulose and hemicellulose to monosaccharides. They invade, and possibly physically disrupt, fragments of lignified plant material which are inaccessible to bacteria or protozoa (Bauchop, 1984).

Table 7.5. Polysaccharide substrates of the main cell wall-digesting bacteria of the rumen. (From Cheng *et al.*, 1984.)

Cellulose	Hemicellulose	Pectic substances
Fibrobacter succinogenes	*B. fibrisolvens*	All these species, plus
Ruminococcus flavefaciens	*R. flavefaciens*	*Lachnospira multiparus*
Ruminococcus albus	*R. albus*	*Streptococcus bovis*
Butyrivibrio fibrisolvens	*Prevotella ruminicola*	*Succinivibrio dextrinosolvens*
Cillobacterium cellulosolvens		
Clostridium lockheadii		
Cellulomonas fimi		
Eubacterium spp.		

Cellulose Is Digested by Endoglucanases and Exoglucanases

Orpin (1984) proposed three types of bacterial cellulases. A modification of his scheme is shown in Fig. 7.8. The three enzymes are endo-β1-4 glucanase, which acts at random within the cellulose molecule; exo-β1-4 glucanase, which acts at the ends of the molecule and releases cellobiose units; and β-D-glucoside glucohydrolase, which splits the cellobiose molecule into its constituent glucoses.

Doerner and White (1990) have described the endoglucanase multienzyme complex in *R. flavefaciens*. There are two polymeric units. Endoglucanase A has 13 protein components, while there are 5 in endoglucanase B. Not all of these components are enzymatically active, but they appear to be all needed for the enzyme complex to be most active. This arrangement is also present in endoglucanases from *R. albus* and *F. succinogenes*. For more detailed descriptions of the bacterial cellulosome and cellulose digestion read Gal *et al.* (1997), Schwarz (2001) and Desvaux (2005). Bacterial hemicelluloses probably act in the same way as the cellulases. Čpeljnik *et al.* (2004) have recently isolated a xylanase from *Pseudobutyrivibrio xylanivorans*.

Volatile Fatty Acids Are Produced by the Microbial Metabolism of Pyruvic Acid

Bacterial metabolism of carbohydrate is conventionally divided into primary and secondary fermentation. Primary fermentation involves the metabolism of digestion products to pyruvate and sometimes is considered to include the digestion of polysaccharides to glucose. Primary fermentation is an old term, and it is best to separate (extracellular) digestion and (intracellular) metabolism. Figure 7.9 shows the intracellular metabolism of glucose to pyruvate. Both glycolysis and the pentose phosphate pathways of glucose oxidation are used by rumen bacteria. Secondary fermentation is the metabolism of pyruvate as an energy source with the ultimate production of VFAs, typically acetic, propionic and butyric acids, as waste products (Fig. 7.10). Formic acid is also produced but is used as a substrate for methane production by the methanogens. Long-chain VFAs (e.g. valeric acid) and some branched-chain VFAs (e.g. iso-butyric and iso-valeric acids) are also synthesized.

Why Volatile Fatty Acids, and How Do Bacteria Get Their Energy?

Bacteria ferment the animal's food to generate energy. The energy yields in the rumen are described in detail by Thauer *et al.* (1977). In an anaerobic environment, most ATP generation is likely to be by substrate phosphorylation, but oxidative phosphorylation is also used by rumen bacteria (Demeyer, 1981; Erfle *et al.*, 1984; Hino and Russell, 1985) and H is produced in large amounts in the rumen. However, it does not accumulate; instead it is sequestered in 'H sinks'. Methane is one of these sinks, produced by methanogens such as *M. ruminantium*. About 1 mole of ATP is generated for each

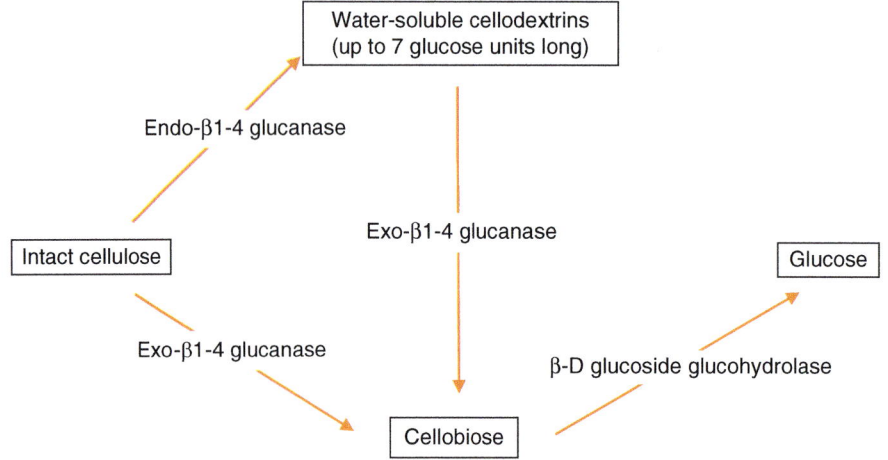

Fig. 7.8. A model of the action of bacterial cellulases.

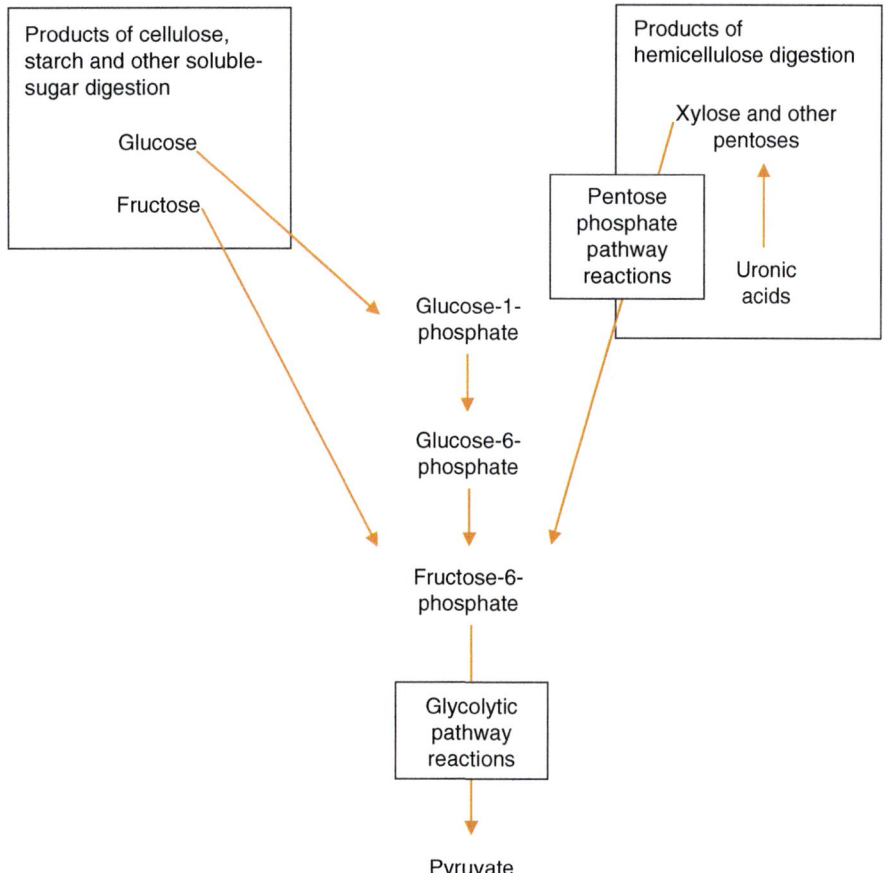

Fig. 7.9. 'Primary fermentation' of carbohydrates by rumen bacteria. (From Cotta and Hespell, 1984; Wallace and McKain, 1991; Wallace, 1996.)

mole of methane produced (Demeyer, 1981). Other H users are the acetogenic bacteria, which incorporate four atoms of H and one molecule of CO_2 into a molecule of acetate.

One of the more interesting, and practically important, questions of rumen microbial metabolism is to understand why different feeding systems result in different rumen VFA profiles. There are two reasons. First, the bacteria which ferment starch and soluble sugars tend to produce a wider range of VFA end products than the fibrolytic species (Table 7.3), so starch-rich (or concentrate-rich) diets are associated with a more propionate-rich VFA profile. Second, VFA production is influenced by the amount of H in the rumen. If the H partial pressure in the rumen increases (e.g. if we feed antibiotics which target methanogens) there is an accumulation of reduced NAD and an accompany-

ing shift in ATP synthesis from oxidative to substrate phosphorylation (Demeyer, 1981). In this situation, some species (e.g. *S. ruminantium* and *R. flavefaciens*; Wolin *et al.*, 1997) move their H into succinate or the longer-chain VFA, and this leads to increased propionate and butyrate production.

Protein Digestion by Bacteria and Protozoa

Bacterial protein is the main type of microbial protein flowing into the small intestine. Faichney *et al.* (1997) reported that in sheep fed forages, fungi and protozoa, respectively, contributed only 0.7–2.7%, and 4–15% of the microbial protein flowing into the duodenum.

A variety of bacterial species digest protein, acting at one of the three stages in the process of digestion and metabolism (Fig. 7.11). *P. ruminicola* occupies a

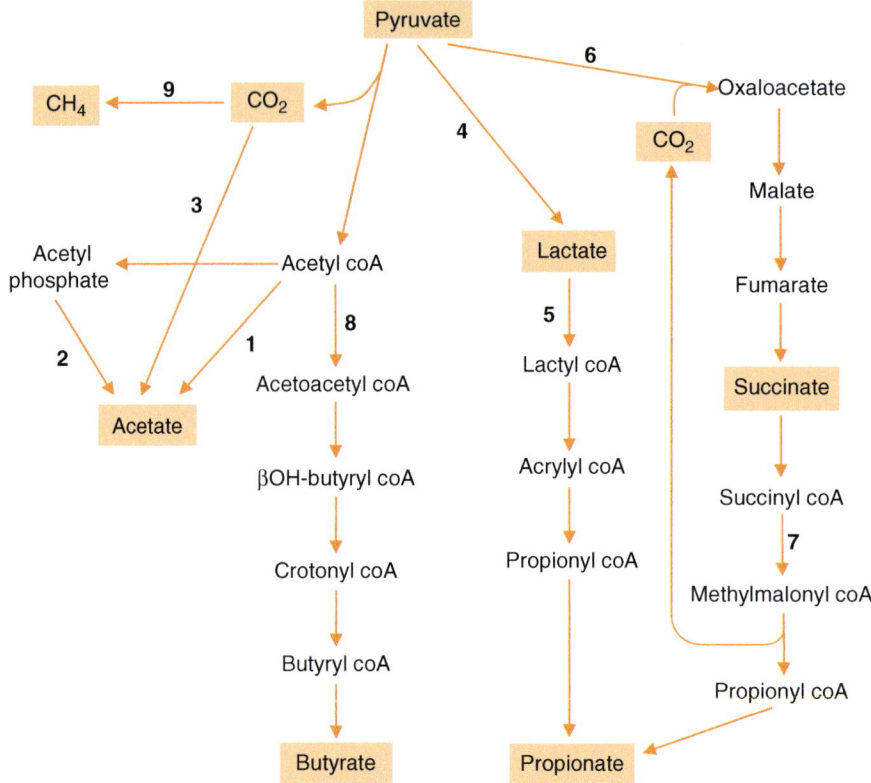

1. Direct synthesis of acetate (e.g. *Ruminococcus flavefaciens* and *Ruminococcus albus* in the presence of methanogens);
2. Acetate phosphate pathway (e.g. *Streptococcus* spp. in situations of extreme carbohydrate shortage);
3. Direct synthesis of acetate from CO_2 (e.g. by the acetogen *Acetitomaculum ruminis*);
4. Lactate synthesis (e.g. *Streptococcus bovis* and *Lactobacillus* spp.);
5. Direct synthesis of propionate (e.g. *Megasphaera elsdenii*);
6. Succinate synthesis (e.g. *Fibrobacter succinogenes*);
7. Propionate synthesis from succinate (e.g. *Selenomonas ruminantium*);
8. Butyrate synthesis (e.g. *Butyrivibrio fibrisolvens*);
9. Methanogenesis (methanogens).

Fig. 7.10. 'Secondary fermentation' of carbohydrates by rumen bacteria. (From Hungate, 1966; Leng, 1970; Harfoot, 1978; Wolin *et al.*, 1997; Konings *et al.*, 1997; Joblin, 1999.)

central position in protein digestion (Wallace and McKain, 1991; Wallace *et al.*, 1995) in part because it has peptidase enzymes which are active against a wide range of substrates (Wallace *et al.*, 1997).

The bacterial proteases are mainly cell-bound (located in the cell membrane or in the glycocalyx surrounding it), while the protozoal enzymes are intracellular (Kopecny and Wallace, 1982; Jouany, 1996). The pH optima for these enzymes range from 6 to 7. The action of many of these proteases is like that of the pancreatic proteases (Cotta and

Hespell, 1984). There is evidence that the proteases are constitutive (i.e. their activity does not change in response to changes in substrate concentration).

Deaminases are probably inducible enzymes, with an optimum pH of about 6.9. The reaction products are VFAs, CO_2 and NH_3, suggesting that the Stickland reaction (Fig. 7.12) is important, as we would expect it to be in an anaerobic environment. In this reaction, two amino acids participate in a reaction in which one acts as an H donor and the other as an H acceptor, e.g. if our amino

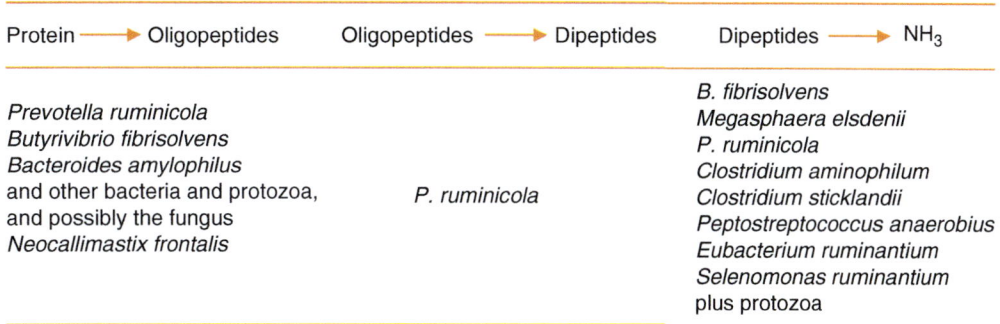

Protein → Oligopeptides	Oligopeptides → Dipeptides	Dipeptides → NH₃
Prevotella ruminicola *Butyrivibrio fibrisolvens* *Bacteroides amylophilus* and other bacteria and protozoa, and possibly the fungus *Neocallimastix frontalis*	*P. ruminicola*	*B. fibrisolvens* *Megasphaera elsdenii* *P. ruminicola* *Clostridium aminophilum* *Clostridium sticklandii* *Peptostreptococcus anaerobius* *Eubacterium ruminantium* *Selenomonas ruminantium* plus protozoa

Fig. 7.11. Organisms involved in protein digestion in the rumen. (From Cotta and Hespell, 1984; Wallace and McKain, 1991; Wallace, 1996.)

acids are proline and alanine, the reaction will yield δ-aminovaleric acid and acetic acid.

It was recognized by Russell *et al.* (1991) that the amount of ammonia produced in the rumen was greater than could be accounted for by protein digestion and metabolism by the main rumen bacterial species. Russell *et al.* (1988), Attwood *et al.* (1998) and Eschenlauer *et al.* (2002) discovered 'hyper-ammonia-producing' bacteria which use peptides and amino acids as their energy source and which appear to be responsible for most of the ammonia produced in the rumen. They are Gram positive organisms which are sensitive to monensin.

Dietary Fatty Acids Can Be Greatly Changed by Rumen Metabolism

It is well recognized that the chemistry of the fatty acids in dietary lipids can be markedly altered in the rumen. Bacteria hydrolyse triacylglycerols, ferment the glycerol and use unsaturated fatty acids as H sinks, so these are saturated (i.e. all the available C bonds are filled with H atoms) as they pass through the rumen (Doreau and Ferlay, 1994; Fig. 7.13A). Saturation can reach 95–100% for linolenic acid and 90% for linoleic acid (Tamminga and Doreau, 1991). Consequently, ruminants have a more saturated body fat than other herbivores

when these fatty acids are incorporated into the animal's body tissue.

Most plant fatty acids exist in the *cis* form (Fig. 7.13B). Rumen bacteria transform some of these into the *trans* form. This appears to give the bacteria a means of regulating the permeability of their cell membranes, especially during resting (i.e. non-growing) phases (Keweloh and Heipieper, 1996; Trevors, 2003).

Some Final Comments on Ruminant Digestive Function

The ruminant animal is an ecosystem in which there are two complementary parts: the animal and the rumen microbiota. This may seem quite obvious, but we must explicitly recognize this because to get optimum performance from the animal/microbial ecosystem we have to provide for the nutritional needs of both parts, as well as the effects of rumen bacterial metabolism on animal performance and product quality. Bacteria are the important part of the rumen microbiota as far as digestion and metabolism of food is concerned, although the protozoa have a beneficial role in fibre digestion. When we design feeding systems for ruminants, we must provide for the bacterial requirements for energy, ammonia, phosphorus, sulfur and CO_2, among other nutrients, and design diets which will maintain an acceptable rumen pH.

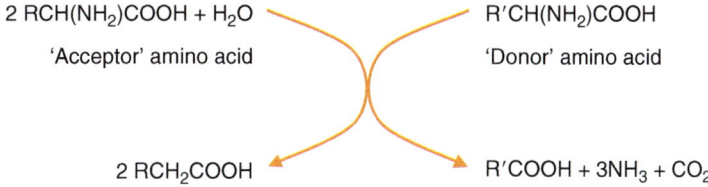

$$2\ RCH(NH_2)COOH + H_2O \qquad\qquad R'CH(NH_2)COOH$$

'Acceptor' amino acid   'Donor' amino acid

$$2\ RCH_2COOH \qquad\qquad\qquad R'COOH + 3NH_3 + CO_2$$

Fig. 7.12. The Stickland reaction for deaminating amino acids in the rumen.

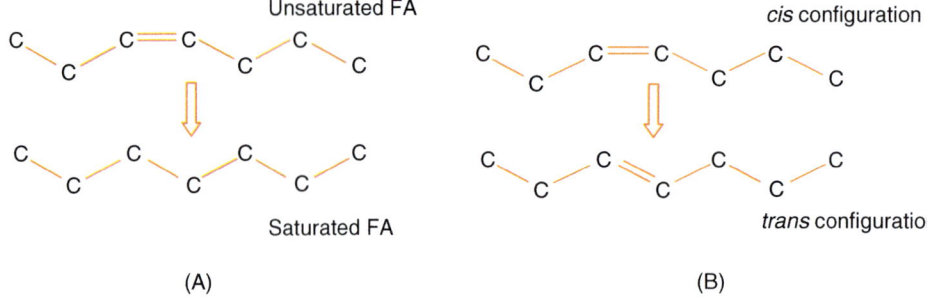

Fig. 7.13. Changes in the chemistry of dietary fatty acids as a result of rumen metabolism: (A) saturated and unsaturated forms of fatty acids; (B) cis and trans isomeric forms of unsaturated fatty acids.

References

Agricultural Research Council (ARC) (1980) *Nutrient Requirements of Ruminant Livestock.* CAB, Farnham Royal, UK.

Akin, D.E. and Rigsby, L.L. (1987) Mixed fungal populations and lignocellulosic tissue degradation in the bovine rumen. *Applied and Environmental Microbiology* 53, 1987–1995.

Akin, D.E., Borneman, W.S. and Windham, W.R. (1988) Rumen fungi: morphological types from Georgia cattle and the attack on forage cell walls. *BioSystems* 21, 385–391.

Al Jassim, R.A.M., Scott, P.T., Trebbin, A.L., Trott, D. and Pollitt, C.C. (2005) The genetic diversity of lactic acid producing bacteria in the equine gastrointestinal tract. *FEMS Microbiology Letters* 248, 75–81.

Altmann, P.L. and Dittmer, D.S. (1968) *Metabolism.* Federation of American Societies for Experimental Biology, Bethesda, Maryland.

Andries, J.I., Buysse, F.X., de Brabander, D.L. and Cottyn, B.G. (1987) Isoacids in ruminant nutrition: their role in ruminal and intermediary metabolism and possible influences on performances – a review. *Animal Feed Science and Technology* 18, 169–180.

Attwood, G.T., Klieve, A.V., Ouwerkerk, D. and Patel, B.K.C. (1998) Ammonia-hyperproducing bacteria from New Zealand ruminants. *Applied and Environmental Microbiology* 64, 1796–1804.

Bauchop, T. (1979) Rumen anaerobic fungi of sheep and cattle. *Applied and Environmental Microbiology* 38, 148–158.

Bauchop, T. (1984) Rumen anaerobic fungi and the utilisation of fibrous feeds. *Reviews in Rural Science* 6, 118–123.

Beal, A.M. (1998) Enzyme activity in parotid and mandibular saliva from red kangaroos, *Macropus rufus. Archives of Oral Biology* 43, 695–699.

Beever, D.E., Osbourn, D.F., Cammell, S.B. and Terry, R.A. (1981) The effect of grinding and pelleting on the digestion of Italian ryegrass and timothy by sheep. *British Journal of Nutrition* 46, 357–370.

Belenguer, A., Balcells, J., Guada, J.A., Decoux, M. and Milne, E. (2005) Protein recycling in growing rabbits: contribution of microbial lysine to amino acid metabolism. *British Journal of Nutrition* 94, 763–770.

Béra-Maillet, C., Estelle Devillard, E., Cezette, M., Jouany, J.-P. and Forano, E. (2005) Xylanases and carboxymethylcellulases of the rumen protozoa *Polyplastron multivesiculatum, Eudiplodinium maggii* and *Entodinium* sp. *FEMS Microbiology Letters* 244, 149–156.

Bird, S.H. and Leng, R.A. (1984) Further studies on the effects of the presence or absence of protozoa in the rumen on live-weight gain and wool growth of sheep. *British Journal of Nutrition* 52, 607–611.

Bird, S.H., Hill, M.K. and Leng, R.A. (1979) The effects of defaunation of the rumen on the growth of lambs on low-protein–high-energy diets. *British Journal of Nutrition* 42, 81–87.

Bird, S.H., Romulo, B. and Leng, R.A. (1994) Effects of lucerne supplementation and defaunation on feed intake, digestibility, N retention and productivity of sheep fed straw based diets. *Animal Feed Science and Technology* 45, 119–129.

Breves, G., Hoeller, H., Harmeyer, J. and Martens, H. (1980) Thiamin balance in the gastrointestinal tract of sheep. *Journal of Animal Science* 51, 1177–1181.

Bryant, M.P. and Robinson, I.M. (1961) An improved nonselective culture medium for ruminal bacteria and its use in determining diurnal variation in numbers of bacteria in the rumen. *Journal of Dairy Science* 44, 1446–1456.

Bryant, M.P. and Robinson, I.M. (1962) Some nutritional characteristics of predominant culturable ruminal bacteria. *Journal of Bacteriology* 84, 605–614.

Bugaut, M. (1987) Occurrence, absorption and metabolism of short chain fatty acids in the digestive tract of mammals. *Comparative Biochemistry and Physiology* 86B, 439–472.

Carey, M.C., Small, D.M. and Bliss, C.M. (1983) Lipid digestion and absorption. *Annual Review of Physiology* 45, 651–677.

Čepeljnik, T., Križaj, I. and Marinšek-Logar, R. (2004) Isolation and characterization of the *Pseudobutyrivibrio xylanivorans* Mz5^T xylanase XynT – the first family 11 endoxylanase from rumen *Butyrivibrio*-related bacteria. *Enzyme and Microbial Technology* 34, 219–227.

Cheng, K.-J. and Costerton, J.W. (1984) Benefitting ruminants by manipulation of bacteria that degrade fibrous feeds and adhere to digestive tract surfaces. Paper presented to *IAEA Research Co-ordination Meeting*, Vienna.

Cheng, K.-J., Stewart, C.S., Dinsdale, D. and Costerton, J.W. (1984) Electron microscopy of bacteria involved in the digestion of plant cell walls. *Animal Feed Science and Technology* 10, 93–120.

Cheng, K.-J., Forsberg, C.W., Minato, H. and Costerton, J.W. (1991) Microbial ecology and physiology of feed degradation within the rumen. In: Tsuda, T., Sasaki, Y. and Kawashima, R. (eds) *Physiological Aspects of Digestion and Metabolism in Ruminants*. Academic Press, Toronto, Ontario, Canada, pp. 595–624.

Chesson, A. and Forsberg, C.W. (1997) Polysaccharide degradation by rumen microorganisms. In: Hobson, P.N. and Stewart, C.S. (eds) *The Rumen Microbial Ecosystem*. Blackie Academic and Professional, London, pp. 329–381.

Chilcott, M.J. and Hume, I.D. (1985) Coprophagy and selective retention of fluid digesta: their role in the nutrition of the common ringtail possum, *Pseudocheirus peregrinus*. *Australian Journal of Zoology* 33, 1–15.

Christiansen, W.C., Woods, W. and Burroughes, W. (1964) Ration characteristics influencing rumen protozoal populations. *Journal of Animal Science* 23, 984–988.

Clarke, R.T.J. (1964) Ciliates of the rumen of domesticated cattle (*Bos taurus* L.). *New Zealand Journal of Agricultural Research* 7, 248–257.

Coleman, G.S. (1986) The metabolism of rumen ciliate protozoa. *FEMS Microbiology Reviews* 39, 321–344.

Coleman, G.S. and Hall, F.J. (1984) The uptake and utilization of *Entodinium caudatum*, bacteria, free amino acids and glucose by the rumen ciliate *Entodinium bursa*. *Journal of Applied Bacteriology* 56, 283–294.

Coleman, G.S. and Sandford, D.C. (1979) The engulfment and digestion of mixed rumen bacteria and individual bacterial species by single and mixed species of rumen ciliate protozoa grown *in vivo*. *Journal of Agricultural Science, Cambridge* 92, 729–742.

Collatz, K.G. and Hoeger, U. (1980) Age-related changes in the body composition of mated and unmated blowflies *Phormia terrae novae*. *Experimental Gerontology* 15, 433–441.

Cookson, A.L., Noel, S.J., Kelly, W.J. and Attwood, G.T. (2004) The use of PCR for the identification and characterisation of bacteriocin genes from bacterial strains isolated from rumen or caecal contents of cattle and sheep. *FEMS Microbiology Ecology* 48, 199–207.

Cork, S.J. and Warner, A.C.I. (1983) The passage of digesta markers through the gut of a folivorous marsupial the koala (*Phascolarctos cinereus*). *Journal of Comparative Physiology, B* 152, 43–52.

Corring, T. (1982) Enzyme digestion in the proximal digestive tract of the pig: a review. *Livestock Production Science* 9, 581–590.

Cosgrove, M. (1998) Nucleotides. *Nutrition* 14, 748–751.

Cotta, M.A. (1988) Amylolytic activity of selected species of rumen bacteria. *Applied and Environmental Microbiology* 54, 772–776.

Cotta, M.A. and Hespell, R.B. (1984) Protein and amino acid metabolism of rumen bacteria. In: Milligan, L.P. *et al.* (eds) *Control of Digestion and Metabolism in Ruminants*. Reston, Englewood Cliffs, New Jersey, pp. 122–136.

Cranford, J.A. and Johnson, E.O. (1989) Effects of coprophagy and diet quality on two microtine rodents (*Microtus pennsylvanicus* and *Microtus pinetorum*). *Journal of Mammalogy* 70, 494–502.

Crawford, A.C., Kricker, J.A., Anderson, A.J., Richardson, N.R. and Mather, P.B. (2004) Structure and function of a cellulase gene in redclaw crayfish. *Cherax quadricarinatus*. *Gene* 340, 267–274.

Daly, K., Stewart, C.S., Flint, H.J. and Shirazi-Beechey, S.P. (2001) Bacterial diversity within the equine large intestine as revealed by molecular analysis of cloned 16S rRNA genes. *FEMS Microbiology Ecology* 38, 141–151.

Dehority, B.A. (1971) Carbon dioxide requirement of various species of rumen bacteria. *Journal of Bacteriology* 105, 70–76.

Dehority, B.A. (2003) *Rumen Microbiology*. Nottingham University Press, Nottingham, UK.

Demeyer, D.I. (1981) Rumen microbes and digestion of plant cell walls. *Agriculture and Environment* 6, 295–337.

Dennis, E.A. (1994) Diversity of group types, regulation, and function of phospholipase A2. *Journal of Biological Chemistry* 269, 13057–13060.

Desvaux, M. (2005) The cellulosome of *Clostridium cellulolyticum*. *Enzyme and Microbial Technology* 37, 373–385.

Devillard, E., Newbold, C.J., Scott, K.P., Forano, E., Wallace, R.J., Jouany, J.-P. and Flint, H.J. (1999) A xylanase produced by the rumen anaerobic protozoan *Polyplastron multivesiculatum* shows close sequence similarity to family 11 xylanases from Gram-positive bacteria. *FEMS Microbiology Letters* 181, 145–152.

Dobson, D.E., Prager, E.M. and Wilson, A.C. (1984) Stomach lysozomes of ruminants. I. Distribution and catalytic properties. *Journal of Biological Chemistry* 259, 11607–11616.

Doerner, K.C. and White, B.A. (1990) Assessment of the endo-1,4-beta-glucanase components of *Ruminococcus* flavefaciens FD-1. *Applied and Environmental Microbiology* 56, 1844–1850.

Dolezal, H.G., Tatum, J.D. and Williams, F.L. Jr (1993) Effects of feeder cattle frame size, muscle thickness, and age class on days fed, weight, and carcass composition. *Journal of Animal Science* 71, 2975–2985.

Domínguez-Bello, M.G., Pacheco, M.A., Ruiz, M.C., Michelangeli, F., Leippe, M. and de Pedro, M.A. (2004) Resistance of rumen bacteria murein to bovine gastric lysozyme. *BMC Ecology* 4, 7. Available at: http://www.biomedcentral.com/1472-6785/4/7

Doreau, M. and Ferlay, A. (1994) Digestion and utilisation of fatty acids by ruminants. *Animal Feed Science and Technology* 45, 379–396.

Dryden, G.McL., Stafford, K.J., Waghorn, G.C. and Barry, T.N. (1995) Comminution of roughages by red deer (*Cervus elaphus*) during the prehension of feed. *Journal of Agricultural Science, Cambridge* 125, 407–414.

Eadie, J.M. (1962) The development of rumen microbial populations in lambs and calves under various conditions of management. *Journal of General Microbiology* 29, 563–578.

Embleton, J.K. and Pouton, C.W. (1997) Structure and function of gastro-intestinal lipases. *Advanced Drug Delivery Reviews* 25, 15–32.

Emmanuel, B., Lawlor, M.J. and McAleese, D. (1969) The rumen buffering system of sheep fed pelleted roughage-concentrate rations. *British Journal of Nutrition* 23, 805–811.

Emmanuel, B., Lawlor, M.J. and McAleese, D. (1970) The effect of phosphate and carbonate bicarbonate supplements on the rumen buffering systems of sheep. *British Journal of Nutrition* 24, 653–660.

Erfle, J.D., Sauer, F.D. and Mahadevan, S. (1984) Energy metabolism in rumen microbes. In: Milligan, L.P. *et al.* (eds) *Control of Digestion and Metabolism in Ruminants*. Reston, Englewood Cliffs, New Jersey, pp. 81–99.

Eschenlauer, S.C.P., McKain, N., Walker, N.D., McEwan, N.R., Newbold, C.J. and Wallace, R.J. (2002) Ammonia production by ruminal microorganisms and enumeration, isolation, and characterization of bacteria capable of growth on peptides and amino acids from the sheep rumen. *Applied and Environmental Microbiology* 68, 4925–4931.

Eugène, M., Archimède, H. and Sauvant, D. (2004) Quantitative meta-analysis on the effects of defaunation of the rumen on growth, intake and digestion in ruminants. *Livestock Production Science* 85, 81–97.

Evans, J.D. and Martin, S.A. (1997) Factors affecting lactate and malate utilization by *Selenomonas ruminantium*. *Applied and Environmental Microbiology* 63, 4853–4858.

Faichney, G.J., Poncet, C., Lassalas, B., Jouany, J.P., Millet, L., Doré, J. and Brownlee, A.G. (1997) Effect of concentrates in a hay diet on the contribution of anaerobic fungi, protozoa and bacteria to nitrogen in rumen and duodenal digesta in sheep. *Animal Feed Science and Technology* 64, 193–213.

Finlay, B.J., Esteban, G., Clarke, K.J., Williams, A.G., Embley, T.M. and Hirt, R.P. (1994) Some rumen ciliates have endosymbiotic methanogens. *FEMS Microbiology Letters* 117, 157–162.

Fonty, G. and Gouet, Ph. (1989) Fibre-degrading microorganisms in the monogastric digestive tract. *Animal Feed Science and Technology* 23, 91–107.

Gal, L., Pagés, S., Gaudin, C., Belaich, A., Reverbel-Leroy, C., Tardif, C. and Belaich, J.P. (1997) Characterization of the cellulolytic complex (cellulosome) produced by *Clostridium cellulolyticum*. *Applied and Environmental Microbiology* 63, 903–909.

Gray, G.M. (1992) Starch digestion and absorption in nonruminants. *Journal of Nutrition* 122, 172–177.

Harfoot, C.G. (1978) Anatomy, physiology and microbiology of the ruminant digestive tract. *Progress in the Chemistry of Fats and Other Lipids* 17, 1–19.

Harmon, D.L. (1993) Nutritional regulation of postruminal digestive enzymes in ruminants. *Journal of Dairy Science* 76, 2102–2111.

Harmon, D.L., Yamka, R.M. and Elam, N.A. (2004) Factors affecting intestinal starch digestion in ruminants: a review. *Canadian Journal of Animal Science* 84, 309–318.

Hayes, J.R., Pence, D.H., Scheinbach, S., D'Amelia, R.P., Klemann, L.P., Wilson, N.H. and Finley, J.W. (1994) Review of triacylglycerol digestion, absorption, and metabolism with respect to SALTRIM triacylglycerols. *Journal of Agricultural and Food Chemistry* 42, 474–483.

Hino, T. and Russell, J.B. (1985) Effect of reducing-equivalent disposal and NADH/NAD on deamination of amino acids by intact rumen microorganisms and their cell extracts. *Applied and Environmental Microbiology* 50, 1368–1374.

Hinsberger, A. and Sandhu, B.K. (2004) Digestion and absorption. *Current Paediatrics* 14, 605–611.

Hobson, P.N. and Stewart, C.S. (eds) (1997) *The Rumen Microbial Ecosystem*, 2 edn. Blackie Academic and Professional, London.

Hoffman, R.R. (1985) Digestive physiology of the deer: their morphological specialisations and adaptations. *Royal Society of New Zealand Bulletin* 22, 393–407.

Hornicke, H. (1981) Utilization of caecal digesta by caecotrophy (soft faeces ingestion) in the rabbit. *Livestock Production Science* 8, 361–366.

Hristov, A.N., Ivan, M., Rode, L.M. and McAllister, T.A. (2001) Fermentation characteristics and ruminal ciliate protozoal populations in cattle fed medium- or high-concentrate barley-based diets. *Journal of Animal Science* 79, 515–524.

Hu, Z.-H., Yu, H.-Q. and Zhu, R.-F. (2005) Influence of particle size and pH on anaerobic degradation of cellulose by ruminal microbes. *International Biodeterioration and Biodegradation* 55, 233–238.

Hudson, R. (undated) *Back to Basics: Deer Taxonomy*. Available at: http://www.deer.rr.ualberta.ca/library/deertaxonomy/deerbasics.htm

Hungate, R.E. (1966) *The Rumen and Its Microbes*. Academic Press, New York.

Hungate, R.E. (1984) Microbes of nutritional importance in the alimentary tract. *Proceedings of the Nutrition Society* 43, 1–11.

Hungate, R.E., Bryant, M.P. and Mah, R.A. (1964) The rumen bacteria and protozoa. *Annual Review of Microbiology* 18, 131–166.

Ivan, M., de Dayrell, M., Mahadevan, S. and Hidiroglou, M. (1992) Effect of bentonite on wool growth and nitrogen metabolism in fauna free and faunated sheep. *Journal of Animal Science* 70, 3194–3202.

Jamry, W., Sasser, T. and Kumar, B.V. (1995) Purification and identification of two distinct isoforms of rabbit pancreatic cholesterol esterase. *International Journal of Biochemistry and Cell Biology* 27, 415–423.

Janis, C.M. and Erhardt, D. (1988) Correlation of relative muzzle width and relative incisor width with dietary preference in ungulates. *Zoological Journal of the Linnean Society* 92, 267–284.

Joblin, K.N. (1999) Ruminal acetogens and their potential to lower ruminant methane emissions. *Australian Journal of Agricultural Research* 50, 1307–1313.

Johansen, H.N., Knudsen, K.N.B., Wood, P.J. and Fulcher, R.G. (1997) Physico-chemical properties and the degradation of oat bran polysaccharides in the gut of pigs. *Journal of the Science of Food and Agriculture* 73, 81–92.

Jones, R.J. and Lowry, J.B. (1984) Australian goats detoxify the goitrogen 3-hydroxy-4(1H) pyridine (DHP) after rumen infusion from an Indonesian goat. *Experientia* 40, 1435–1436.

Jouany, J.-P. (1996) Effect of rumen protozoa on nitrogen utilization by ruminants. *Journal of Nutrition* 126(4S), 1335S–1346S.

Kalmokoff, M.L. and Teather, R.M. (1997) Isolation and characterization of a bacteriocin Butyrivibriocin AR10 from the ruminal anaerobe *Butyrivibrio fibrisolvens* AR10: evidence in support of wide spread occurrence of bacteriocin-like activity among ruminal isolates of *B. fibrisolvens*. *Applied and Environmental Microbiology* 63, 394–402.

Kempster, A.J., Cook, G.L. and Grantley-Smith, M. (1986) National estimates of the body composition of British cattle, sheep and pigs with special reference to trends in fatness. A review. *Meat Science* 17, 107–138.

Kertin, A., Bardos, L., Deli, J. and Olah, P. (2005) Relationship of retinoid and carotenoid metabolism with caecotrophy in rabbits. *Acta Veterinaria Hungarica* 53, 309–318.

Keweloh, H. and Heipieper, H.J. (1996) Trans unsaturated fatty acids in bacteria. *Lipids* 31, 129–137.

Klieve, A.V. and Swain, R.A. (1993) Estimation of ruminal bacteriophage numbers by pulsed-field gel electrophoresis and laser densitometry. *Applied and Environmental Microbiology* 59, 2299–2303.

Klieve, A.V., Bain, P.A., Yokoyama, M.T., Ouwerkerk, D., Forster, R.J. and Turner, A.F. (2004) Bacteriophages that infect the cellulolytic ruminal bacterium *Ruminococcus albus* AR67. *Letters in Applied Microbiology* 38, 333–338.

Knudsen, K.E., Jensen, B.B. and Hansen, I. (1993) Digestion of polysaccharides and other major components in the small and large intestine of pigs fed on diets consisting of oat fractions rich in beta-D-glucan. *British Journal of Nutrition* 70, 537–556.

Konings, W.N., Lolkema, J.S., Bolhuis, H., van Veen, H.W., Poolman, B. and Driessen, A.J.M. (1997) The role of transport processes in survival of lactic acid bacteria. Energy transduction and multidrug resistance. *Antonie van Leeuwenhoek* 71, 117–128.

Kopecny, J. and Wallace, R.J. (1982) Cellular location and some properties of proteolytic enzymes of rumen bacteria. *Applied and Environmental Microbiology* 43, 1026–1033.

Kreikemeier, K.K., Harmon, D.L., Peters, J.P., Gross, K.L., Armendark, C.K. and Krehbiel, C.R. (1990) Influence of dietary forage and feed intake on carbohydrase activities and small intestinal morphology of calves. *Journal of Animal Science* 68, 2916–2929.

Krumholz, L.R., Forsberg, C.W. and Veira, D.M. (1983) Association of methanogenic bacteria with rumen protozoa. *Canadian Journal of Microbiology* 29, 676–680.

Latham, M.J., Sharpe, M.E. and Sutton, J.D. (1971) The microflora of the rumen of cows fed hay and high cereal rations and its relationship to the rumen fermentation. *Journal of Applied Bacteriology* 34, 425–434.

Lee, S.S., Ha, J.K. and Cheng, K.-J. (2000) Influence of an anaerobic fungal culture administration on *in vivo* ruminal fermentation and nutrient digestion. *Animal Feed Science and Technology* 88, 201–217.

Leng, R.A. (1970) Formation and production of volatile fatty acids in the rumen. In: Phillipson, A.T. *et al.* (eds) *Physiology of Digestion and Metabolism in the Ruminant*. Oriel Press, Newcastle upon Tyne, UK, pp. 406–421.

Mackie, R.I., Gilchrist, F.M.C., Robberts, A., Hannah, P.E. and Schwartz, H.M. (1978) Microbiological and chemical changes in the rumen during the stepwise adaptation of sheep to high concentrate diets. *Journal of Agricultural Science, Cambridge* 90, 241–254.

Maroux, S., Coudrier, E., Feracci, H., Gorvel, J.-P. and Louvard, D. (1988) Molecular organisation of the intestinal brush border. *Biochimie* 70, 1297–1306.

McAllister, T.A. and Cheng, K.-J. (1996) Microbial strategies in the ruminal digestion of cereal grains. *Animal Feed Science and Technology* 62, 29–36.

McCowan, R.P., Cheng, K.-J. and Costerton, J.W. (1980) Adherent bacterial populations on the bovine rumen wall: distribution patterns of adherent bacteria. *Applied and Environmental Microbiology* 39, 233–241.

McSweeney, C.S., Dulieu, A., Katayama, Y. and Lowry, J.B. (1994) Solubilization of lignin by the ruminal anaerobic fungus *Neocallimastix patriciarum*. *Applied and Environmental Microbiology* 60, 2985–2989.

McSweeney, C.S., Blackall, L.L., Collins, E., Conlan, L.L., Webb, R.I., Denman, S.E. and Krause, D.O. (2005) Enrichment, isolation and characterisation of ruminal bacteria that degrade non-protein amino acids from the tropical legume *Acacia angustissima*. *Animal Feed Science and Technology* 121, 191–204.

Mendoza, G.D., Britton, R.A. and Stock, R.A. (1993) Influence of ruminal protozoa on site and extent of starch digestion and ruminal fermentation. *Journal of Animal Science* 71, 1572–1578.

Michalet-Doreau, B., Fernandez, I., Peyron, C., Millet, L. and Fonty, G. (2001) Fibrolytic activities and cellulolytic bacterial community structure in the solid and liquid phases of rumen contents. *Reproduction Nutrition Development* 41, 187–194.

Miled, N., Canaan, S., Dupuis, L., Roussel, A., Rivière, M., Carrière, F., de Caro, A., Cambillau, C. and Verger, R. (2000) Digestive lipases: from three-dimensional structure to physiology. *Biochimie* 82, 973–986.

Miron, J., Ben-Ghedalia, D. and Morrison, M. (2001) Invited review: adhesion mechanisms of rumen cellulolytic bacteria. *Journal of Dairy Science* 84, 1294–1309.

Moir, R.J. (1951) The seasonal variation in the ruminal micro-organisms of grazing sheep. *Australian Journal of Agricultural Research* 2, 322–330.

Morvan, B., Dore, J., Rieu-Lesme, F., Foucat, L., Fonty, G. and Gouet, P. (1994) Establishment of hydrogen-utilizing bacteria in the rumen of the newborn lamb. *FEMS Microbiology Letters* 117, 249–256.

Mould, F.L., Kliem, K.E., Morgan, R. and Mauricio, R.M. (2005) *In vitro* microbial inoculum: a review of its function and properties. *Animal Feed Science and Technology* 123–124, 31–50.

Narita, Y., Oda, S., Moriyama, A. and Kageyama, T. (2002) Primary structure, unique enzymatic properties, and molecular evolution of pepsinogen B and pepsin B. *Archives of Biochemistry and Biophysics* 404, 177–185.

Newbold, C.J., Wallace, R.J. and McIntosh, F.M. (1996) Mode of action of the yeast *Saccharomyces cerevisiae* as a feed additive for ruminants. *British Journal of Nutrition* 76, 249–261.

Nielsen, B.B., Zhu, W.-Y., Dhanoa, M.S., Trinci, A.P.J. and Theodorou, M.K. (2002) Comparison of the growth kinetics of anaerobic gut fungi on wheat straw in batch culture. *Anaerobe* 8, 216–222.

Nielsen, P.K. and Foltmann, B. (1995) Purification and characterization of porcine pepsinogen B and pepsin B. *Archives of Biochemistry and Biophysics* 322, 417–422.

Obispo, N.E. and Dehority, B.A. (1992) A most probable number method for enumeration of rumen fungi with studies on factors affecting their concentration in the rumen. *Journal of Microbiological Methods* 16, 259–270.

Ogimoto, K. and Imai, S. (1981) *Atlas of Rumen Microbiology*. Japan Scientific Societies Press, Tokyo.

Orpin, C.G. (1977) Invasion of plant tissue in the rumen by the flagellate *Neocallimastix frontalis*. *Journal of General Microbiology* 98, 423–430.

Orpin, C.G. (1984) The role of ciliate protozoa and fungi in the rumen digestion of plant cell walls. *Animal Feed Science and Technology* 10, 121–143.

Paul, S.S., Kamra, D.N., Sastry, V.R.B., Sahu, N.P. and Agarwal, N. (2004) Effect of administration of an anaerobic gut fungus isolated from wild blue bull (*Boselaphus tragocamelus*) to buffaloes (*Bubalus bubalis*) on *in vivo* ruminal fermentation and digestion of nutrients. *Animal Feed Science and Technology* 115, 143–157.

Pei, Y.-X., Wang, D.-H. and Hume, I.D. (2001) Selective digesta retention and coprophagy in Brandt's vole (*Microtus brandti*). *Journal of Comparative Physiology* 171B, 457–464.

Petkov, A. (1976) Number and generic composition of ciliates in the rumen and caecum of lambs in relation to age. *Zhivotnov'dni Nauki* 13, 64–69.

Prins, R.A. (1988) Microbial digestion in herbivores: overall fermentation and reactor design. In: *Rumen Microbiology*, Proceedings of Caput College 1988, Biologisch Centrum, Haren, The Netherlands.

Purser, D.B. and Moir, R.J. (1959) Ruminal flora studies in the sheep. IX. The effect of pH on the ciliate population of the rumen *in vivo*. *Australian Journal of Agricultural Research* 10, 555–564.

Regensbogenova, M., McEwan, N.R., Javorsky, P., Kisidayova, S., Michalowski, T., Newbold, C.J., Hackstein, J.H.P. and Pristas, P. (2004) A re-appraisal of the diversity of the methanogens associated with the rumen ciliates. *FEMS Microbiology Letters* 238, 307–313.

Rezaeian, M., Beakes, G.W. and Parker, D.S. (2004) Distribution and estimation of anaerobic zoosporic fungi along the digestive tracts of sheep. *Mycological Research* 108, 1227–1233.

Richardson, S. and Gorton, L. (2003) Characterisation of the substituent distribution in starch and cellulose derivatives. *Analytica Chimica Acta* 497, 27–65.

Rieu-Lesme, F., Dauga, C., Morvan, B., Bouvet, O.M.M., Grimont, P.A.D. and Doré, J. (1996) Acetogenic coccoid spore-forming bacteria isolated from the rumen. *Research in Microbiology* 147, 753–764.

Ritchie, A.E., Robinson, I.M. and Allison, M.J. (1970) Rumen bacteriophage: survey of morphological types. In: Favard, P. (ed.) *Microscopie Electronique*, Vol. 3. Société Francaise de Microscopie Electronique, Paris, pp. 333–334.

Russell, J.B. and Rychlik, J.L. (2001) Factors that alter rumen ecology. *Science* 292, 1119–1122.

Russell, J.B., Strobel, H.J. and Chen, G. (1988) Enrichment and isolation of a ruminal bacterium with a very high specific activity of ammonia production. *Applied and Environmental Microbiology* 54, 872–877.

Russell, J.B., Onodera, R. and Hino, T. (1991) Ruminal protein fermentation: new perspectives on previous contradictions. In: Tsuda, T., Sasaki, Y. and Kawashima, R. (eds) *Physiological Aspects of Digestion and Metabolism in Ruminants*. Academic Press, Tokyo, pp. 681–697.

Salmon, H. (1999) The mammary gland and neonate mucosal immunity. *Veterinary Immunology and Immunopathology* 72, 143–155.

Santra, A. and Karim, S.A. (2000) Growth performance of faunated and defaunated Malpura weaner lambs. *Animal Feed Science and Technology* 86, 251–260.

Santra, A. and Karim, S.A. (2002) Influence of ciliate protozoa on biochemical changes and hydrolytic enzyme profile in the rumen ecosystem. *Journal of Applied Microbiology* 92, 801–811.

Schwarz, W. (2001) *The Cellulose and Cellulosome Page*. Available at: http://www.wzw.tum.de/mbiotec/cellpage.htm

Schwartz, H.M. and Gilchrist, F.M.C. (1975) Microbial interactions with the diet and the host animal. In: McDonald, I.W. and Warner, A.C.I. (eds) *Digestion and Metabolism in the Ruminant*. University of New England Publishing Unit, Armidale, New South Wales, Australia, pp. 165–179.

Siddons, R.C. (1968) Carbohydrase activities in the bovine digestive tract. *Biochemical Journal* 108, 839–844.

Sjaastad, O.V., Hove, K. and Sand, O. (2003) *Physiology of Domestic Animals*. Scandinavian Veterinary Press, Oslo, pp. 543, 548.

Skillman, L.C., Evans, P.N., Naylor, G.E., Morvan, B., Jarvis, G.N. and Joblin, K.N. (2004) 16S ribosomal DNA-directed PCR primers for ruminal methanogens and identification of methanogens colonising young lambs. *Anaerobe* 10, 277–285.

Sniffen, C.J., O'Connor, J.J.D., Van Soest, P.J., Fox, D.G. and Russell, J.B. (1992) A net carbohydrate and protein system for evaluating cattle diets: II. Carbohydrate and protein availability. *Journal of Animal Science* 70, 3562–3577.

Steffanson, B., Nielsen, C.U., Brodin, B., Eriksson, A.H., Andersen, R. and Frokjaer, S. (2004) Intestinal solute carriers: an overview of trends and strategies for improving oral drug absorption. *European Journal of Pharmaceutical Sciences* 21, 3–16.

Stewart, C.S., Fonty, G. and Gouet, P. (1988) The establishment of rumen microbial communities. *Animal Feed Science and Technology* 21, 69–97.

Stewart, C.S., Flint, H.J. and Bryant, M.P. (1997) The rumen bacteria. In: Hobson, P.N. and Stewart, C.S. (eds) *The Rumen Microbial Ecosystem*. Blackie Academic and Professional, London, pp. 10–72.

Strobel, H.J. and Russell, J.B. (1991) Succinate transport by a ruminal selenomonad and its regulation by carbohydrate availability and osmotic strength. *Applied and Environmental Microbiology* 57, 248–254.

Sun, C.Q., O'Connor, C.J. and Roberton, A.M. (2002) The antimicrobial properties of milkfat after partial hydrolysis by calf pregastric lipase. *Chemico-Biological Interactions* 140, 185–198.

Suzuki, M., Fujimoto, W., Goto, M., Morimatsu, M., Syuto, B. and Iwanaga, T. (2002) Cellular expression of gut chitinase mRNA in the gastrointestinal tract of mice and chickens. *Journal of Histochemistry and Cytochemistry* 50, 1081–1089.

Tagg, J.R., Dajani, A.S. and Wannamaker, L.W. (1976) Bacteriocins of gram-positive bacteria. *Bacteriological Reviews* 40, 722–756.

Takenaka, A., D'Silva, C.G., Kudo, H., Itabashi, H. and Cheng, K.J. (1999) Molecular cloning, expression and characterization of an endo-beta-1,4-glucanase cDNA from *Epidinium caudatum. Journal of General and Applied Microbiology* 45, 57–61.

Tamminga, S. and Doreau, M. (1991) Lipids and rumen digestion. In: Jouany, J.P. (ed.) *Rumen Microbial Metabolism and Ruminant Digestion*. INRA, Paris, pp. 151–160.

Thauer, R.K., Jungermann, K. and Decker, K. (1977) Energy conservation in chemotrophic anaerobic bacteria. *Bacteriological Reviews* 41, 100–180.

Theodorou, M.K., Lowe, S.E. and Trinei, A.P.J. (1988) The fermentative characteristics of anaerobic rumen fungi. *BioSystems* 21, 371–376.

Thomson, A.B.R., Schoeller, C., Keelan, M., Smith, L. and Clandinin, M.T. (1993) Lipid absorption: passing through the unstirred layers, brush-border membrane, and beyond. *Canadian Journal of Physiology and Pharmacology* 71, 531–555.

Thorley, C.M., Sharpe, M.E. and Bryant, M.P. (1968) Modification of the rumen bacterial flora by feeding cattle ground and pelleted roughages as determined with culture media with and without rumen fluid. *Journal of Dairy Science* 51, 1811–1816.

Toshio, I. and Fraenkel, G. (1966) The effect of nitrogen starvation on *Tenebrio molitor* L. *Journal of Insect Physiology* 12, 803–817.

Trabalza-Marinucci, M., Poncet, C., Delval, E. and Fonty, G. (2006) Evaluation of techniques to detach particle-associated microorganisms from rumen contents. *Animal Feed Science and Technology* 125, 1–16.

Trevors, J.T. (2003) Fluorescent probes for bacterial cytoplasmic membrane research. *Journal of Biochemical and Biophysical Methods* 57, 87–103.

Udén, P. (1988) The effect of grinding and pelleting hay on digestibility, fermentation rate, digesta passage and rumen and faecal particle size in cows. *Animal Feed Science and Technology* 19, 145–157.

Vance, R.D., Preston, R.L., Klosterman, E.W. and Cahill, V.R. (1972) Utilization of whole shelled and crimped corn grain with varying proportions of corn silage by growing-finishing steers. *Journal of Animal Science* 35, 598–605.

Van Soest, P.J. (1975) Physico-chemical aspects of fibre digestion. In: McDonald, I.W. and Warner, A.C.I. (eds) *Digestion and Metabolism in the Ruminant*. University of New England Publishing Unit, Armidale, New South Wales, Australia, pp. 352–365.

Wallace, R.J. (1996) Ruminal microbial metabolism of peptides and amino acids. *Journal of Nutrition* 126(4S), 1326S–1334S.

Wallace, R.J. and McKain, N. (1991) A survey of peptidase activity in rumen bacteria. *Journal of General Microbiology* 137, 2259–2264.

Wallace, R.J., Kopecny, J., Broderick, G.A., Walker, N.D., Sichao, L., Newbold, C.J. and McKain, N. (1995) Cleavage of di- and tripeptides by *Prevotella ruminicola*. *Anaerobe* 1, 335–343.

Wallace, R.J., McKain, N., Broderick, G.A., Rode, L.M., Walker, N.D., Newbold, C.J. and Kopecny, J. (1997) Peptidases of the rumen bacterium *Prevotella ruminicola*. *Anaerobe* 3, 35–42.

Watanabe, H. and Tokuda, G. (2001) Animal cellulases. *Cellular and Molecular Life Sciences* 58, 1167–1178.

Wilkie, D.P.D. (1926) An address on the functions of the biliary passages in relation to their pathology. *The Lancet* 208, 689–690.

Williams, A.G. and Withers, S.E. (1991) Effect of ciliate protozoa on the activity of polysaccharide-degrading enzymes and fibre breakdown in the rumen ecosystem. *Journal of Applied Bacteriology* 70, 144–155.

Windham, W.R. and Akin, D.E. (1984) Rumen fungi and forage fiber degradation. *Applied and Environmental Microbiology* 48, 473–476.

Wolin, M.J., Miller, T.L. and Stewart, C.S. (1997) Microbe–microbe interactions. In: Hobson, P.N. and Stewart, C.S. (eds) *The Rumen Microbial Ecosystem*. Chapman & Hall, London, pp. 329–381.

Wolstrup, J., Jensen, V. and Jensen, K. (1974) The microflora and concentrations of volatile fatty acids in the rumen of cattle fed on single component rations. *Acta Veterinaria Scandinavica* 15, 244–255.

Wright, A.-D.G., Williams, A.J., Winder, B., Christophersen, C.T., Rodgers, S.L. and Smith, K.D. (2004) Molecular diversity of rumen methanogens from sheep in Western Australia. *Applied and Environmental Microbiology* 70, 1263–1270.

8 Water Use and Requirements

It is axiomatic that animals need water, but we cannot state with the same certainty how much water an individual animal will need in a particular situation, or how it will respond to changes in water quality. For welfare and performance in farm and companion animals, we must ensure that they have an adequate supply of good-quality water. To do this, we need information on water requirements so that we can design watering facilities and storages. This is particularly important in arid and semi-arid regions where water supplies are often limited and may have high concentrations of dissolved solids.

Animals Use Water for a Wide Variety of Functions

Water has many roles in animal physiology and metabolism, and so the amounts of water needed vary as the animal's physiological and metabolic states change. This makes it difficult to state precise water requirements, and we need to understand how changes in physiology, metabolism and water and feed quality affect an animal's water requirements.

Animals use water:

- as a medium for carrying digesta through the digestive tract;
- as a carrier for body secretions (e.g. bile, pancreatic juice and saliva);
- to control body temperature by evaporation of water through transepidermal water loss, sweating or panting;
- as a lubricant and cushion against shocks (e.g. tears, synovial fluid in joints);
- as a constituent of cellular material;
- to dissolve oxygen in the lungs;
- for transportation (e.g. of absorbed nutrients in blood and lymph);
- as a solvent or dispersant;
- as a medium for excretion of wastes (e.g. urine and faeces);

- in the reactions which are carried out within cells when organic molecules are metabolized (examples include the oxidation of fatty acids, the tricarboxylic acid cycle, and the pentose phosphate and uronic acid pathways); and
- as a constituent of milk and other products.

Water Requirements Are Determined by the Sum of Water Losses from the Body and Water Sequestered in Animal Products

Clearly, anything which increases or reduces the rate at which water is lost from the body will change the total water requirement. Animals lose water by excretion in urine and faeces, in sweat and saliva and by incorporating it into products like milk and eggs (these are the 'sensible' losses); and in water vaporized from the lungs and by diffusion through the skin (the 'insensible' losses). Factors that may influence the drinking water requirements of animals include the water content, chemical composition and quantity of food, the amount and nature of ions dissolved in the drinking water, water temperature and the ambient temperature and humidity. Water sequestered in animal products such as a fetus, eggs or milk, cannot be withdrawn and reused for metabolic purposes, although body tissue can become dehydrated when there is a high rate of water loss (usually by evaporation) which cannot be immediately replaced. The balance between water outflow and that available from metabolic and food sources has to be made up from drinking water (Fig. 8.1).

Animals Have Three Sources of Water

The first source of water is that which is present within or adhering to food. Forages may contain up to 85% moisture, and immature pasture can easily supply much of the water needed by grazing animals.

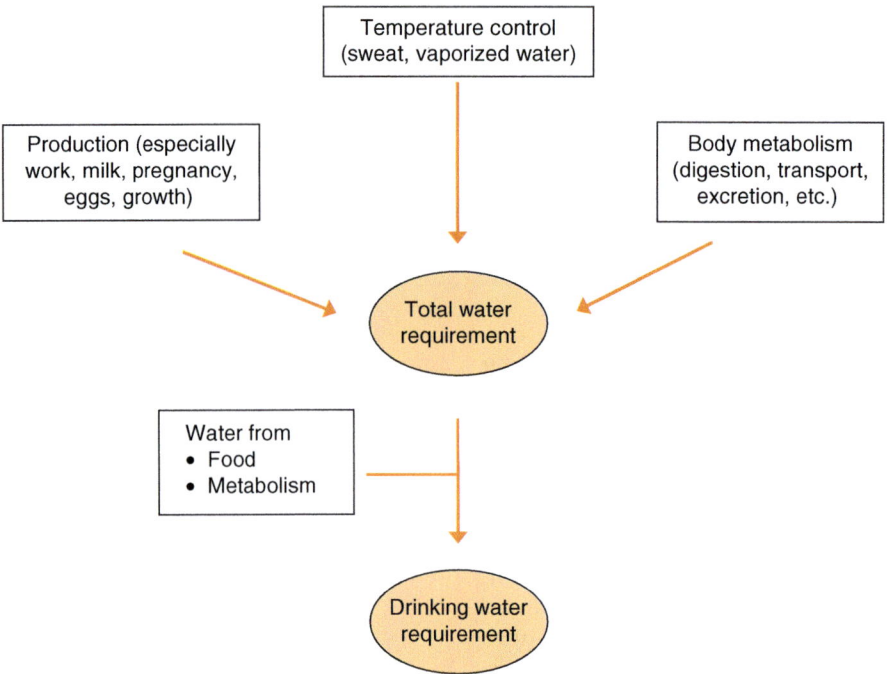

Fig. 8.1. Factors which influence the drinking water requirements of animals.

This water is found in three main forms: the most quantitatively important is intracellular water, i.e. that found in the cytosol. Other water is trapped in pores in the cell walls and occurs as water of crystallization. Most grasses have 15–25% dry matter (DM), and thus contain 75–85% water (NRC, 1982). The amounts of water in forages decrease as the plants mature. A dairy cow eating 20 kg of DM each day, while grazing a temperate grass pasture, will therefore ingest over 100 kg/day of plant cell water. This amount of feed water will be increased if the pasture has been irrigated recently or if it has rained, as the adhering water will contribute to the water ingested with the feed. On the other hand, air-dry feeds such as hay, cereal grains and protein meals contain little water, characteristically 5–10%. Thus, a growing pig eating 2.25 kg of a typical commercial feed daily will ingest only some 200 g of food water.

A second source of water is that produced within the animal's cells when it oxidizes carbohydrates, amino acids and fatty acids as energy sources, and when it synthesizes protein and fat. Chemical energy removed from the substrate molecules during this process is stored as ATP (adenosine triphosphate). ATP synthesis begins with the removal of H atoms from the carbohydrate, amino acid or fatty acid sub-strate by dehydrogenation reactions which are cofactored by various H-carrying substances, often nicotinamide adenine dinucleotide (NAD$^+$). Cells have a limited amount of NAD$^+$ and use oxidative phosphorylation to both re-oxidize NADH and produce ATP. The H removed by dehydrogenation reactions is combined with O_2 at the end of the electron transfer pathway, forming water. Van Es (1969) calculated that 0.56, 1.07 and 0.42 g water are synthesized for each gram of carbohydrate, fat and protein oxidized, respectively. Inadequate food intake will cause body fat and protein to be used to meet the animal's energy and protein requirements for maintenance. Pigs which are subjected to high ambient temperatures eat less and lose more body weight, but this may contribute positively to their water balance. Lactating sows which lost live weight at 600–1100 g/day mobilized between 200 and 600 g water daily as a result of this body weight loss (Renaudeau et al., 2001).

Schiavon and Emmans (2000) suggest that 0.16, 0.07 and 0.60 g water are produced, respectively, for each gram of protein retained in the body from dietary protein, lipid retained from dietary lipid and lipid retained from dietary carbohydrate. This water is produced when peptide bonds are formed during protein synthesis, during the synthesis of

Chapter 8

fatty acids from carbohydrates and when fatty acids and glycerol esterify to form fats.

A lactating cow eating 15 kg daily of DM from a mixed forage/concentrate diet will generate about 6 kg of metabolic water each day, while a growing pig eating 2000 g daily of DM from a typical mixture of cereal grains, protein meals and minerals will generate some 800 g metabolic water.

There is some disagreement about the physiological significance of metabolic water. When nutrients are metabolized as energy sources there is an increased requirement for oxygen, which then generates an increased water loss as air is exhaled (this air is saturated with water vapour). Further, the chemical energy produced by oxidation is not used with complete efficiency and some is lost as heat. This increases the amount of water needed for heat dissipation. Also, when amino acids are deaminated and oxidized for energy, there is an increased urinary urea excretion. Schmidt-Nielsen (1964) and Chew (1965) have argued that these water losses, especially in arid environments, may be similar to, or more than, the amount of metabolic water produced. This point remains undecided.

Under some circumstances, though, the contribution of feed and metabolic water to the animal's water economy may be significant. In a notable series of experiments, sheep were reported to survive and reproduce without access to drinking water for more than a year (Brown and Lynch, 1972; Lynch *et al.*, 1972). These sheep grew the same amount of wool and had similar lambing rates as ewes which were given drinking water. However, their lambs, which were lighter at birth, and the ewes did not tolerate dry summer pasture conditions. In practical animal feeding we should not assume that metabolic and feed water will meet an animal's total water requirement.

Finally, we need to consider drinking water. Dealing with this most obvious water source as the last point in this discussion may seem perverse, but actually animals drink to make up any deficit between their total water requirements and that obtained from feed and any net gain from the production of water during energy metabolism (Winchester and Morris, 1956; Birrell, 1992).

Total Body Water, Water Turnover Rate and Water Balance Are the Basic Physiological Parameters of Water Use

Isotope dilution studies enable us to measure the total body water (TBW) content, and the rate at which this is exchanged with 'new' water. TBW contents vary inversely with body fat content, from 55% to 80% (Murphy, 1992). The water turnover rate (WTR) is the amount of body water which is replaced each day from drinking water, feed water and metabolic water. WTR is predicted by the relationship (MacFarlane and Howard, 1966):

$$WTR = fWTR \times W^{0.82} \qquad (8.1)$$

where
WTR = water turnover rate (ml/day)
fWTR = fractional water turnover rate (ml/kg body mass daily)
W = body mass (kg)

The allometric relationship ($W^{0.82}$) is intended to correct for the body's fat content. Other workers have related WTR to $TBW^{0.82}$. The fractional water turnover rate (fWTR) is the proportion of the TBW which is replaced each day. WTR is measured by observing the rate at which a marker concentration decreases in the water of a selected body fluid. Typically, an amount of labelled water, either ^{2}HHO (deuterated water) or ^{3}HHO (tritiated water), is injected, and the isotope enrichment of a body fluid is recorded. Plasma, serum, urine and milk are suitable fluids. The enrichment in the fluid (e.g. urine) increases as the isotope spreads through the body water, then declines as the injected isotope becomes diluted as 'new' water exchanges with the body water (Fig. 8.2A). The rate of decline gives the fWTR, and if the enrichment is extrapolated back to zero time, the animal's TBW content can be estimated (Fig. 8.2B).

The WTR is influenced by all of those factors which influence the rate of water use, because to maintain the body water content, lost or used water has to be replaced with new water. WTR is influenced by ambient temperature, increasing in hot conditions because of sweating (MacFarlane and Howard, 1966; El-Nouty *et al.*, 1991) and sometimes in cold conditions because of thermogenesis through increased metabolic rate (Deavers and Hudson, 1977). Animal breed influences the animal's response to high temperatures, and thus the WTR (Kamal *et al.*, 1972; Ismail *et al.*, 1995), as does exercise (Khan and Ghosh, 1989) and the ingestion of compounds such as sodium chloride that increase urinary or faecal water excretion (Meintjes and Olivier, 1992). WTR is higher in pregnant than in non-pregnant animals (Hassan *et al.*, 1988), and higher in lactation than in pregnancy (Murphy, 1992; Chaiyabutr *et al.*, 1997). Indicative

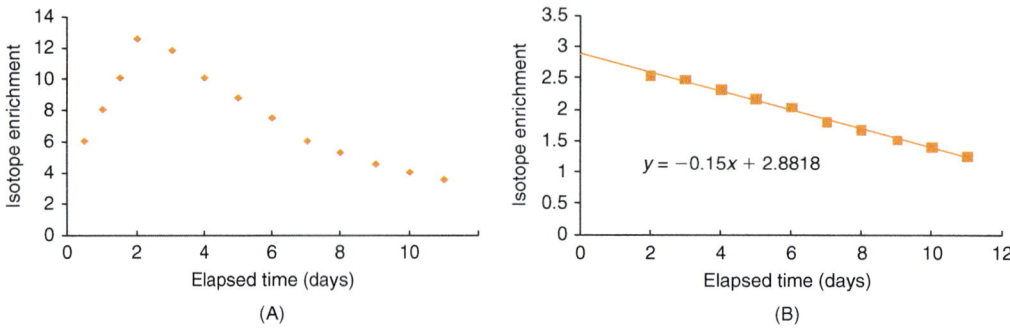

Fig. 8.2. Body water enrichment with time after injection of an isotope: (A) normal plot of enrichment versus time; (B) plot of ln (enrichment) versus time, the linear regression coefficient is the fractional water turnover rate (0.15/day in this case), and the value of enrichment at the zero time intercept is used to calculate the total body water content.

Table 8.1. Indicative water turnover rates (WTR) in domestic animals.

Species	WTR (ml/$W_{kg}^{0.82}$)	Animal type, maximum ambient temperature, feed type
Pig[a]	250	Growers, 25°C, dry pelleted feed
Horse[b]	198–279	Australian Stock Horse, 25°C, lucerne hay
Cow[c]	295	Friesian, normal farm rations
Cow[c]	259	Friesian × Haryana, normal farm rations
Sheep[d]	152	Barki and Rahmani, 29°C

[a]Yang *et al.* (1981).
[b]Van der Aa Kuhle *et al.* (2004).
[c]Sengar *et al.* (1983).
[d]El-Nouty *et al.* (1988).

Table 8.2. Water turnover rate parameters in free-living and captive animals. (From Nagy and Peterson, 1988.)

Animal type	Habitat	Equation parameters[a]	
		K	*X*
Eutherian mammals			
All	Captive	0.159	0.946
Herbivores	Free-living	0.708	0.795
Carnivores and granivores	Free-living	0.248	0.795
Marsupial mammals			
All	Captive	0.547	0.771
Herbivores	Free-living	0.874	0.711
Carnivores	Free-living	1.862	0.711

[a]The prediction equation fitted was WTR = $k \times W^x$, where WTR is in ml/day, and k is in ml/W_g^x.

WTR for domestic animals kept under 'normal' situations are summarized in Table 8.1.

Nagy and Peterson (1988) reviewed the WTR of different animal species, both free-living and in captivity. They noted that free-living animals often had greater WTR than similar animals in captivity, but that it is possible to manipulate the conditions of animals in captivity so that their WTR changes markedly, 'up to 70-fold' (Nagy and Peterson, 1988). Average values for these data are summarized in Table 8.2.

Table 8.3. Water balance of lactating Friesian cows under thermoneutral conditions. (From Richards, 1985.)

Water intake				Water excretion					
Drunk	Food	Metabolic	Total	Faeces	Urine	Saliva	Sweat	Milk	Total[b]
Amounts per day (kg)									
78.6	2.2	6[a]	86.8	32.7	23.4	0.2	9.8	13.9	80
Proportion of total water (%)									
88.5	2.5	6.9	100	37.7	27.0	0.2	1.1	16.0	92.1

[a]Estimated.
[b]Excluding water vaporized from lungs (estimated to be 6.8 kg/day).

The difference between the amounts of ingested and excreted water is the water balance. Generally, this is in equilibrium, i.e. the amount of ingested water equals that excreted. A water balance table for a lactating cow eating 18 kg DM/day of an air-dry food (30% barley straw, 70% barley/wheat/molasses concentrate) is given in Table 8.3.

Drinking Water Requirements Are Influenced by Animal Genotype, the Environment, and Interactions Between Them

Evaporation is an important route of water loss which increases in hot temperatures (Shebaita *et al.*, 1992; Shafie *et al.*, 1994). Approximately 400 g water is dissipated as vapour for each MJ of heat removed from the body (Blaxter, 1989). Under hot conditions, or at high rates of production, animals may lose as much or more water by evaporation than is excreted in urine and faeces combined (Taylor, 1970; Silanikove *et al.*, 1997). Consequently, drinking water intakes are higher in high ambient temperatures (Vercoe *et al.*, 1972; Richards, 1985; El-Nouty *et al.*, 1991; Olsson *et al.*, 1995). Different breeds of cattle, sheep and goats have differing abilities to cope with high temperatures. Schoeman and Visser (1995) and Ferreira *et al.* (2002) have shown differences of over 100% in the voluntary water consumption of goats and sheep, and various breeds of sheep. Similarly, Winchester and Morris (1956) and Siebert and MacFarlane (1969) have reported different water intakes by Santa Gertrudis and Shorthorn cattle. The response to heat stress may involve either respiratory water loss or sweating or both, and the relative importance of these varies between species (Macfarlane *et al.*, 1963; Baccari *et al.*, 1997; De Lamo *et al.*, 2001). Sweating is a more water-efficient mechanism than respiratory loss (panting) (Macfarlane *et al.*, 1963). Of particular interest is the source of water for respiratory temperature control. The camel draws on water in the digesta, while sheep use intracellular water and this makes them more susceptible to heat stress (Macfarlane *et al.*, 1963).

Air exhaled from the lungs is saturated and so breathing humid air can be expected to reduce the drinking water requirement. Air humidity and drinking water intake by lactating Friesian cows are negatively correlated (Cowan *et al.*, 1978).

Water intake can be influenced by its temperature. Horses recovering from strenuous exercise are encouraged to drink if they are given water at near ambient temperature, rather than water which is cooler or warmer (Schott *et al.*, 2003). Dairy cattle prefer water at about ambient temperature (NRC, 2001). In warm temperatures, providing cool water will decrease the amount cattle drink. Dairy cattle and goats drink significantly less cold (1°C) than warm (39°C) water (Cunningham *et al.*, 1964; Olsson *et al.*, 1997) in normal conditions. The ingestion of cold water significantly lowers ruminal and often rectal temperatures of sheep, goats and dairy cattle and it can take the rumen up to 2 h to return to normal temperature (Dillon and Nichols, 1955; Dracy *et al.*, 1963; Cunningham *et al.*, 1964).

Food Consumption and Food Type Both Influence Water Consumption

There is a rather constant relationship between feed intake and water consumption (Ritzman and Benedict, 1924). Animals need about 2.5–4 l of water/kg of feed DM, provided that they are not heat stressed. This water/DM ratio has been reported consistently, for cattle, sheep, goats, deer and pigs (ARC, 1980; Yang *et al.*, 1981; Silanikove, 1992; Misra and Singh, 2002; Yape Kii and Dryden, 2005a). We cannot apply this relationship to a water/*food* ratio. Air-dry foods such as mature pasture or hay, or the concentrate foods typically used in pig production and sometimes given to companion animals, have low water contents. Animals eating this type of food need more drinking water than when they consume succulent foods like fresh, immature pasture, or moisture-rich foods such as canned pet foods. Thus, the water/food ratio changes although the water/DM ratio stays quite constant.

It has been claimed that a high dietary protein content may also increase the drinking water requirement. However, the data indicate that this is so only if the food supplies amounts of protein, or ruminally degradable protein in the case of ruminants, in excess of the animal's requirements or capacities to use these nutrients. Any increased water requirement is associated with a need to excrete urea produced from the deamination of excess amino acids or to excrete urea synthesized when excess rumen NH_3 is produced. Thus, the relationship between dietary protein content and water intake depends on the animal's protein requirements as much as the diet protein content, and so the water intake response of an animal to eating a protein-rich diet may depend on its physiological status. For example, pigs which drink less water when given foods of lower protein content (Zuzuki *et al.*, 1998; Le Bellego and Noblet, 2002) have nevertheless grown and deposited protein at similar rates as the control animals. Positive relationships between protein content and water intake have been reported for non-lactating and lactating dairy cattle (Holter and Urban, 1992; Huber *et al.*, 1994), and when excess ruminally degradable protein was fed to dairy cattle (Higginbotham *et al.*, 1989).

There is some evidence that the water intakes of herbivores increase when they are given fibre-rich diets. Eating a fibre-rich diet may increase the amount of saliva produced (and swallowed, so it is not clear that this will consequently increase an animal's drinking water requirement) and may increase the amount of water excreted in faeces. Holzer *et al.* (1976), Warren *et al.* (1999) and Yape Kii and Dryden (2005a) have reported that steers, horses and deer drink more water when given a fibre-rich diet. On the other hand, Al-Homidan and Ahmed (2000) found no effect of dietary fibre content on the water intake of rabbits.

Foods which have high contents of absorbable minerals will increase the amount of water which is needed to excrete wastes in the urine. Examples include the increased water intake by cattle given diets containing the rumen buffers MgO and NaHCO$_3$ (Teh *et al.*, 1987), the inclusion of salt in the food of sheep and deer (Waghorn *et al.*, 1994; Ru *et al.*, 2004a) and a mineral supplement in pig food (Zheng-Li *et al.*, 2001).

The water consumption of pigs is often greater when they are given a meal feed, rather than pelleted feed (Laitat *et al.*, 1999). Offering food as slurry reduces drinking water intake, as would be expected (Miyawaki *et al.*, 1998).

Animals may also use water to allay hunger. Pigs have been reported to drink more when they are underfed (Yang *et al.*, 1981) and this behaviour has also been reported in rodents.

Producing Animals Need More Water than Animals at Maintenance

Water is 73% of the fat-free body tissue (Wang *et al.*, 1999), and this is almost an interspecies constant. However, when we include the fat typically in animal carcasses we find that the TBW content is less in rapidly growing animals or in fat, mature animals (Weis *et al.*, 2004). Gilts growing to between 45 and 85 kg live weight and with 55–63% water in their tissues (Bikker *et al.*, 1996) had water deposition rates of 0.26–0.62 kg/day. Horses have about 58–68% water (Robb *et al.*, 1972; Gee *et al.*, 2003) and this is influenced by sex and health status. Sheep and goats have been reported to have 47–72% water in their total body tissue (Bocquier *et al.*, 1999; Wuliji *et al.*, 2003).

Pregnancy and lactation both increase the water consumption of animals. Pregnant animals drink more, because water is needed for fetal growth. Faichney *et al.* (2004) have published values of 82 ml/day for the average net water uptake by Merino ewe fetuses and a total water flux between the fetus and the dam of 16–43 l/h. The average water contents of bovine, equine and porcine fetuses are 72–85% (Meyer and Ahlswede, 1976; ARC, 1980; McPherson *et al.*, 2004), so these species need to sequester 40, 43 and 7.5 kg of water, respectively, in at-term fetuses.

Milk contains about 85% water (e.g. 82% in buffalo, 85.5–88% in dairy cattle, 86–88% in goats and 90% in horses; Altmann and Dittmer, 1968). Mares may produce between 12 and 20 kg of milk daily (Doreau and Dussap, 1980), and dairy goats produce 3–7 kg daily (Knights and Garcia, 1997). The milk production of dairy cattle is very variable, from less than 5 to more than 45 kg/day. Simple calculations show that goats, mares and dairy cows may sequester up to 6, 17 and 38 kg of water in milk each day, respectively. The water intake of lactating cattle is increased over that of dry cows (water consumption by Friesian cows increased from 36 to 106 l/day; Silanikove et al., 1997) and lactation is associated with a marked increase in the WTR (Chaiyabutr et al., 1997). Dairy cattle use about 2.5 kg of water to produce 1 kg of milk (Silanikove et al., 1997). A safe practice is to allow 3–4 l of drinking water for each litre of milk produced. This allows for the extra water needed to facilitate digestion of the extra food eaten, and to allow for the secretion of water in milk.

Working horses may use very large amounts of water for cooling. Sweating rates of exercising horses may be as much as 55 ml/m^2/min (McCutcheon et al., 1999). A horse may lose 20 kg of sweat during a component of an Olympic 3-day event (McCutcheon and Geor, 1996). The amount of water drunk after exercise clearly depends on the intensity of the work, the ambient temperature and individual animal variation, but as a guide, Arabian horses given long-distance rides (45–60 km)

may drink 10–25 l immediately after the exercise (Düsterdieck et al., 1999; Butudom et al., 2003).

Estimates of Drinking Water Requirements

The volume of drinking water consumed is influenced by the animal's genotype (species and/or breed), and its age, physiological status, activity or productivity, food intake and food quality, and climatic conditions. These points are reiterated because there is an understandable tendency to accept published recommendations of water requirements as unvarying. The estimates of the requirements of domestic animals for drinking water, and of tolerance to salinity in water, given in Table 8.4 are guides only.

Better estimates of water requirements can be obtained from prediction equations which relate water intakes to measurable aspects of the animal's diet, productivity and environment. As an example, the milk-free water balance (WB, kg/day; i.e. total water turnover less the amount excreted in milk) in lactating dairy

Table 8.4. Water requirements and salinity tolerances of domestic animals.

Animal	Water requirement[a] (l/day)	Maximum salinity[a] (mg TDS/l)
Beef cattle		
Lactating cow	43–60	4,500–6,000
Growing heifer, steer	15–57	6,000–10,000
Mature dry cow	22–78	6,000–10,000
Dairy cattle		
Lactating cow	80–105	3,000–5,000
Calf (1–4 weeks)	1–4.5	3,000–7,000
Deer		
Red deer hinds	4.5	9,200
Rusa deer stags	7	8,500
Goats		
Dry adults	1.5–4	15,000–20,000
Lactating does	6–15	10,000–14,000
Horses		
Working	27–68	1,000–5,000
Resting	20–27	5,000–7,000
Foals (6–10 weeks)	4–5.5	1,000–5,000
Pigs		
Dry and breeding sows	10–40	5,000–6,000
Growers and finishers	3.5–14	5,000–6,000
Weaners	0.5–9	5,000–6,000
Sheep		
Lactating ewe	14	4,500–6,000
Dry ewe, wether	7	10,000–15,000
Lamb	2	3,500–4,000

[a]Water requirements and salinity tolerances are adapted from NRC (1974, 1981, 1985, 1989, 1998, 2000, 2001), Morand-Fehr and Sauvant (1978), SCA (1987), Barrell and Topp (1989), Kohnke et al. (1999), Ru et al. (2004b) and Yape Kii and Dryden (2005a,b).

cows can be predicted very accurately ($R^2 = 0.97$) by the following relationship from Silanikove *et al.* (1997):

$$WB = (0.14\ DEI) + (0.97\ EVAP) \qquad (8.2)$$

where
DEI = digestible energy intake (MJ/day)
EVAP = respiratory and cutaneous water losses (kg/day)

With relationships such as these we can calculate water requirements which are tailored to each individual animal's size, productivity and diet, and also identify the important factors which influence water requirements, e.g. food intake (metabolizable energy (ME) intake or DM intake), ambient temperature (also indicated by the evaporative water loss), Na content of the food and level of productivity.

Other prediction equations are:

Sheep

(McMeniman and Pepper, 1982):
$$W = 0.429 + (0.073\ T) - (0.013\ R) \qquad (8.3a)$$

(Forbes, 1968): $W = (3.86\ DMI) - 0.99 \quad (8.3b)$

Lactating dairy cows

(Murphy *et al.*, 1983): $W = 15.99 + (1.58\ DMI) + (0.09\ MY) + (0.05\ Na) + (1.2\ T) \qquad (8.3c)$

(ARC, 1980): $W = (2.15\ DMI) + (0.73\ MY) + 12.3 \qquad (8.3d)$

Non-lactating dairy cows

(Holter and Urban, 1992): $W = (0.2296\ DMC) + (2.212\ DMI) + (0.03944\ CP^2) \qquad (8.3e)$

Lot-fed beef cattle

(Hicks *et al.*, 1988): $W = (0.76\ T) + (0.13\ DMI) - (0.66\ R) - (0.29\ Na/DMI) - 6.31 \qquad (8.3f)$

Weaner pigs

(Brooks *et al.*, 1984): $W = 0.149 + (3.053\ food\ intake) \qquad (8.3g)$

where
W = water intake (kg/day)
T = maximum ambient temperature (°C)
R = rainfall (mm/day)
DMI = food DM intake (kg/day)
DMC = food DM content (%)
MY = milk yield (kg/day)
Na = sodium intake (g/day)
CP = crude protein (%)

Water Salinity Affects Water and Food Consumption and Animal Health

'Saline' water is water which contains dissolved solids. Often these are the salts of common soil minerals such as Na, Mg and Ca chlorides, carbonates, bicarbonates, sulfates and phosphates, but other substances such as nitrates and nitrites, and trace minerals, may also be present. Saline water must be consumed in greater amounts than pure water because some of the water drunk must be used to excrete its dissolved salts. This response has been reported, up to the point of maximum salinity tolerance, in most animal species including goats (El Gawad, 1997), sheep (Pierce, 1957; Wilson and Dudzinski, 1973; El Sherif and El Hassanein, 1996) and deer (Ru *et al.*, 2004b), but not always with cattle (Saul and Flinn, 1985; Ray, 1989; Kattnig *et al.*, 1992; Bahman *et al.*, 1993). The small amount of information on the behaviour of pigs and horses given saline water is consistent with the responses observed in ruminants (Anderson and Stothers, 1978; Seynaeve *et al.*, 1996; Butudom *et al.*, 2002). The increased water intake is reflected in greater urinary and sometimes faecal water excretion (Laredo *et al.*, 1996; Ganong, 1997; Stricker *et al.*, 2001).

The reported feed intake responses of ruminants given saline water vary, and are influenced by the salinity level, animal species and physiological status, and environmental conditions. For example, the feed intake of sheep increased concomitantly with water intake when they were given drinking water with moderate salt concentrations, but feed intakes were slightly reduced when these animals were given high-saline water (Wilson and Dudzinski, 1973). The published data on changes in N retention and feed digestibility are ambiguous, but authors working with deer, cattle and sheep generally have reported no changes in these (Hadjipanayiotou, 1984; Kattnig *et al.*, 1992; Yape Kii and Dryden, 2005b).

Mammals can concentrate their urine to osmolarities much greater than that of their blood. This ability helps to protect animals against the effects of drinking saline water. However, animals of larger species are less able to concentrate their urine than smaller ones. The relationship between live weight and maximal urinary concentration (U_{osm}, mmol/kg water) is (Beuchat, 1996):

$$U_{osm} = 2667\ W^{-0.09} \qquad (8.4)$$

where
W = live weight (kg)

The maximum urine concentration achieved by mammals is 6500 mOsm/l in the jerboa; Merino sheep, cats and dogs fall in the mid-range (3200, 3200 and 2400 mOsm/l, respectively), while man and the pig have only limited capacities to concentrate their urine (1200 and 1100 mOsm/l, respectively) (Chew, 1965).

The ability of animals to concentrate their urine is closely associated with the concept of obligatory urine water loss, introduced by Chew (1965). When drinking water becomes saline, the animal has to use some of the water it drinks to excrete the salts which are dissolved in the drinking water. This leads to the sequence of events in Fig. 8.3. As the load of dissolved solids increases there is progressively less 'free' water for the animal's metabolism and temperature control. One of the first responses to saline water is to drink more water. This persists until the salt load becomes too great, at which point water intake falls to very low levels, imposing severe physiological stress.

Tolerance of saline water (i.e. the amount an animal can drink without becoming sick or dying) depends on its physiological state (whether it is growing, lactating, pregnant, etc.), the climatic conditions (air temperature, wind speed, humidity, access to shade), the type of food eaten (especially the salt, and possibly protein, contents which influence the amounts of wastes needed to be excreted in the urine) and the animal's ability to concentrate its urine.

There are many different behavioural and physiological responses to saline water, and the expression of these varies between individuals and between species, and in the way in which animals are introduced to saline water. Pierce (1959) showed that Merino sheep were able to adapt well to high-saline water when the salt concentration was gradually increased, but that their food consumption markedly declined if they were abruptly changed from low-salinity to higher-salinity water. A similar adaptation by pigs to sulfate-rich water was reported by Anderson and Stothers (1978). Salinity tolerance is also affected by the

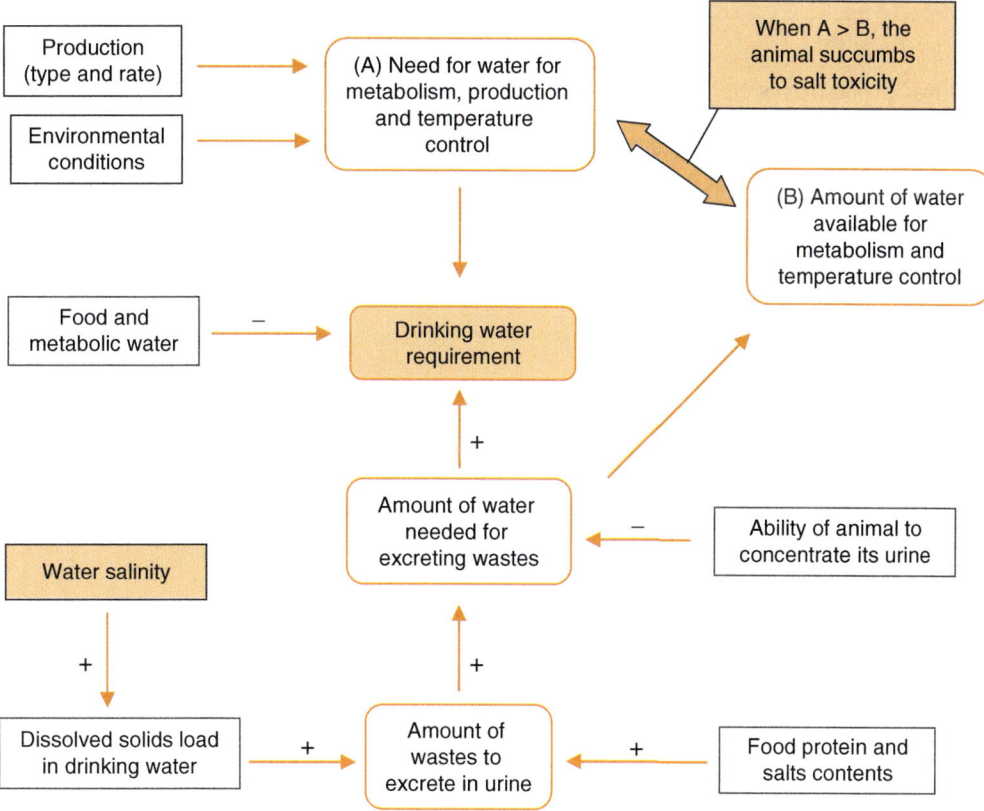

Fig. 8.3. Factors which affect an animal's tolerance of saline water.

physiological status of the animal. Mature animals are more resistant to salt toxicity than young ones (NRC, 1974). Pierce (1968) and Wilson and Dudzinski (1973) suggested that young sheep or those with no experience of saline water appeared to be less tolerant of high-saline water than mature sheep.

There are several symptoms of saline water poisoning. The more common are reduced growth (or a decline in body weight), and occasional diarrhoea and general weakness, seen when animals drink high-salinity water near to their upper limit of tolerance (Pierce, 1957, 1959; Saul and Flinn, 1985; Ru et al., 2004b). Some neurological symptoms can occur. Bobak and Salm (1996) and Ayoub and Salm (2003) reported changes in the hypothalamic supraoptic nucleus and the ventral glial limitans in rats given a salt level of 20,000 mg/l. Polioencephalomalacia-like symptoms were observed in cattle given sulfur-rich water containing 3875 mg TDS/kg by Beke and Hironaka (1991).

Prudent animal managers will periodically test the salinity of their animals' water. The simplest way of doing this is by an electrical conductivity measurement. The result is in Siemens (either mS or μS). The relationship between conductivity and total dissolved salts concentration (TDS, mg/l) is:

$$TDS = 0.68\ C \qquad (8.5)$$

where
C = conductivity (μS/cm)

Results of salinity testing should be interpreted in the light of several factors. First, individual animal species vary in their ability to tolerate saline water, because they have different capacities to concentrate their urine. For this reason, sheep are more tolerant of saline water than cattle, and both are more tolerant than pigs or horses (Table 8.4). Further, not all water contains the same types of dissolved solids. General, i.e. not species-specific,

Table 8.5. Recommended maximum levels of minerals and other constituents in drinking water for animals.

Constituent	Units	ANZECC (2000)	NRC (1974)	CCME (1987)	NRC (1989)
Faecal coliforms	cells/100 ml	1,000			
Algae	cells/ml	10,000[a]			
TDS[b]	mg/l	3,000–13,000		3,000	6,500
Nitrate-N	mg/l	30.0	99.5	22.5	100.0
Nitrite-N	mg/l	10.0	10.0	3.0	10.0
Aluminium	mg/l	5.0		5.0	
Arsenic	mg/l	0.5	0.2	0.5	0.2
Beryllium	mg/l	0.1		0.1	
Boron	mg/l	5.0		5.0	
Cadmium	mg/l	0.01	0.05	0.02	0.05
Calcium	mg/l	1,000		1,000	
Chromium	mg/l	1.0	1.0	1.0	1.0
Cobalt	mg/l	1.0	1.0	1.0	1.0
Copper[c]	mg/l	0.5	0.5	0.5–5.0	0.5
Fluoride	mg/l	2.0	2.0	2.0	2.0
Lead	mg/l	0.1	0.1	0.1	0.1
Magnesium	mg/l	600.0			
Mercury	mg/l	0.002	0.01	0.003	0.01
Molybdenum	mg/l	0.01		0.05	
Nickel	mg/l	1.0	1.0	1.0	1.0
Selenium	mg/l	0.02		0.05	
Sulfate	mg/l			1,000	
Uranium	mg/l	0.2			
Vanadium	mg/l	0.1	0.01	0.01	0.1
Zinc	mg/l	20.0	25.0	25.0	25.0

[a]Depends on species of algae.
[b]Depends on animal species.
[c]Depends on food Cu levels, the Mo and sulfate intakes and animal species.

recommendations on the maximum levels of dissolved chemicals and pathogenic organisms in drinking water are in Table 8.5. Sulfates and chlorides are better tolerated than carbonates, while sodium salts are better tolerated than magnesium salts. NRC (2001) suggests that levels of nitrate greater than 220 mg/l may cause ill health if given to cattle over a long period. On the other hand, pigs are quite resistant to nitrate, and SCA (1987) recommends maximum concentrations in drinking water of 500 mg nitrate/l and 100 mg nitrite/l. Data on total salt contents should be evaluated in the light of the types of dissolved solids present.

Water offered to grazing animals may become more saline in dry seasons because surface water becomes concentrated, or is lost, through evaporation. This may be accompanied by a greater reliance on mineral-rich underground water. In addition, mature, dry pastures contain less water and the animal's requirements for drinking water will correspondingly increase.

References

Agricultural Research Council (ARC) (1980) *Nutrient Requirements of Ruminant Livestock.* CAB, Farnham Royal, UK.

Altmann, P.L. and Dittmer, D.S. (1968) *Metabolism.* Federation of American Societies for Experimental Biology, Bethesda, Maryland.

Al-Homidan, A. and Ahmed, B.M. (2000) Productive performance and digestion kinetics of California rabbits as affected by combinations of ambient temperature and dietary crude fiber level. *Egyptian Journal of Rabbit Science* 10, 281–294.

Anderson, D.M. and Stothers, S.C. (1978) Effects of saline water high in sulfates, chlorides and nitrates on the performance of young weanling pigs. *Journal of Animal Science* 47, 900–907.

ANZECC (2000) *Australian and New Zealand Guidelines for Fresh and Marine Water Quality.* Australian and New Zealand Environment and Conservation Council, Canberra, Australia.

Ayoub, A.E. and Salm, A.K. (2003) Increased morphological diversity of microglia in the activated hypothalamic supraoptic nucleus. *Journal of Neuroscience* 23, 7759–7766.

Baccari, F. Jr, Brasil, L.H.A., Teodoro, S.M., Goncalves, H.C., Wechsler, F.S. and Aguiar, I.S. (1997) Thermoregulatory responses of Alpine goats during thermal stress. In: Bottcher, R.W. and Hoff, S.J. (eds) *Livestock Environment 5*, Vol. 2. Proceedings of the Fifth International Symposium, Bloomington, Minnesota, 29–31 May, 1997, pp. 789–794.

Bahman, A.M., Rooke, J.A. and Topps, J.H. (1993) The performance of dairy cows offered drinking water of low or high salinity in a hot arid climate. *Animal Production* 57, 23–28.

Barrell, G.K. and Topp, D.F. (1989) Water intake of red deer stags consuming dryland pasture or indoors on concentrated feeds. *Proceedings of the New Zealand Society of Animal Production* 49, 21–24.

Beke, G.J. and Hironaka, R. (1991) Toxicity to beef cattle of sulfur in saline well water: a case study. *The Science of the Total Environment* 101, 281–290.

Beuchat, C.A. (1996) Structure and concentrating ability of the mammalian kidney: correlations with habitat. *American Journal of Physiology, Regulatory Integrative and Comparative Physiology* 271, R157–R179.

Bikker, P., Verstegen, M.W.A. and Campbell, R.G. (1996) Performance and body composition of finishing gilts (45 to 85 kg) as affected by energy intake and nutrition in earlier life. II. Protein and lipid accretion in body compartments. *Journal of Animal Science* 74, 817–826.

Birrell, H.A. (1992) Estimates of the water intake of grazing and housed sheep under normal summer conditions as well as for sheep and cattle during a drought in south-western Victoria. *Proceedings of the Australian Society of Animal Production* 19, 409–411.

Blaxter, K.L. (1989) *Energy Metabolism in Animals and Man.* Cambridge University Press, Cambridge, UK.

Bobak, J.B. and Salm, A.K. (1996) Plasticity of astrocytes of the ventral glial limitans subjacent to the supraoptic nucleus. *Journal of Comparative Neurology* 376, 188–197.

Bocquier, F., Guillouet, P., Barillet, F. and Chilliard, Y. (1999) Comparison of three methods for the *in vivo* estimation of body composition in dairy ewes. *Annales de Zootechnie* 48, 297–308.

Butudom, P., Schott, H.C., Davis, C.A., Nielsen, B.D. and Eberhart, S.W. (2002) Drinking salt water enhances rehydration in horses dehydrated by frusemid administration and endurance exercise. *Equine Veterinary Journal* 9(Suppl), 513–518.

Butudom, P., Axiak, S.M., Nielsen, B.D., Eberhart, S.W. and Schott, H.C. (2003) Effect of varying initial drink volume on rehydration of horses. *Physiology and Behavior* 79, 135–142.

Brooks, P.H., Russel, S.J. and Carpenter, J.L. (1984) Water intake of weaned piglets from three to seven weeks old. *Veterinary Record* 115, 513–515.

Brown, G.D. and Lynch, J.J. (1972) Some aspects of the water balance of sheep at pasture when deprived of drinking water. *Australian Journal of Agricultural Research* 23, 669–684.

Canadian Council of Ministers of the Environment (CCME) (1987) *Canadian Water Quality Guidelines.* Water Quality Branch, Inland Waters Directorate, Environment Canada, Ottawa.

Chaiyabutr, N., Komolvanich, S., Sawangkoon, S., Preuksagorn, S. and Chanpongsang, S. (1997) The regulation of body fluids and mammary circulation during late pregnancy and early lactation of cross-bred Holstein cattle feeding on different types of roughage. *Journal of Animal Physiology and Animal Nutrition* 77, 167–179.

Chew, R.M. (1965) Water metabolism of mammals. In: Meyer, W.W. and van Gelder, R.G. (eds) *Physiological Mammalogy*, Vol. 2. Academic Press, New York, pp. 44–178, 112–113.

Cowan, R.T., Shackel, D. and Davison, T.M. (1978) Water intakes, milk yield and grazing behaviour of Friesian cows with restricted access to water in a tropical upland environment. *Australian Journal of Experimental Agriculture and Animal Husbandry* 18, 190–195.

Cunningham, M.D., Martz, F.A. and Merilan, C.P. (1964) Effect of drinking-water temperature upon ruminant digestion, intraruminal temperature and water consumption of non-lactating cows. *Journal of Dairy Science* 47, 382–385.

Deavers, D.R. and Hudson, J.W. (1977) Effect of cold exposure on water requirements of three species of small mammals. *Journal of Applied Physiology* 43, 121–125.

De Lamo, D.A., Lacolla, D. and Heath, J.E. (2001) Sweating in the guanaco (*Lama guanicoe*). *Journal of Thermal Biology* 26, 77–83.

Dillon, R.D. and Nichols, R.E. (1955) Changes in temperature of reticulo-ruminal content following the drinking of water. *American Journal of Veterinary Research* 16, 69–70.

Doreau, M. and Dussap, G. (1980) Estimation of milk production in the nursing mare by labeling the body water of the foal. *Reproduction, Nutrition and Development* 20, 1883–1892.

Dracy, A.E., Jahn, J.R. and Essler, W.O. (1963) Recording intrareticular temperatures by radiosonde equipment. *Journal of Dairy Science* 46, 241–242.

Düsterdieck, K.F., Schott, H.C., Eberhart, S.W., Woody, K.A. and Coenen, M. (1999) Electrolyte and glycerol supplementation improve water intake by horses performing a simulated 60 km endurance ride. *Equine Veterinary Journal* 30(Suppl.), 418–424.

El Gawad, E.I.A. (1997) Physiological responses of Barki and Damascus goats and their crossbreed to drinking saline water. *Alexandria Journal of Agricultural Research* 42, 23–36.

El-Nouty, F.D., Hassan, G.A., Taher, T.H., Samak, M.A., Abo-Elezz, Z. and Salem, M.H. (1988) Water requirements and metabolism in Egyptian Barki and Rahmani sheep and Baladi goats during spring, summer and winter seasons. *Journal of Agricultural Science, Cambridge* 111, 27–34.

El-Nouty, F.D., El-Naggar, M.I., Hassan, G.A. and Salem, M.H. (1991) Effect of lactation on water requirements and metabolism in Egyptian sheep and goats. *World Review of Animal Production* 26, 39–43.

El Sherif, M.A. and El Hassanein, E.E. (1996) Influence of drinking saline water on growth and distribution of

body fluids in sheep. *Alexandria Journal of Agricultural Research* 41, 1–9.

Faichney, G.J., Fawcett, A.A. and Boston, R.C. (2004) Water exchange between the pregnant ewe, the fetus and its amniotic and allantoic fluids. *Journal of Comparative Physiology. B. Biochemical, Systemic, and Environmental Physiology* 174, 503–510.

Ferreira, A.V., Hoffman, L.C., Schoeman, S.J. and Sheridan, R. (2002) Water intake of Boer goats and Mutton merinos receiving either a low or high energy feedlot diet. *Small Ruminant Research* 43, 245–248.

Forbes, J.M. (1968) The water intake of ewes. *British Journal of Nutrition* 22, 33–43.

Ganong, W.F. (1997) *Review of Medical Physiology*, 18th edn. Appleton and Lange, London, p. 671.

Gee, E.K., Fennessy, P.F., Morel, P.C.H., Grace, N.D., Firth, E.C. and Mogg, T.D. (2003) Chemical body composition of 20 thoroughbred foals at 160 days of age, and preliminary investigation of techniques used to predict body fatness. *New Zealand Veterinary Journal* 51, 125–131.

Hadjipanayiotou, M. (1984) Influence of type of water on diet digestibility, mineral balance and rumen metabolites of Chios sheep. *Zeitschrift Fuer Tierphysiologie Tierernaehrung und Futtermittelkunde* 52, 194–200.

Hassan, G.A., El-Nouty, F.D., Salem, M.H., Latif, M.G. and Badawy, A.M. (1988) Water requirements and metabolism of Egyptian sheep and goats as affected by breed, season and physiological status. In: *Isotope Aided Studies on Livestock Productivity in Mediterranean and North African Countries*. Proceedings of the Final Research Co-ordination Meeting, Rabat, 23–27 March 1987. International Atomic Energy Agency, Vienna, pp. 65–83.

Hicks, R.B., Owens, F.N., Gill, D.R, Martin, J.J. and Strasia, C.A. (1988) Water intake by feedlot steers. *Oklahoma Animal Science Report* Mp-125, 208.

Higginbotham, G.E., Torabi, M. and Huber, J.T. (1989) Influence of dietary protein concentration and degradability on performance of lactating cows during hot environmental temperatures. *Journal of Dairy Science* 72, 2554–2564.

Holter, J.B. and Urban, W.E. (1992) Water partitioning and intake prediction in dry and lactating Holstein cows. *Journal of Dairy Science* 75, 1472–1479.

Holzer, Z., Tagari, H., Levy, D. and Volcani, R. (1976) Soaking of complete fattening ration high in poor roughage. 2. The effect of moisture content and of particle size of the roughage component on the performance of male cattle. *Animal Production* 22, 41–53.

Huber, J.T., Higginbotham, G., Gomez-Alarcon, R.A., Taylor, R.B., Chen, K.H., Chan, S.C. and Wu, Z. (1994) Heat stress interactions with protein, supplemental fat, and fungal cultures. *Journal of Dairy Science* 77, 2080–2090.

Ismail, E., Abd-El-Latif, H., Hassan, G.A. and Salem, M.H. (1995) Water metabolism and requirements of sheep as affected by breed and season. *World Review of Animal Production* 30, 95–105.

Kamal, T.H., Shehata, O. and Elbanna, I.M. (1972) Effect of heat and water restriction on water metabolism and body fluid compartments in farm animals. *Isotope Studies on the Physiology of Domestic Animals.* International Atomic Energy Agency, Vienna, pp. 95–102.

Kattnig, R.M., Pordomingo, A.J., Schneberger, A.G., Duff, G.C. and Wallace, J.D. (1992) Influence of saline water on intake, digesta kinetics, and serum profiles of steers. *Journal of Range Management* 45, 514–518.

Khan, M.S. and Ghosh, P.K. (1989) Physiological responses of desert sheep and goats to grazing during summer and winter. *Indian Journal of Animal Sciences* 59, 600–603.

Knights, M. and Garcia, G.W. (1997) The status and characteristics of the goat (*Capra hircus*) and its potential role as a significant milk producer in the tropics: a review. *Small Ruminant Research* 26, 203–215.

Kohnke, J., Kelleher, F. and Trevor-Jones, P. (1999) *Feeding Horses in Australia.* RIRDC Publication No. 99/49. Rural Industries Research and Development Corporation, Canberra, Australia.

Laitat, M., Vandenheede, M., Desiron, A., Canart, B. and Nicks, B. (1999) Comparison of performance, water intake and feeding behaviour of weaned pigs given either pellets or meal. *Animal Science* 69, 491–499.

Laredo, S., Yuen, K., Sonnenberg, B. and Halperin, M.L. (1996) Coexistence of central diabetes insipidus and salt wasting: the difficulties in diagnosis, changes in natremia, and treatment. *Journal of the American Society of Nephrology* 7, 2527–2532.

Le Bellego, L. and Noblet, J. (2002) Performance and utilization of dietary energy and amino acids in piglets fed low protein diets. *Livestock Production Science* 76, 45–58.

Lynch, J.J., Brown, G.D., May, P.F. and Donnelly, J.B. (1972) The effect of withholding drinking water on wool growth and lamb production of grazing Merino sheep in a temperate climate. *Australian Journal of Agricultural Research* 23, 659–668.

MacFarlane, W.V. and Howard, G. (1966) Water content and turnover of identical twin *Bos indicus* and *Bos taurus* in Kenya. *Journal of Agricultural Science, Cambridge* 66, 297–302.

MacFarlane, W.V., Morris, R.J.H. and Howard, B. (1963) Turn-over and distribution of water in desert camels, sheep, cattle and kangaroos. *Nature* 197, 270–271.

McCutcheon, L.J. and Geor, R.J. (1996) Sweat fluid and ion losses in horses during training and competition in cool vs. hot ambient conditions: implications for ion supplementation. *Equine Veterinary Journal* 22(Suppl.), 54–62.

McCutcheon, L.J., Geor, R.J., Ecker, G.L. and Lindinger, M.I. (1999) Equine sweating responses to submaximal exercise during 21 days of heat acclimation. *Journal of Applied Physiology* 87, 1843–1851.

McMeniman, N.P. and Pepper, P.M. (1982) The influence of environmental temperature and rainfall on the water intake of sheep consuming mulga (*Acacia aneura*). *Proceedings of the Australian Society of Animal Production* 14, 443–446.

McPherson, R.L., Ji, F., Wu, G., Blanton, J.R. and Kim, S.W. (2004) Growth and compositional changes of fetal tissues in pigs. *Journal of Animal Science* 82, 2534–2540.

Meintjes, R.A. and Olivier, R. (1992) The effects of salt loading via two different routes on feed intake, body water turnover rate and electrolyte excretion in sheep. *Onderstepoort Journal of Veterinary Research* 59, 91–96.

Meyer, H. and Ahlswede, L. (1976) Uber das intrauterine Wachstum und die Korperzusammensetzung von Fohlen sowie den Nahrstoffbedarf tragender Stuten. *Ubersichten zur Tierernahrung* 4, 263–292.

Misra, A.K. and Singh, K. (2002) Effect of water deprivation on dry matter intake, nutrient utilization and metabolic water production in goats under semiarid zone of India. *Small Ruminant Research* 46, 159–165.

Miyawaki, K., Hoshina, K. and Itoh, S. (1998) Effects of wet-dry feeding for postweaning pigs on growth, feed intake, water consumption and eating behaviour. *Japanese Journal of Swine Science* 35, 9–17.

Morand-Fehr, P. and Sauvant, D.S. (1978) Caprins. In: *Alimentation des Ruminant.* INRA, Paris.

Murphy, M.R. (1992) Water metabolism of dairy cattle. *Journal of Dairy Science* 75, 326–333.

Murphy, M.R., Davis, C.L. and McCoy, G.C. (1983) Factors affecting water consumption by Holstein cows in early lactation. *Journal of Dairy Science* 66, 35–38.

Nagy, K.A. and Peterson, C.C. (1988) *Scaling of Water Flux Rate in Animals.* University of California Press, Berkeley/Los Angeles, California.

National Research Council (NRC) (1974) *Nutrients and Toxic Substances in Water for Livestock and Poultry.* National Academy Press, Washington, DC.

National Research Council (NRC) (1981) *Nutrient Requirements of Goats.* National Academy Press, Washington, DC.

National Research Council (NRC) (1982) *United States-Canadian Tables of Feed Composition: Nutritional Data for United States and Canadian Feeds*, 3rd rev. edn. National Academy Press, Washington, DC.

National Research Council (NRC) (1985) *Nutrient Requirements of Sheep*, 6th rev. edn. National Academy Press, Washington, DC.

National Research Council (NRC) (1989) *Nutrient Requirements of Horses*, 5 rev. edn. National Academy Press, Washington, DC.

National Research Council (NRC) (1998) *Nutrient Requirements of Swine*, 10 rev. edn. National Academy Press, Washington, DC.

National Research Council (NRC) (2000) *Nutrient Requirements of Beef Cattle*, 7th rev. edn. Update. National Academy Press, Washington, DC.

National Research Council (NRC) (2001) *Nutrient Requirements of Dairy Cattle*, 7 rev. edn. National Academy Press, Washington, DC.

Olsson, K., JoGiter-Hermelin, M., Hossaini-Hilali, J., Hydbring, E. and Dahlborn, K. (1995) Heat stress causes excessive drinking in fed and food deprived pregnant goats. *Comparative Biochemistry and Physiology* 110A, 309–317.

Olsson, K., Cvek, K. and Hydbring, E. (1997) Preference for drinking warm water during heat stress affects milk production in food-deprived goats. *Small Ruminant Research* 25, 69–75.

Pierce, A.W. (1957) Studies on salt tolerance of sheep I. The tolerance of sheep for sodium chloride in the drinking water. *Australian Journal of Agriculture Research* 8, 711–722.

Pierce, A.W. (1959) Studies on salt tolerance of sheep II. The tolerance of sheep for mixture of sodium chloride and magnesium chloride in the drinking water. *Australian Journal of Agriculture Research* 10, 725–735.

Pierce, A.W. (1968) Studies on salt tolerance of sheep. VIII. The tolerance of grazing ewes and their lambs for drinking waters of the types obtained from underground sources in Australia. *Australian Journal of Agricultural Research* 19, 589–595.

Ray, D.E. (1989) Interrelationships among water quality, climate and diet on feedlot performance of steer calves. *Journal of Animal Science* 67, 357–363.

Renaudeau, D., Quiniou, N. and Noblet, J. (2001) Effects of exposure to high ambient temperature and dietary protein level on performance of multiparous lactating sows. *Journal of Animal Science* 79, 1240–1249.

Richards, J.I. (1985) Effect of high daytime temperatures on the intake and utilisation of water in lactating Friesian cows. *Tropical Animal Health and Production* 17, 209–217.

Ritzman, E.G. and Benedict, F.G. (1924) *The Effect of Varying Feed Levels on the Physiological Economy of Steers*. New Hampshire Agricultural Experiment Station Technical Bulletin No 26.

Robb, J., Harper, R.B., Hintz, H.F., Reid, J.T., Lowe, J.E., Schryver, J.E. and Rhee, M.S.S. (1972) Chemical composition and energy value of the body, fatty acid composition of adipose tissue, and liver and kidney size in the horse. *Animal Production* 14, 25–34.

Ru, Y.J., Fischer, M., Glatz, P.C. and Bao, Y.M. (2004a) Effect of salt level in the feed on performance of red and fallow weaner deer. *Asian-Australasian Journal of Animal Sciences* 17, 638–642.

Ru, Y.J., Glatz, P.C. and Bao, Y.M. (2004b) *Effect of Salt Intake on Feed Intake and Growth Rate of Fallow and Red Weaner Deer*. RIRDC Publication No 04/054. Rural Industries Research and Development Corporation, Canberra, Australia.

Saul, G.R. and Flinn, P.C. (1985) Effect of saline drinking water on growth and water and feed intakes of weaner heifers. *Australian Journal of Experimental Agriculture* 25, 734–738.

Schiavon, S. and Emmans, G.C. (2000) A model to predict water intake of a pig growing in a known environment on a known diet. *British Journal of Nutrition* 84, 873–883.

Schmidt-Nielsen, K. (1964) *Desert Animals: Physiological Problems of Heat and Water*. Clarendon Press, Oxford, UK.

Schoeman, S.J. and Visser, J.A. (1995) Comparative water consumption and efficiency in three divergent sheep types. *Journal of Agricultural Science, Cambridge* 124, 139–143.

Schott, H.C., Butudom, P., Nielsen, B. and Eberhart, S.W. (2003) Strategies to increase voluntary drinking after exercise. In: *Proceedings of the 49th AAEP Convention*, 21–25 December, New Orleans, Louisiana, pp. 132–136.

Sengar, S.S., Verma, D.N., Singh, U.B. and Johri, S.B. (1983) Nutrient utilization, body composition and water metabolism in cattle. *Journal of Nuclear Agriculture and Biology* 12, 109–110.

Seynaeve, M., De Wilde, R., Janssens, G. and De Smet, B. (1996) The influence of dietary salt level on water consumption, farrowing, and reproductive performance of lactating sows. *Journal of Animal Science* 74, 1047–1055.

Shafie, M.M., Murad, H.M., El-Bedawy, T.M. and Salem, S.M. (1994) Effect of heat stress on feed intake, rumen fermentation and water turnover in relation to heat tolerance response by sheep. *Egyptian Journal of Animal Production* 31, 317–327.

Shebaita, M.K., Yousri, R.M. and Pfau, A. (1992) Water economy and water pool in animals under heat stress. *International Journal of Animal Sciences* 7, 235–240.

Siebert, B.D. and MacFarlane, W.V. (1969) Body water content and water turnover of tropical *Bos taurus, Bos indicus, Bibos banteng* and *Bos bubalus bubalis*. *Australian Journal of Agricultural Research* 20, 613–622.

Silanikove, N. (1992) Effect of water scarcity and hot environment temperature on appetite and digestion of ruminants. *Livestock Production Science* 30, 175–194.

Silanikove, N., Maltz, E., Halevi, A. and Shinder, D. (1997) Metabolism of water, sodium, potassium, and chlorine by high yielding dairy cows at the onset of lactation. *Journal of Dairy Science* 80, 949–956.

Standing Committee on Agriculture (SCA) (1987) *Feeding Standards for Australian Livestock. Pigs*. CSIRO, Melbourne, Australia.

Stricker, E.M., Craver, C.F., Curtis, K.S., Peacock-Kinzig, K.A., Sved, A.F. and Smith, J.C. (2001) Osmoregulation in water-deprived rats drinking hypertonic saline: effect of area postrema lesions. *American Journal of Physiology. Regulatory, Integrative and Comparative Physiology* 280, R831–R842.

Taylor, C.R. (1970) Strategies of temperature regulations: effect on evaporation in East African ungulates. *American Journal of Physiology* 219, 1131–1135.

Teh, T.H., Hemken, R.W., Bremel, D.H. and Harmon, R.J. (1987) Comparison of buffers on rumen functions, turnover rate and gastric secretions in Holstein steers. *Animal Feed Science and Technology* 17, 257–270.

Van der Aa Kuhle, K., Cawdell-Smith, A.J., Bryden, W.L., Davies, P.S.W. and Dryden, G.McL. (2004) Equine water kinetics as influenced by age and temperament. *Asia Pacific Journal of Clinical Nutrition* 29, S119.

Van Es, A.J.H. (1969) Report to the sub-committee on constants and factors: constants and factors regarding metabolic water. In: Blaxter, K.L., Kielanowski, J. and Thorbek, G. (eds) *Proceedings of the 4th Symposium on Energy Metabolism of Farm Animals.* European Association for Animal Production Publication no. 12, Oriel Press, Newcastle upon Tyne, UK, pp. 513–514.

Vercoe, J.E., Frisch, J.E. and Moran, J.B. (1972) Apparent digestibility, nitrogen utilization, water metabolism and heat tolerance of Brahman cross, Africander cross and Shorthorn × Hereford steers. *Journal of Agricultural Science, Cambridge* 79, 71–74.

Waghorn, G.C., Black, H. and Horsburgh, T. (1994) The effect of salt and bentonite supplementation on feed and water-intake, faecal characteristics and urine output in sheep. *New Zealand Veterinary Journal* 42, 24–29.

Wang, Z.M., Deurenberg, P., Pietrobelli, A., Baumgartner, R.N. and Heymsfield, S.B. (1999) Hydration of fat-free body mass: review and critique of a classic body-composition constant. *American Journal of Clinical Nutrition* 69, 833–841.

Warren, L.K., Lawrence, L.M., Brewster-Barnes, T., Powell, D.M. and Jeffcot, L.B. (1999) The effect of dietary fibre on hydration status after dehydration with frusemide. *Equine Veterinary Journal* 30(Suppl.), 508–513.

Weis, R.N., Birkett, S.H., Morel, P.C. and de Lange, C.F. (2004) Effects of energy intake and body weight on physical and chemical body composition in growing entire male pigs. *Journal of Animal Science* 82, 109–121.

Wilson, A.D. and Dudzinski, M.L. (1973) Influence of the concentration and volume of saline water on the food intake of sheep, and on their excretion of sodium and water in urine and faeces. *Australian Journal of Agriculture Research* 24, 245–256.

Winchester, C.F. and Morris, M.J. (1956) Water intake rates of cattle. *Journal of Animal Science* 15, 722–740.

Wuliji, T., Goetsch, A.L., Puchala, R., Sahlu, T., Merkel, R.C., Detweiler, G., Soto-Navarro, S., Luo, J. and Shenkoru, T. (2003) Relationships between body composition and shrunk body weight and urea space in growing goats. *Journal of Applied Animal Research* 23, 1–24.

Yang, T.S., Howard, B. and Macfarlane, W.V. (1981) Effects of food on drinking behaviour of growing pigs. *Applied Animal Ethology* 7, 259–270.

Yape Kii, W. (2004) *Water Requirements of Rusa Deer.* MPhil thesis, The University of Queensland, Queensland, Australia.

Yape Kii, W. and Dryden, G.McL. (2005a) Water consumption by rusa deer (*Cervus timorensis*) stags as influenced by different types of feed. *Animal Science* 80, 83–88.

Yape Kii, W. and Dryden, G.McL. (2005b) Effect of drinking saline water on feed and water intake, feed digestibility, and nitrogen and mineral balances of rusa deer stags (*Cervus timorensis rusa*). *Animal Science* 81, 99–105.

Zheng-Li, Jiang-ZongYong, Yu-DeQian and Lin-YingCai (2001) The effect of dietary electrolyte balance on the performance and specific blood parameters of growing pigs. *Acta Zoonutrimenta Sinica* 13, 10–14.

Zuzuki, K., Cheng, XuChun, Kano, H., Shimizu, T. and Sato, Y. (1998) Influence of low protein diets on water intake and urine and nitrogen excretion in growing pigs. *Animal Science and Technology* 69, 267–270.

9 Minerals: Their Functions and Animal Requirements

Introduction

The 'minerals', i.e. elements other than C, H, O and N which are required by animals for their normal metabolism, occur in Periods 1–5 of the periodic table (Fig. 9.1). Ideas about why certain minerals are widely distributed in animal tissues and are essential nutrients, while others are not, are discussed by Williams (1996).

Seven minerals, Na, K, Ca, Mg, P, S and Cl, are required by animals in quite large amounts, i.e. in g/day, and are measured in the diet in g/kg or per cent. These are the 'major minerals' or 'macrominerals' (Underwood and Suttle, 1999). Other minerals are present in body tissue at levels of less than 0.01%. Requirements for these are small, typically mg/day or μg/day, or inclusion in the food in mg/kg or μg/kg. These are referred to as the 'trace minerals' or 'microminerals' (Mertz, 1986). Minerals which are required in the diet at less than 50 μg/kg are sometimes called the 'ultra-trace' minerals (Nielsen et al., 1980). Although animals need larger amounts of the major minerals than of the trace minerals this does not mean that they are more metabolically important.

All of the minerals which are involved in the body's metabolism are 'essential nutrients'. Several minerals are potentially 'limiting nutrients' in practical animal feeding because they may not occur in conventional diets in adequate concentrations. This list includes Ca, P, Mg, Na, Cu, Co, I, Fe, Se and Zn.

There are 17 minerals which we know without any doubt to be essential for animal metabolism, and another nine which may possibly be essential (Fig. 9.1, Table 9.3). It can be difficult to demonstrate unequivocally the essentiality of a mineral, especially those in the ultra-trace category. Conventionally, a nutrient (e.g. a mineral) is essential if it performs some physiological or biochemical function which cannot be performed by any other nutrient. But we need to decide if our definition of essentiality can be extended to include substances which have a pharmacological, rather than a nutritional, effect on animal metabolism. There is also a practical problem in investigating trace and ultra-trace minerals – the minute amounts needed of some of these makes it difficult to prepare diets which are deficient in the mineral in question. More complications arise from between-animal species and between-individual (including age and sex) differences in response to a mineral, differences in the availability of minerals from different types of food, interactions between minerals and other nutrients, and mineral recycling within the animal (Frieden, 1978).

In some situations in practical animal feeding an essential mineral (often a trace mineral) can be toxic, and this can pose real problems in nutritional management. Some trace minerals are potentially poisonous because there is a small window between adequacy and a toxic excess. For example, some cereal by-products, oilseed meals and molasses have 20–30, or up to 80 mg Cu/kg (Evans, 1985), whereas the sheep's upper tolerable limit for Cu is about 25 mg/kg, or less in diets which have low levels of Zn and/or Mo (NRC, 1985a). Other essential minerals are toxic in certain oxidation states, e.g. Cr(IV), (V) and (VI); and V(IV) and (V).

Normally safe foods can become poisonous if accidentally loaded with an excessive amount of a mineral (Wilkinson et al., 2003). Pastures and crops adjoining major highways have been contaminated with Pb from tetraethyl lead added to motor fuel (Zhao et al., 2004; Ahamed and Siddiqui, 2007). Following the Chernobyl incident, Scandinavian reindeer pastures were contaminated with ^{134}Cs and other radioactive elements (Staaland et al., 1995). Sheep were poisoned in New Zealand when pastures were loaded with F after the eruption of Mt Ruapehu (Cronin et al., 2003). Hg (Magos and Clarkson, 2006) and Cd (Neathery and Miller, 1975; Meharg and French, 1995) may accumulate in foods and are poisonous. Ge and Al

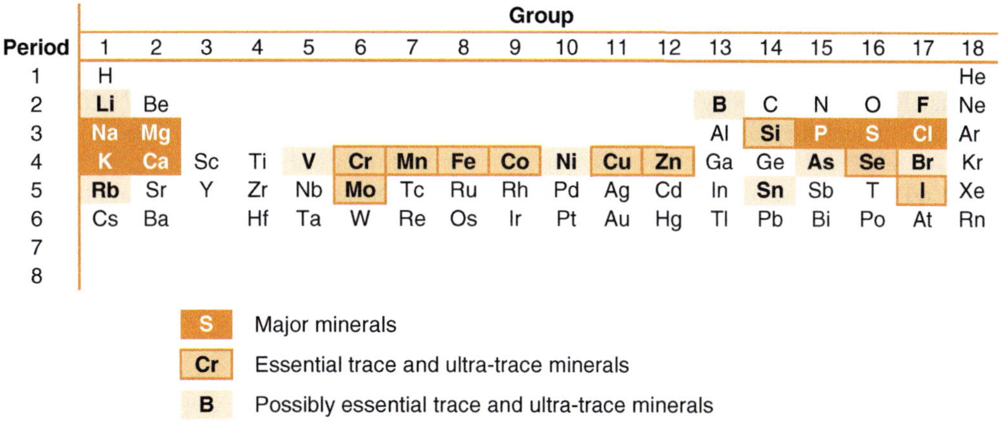

	Group																	
Period	1	2	3	4	5	6	7	8	9	10	11	12	13	14	15	16	17	18
1	H																	He
2	Li	Be											B	C	N	O	F	Ne
3	Na	Mg											Al	Si	P	S	Cl	Ar
4	K	Ca	Sc	Ti	V	Cr	Mn	Fe	Co	Ni	Cu	Zn	Ga	Ge	As	Se	Br	Kr
5	Rb	Sr	Y	Zr	Nb	Mo	Tc	Ru	Rh	Pd	Ag	Cd	In	Sn	Sb	T	I	Xe
6	Cs	Ba		Hf	Ta	W	Re	Os	Ir	Pt	Au	Hg	Tl	Pb	Bi	Po	At	Rn
7																		
8																		

S — Major minerals

Cr — Essential trace and ultra-trace minerals

B — Possibly essential trace and ultra-trace minerals

Fig. 9.1. The periodic table showing elements which are nutritionally essential, or where there is evidence which supports speculation that they are essential.

are commonly found in animal foods. They appear to have no metabolic roles (although Ge possibly has some pharmacological ones) and are toxic in sufficiently high concentrations (Arts *et al.*, 1994; Domingo, 1995; Klein, 2005), although this appears to be rare.

Minerals do not influence metabolism independently of other nutrients. There are mineral–mineral interactions, such as competition for binding sites on transport proteins, chemical bonding to form unabsorbable complexes which influence the availabilities of certain minerals, and in some cases minerals cofactor enzymes which synthesize mineral-containing macromolecules. There are mineral–vitamin interactions as a result of some minerals and vitamins having complementary biochemical functions.

Major Minerals

Sulfur

S is a component of the S-amino acids cysteine, cystine and methionine. The electron configuration of S atoms allows cyst(e)ine to participate in di-thiol (S–S) bridges between peptide strands in molecules such as keratin (the hair fibre protein). S atoms can form thio esters (bonds between S and C); these bonds are unstable and therefore reactive. A di-thiol bond can be reduced to two sulfydryl (–SH) groups, allowing S to act as an acceptor and donor of H atoms, as in lipoic acid, glutathione and coenzyme A. Giles *et al.* (2003) have reviewed the chemistry of S relevant to the capacity of cysteine proteins to act as metal ion

sinks, antioxidants and regulators of metabolic pathways.

S, per se, is rarely deficient in conventional diets, although the S-amino acids can be limiting amino acids in monogastric animals' diets. Some authorities recommend a minimum dietary N/S ratio for ruminants, e.g. 14:1 (ARC, 1980), 12.5:1 to 14.3:1 (SCA, 1990), 10:1 (NRC, 1985a) and others cited by Qi *et al.* (1994). Mature forage can have low protein and S contents, but supplements providing ruminally available N often effectively promote live weight gain, without additional S. Giving cystine and/or methionine to ruminants in a way which bypasses rumen degradation will improve wool growth (Dryden *et al.*, 1969; Reis *et al.*, 1990; Sherlock *et al.*, 2001) and milk protein content (Schwab, 1996; Leonardi *et al.*, 2003). However, this reflects the effects of microbial digestion in the rumen more than the S-amino acid content of the food. S may induce polioencephalomalacia in ruminants through being reduced to H_2S (Gould *et al.*, 1991), and an upper dietary limit of 0.4% S is recommended.

Calcium and phosphorus

Ca and P are often considered together because aspects of their metabolism are very closely related. Although they have different roles in soft tissue metabolism, they are both the main constituents of bone mineral, bone is their major storage reserve and they influence the absorption and metabolism of each other. They are required in large amounts. For example, an adult pig weighing 100 kg has about 950 g Ca and 400 g P in its body, and needs a diet which supplies 13.8 g Ca and 12.3 g P each day (NRC, 1998).

Approximately 99% of the body's Ca and 80% of P is in bone, where they are combined chemically with each other and with other minerals. The basic bone Ca/P complex is hydroxyapatite, with the general formula $Ca_{10}(PO_4)_6(OH)_2$ (Giraud-Guille *et al.*, 2004). Other minerals are found in bone, e.g. in equine phalanx trabecular bone we have 240 mg Ca/g (or 39% of the total mineral), 99 mg P/g and 2 mg Mg/g (Van der Harst *et al.*, 2004). Bone has a variable chemical composition because of substitutions of Mg^{2+}, Ba^{2+} or Sr^{2+} for Ca^{2+}; HPO_4^{2-} or CO_3^{2-} for PO_4^{3-}; and HPO_4^{2-}, CO_3^{2-}, Cl^- or F^- for OH^- (Timlin *et al.*, 2000; Giraud-Guille *et al.*, 2004). Hydroxyapatite confers rigidity and strength to bone, and thus structural strength to the body and protection for organs such as the brain and the heart.

Ca has a central place in soft tissue cell metabolism as a second messenger in muscle, liver and other tissues (Berridge *et al.*, 2000); as a regulator of the movement of K^+ through cell membranes; and as a coenzyme (e.g. Ca/Mg endonuclease; Yakovlev *et al.*, 2000). Ca has important roles in muscle and nerve metabolism (Alberts *et al.*, 2002). At the start of a muscle contraction Ca^{2+} moves out of the sarcoplasmic reticulum and into the muscle cell where it binds to troponin C and exposes the actin fibres to myosin. Nerve impulse transmission across the synaptic gaps between nerve cells and nerve and muscle cells also requires Ca^{2+}. When the action potential reaches the end of the axon it opens voltage-gated Ca^{2+} channels and the influx of Ca^{2+} ruptures the acetylcholine vesicles and allows acetylcholine to diffuse across the synaptic gap and to activate ligand-gated channels in the neighbouring dendrite. In blood clotting, Ca^{2+} is a cofactor to enzymes which work at several stages in fibrin synthesis and clot development (Moran and Viele, 2005).

P also has important roles in soft tissue metabolism. It is a constituent of numerous energy-regulating metabolites (e.g. the adenosine phosphates and creatine phosphate) and participates in many aspects of intermediary metabolism. It is part of the ribose-phosphate backbone of DNA and RNA, and is a constituent of cell membrane phospholipids (Alberts *et al.*, 2002).

Ca^{2+} and inorganic P are absorbed, by passive and active mechanisms, through the rumen wall and the small intestinal epithelium. Vitamin D is involved, promoting the expression of the gene which codes for the Ca binding protein in the small intestinal epithelium (Breves *et al.*, 2007; Schröder and Breves, 2007). There are species differences in Ca and P absorption. Monogastrics (pigs) rely almost entirely on small intestinal absorption and this is important in regulating the animal's Ca status (Partridge, 1978; Schröder and Breves, 2007). Horses appear to regulate Ca status mostly via renal function (Caple, 1990). Ruminants absorb Ca and P from the rumen when the concentrations in rumen liquor are sufficiently high (Beardsworth *et al.*, 1989; Caple, 1989; Schröder and Breves, 2007) as well as from the small intestine.

Sufficient dietary contents are the first, but not the only, criterion of the adequacy of mineral supply. Excess phosphate reduces the concentration of free Ca^{2+} ions and thus interferes with Ca absorption (Soares, 1995a). The recommended ratio of total P/Ca for pigs is 1:1 to 1.5:1 (AFRC, 1990; NRC, 1998), and of available P/Ca is 2:1 to 3:1 (NRC, 1998). Recommended ratios of total P/Ca are, for horses, not less than 1:1 (NRC, 2007) although ratios of up to 6:1 may be tolerated (Jordan *et al.*, 1975); beef cattle, 1:1 to 7:1 (NRC, 2000); dogs, 1.2:1 to 1.6:1 (Dzanis, 1994); and up to 14:1 for the black rhinoceros (Clauss *et al.*, 2007).

Not all of the P in foods can be digested and absorbed. Availability depends on its chemical form in the food, and the dietary Ca/P ratio. Much of the P in plant material, especially in seeds, is present as phytin (*myo*-inositol(1,2,3,4,5,6)*hexakis*-dihydrogen phosphate), and animals with little microbial digestive activity, such as pigs, digest little of this. P in wheat is more available than in other cereal grains; this is partly due to the activity of endogenous wheat phytase (Soares, 1995b; Godoy *et al.*, 2005). The bioavailability of P is improved in high-moisture grain (Ross *et al.*, 1983; Soares, 1995b). Rumen microorganisms digest phytate and release the P (Raun *et al.*, 1956; Yanke *et al.*, 1998). Phytate-P is therefore largely available in the ruminant digestive tract (Clark *et al.*, 1986), although cattle and sheep sometimes respond positively to exogenase phytase supplements (Bravo *et al.*, 2003; Kincaid *et al.*, 2005). Horses absorb about 35% of plant P (NRC, 2007), increasing to 70% if phytase is added to the diet (Van Doorn *et al.*, 2004). The bioavailability of P from non-plant sources varies (Soares, 1995b). Availability is low in undefluorinated rock phosphates

(NRC, 1998), but high (about 90%) in meat meals (Traylor *et al.*, 2005a).

The availability of Ca for ruminants in conventional foods averages about 64% (AFRC, 1991), and the NRC uses 50% for Ca availability for horses and 68% for cattle (NRC, 2000, 2007). For pigs, the Ca in inorganic sources such as dicalcium phosphate and limestone is readily (93–102% compared to calcium carbonate) available (Ross *et al.*, 1984), but Ca is less (51% and 78%) available from dolomitic limestone (Ross *et al.*, 1984). Where Ca is present in plant-derived foods as the oxalate, or bound to phytin, it is much less readily available, for both pigs and ruminants (references in Soares, 1995a). Addition of a phytase to the diet improves the availability of phytate-bound Ca for non-ruminants (Harper *et al.*, 1997; Igbasan *et al.*, 2001; Van Doorn *et al.*, 2004). Ca may be less available if the particle size of the source (e.g. limestone, bone, etc.) is large, but the question is not resolved, e.g. contrasting results of Burnell *et al.* (1989) and Traylor *et al.* (2005b) for meat meals fed to pigs.

Availabilities of Ca and P for pigs, horses and ruminants are summarized in Table 9.1. These are indicative values only. There are several different ways of measuring mineral availabilities which give different values, and different estimates of the endogenous excretion of Ca and P greatly influence the estimated 'true' absorption coefficients (Ammerman, 1995; Bravo *et al.*, 2003).

Ca, P and/or vitamin D deficiencies impair bone mineralization, causing the following conditions:

1. Rickets, osteomalacia and osteoporosis: rickets and osteomalacia are the juvenile and adult forms of defective bone growth where the volume of bone is normal but it is not properly mineralized (Baburaj and Reid, 2006). Osteoporosis occurs when there is loss of bone matrix as well as mineral. These conditions can occur through P or vitamin D deficiencies, as well as Ca deficiency (Nisbet *et al.*, 1970).
2. Osteochondrosis and arthritis: these joint disorders involve either damage to developing or mature cartilage, or impaired mineralization of the articular surface during growth or at maturity. In equine osteodystrophia we have the splitting-off of pieces of cartilage in the maturing joint, which may be associated with necrosis and synovitis (Aldred, 1998). The aetiology of osteochondrosis is not fully understood. Deficiencies of Ca, P, Cu and Zn, and

overfeeding and over-exercising of young horses, may contribute to the condition (Murray *et al.*, 2001).
3. Hind-leg paralysis (posterior paralysis): this occurs in pigs, especially in sows nearing the end of their lactation (NRC, 1998).
4. Osteodystrophia fibrosa (ODF, or 'bighead'): this is a condition in horses associated with the binding of Ca in forages as Ca oxalate, or from feeding low-Ca diets (Blaney *et al.*, 1981; McKenzie *et al.*, 1981). The three stages of ODF are: (i) shifting lameness; (ii) swelling of facial bones, respiratory problems and arthritis; and (iii) death. Several tropical grasses have very high oxalate contents, e.g. 0.5–2% (DM basis) in kikuyu (*Pennisetum clandestinum*), pangola (*Digitaria eriantha*) and

Table 9.1. Estimates of the true availabilities of Ca and P in foods for pigs, horses and ruminants. (From Ross *et al.*, 1983; SCA, 1987; Martz *et al.*, 1990; Soares, 1995a,b; NRC, 1998; Godoy *et al.*, 2002; Bravo *et al.*, 2003; Meschy and Ramirez-Perez, 2005; Traylor *et al.*, 2005a,b; values are directly measured true availabilities except where indicated. True availability is the proportion absorbed, calculated in a way which corrects for any endogenous excretion into the digestive tract.)

	True availability (%)	
Food type	Ca	P
Pigs		
Forages, legumes	6–11[a]	12[b]
Cereal grains	–	8–36
Cereal by-products	–	11–54
Oilseed meals	–	1–25
Fish and meat meals	–	45–72
Mineral sources	41–53[a]	21–84[a]
Horses		
Forages, legumes	76–80	–
Mineral sources	67–73	–
Ruminants		
Forages, legumes	24–56	56–94
Forages, grasses	59	60–75
Cereal grains	–	72–76
Cereal by-products	–	25–78
Oilseed meals	15[c]	49–71
Fish and meat meals	–	85
Mineral sources	49–67[a]	34–100

[a]Using calcium carbonate as the reference substance (true availability = 59% for cattle, 53% for pigs).
[b]Using dicalcium phosphate as the reference substance (true availability = 70% for pigs).
[c]Apparent availability.

green panic (*Panicum maximum*); 4% in buffel grass (*Cenchrus ciliaris*), and up to 10% in setaria (*Setaria* spp.) (Barry and Blaney, 1987).

P deficiency can reduce food intake by ruminants (McLachlan and Ternouth, 1985). Deficient animals become unproductive and, in severe cases, lame and grossly emaciated. Animals suffering from P deficiency often display unusual eating behaviour ('pica'), including eating bones and animal carcasses. These may carry *Clostridium botulinum* with the risk of the animal contracting the fatal condition, botulism (Jansen, 1969; Brizuela, 1996).

Ca deficiencies occur when animals are fed cereal-rich diets without supplements, or when horses (and also ruminants; Panda and Sahu, 2002) graze oxalate-rich pastures. Cereal grains have very little Ca (0.02–0.1%; Evans, 1985). This is insufficient for all animals, and the content of available Ca is further reduced by binding to phytate. Ca is present in high concentrations in some foods, e.g. 1.5% in lucerne forage, and 12% in low-quality meat meals. These levels may interfere with the absorption of other minerals, especially P and Zn. Supplementary Ca can be provided from limestone (36% Ca), dicalcium phosphate (23% Ca) and similar sources. Hypocalcaemia is a metabolic Ca deficiency disorder. For descriptions of the causes and control of the disorder in dairy cattle see McNeill *et al.* (2002) and Thilsing-Hansen *et al.* (2002).

Cereal grains have about 0.35% P (Evans, 1985), but its low availability, especially for monogastric animals, means that this level may not meet animal requirements. Grazing animals are exposed to P deficiency in two ways. Some soils have very low levels of plant-available P (McCosker and Winks, 1994) and this results in P-deficient pasture. Further, the P content of a plant changes through its growth cycle because plants withdraw nutrients into the crown and roots before dry weather (in the case of 'drought-avoiding' perennials), or when the plant dies after setting seed (in the case of annual plants). In a northern Australian native tropical grass pasture P contents varied with season between 0.02% and 0.15%, DM basis (Norman, 1963). Seasonal changes are also reported in temperate pasture species (e.g. Opitz von Boberfeld and Banzhaf, 2006; Miller and Thompson, 2007).

The nature of the response to supplementary P depends on whether it is the first-limiting nutrient. If it is, we can expect a response (Holm *et al.*, 1981). If not, animals given only P may not respond, and may perform worse (for as yet unexplained reasons) than if not supplemented (Cohen, 1978; Wadsworth *et al.*, 1988). As with other nutrients, a decision to give a P supplement is made for either welfare or economic reasons. It may be uneconomic to give a P supplement to a grazing animal, except in areas of low soil-P status where animal production is limited without a supplement. Supplementary P can be provided from several sources, e.g. dicalcium phosphate (18% P), monocalcium phosphate (21% P) and sodium tripolyphosphate (34% P). The usefulness of some P sources is limited, e.g. by excess F in undefluorinated rock phosphate and by the corrosiveness of phosphoric acid. Water-soluble sources are more versatile than insoluble forms because they can be added to drinking water, or included in molasses-based supplement blocks, as well as added to the food.

Magnesium

About half of the body Mg is in bone (Saris *et al.*, 2000) where it influences the size of apatite crystals and thus bone fragility (large crystals give a denser, but more fragile, bone). Mg is a cofactor to more than 300 enzymes. It has essential roles in, for example, energy metabolism (e.g. glycolytic pathway enzymes including pyruvic kinase and decarboxylase, triose phosphate isomerize; Baker and Worthley, 2002); cell membrane stability, nucleic acid and protein synthesis, and regulation of intracellular Ca and K concentrations (Saris *et al.*, 2000); muscle contraction, modulation of nerve impulse passage at synapses (Liu, 2007); and it is bound in the RNA polymerase active site (Cramer *et al.*, 2001). Mg deficiency reduces the cell's capacity to detoxify reactive oxygen species (Kuzniar *et al.*, 2003).

Mg is absorbed from the small intestine in preruminant and monogastric animals, and largely from the rumen in ruminants (Axford *et al.*, 1975; Patience and Zijlstra, 2001; Goff, 2006). Active absorption occurs via an Na^+-dependent transporter, and there is also some passive diffusion. Active absorption from the rumen is impaired by high diet K and low diet Na concentrations (Goff, 2006). Bone is not a useful Mg reserve and does not contribute Mg in times of deficiency.

The dietary Mg requirement for pigs is 0.04% or less (NRC, 1998). Although pigs are rarely

Mg-deficient because most concentrate foods have at least this content, a short period of supplementation before slaughter can improve pig meat quality (Frederick *et al.*, 2006). Recommended dietary Mg concentrations for lactating dairy cattle are between 0.18% and 0.21% of the diet DM (NRC, 2001) depending on stage of lactation and milk yield. Many plant-derived foods contain at least this much Mg. Nevertheless, lactating dairy cattle may experience a Mg deficiency resulting from low food intake, low pasture Na content, excessive K or non-protein N intakes, the formation of unavailable salts by combination of Mg^{2+} with plant fatty acids or when the rumen pH rises above 6.5 (NRC, 2001; Goff, 2006). In certain circumstances, temperate pastures may not have sufficient Mg. Agronomic factors which are important in maintaining the Mg content of temperate pastures are discussed by Edmeades (2004). Hypomagnesaemia rarely occurs in cattle grazing tropical pastures because these generally contain more Mg than temperate plants (Minson and Norton, 1982).

Signs of deficiency in the cow and the pig are initially depressed appetite and reduced growth (especially in the pig), nervousness or excitement, ataxia and finally convulsions followed by death (NRC, 2000; Patience and Zijlstra, 2001). Mg excess reduces diet digestibility and induces lethargy and severe diarrhoea in cattle (Chester-Jones *et al.*, 1990).

Sodium and potassium

These minerals are electrolytes, i.e. elements which are normally found as the dissociated ion in dilute aqueous solution. K^+ and Na^+ are involved in the regulation of intracellular and extracellular osmotic pressure, respectively, the pH of body fluids (together with Cl^-, HCO_3^- and other anions) and programmed cell death (Bortner and Cidlowski, 2007); and are enzyme cofactors. Na^+ is necessary for the action of some cell membrane carrier proteins, and the transmission of nerve action potentials by virtue of its passage through voltage-gated Na^+ channels in nerve cell membranes (Alberts *et al.*, 2002). More than 50% of body tissue Na^+ occurs in bone (Coenen, 2005). K^+ is involved in protein synthesis in the ribosome (Cahn and Lubin, 1978), neurotransmitter release, nerve impulse transmission at synapses, nerve and muscle cell excitability, insulin secretion and smooth muscle tone (Ghatta *et al.*, 2006).

Na deficiency causes inappetance, uncoordination, shivering, general weakness, hair loss and fatigue, and can be associated with stereotypical licking behaviour, low milk yield and emaciation (Phillips *et al.*, 1999; NRC, 2000). Deficiencies in grazing animals occur when they graze Na-deficient pastures – Na content varies with plant species, and seasonally. The risk of deficiency is greater when plant moisture contents are high (because animals drink less water) and when the drinking water is relatively free of dissolved ions. Na occurs in very low concentrations in cereal grains (0.01–0.03%, DM basis; Evans, 1985; NRC, 2000) and some forages (e.g. 0.01% DM basis in ryegrass and similar or lower values in other temperate grasses; NRC, 2000; Arizmendi-Maldonado *et al.*, 2002; Tremblay *et al.*, 2006). Pastures distant from the coast may be Na-deficient. K is rarely deficient in animal diets as most ingredients have at least the concentration required in diets for pigs (NRC, 1998) and cattle (NRC, 2000, 2001). Molasses, in particular, has a high K content (Stewart, 1976).

Requirements for Na and K are increased when animals sweat. Working horses need more Na and K than are provided in conventional diets, especially in hot conditions (Kronfeld, 2001). Large amounts of these minerals can be lost in sweat. For example, a 550 kg horse performing 'heavy' work and excreting 20 l sweat daily needs an extra 57.8 g Na daily above its maintenance requirement, as well as an extra 12.5 g K and 66 g Cl (Coenen, 2005).

Supplementary Na is most commonly provided as salt (NaCl), often given as a compressed salt block. Commercial salt is not chemically pure; it may contain small amounts (less than 1 g/kg) of Ca, Mg and S, and traces of Cu and Fe. Excess salt is sometimes deliberately added to foods to reduce consumption. There is little risk of Na toxicity, as an excess is rapidly excreted. However, foods with very high salt contents may adversely affect animal performance. Australian saltbush (*Atriplex* spp.) has over 40% Na in the DM (Warren *et al.*, 1990) and sheep given diets rich in this ate less food and drank more water, and lost live weight and body condition.

The dietary cation anion difference

Optimizing the dietary electrolyte balance, or 'dietary cation anion difference' (DCAD, in Eq/kg), may improve animal performance (Austic, 1983). The index can be calculated in several ways (Charbonneau *et al.*, 2006), e.g. as suggested by Ender *et al.* (1971), Equation 9.1a, and by Goff *et al.* (2004), Equation 9.1b.

$$DCAD = [Na^+ + K^+] - [Cl^- + S^{2-}] \qquad (9.1a)$$

$$DCAD = [Na^+ + K^+] - [Cl^- + 0.6\,S^{2-}] \qquad (9.1b)$$

where
$[Na^+]$, $[K^+]$, $[Cl^-]$, $[S^2]^-$ = concentrations of these ions (mEq/kg)

Goff *et al.* (2004) and Goff (2006) have discussed the chemical and physiological bases for these equations. Other DCAD equations include terms for Ca^{2+}, Mg^{2+} and phosphate (PO_4^{3-}) but meta-analyses (systematic statistical examinations of combined data from several sources) by Charbonneau *et al.* (2006) and Lean *et al.* (2006) suggest that the two listed above are the best predictors of hypocalcaemia, when applied to diets fed before calving. Horst *et al.* (1997) discuss the agronomic and feeding aspects of DCAD application in the dairy industry.

DCAD of 40 mEq/100 g appears to improve growth in pigs (Haydon *et al.*, 1990). DCAD of 15–35 mEq/100 g improved growth in lot-fed cattle (Ross *et al.*, 1994). Balances of up to 32 mEq/100 g increased milk yield in dairy cattle (West *et al.*, 1991) and food intake and growth of dairy calves (Jackson *et al.*, 1992). Lactating cows which are fed diets rich in cations (particularly Na^+ and K^+) are at greater risk of hypocalcaemia ('milk fever'), a potentially fatal disorder. These animals synthesize less 1,25-dihydroxycholecalciferol (Goff *et al.*, 1991) and so are less able to mobilize Ca from bone. Thus, with dairy cows, we aim to optimize, rather than maximize, the DCAD (Goff *et al.*, 2004).

Requirements for the Major Minerals

Estimates of requirements are provided by a number of authorities. Table 9.2 summarizes the British (AFRC, 1991), French (Jarridge, 1989), Australian (SCA, 1987, 1990) and American (NRC, 1981, 1985a,b, 1998, 2000, 2001, 2007; Dzanis, 1994) recommendations. The data are presented as recommended minimum dietary concentrations. This form is useful for ration formulations, but it also introduces the problem of accuracy in estimates of food intake, as these data are the recommended daily

Table 9.2. Recommended minimum dietary concentrations for the major minerals.

| Animal species | Animal type | Source | Recommendation (g/kg DM) | | | | |
			Ca	P[a]	Mg	Na	K
Pigs	Baby (up to 10 kg)	SCA, 1987	10.3	8.2	0.4	1.2	2.3
		NRC, 1998	8.9	4.4	0.4	2.2	3.1
	Grower (>45 kg)	SCA, 1987	7.2	5.7	0.2	0.7	2.3
		NRC, 1998	5.5	2.1	0.4	1.1	2.1
	Pregnant sow[b]	SCA, 1987	8.2	7.3	0.4	1.2	–
		NRC, 1998	8.3	3.9	0.4	1.7	2.2
	Lactating sow[b]	SCA, 1987	8.2	7.3	0.4	1.2	–
		NRC, 1998	8.3	3.9	0.4	2.2	2.2
Horses	Growing yearlings	NRC, 2007	3.8	2.1	0.5	0.7	1.7
(standard-bred,	Adult, light work[c]	NRC, 2007	3.0	1.8	1.0	1.4	2.9
500 kg adult	Pregnant mare[b]	NRC, 2007	3.6	2.6	0.8	1.1	2.6
weight)	Lactating mare[b]	NRC, 2007	4.7	3.1	0.9	1.0	3.8
Goats (dairy	Mature (maintenance)	NRC, 1981	2.8	1.9	–	–	5.0
type, 60 kg)	Lactating doe (5 kg	NRC, 1981	4.2	2.9	–	–	8.0
	FCM/day)[b]	Jarridge, 1989	7.5–8.5	3.5	1.4	1.4	–
Sheep (wool type,	Mature	SCA, 1990	1.3	1.4	0.2	0.7	–
60 or 70 kg)	(maintenance)[d]	NRC, 1985a	2.0	2.0	1.2–1.8	0.9–1.8	5.0–8.0
		AFRC, 1991	1.2	1.6	–	–	–
	Lactating ewe	SCA, 1990	6.0	4.3	0.4	0.9	–
	(3 kg milk/day)[b,d]	NRC, 1985a	3.2	2.6	1.2–1.8	0.9–1.8	5.0–8.0
		Jarridge, 1989	7.0–8.0	4.0	1.4	1.5	–
		AFRC, 1991	4.9	4.9	–	–	–

Continued

Table 9.2. Continued

Animal species	Animal type	Source	Recommendation (g/kg DM)				
			Ca	P[a]	Mg	Na	K
Dairy cattle	Growing heifer	SCA, 1990[d]	4.2	2.7	0.2	0.8	–
(Friesian)	(250 kg, 1 kg	NRC, 2001[e]	4.1	2.3	1.1	0.8	4.8
	gain/day)	Jarridge, 1989	5.5–6.5	3.2–3.8	1.2	1.5	–
		AFRC, 1991	3.5	3.5	–	–	–
	Lactating cow	SCA, 1990[d]	4.0	2.9	0.3	1.2	–
	(600 kg, 25 kg	NRC, 2001[e]	6.2	3.2	1.8	2.2	10.0
	FCM/day)[b]	Jarridge, 1989	6.5–7.2	3.8–4.0	1.5–2.0	1.5	–
		AFRC, 1991	3.8	4.1	–	–	–
Beef cattle	Growing yearling	SCA, 1990[d]	5.2	3.2	0.2	0.8	–
(*Bos taurus*,	(250 kg, 1.2–1.5 kg	NRC, 2000	2.1	1.3	1.0	0.6–0.8	6.0
500 kg adult	gain/day)	Jarridge, 1989	8.0–9.5	4.5–5.0	1.4	1.5	–
weight)		AFRC, 1991	4.3	3.5	–	–	–
	Mature bull	SCA, 1990[d]	2.3	2.4	0.3	–	–
	(maintenance)	AFRC, 1991	1.8	1.9	–	–	–
		NRC, 2000[f]	1.4	1.1	0.1	0.6–0.8	3.0–4.0
Dogs	Growing puppy[g]	NRC, 1985b	5.9	4.4	0.4	0.6	4.4
		Dzanis, 1994	10.0	8.0	–	3.0	6.0
	Lactating bitch[g]	NRC, 1985b					
		Dzanis, 1994	10.0	8.0	–	3.0	6.0
	Adult (maintenance)	NRC, 1985b					
		Dzanis, 1994	6.0	5.0	–	0.6	6.0
Cats	Growing kitten[g]	NRC, 1986					
		Dzanis, 1994	10.0	8.0	–	2.0	6.0
	Lactating queen[g]	NRC, 1986					
		Dzanis, 1994	10.0	8.0	–	2.0	6.0
	Adult (maintenance)	NRC, 1986					
		Dzanis, 1994	6.0	5.0	–	2.0	6.0

[a]Available P for pigs, total P for all other species.
[b]The requirement is for a female in the eleventh month of gestation or in the first month of lactation FCM is milk with 4% fat.
[c]See NRC (2007) for definitions of work intensities.
[d]The SCA (1990) recommendations for Ca and P are calculated for a sheep eating 1.2 (at maintenance) or 1.87 (lactating) kg DM/day (9 MJ ME/kg DM), growing yearlings eating 6.9 (beef) or 7.5 (dairy) kg DM/day (10.5 MJ ME/kg DM), and a lactating cow eating 17.8 kg DM/day (calculations based on the potential intake of a cow with a peak milk yield of 25 kg/day); the AFRC (1991) recommendations are calculated using their equations but the same DM intakes as for the SCA (1990) values.
[e]When fed the diets nominated in NRC (2001).
[f]Maintenance requirements for Mg, K and Na are not given; these values are the amounts needed to replace endogenous losses (Mg) or for low levels of live weight gain (Na and K) and assuming a DM intake of 10.8 kg/day.
[g]The requirement is stated to be for 'growing and reproducing' animals.

amounts of each nutrient, divided by the expected daily dry matter (DM) intake. There are some quite large variations between the different recommendations. These reflect the different data sets and assumptions on which the recommendations are based, including the age and physiological status of the animals involved, improvements in experimental technique, the criteria adopted for adequacy, differences in estimating endogenous secretions (for estimating true availabilities) and whether factorial or empirical methods are used to estimate requirements.

The Trace Minerals

Co, Cu, Cr, Fe, I, Mn, Mo, Se and Zn are trace minerals which are unequivocally essential for animal metabolism. Many of them are transition metals which have more than one oxidation state. This makes them suitable as cofactors to reductase and oxidase enzymes, and in situations where they function as electron carriers, e.g. in the cytochrome pathway. They may also be constituents of metabolically important molecules, such as cobalamin (Co), haem (Fe), and thyroxine (I). Some characteristics of these minerals are summarized in Table 9.3.

Table 9.3. Metabolic roles of the trace and ultra-trace minerals.

Mineral	Function
Co	The central constituent of the cobalamin moiety of cyanocobalamin (vitamin B12). This vitamin cofactors methylmalonyl coA mutase and so promotes the introduction of propionate C into the tricarboxylic acid cycle. Bioavailability 10–25% for monogastric animals (Thompson *et al.*, 2004) Review: Hostetler *et al.* (2003)
Cu	An enzyme cofactor, e.g. in keratin synthesis (hair fibre protein); to tyrosinase (melanin synthesis); to lysyl oxidase (cross-linking of collagen macromolecules, and so in angiogenesis and bone cartilage growth); cytochrome c oxidase (cycles between Cu^{2+} and Cu^{3+} in electron transport); ceruloplasmin (Fe absorption and transport and uptake of Fe by cells synthesizing haemoglobin); Cu/Zn superoxide dismutase (detoxification of peroxide radicals and protection of biomembranes), and others (Deschamps *et al.*, 2005). Bioavailability 1–10% for ruminants, 70–85% for non-ruminants and unweaned ruminants. High Mo and sulfate, and very high Zn, impair Cu absorption (ARC, 1980; Spears, 2003). CuO is very poorly absorbed by all species Reviews: Linder and Hazegh-Azam (1996), Grace and Wilson (2002) and Hostetler *et al.* (2003)
Cr	As part of the oligopeptide chromodulin (Vincent, 2004), Cr increases the animal's sensitivity to insulin and improves glucose tolerance. But Cr(VI) is toxic and its reduction intermediates (Cr(V), (IV)) involve the production of reactive oxygen species. Bioavailability is 5% or less (Spears, 1999; Thompson *et al.*, 2004) Reviews: Spears (1999) and Vincent (2000)
Fe	A constituent of haem and Fe–S proteins. Fe is thus important in haemoglobin and myoglobin (i.e. carrying O in the blood and storing it in muscle cells, respectively), and in the electron transport pathway where it cycles between Fe^{2+} and Fe^{3+} as it accepts and passes on electrons. Excess Co and Zn interfere with Fe absorption (Henry and Miller, 1995) Review: Hentze *et al.* (2004)
I	The central mineral constituent of thyroxine and triiodothyronine. A deficiency is associated with reduced metabolic rate, vigour and cold tolerance. Deficient animals may develop goitre (an enlargement of the thyroid gland) and there may be retarded development of fetuses. Excessive intakes of I (e.g. from seaweed meals) will reduce the uptake of I by the thyroid gland and can cause hypothyroidism and goitre (Baker, 2004) Review: Hostetler *et al.* (2003)
Mn	Mn-superoxide dismutase; deficiency reduces growth and interferes with bone remodelling (reduced osteoclast and osteoblast activity, brittle bones) and is associated with dermatitis and nervous conditions. Excessive intakes of Mn can be toxic. Bioavailability for ruminants is 1% (Spears, 2003). Excess Ca, Co, Cu, Fe and P may reduce Mn absorption (Henry, 1995b) Reviews: Freeland-Graves and Turnlund (1996) and Hostetler *et al.* (2003)
Mo	A constituent of molybdopterin, a cofactor to enzymes such as xanthine oxidase. Deficiency is associated with inability to metabolize amino acids and purines. Bioavailability about 95% in monogastric animals (Thompson *et al.*, 2004) Review: Freeland-Graves and Turnlund (1996)
Se	Se acts as an antioxidant by virtue of its being a cofactor to glutathione peroxidase and other enzymes. It is involved in the de-iodination of thyroxine to its active form as it is a cofactor to the deiodinases (Levanger and Whanger, 1996). Selenate may mimic the action of insulin (Stapleton, 2000). Se forms complexes with Cd and Hg, and protects against these poisonous minerals. Selenoproteins are involved in sperm and muscle function, and Se is involved in immune function (Brown and Arthur, 2001). Se status is reduced by increased S intake and the presence of cyanogenetic glucosides (Spears, 2003) Reviews: Reilly (1996), Brown and Arthur (2001) and Hostetler *et al.* (2003)
Zn	An enzyme cofactor: e.g. alcohol dehydrogenase, alkaline phosphatase, Cu/Zn superoxide dismutase (Hill and Spears, 2001); RNA polymerase (assembly of the enzyme protein and the termination of transcription) (Markov *et al.*, 1999; King *et al.*, 2004). Zn-containing synapses are abundant in the cerebral cortex, where Zn may modulate neuronal excitability (Frederickson *et al.*, 2000). Zn protects against cell death by suppressing a Ca/Mg endonuclease which severs DNA strands (Truong-Tran *et al.*, 2000). Bone (especially trabecular bone) acts as a storage organ for Zn (Windisch *et al.*, 2002). Bioavailability for monogastric animals is 10–40% (Thompson *et al.*, 2004). A variable proportion of dietary Zn is absorbed, the proportion declines at high dietary Zn levels (Spears, 2003), and if high levels of Ca are in the food (e.g. 1.6% Ca in pig foods) Reviews: Hambidge *et al.* (2000) and Hostetler *et al.* (2003)

Another group of trace and ultra-trace minerals has well-defined effects which are useful for optimizing animal performance even though the function appears to be pharmacological rather than physiological. These minerals are listed in Table 9.4 (p. 142).

Practical Implications of Trace Mineral Deficiencies and Excesses

Trace mineral deficiencies can often be traced back to low soil mineral status; this applies to grazing animals and also to intensively fed animals which are given cereal grains grown on mineral-deficient soils. Deficiencies can adversely affect animal performance, but in some cases supplementation is not economic, or if done without care can induce deficiencies or excesses in the supplemented, or another, mineral. Excesses of some trace minerals, again usually related to soil mineral status, can have serious effects on animal health. These aspects of trace mineral nutrition are well reviewed by Reid and Horvath (1980).

Cobalt

It is not required by plants, and so pastures can appear healthy and nutritious while the animals grazing them suffer from the deficiency. Deficient ruminants become anaemic, and their appetite is depressed with resulting weight loss and, in severe cases, emaciation (NRC, 2000). Similar signs occur in deficient pigs, as well as a humped back, uncoordination and muscle tremors (Hill and Spears, 2001). Milk has little vitamin B12, and if the neonate's liver stores of the vitamin are depleted by the dam being fed a Co-deficient food during pregnancy, the young animal will show signs of deficiency while the dam may appear to be normal (Quirk and Norton, 1982). The dietary requirement is very small, 0.1 mg/kg for ruminants, and Co becomes toxic at about 10 mg/kg (Ellison, 1995; NRC, 2000). Pigs should be given vitamin B12 rather than a Co supplement as excess dietary Co is toxic (at about 150 mg/kg; NRC, 1998).

Co can be given to grazing animals by a 'cobaltized' salt block (a salt block with added Co salts). Dewey et al. (1958) developed a Co rumen 'bullet', which is given orally and remains in the rumen where the Co gradually leaches out. The current formulation is a compressed pellet of (up to 60%) CoO and Fe powder which is usually given with a small corrugated steel block which rubs against the bullet to dislodge calcium phosphate accretions; this will last for about 20 months (McFarlane et al., 1992). Excess Co (as well as the deficiency) reduces the animal's appetite (Henry, 1995a).

Copper

Cu is absorbed from the stomach and colon, transported in the blood while bound to ceruloplasmin and albumin and stored in the liver and kidney. There are substantial between-animal differences in the efficiency of absorption; e.g. Suttle (1974) fed a semi-purified diet and measured absorption coefficients in individual sheep ranging from 4.5% to 11.4%. Pre-ruminant lambs and calves absorb much more Cu than ruminating animals. Suttle (1975) found that absorbability fell by 1.18% each day between birth and weaning, and from 47% to 11% within 10 days of weaning. Absorption is compromised by excess sulfate and Mo as these form a complex of Cu and tetrathiomolybdate. Excesses of Zn and Fe (and ingestion of Fe-rich soil) will also precipitate a Cu deficiency (Hill and Spears, 2001). Zn excess promotes the synthesis of metallothionein, which has a higher affinity for Cu than Zn, and the Cu-metallothionein complex is not absorbed (Bremner and Beattie, 1995). Cu is excreted via the bile, bound to metallothionein.

Animal performance can be seriously compromised by Cu deficiency. Deficiencies occur in a wide variety of soil types but animals grazing pastures growing on peat and poorly drained soils, low-fertility sandy soils, and Mo- and/or S-rich soils are at risk (ARC, 1980; Reid and Horvath, 1980).

The multiple enzymic and other roles of Cu lead to a large number and variety of symptoms in deficient animals. Ruminants suffer from deformed and weak legs and possible arthritis, failure of the myelin sheath of the spinal cord to form correctly (this condition is known as 'swayback' or 'enzootic ataxia' and is characterized by animals, especially lambs and kids, which are born unable to use their hind legs), diarrhoea, pale-coloured (depigmented) hair, wool with low tensile strength and little crimp, reduced spermatogenesis and depressed oestrus and anaemia (Langlands, 1987; Grace, 1988; NRC, 2000). Suttle and Jones (1989) concluded that immune function (against bacterial infection) of sheep is impaired by Cu deficiency but that the reverse may hold for internal parasite infection. Increased susceptibility to diseases was reported by Smart *et al.* (1992) in Canadian cattle. Cu-deficient pigs suffer from leg deformities, anaemia, cardiac hypertrophy, rupture of blood vessels, impaired growth and depigmentation (NRC, 1998; Hill and Spears, 2001), and neonatal ataxia has been associated with low liver Cu levels (McGavin *et al.*, 1962).

Cu has some pharmacological applications. It has been used (usually as the sulfate), at 125–250 mg/kg of diet, as a growth promoter in pigs (Braude, 1967; Coffey *et al.*, 1994). CuO wire particles have been used successfully (albeit very carefully) to control intestinal nematodes in sheep (Bang *et al.*, 1990; Burke *et al.*, 2007).

Cu deficiencies can be a particularly intractable problem, especially when the deficiency is only marginal or occurs sporadically. For sheep, at least, an injudicious supplement can convert a marginal Cu status into toxicity. Responses to supplements are varied. There are many reports of Cu supplements improving biochemical measures of animals' Cu status, and some of positive effects on growth and reproduction (Vanniekerk *et al.*, 1995; Bailey *et al.*, 2001), but reports of no improvements in production are also common (Lloyd Davies and Chandrasekaran, 1980; Wesley-Smith and Schlink, 1990; Nicol *et al.*, 2003). When necessary, animals can be given Cu by dosing with capsules containing CuO wire particles – the Cu leaches out over 6–8 months (Rogers and Poole, 1988), Cu glycinate or preferably Cu-EDTA given as subcutaneous injections (but these have to be repeated frequently), oral dosing with an aqueous $CuSO_4$ solution or a copperized salt block. The use of glass pellets and injections to give supplementary Cu are described by McFarlane *et al.* (1991) and Baker and Ammerman (1995).

Because individual animals vary in their abilities to absorb, store and mobilize Cu, Cu toxicity is a risk whenever supplements are given in a way that does not control the individual animal dose. The risk is large in sheep (compared to cattle, horses and pigs) because the difference between adequacy and toxicity is quite small (NRC, 1985a,b; SCA, 1990). Cu toxicity in sheep begins with a slow accumulation of Cu in the liver, followed by a 'haemolytic crisis' (jaundice, anorexia, thirst and haemolysis of red blood cells with resulting haemo-

globinuria) which usually results in death (Bremner, 1998). Excess Cu can promote the formation of reactive oxygen species causing oxidative damage to cells (Gaetke and Chow, 2003), e.g. lipid peroxidation in cell membranes, DNA damage and tissue necrosis (Bremner, 1998; NRC, 2000). Pigs bind more Cu to metallothionein in the liver than sheep do (Bremner, 1998) and this may explain the difference in susceptibility to Cu poisoning in these species.

Recommended dietary Cu levels (NRC, 1985a,b, 1998, 2000, 2007; SCA, 1990; Dzanis, 1994; Ellison, 1995) are (mg/kg): pigs, 3.5 (finishers), 6.0 (neonates); horses and beef cattle, 10; sheep, 5–8 (adults), 11 (pregnant ewes); deer, 7–20; dogs, 2.6. Cu becomes toxic at (mg/kg): pigs, 250; horses, 800; beef cattle, 100; and sheep, 8–25. Toxic levels vary with the S and Mo contents in the food.

Iodine

I is a central constituent of thyroxine (tetraiodothyronine) and triiodothyronine. The principal deficiency symptom is goitre, an enlarged thyroid gland. Because of the metabolic roles of thyroxine, I deficiency causes a variety of other effects. Deficient piglets are born hairless and with skin oedema, and there is a high mortality (Hill and Spears, 2001); early brain development is impaired (Delange, 2000); and ruminants suffer from irregular oestrus and low conception rates (NRC, 2000). Provision of additional I to apparently normal animals can sometimes improve performance. Supplemented ewes gave birth to lambs with greater follicle density and a higher ratio of secondary to primary follicles (Knights *et al.*, 1979). Supplementary I can be given orally (e.g. an iodized salt block) using KI or a similar substance, or by an intramuscular injection of iodized oil (Myers and Ross, 1959).

Iron

Animals ingest Fe from soil adhering to pasture, and in water, as well as that in foods, so a deficiency is very rarely reported in grazing animals. However, as pointed out by Henry and Miller (1995), helminth infections lead to the loss of haem Fe through blood loss into the damaged digestive tract. Sows' milk has very little Fe and newborn pigs are given a supplement to avoid a deficiency (indicated by spasmodic and laboured breathing, loss of appetite and death). The usual method is a single injection of Fe-dextran (200 mg) or a similar preparation, given before the piglet is 72 h old (this approach may lead to abscesses or staining at the injection site), or an oral dose of, for example, Fe-galactan within 18 h of birth (Egeli and Framstad, 1999; Hill and Spears, 2001; Webster *et al.*, 2005). Attempts to increase liver Fe stores in the fetus or milk Fe content, by feeding more Fe to the sow, have generally been unsuccessful (NRC, 1998). There are reports of improved performance from the use of Fe-chelates (Fe bound to amino acids) in which the Fe is more available than in inorganic forms. For example, Kegley *et al.* (2002) and Svoboda and Drabek (2003) reported partial success in replacing an Fe-dextran injection with oral Fe-glycine or Fe-methionine given 1–2 days within birth. Fe toxicities are generally due to inappropriate supplementation. The toxic dose for a newborn pig is six times the usual administered dose of 200 mg (NRC, 1998).

Selenium

Se may be deficient in pastures growing on volcanic soils, e.g. in New Zealand, western Australia, and parts of the USA, the UK, France and Scandinavia (Reid and Horvath, 1980). Se concentrations in cereal grains grown on these soils may also be low. Recommended dietary Se levels are (mg/kg): pigs, 0.35 (neonate), 0.15 (adults); horses and beef cattle, 0.1; sheep, 0.1–0.2; deer, 0.03–0.05 (NRC, 1985a, 1998, 2000, 2007; Ellison, 1995).

Deficiency effects vary between animal species. Pigs suffer from reduced conception rate and milk production, oedema, muscle and liver damage, impaired ability of blood to transport O, impaired immune response and a sudden death syndrome in fast-growing young pigs (NRC, 1998; Mahan, 2001). In ruminants the deficiency causes reduced fertility (embryo resorption), muscular dystrophy (white muscle disease) and ill-thrift in young sheep and cattle (Grace, 1988; NRC, 1998). Se supplements include a rumen 'bullet' containing Se metal, oral aqueous solution of Na selenate (sometimes incorporated with anthelmintic drenches) and 'selenized' salt blocks (reviewed by Langlands, 1987). As for Cu, there are risks of toxic Se intakes from supplements which can be consumed ad lib.

Se toxicities are generally due to animals eating Se-accumulating plants such as *Astragalus bisulca-* *tus* and *Neptunia amplexicaulis*, or, because Se is potentially very poisonous, overdosing a Se supplement. Toxicity can be both acute and chronic. Symptoms are difficulty in breathing, then death: lameness, emaciation, laminitis, liver cirrhosis, nephritis, hair loss from the mane and/or tail or shedding of the fleece in sheep; loss of weight in cattle (NRC, 2000); pigs also lose hair, reduce food intake and suffer fatty liver degeneration (NRC, 1998). Toxicity may occur at 2 mg/kg of the diet for sheep, horses and beef cattle, and at 5 mg/kg for pigs.

Zinc

Zn is absorbed via a saturable Zn carrier mechanism and by passive diffusion (Lönnerdal, 2000). Availability is reduced by binding to phytin in seed-rich diets, and by excess amounts of Ca in the diet possibly as a result of the formation of an insoluble Ca/Zn/phytate complex (Fordyce *et al.*, 1987).

A deficiency of Zn is associated with a loss of appetite, growth depression, slow wound healing, hair loss, impaired immune status and impaired reproduction and testicular development (Underwood and Somers, 1969; Martin *et al.*, 1994; NRC, 1998; Hill and Spears, 2001). The recommended dietary Zn level for cattle is 30 mg/kg DM (but up to 100 mg/kg DM for stressed cattle). The NRC (1998) recommendations for pigs fed conventional diets are (mg/kg): 100 for baby pigs, and 50 for finishers and breeding animals. Deer require 20–35 mg Zn/kg DM (Ellison, 1995).

Breeding males may have a greater Zn requirement than other types of animal, and certainly suffer more severe consequences of a deficiency. Phytate- and/or Ca-rich foods increase the dietary Zn levels required by pigs because they interfere with Zn absorption. Toxicities are rare; cattle will tolerate at least 500 mg Zn/kg DM (NRC, 2000) and pigs have been fed up to 3000 mg Zn/kg diet without adverse effects. However, the chemical form of the Zn supplement, and thus the availability of Zn, influences the response to these levels of supplementation (NRC, 1998). At high dietary levels, Zn acts as an antagonist to Cu absorption (Hill and Spears, 2001).

Minerals with Unresolved Status

The physiological roles and essentiality of the nine minerals in Table 9.4 is unresolved. In most cases

Table 9.4. Minerals with unresolved status as essential nutrients.

Mineral	Function
As	Limited studies with animals suggest that As is involved in gene expression and methionine metabolism Review: Food and Nutrition Board (2000)
B	There is little direct evidence, but studies in plants and animals suggest that B is needed for cell membrane functioning (possibly involved in hormone action (transmembrane signalling)), may be involved in immune function, increases plasma vitamin D in animals with low vitamin D status, improves Ca and Mg retention in lambs and increases chondrocyte density when given as a supplement to deficient chicks; potentially toxic Reviews: Spears (1999) and Nielsen (2000)
Br	No requirement has been demonstrated, but Br therapy can inhibit the ill-effects of excess I intake (Baker, 2004). Br is recognized as a toxic element and most of the research regarding Br in animal foods focuses on this aspect
F	May stimulate the proliferation of osteoblasts. In a pharmacological sense, it hardens bones and teeth (by substituting for CO_3^{2-} and OH^- ions) when incorporated into these tissues. Excessive intakes cause mottling of teeth and overgrowth of bones, and are poisonous Review: Cerklewski (1997)
Li	In the goat, a Li deficiency has been associated with depressed oxidase and dehydrogenase enzyme activities, low birthweights, depressed fertility and low milk production. Li has pharmacological effects (e.g. depression of dopamine synthesis) and a neuroprotective function Reviews: Schrauzer (2002) and Yuan *et al.* (2004)
Ni	May be necessary for Fe absorption (depressed haematocrit is a consistent observation in deficient animals), and in the action of enzymes involved in amino acid, carbohydrate and lipid metabolism (but these effects have not been proven). At high intakes, Ni is teratogenic and possibly carcinogenic Review: Denkhaus and Salnikow (2002)
Rb	Chemically very similar to K and is transported through cell membranes via the Na^+/K^+-ATPase system. May have a pharmacological effect similar to that of Li (Gambarana *et al.*, 1999) and may influence the tissue concentrations of Ca, P, Mg, Zn, Cu, Fe, Na and K (Yokoi *et al.*, 1996)
Si	Acts as an antagonist to Al absorption and transport and by this may improve the absorption of Cu (Al interferes with Cu uptake). This effect may explain the apparent requirement for Si for connective tissue synthesis Review: Birchall *et al.* (1996)
Sn	A potent inducer of haemoglobin oxidase-1 (Barrera *et al.*, 2003) and thus has an antioxidant role
V	Has an insulin-like effect (reduces blood glucose levels; Shechter *et al.*, 2003; Thompson *et al.*, 2004) and anti-tumour effect. It is present in bone but is probably not a required constituent for that tissue (Shukla *et al.*, 2004). Bioavailability is about 5% (Thompson *et al.*, 2004) Reviews: French and Jones (1993) and Baran (2000)

they are poisonous in excess, but some have beneficial pharmacological roles. Nielsen (1996) and Uthus and Seaborn (1996) list the symptoms in animals which have been associated with deficiencies of several of these minerals.

References

Agricultural Research Council (ARC) (1980) *The Nutrient Requirements of Ruminant Livestock.* CAB, Farnham Royal, UK.

Agriculture and Food Research Council (AFRC) (1990) Nutrient requirements of sows and boars. *Nutrition Abstracts and Reviews* 60, 383–406.

Agriculture and Food Research Council (AFRC) (1991) A reappraisal of the calcium and phosphorus requirements of sheep and cattle. *Nutrition Abstracts and Reviews* 61, 573–612.

Ahamed, M. and Siddiqui, M.K.J. (2007) Environmental lead toxicity and nutritional factors. *Clinical Nutrition* 26, 400–408.

Alberts, B., Johnson, A., Lewis, J., Raff, M., Roberts, K. and Walter, P. (2002) *Molecular Biology of the Cell*, 4th edn. Garland Science, New York.

Aldred, J. (1998) *Developmental Orthopaedic Disease in Horses.* RIRDC, Canberra. Available at: http://www.rirdc.gov.au/reports/HOR/US-45A.doc

Ammerman, C.B. (1995) Methods for estimation of mineral bioavailability. In: Ammerman, C.B., Baker, D.H.

and Lewis, A.J. (eds) *Bioavailability of Nutrients for Animals*. Academic Press, San Diego, California, pp. 84–94.

Arizmendi-Maldonado, D., McDowell, L.R., Sinclair, T.R., Mislevy, P., Martin, F.G. and Wilkinson, N.S. (2002) Mineral concentrations in four tropical forages as affected by increasing day length. I. Macrominerals. *Communications in Soil Science and Plant Analysis* 33, 1991–2000.

Arts, J.H.E., Til, H.P., Kuper, C.F., de Neve, R. and Swennen, B. (1994) Acute and subacute inhalation toxicity of germanium dioxide in rats. *Food and Chemical Toxicology* 32, 1037–1046.

Austic, R.E. (1983) Interaction of dietary electrolytes in the nutrition of poultry and swine. *Recent Advances in Animal Nutrition in Australia* 7, 334–339.

Axford, R.F., Tas, M.V., Evans, R.A. and Offer, N.W. (1975) The absorption of magnesium from the fore-stomachs, stomach and small intestine of sheep. *Research in Veterinary Science* 19, 333–334.

Baburaj, K. and Reid, D.M. (2006) Osteomalacia. *Surgery (Oxford)* 24, 388–389.

Bailey, J.D., Ansotegui, R.P., Paterson, J.A., Swenson, C.K. and Johnson, A.B. (2001) Effects of supplementing combinations of inorganic and complexed copper on performance and liver mineral status of beef heifers consuming antagonists. *Journal of Animal Science* 79, 2926–2934.

Baker, D.H. (2004) Iodine toxicity and its amelioration. *Experimental Biology and Medicine* 229, 473–478.

Baker, D.H. and Ammerman, C.B. (1995) Copper bioavailability. In: Ammerman, C.B., Baker, D.H. and Lewis, A.J. (eds) *Bioavailability of Nutrients for Animals*. Academic Press, San Diego, California, pp. 127–156.

Baker, S.B. and Worthley, L.I. (2002) The essentials of calcium, magnesium and phosphate metabolism: Part I. Physiology. *Critical Care and Resuscitation* 4, 301–306.

Bang, K.S., Familton, A.S. and Sykes, A.R. (1990) Effect of copper oxide wire particle treatment on establishment of major gastrointestinal nematodes in lambs. *Research in Veterinary Science* 49, 132–137.

Baran, E.J. (2000) Oxovanadium(IV) and oxovanadium(V) complexes relevant to biological systems. *Journal of Inorganic Biochemistry* 80, 1–10.

Barrera, D., Maldonado, P.D., Medina-Campos, O.N., Hernández-Pando, R., Ibarra-Rubio, M.E. and Pedraza-Chaverrí, J. (2003) HO-1 induction attenuates renal damage and oxidative stress induced by $K_2Cr_2O_7$. *Free Radical Biology and Medicine* 34, 1390–1398.

Barry, T.N. and Blaney, B.J. (1987) Secondary compounds of forages. In: Hacker, J.B. and Ternouth, J.H. (eds) *The Nutrition of Herbivores*. Academic Press, Sydney, Australia, pp. 91–119.

Beardsworth, L.J., Beardsworth, P.M. and Care, A.D. (1989) The effect of ruminal phosphate concentration on the absorption of calcium, phosphorus and magnesium from the reticulo-rumen of the sheep. *British Journal of Nutrition* 61, 715–723.

Berridge, M.J., Lipp, P. and Bootman, M.D. (2000) The versatility and universality of calcium signalling. *Nature Reviews Molecular Cell Biology* 1, 11–21.

Blaney, B.J., Gartner, R.J.W. and McKenzie, R.A. (1981) The inability of horses to absorb calcium from calcium oxalate. *Journal of Agricultural Science, Cambridge* 97, 639–641.

Birchall, J.D., Bellia, J.P. and Roberts, N.B. (1996) On the mechanisms underlying the essentiality of silicon – interactions with aluminium and copper. *Coordination Chemistry Reviews* 149, 231–240.

Bortner, C.D. and Cidlowski, J.A. (2007) Cell shrinkage and monovalent cation fluxes: role in apoptosis. *Archives of Biochemistry and Biophysics* 462, 176–188.

Braude, R. (1967) Copper as a stimulant in pig feeding (cuprum propecunia). *World Review of Animal Production* 3(11), 69–82.

Bravo, D., Meschy, F., Bogaert, C. and Sauvant, D. (2003) Phosphorus availability of oilseed meals determined by the apparent faecal digestibility technique. *Animal Feed Science and Technology* 108, 43–60.

Bremner, I. (1998) Manifestations of copper excess. *American Journal of Clinical Nutrition* 67 (Suppl.), 1069S–1073S.

Bremner, I. and Beattie, J.H. (1995) Copper and zinc metabolism in health and disease: speciation and interactions. *Proceedings of the Nutrition Society* 54, 489–499.

Breves, G., Kock, J. and Schroder, B. (2007) Transport of nutrients and electrolytes across the intestinal wall in pigs. *Livestock Science* 109, 4–13.

Brizuela, C.M. (1996) Bovine botulism in Paraguay. *Tropical Animal Health and Production* 28, 221–222.

Brown, K.M. and Arthur, J.R. (2001) Selenium, selenoproteins and human health: a review. *Public Health and Nutrition* 4, 593–599.

Burke, J.M., Morrical, D. and Miller, J.E. (2007) Control of gastrointestinal nematodes with copper oxide wire particles in a flock of lactating Polypay ewes and offspring in Iowa, USA. *Veterinary Parasitology* 146, 372–375.

Burnell, T.W., Cromwell, G.L. and Stahly, T.S. (1989) Bioavailability of phosphorus in meat and bone meal for pigs. *Journal of Animal Science* 67(Suppl. 2), 38 (Abstr.)

Cahn, F. and Lubin, M. (1978) Inhibition of elongation steps of protein synthesis at reduced potassium concentrations in reticulocytes and reticulocyte lysate. *Journal of Biological Chemistry* 253, 7798–7803.

Caple, I.W. (1989) Nutritional problems affecting calcium and magnesium metabolism in grazing ruminants. *Recent Advances in Animal Nutrition in Australia* 10, 37–46.

Caple, I.W. (1990) Vitamin and mineral nutrition of the horse. *Proceedings of the Australian Society of Animal Production* 18, 117–120.

Cerklewski, F.L. (1997) Fluoride bioavailability – nutritional and clinical aspects. *Nutrition Research* 17, 907–929.

Charbonneau, E., Pellerin, D. and Oetzel, G.R. (2006) Impact of lowering dietary cation–anion difference in nonlactating dairy cows: a meta-analysis. *Journal of Dairy Science* 89, 537–548.

Chester-Jones, H., Fontenot, J.P. and Veit, H.P. (1990) Physiological and pathological effects of feeding high levels of magnesium to steers. *Journal of Animal Science* 68, 4400–4413.

Clark, W.D. Jr, Wohlt, J.E., Gilbreath, R.L. and Zajac, P. K. (1986) Phytate phosphorus intake and disappearance in the gastrointestinal tract of high producing dairy cows. *Journal of Dairy Science* 69, 3151–3155.

Clauss, M., Castell, J.C., Kienzle, E., Schramel, P., Dierenfeld, E.S., Flach, E.J., Behlert, O., Streich, W.J., Hummel, J. and Hatt, J.M. (2007) Mineral absorption in the black rhinoceros (*Diceros bicornis*) as compared with the domestic horse. *Journal of Animal Physiology and Animal Nutrition* 91, 193–204.

Coenen, M. (2005) Exercise and stress: impact on adaptive processes involving water and electrolytes. *Livestock Production Science* 92, 131–145.

Coffey, R.D., Cromwell, G.L., and Monegue, H.J. (1994) Efficacy of a copper-lysine complex as a growth promotant for weanling pigs. *Journal of Animal Science* 72, 2880–2886.

Cohen, R.D.H. (1978) Phosphorus nutrition of ruminants. *Recent Advances in Animal Nutrition in Australia* 4, 120–127.

Cramer, P., Bushnell, D.A. and Kornberg, R.D. (2001) Structural basis of transcription: RNA polymerase II at 2.8 Ångstrom resolution. *Science* 292, 1863–1876.

Cronin, S.J., Neall, V.E., Lecointre, J.A., Hedley, M.J. and Loganathan, P. (2003) Environmental hazards of fluoride in volcanic ash: a case study from Ruapehu volcano, New Zealand. *Journal of Volcanology and Geothermal Research* 121, 271–291.

Delange, F. (2000) The role of iodine in brain development. *Proceedings of the Nutrition Society* 59, 75–79.

Denkhaus, E. and Salnikow, K. (2002) Nickel essentiality, toxicity, and carcinogenicity. *Critical Reviews in Oncology/Hematology* 42, 35–56.

Deschamps, P., Kulkarni, P.P., Gautam-Basak, M. and Sarkar, B. (2005) The saga of copper(II)–L-histidine. *Coordination Chemistry Reviews* 249, 895–909.

Dewey, D.W., Lee, H.J. and Marston, H.R. (1958) The provision of cobalt to ruminants by means of heavy pellets. *Nature* 181, 1367–1371.

Domingo, J.L. (1995) Reproductive and developmental toxicity of aluminum: a review. *Neurotoxicology and Teratology* 17, 515–521.

Dryden, G.M., Wickham, G.A. and Cockrem, F. (1969) Intravenous infusion of cysteine and wool growth of Romney sheep. *New Zealand Journal of Agricultural Research* 12, 580–587.

Dzanis, D.A. (1994) The Association of American Feed Control Officials dog and cat food nutrient profiles: substantiation of nutritional adequacy of complete and balanced pet foods in the United States. *Journal of Nutrition* 124, 2535S–2539S.

Edmeades, D.C. (2004) The magnesium requirements of pastures in New Zealand: a review. *New Zealand Journal of Agricultural Research* 47, 363–380.

Egeli, A.K. and Framstad, T. (1999) An evaluation of iron-dextran supplementation in piglets administered by injection on the first, third or fourth day after birth. *Research in Veterinary Science* 66, 179–184.

Ellison, R.S. (1995) Trace elements in deer. *Proceedings of the Deer Branch of the New Zealand Veterinary Association* 12, 57–68.

Ender, F., Dishington, I.W. and Helgebostad, A. (1971) Calcium balance studies in dairy cows under experimental induction and prevention of hypocalcaemic paresis puerperalis. The solution of the aetiology and the prevention of milk fever by dietary means. *Zeitschrift für Tierphysiologie, Tierernährung und Futtermittelkunde* 28, 233–256.

Evans, M. (1985) *Nutrient Composition of Feedstuffs for Pigs and Poultry*. Queensland Department of Primary Industries, Brisbane, Australia.

Food and Nutrition Board (2000) *Dietary Reference Intakes for Vitamin A, Vitamin K, Arsenic, Boron, Chromium, Copper, Iodine, Iron, Manganese, Molybdenum, Nickel, Silicon, Vanadium, and Zinc*. National Academies Press, Washington, DC.

Fordyce, E.J., Forbes, R.M., Robbins, K.R. and Erdman, J.W. Jr (1987) Phytate × Ca/Zn molar ratios: are they predictive of Zn bioavailability? *Journal of Food Science* 52, 440–444.

Frederick, B.R., van Heugten, E., Hanson, D.J. and See, M.T. (2006) Effects of supplemental magnesium concentration of drinking water on pork quality. *Journal of Animal Science* 84, 185–190.

Frederickson, C.J., Suh, S.W., Silva, D., Frederickson, C.J. and Thompson, R.B. (2000) Importance of zinc in the central nervous system: the zinc-containing neuron. *Journal of Nutrition* 130(5S Suppl.), 1471–1483.

Freeland-Graves, J.H. and Turnlund, J.R. (1996) Deliberations and evaluations of the approaches, endpoints and paradigms for manganese and molybdenum dietary recommendations. *Journal of Nutrition* 126, 2435S–2440S.

French, R.J. and Jones, P.J.H. (1993) Role of vanadium in nutrition: metabolism, essentiality and dietary considerations. *Life Sciences* 52, 339–346.

Frieden, E. (1978) Modes of metal metabolism in mammals. In: Kirchgessner, M. (ed.) *Trace Element Metabolism in Man and Animals 3*. Arbeitskreis fur Tierernahrungsforschung. Weihenstephan, Germany, pp. 8–14.

Gaetke, L.M. and Chow, C.K. (2003) Copper toxicity, oxidative stress, and antioxidant nutrients. *Toxicology* 189, 147–163.

Gambarana, C., Ghiglieri, O., Masi, F., Simona Scheggi, S., Tagliamonte, A. and De Montis, M.G. (1999) The effects of long-term administration of rubidium or lithium on reactivity to stress and on dopamine output in the nucleus accumbens in rats. *Brain Research* 826, 200–209.

Ghatta, S., Nimmagadda, D., Xu, X. and O'Rourke, S.T. (2006) Large-conductance, calcium-activated potassium channels: structural and functional implications. *Pharmacology and Therapeutics* 110, 103–116.

Giles, N.M., Watts, A.B., Giles, G.I., Fry, F.H., Littlechild, J.A. and Jacob, C. (2003) Metal and redox modulation of cysteine protein function. *Chemistry and Biology* 10, 677–693.

Giraud-Guille, M.-M., Belamie, E. and Mosser, G. (2004) Organic and mineral networks in carapaces, bones and biomimetic materials. *Comptes Rendues Palevol* 3, 503–513.

Godoy, S., Meschy, F., Chicco, C. and Requena, F. (2002) Phosphorus digestibility of different feeds in sheep. *Revista Cientifica, Facultad de Ciencias Veterinarias, Universidad del Zulia* 12 (Suppl. 2), 412–415.

Godoy, S., Chicco, C., Meschy, F. and Requena, F. (2005) Phytic phosphorus and phytase activity of animal feed ingredients. *Interciencia* 30, 24.

Goff, J.P. (2006) Macromineral physiology and application to the feeding of the dairy cow for prevention of milk fever and other periparturient mineral disorders. *Animal Feed Science and Technology* 126, 237–257.

Goff, J.P., Ruiz, R. and Horst, R.L. (2004) Relative acidifying activity of anionic salts commonly used to prevent milk fever. *Journal of Dairy Science* 87, 1245–1255.

Goff, J.P., Reinhardt, T.A. and Horst, R.L. (1991) Enzymes and factors controlling vitamin D metabolism and action in normal and milk fever cows. *Journal of Dairy Science* 74, 4022–4032.

Gould, D.H., McAllister, M.M., Savage, J.C. and Hamar, D.W. (1991) High sulfide concentrations in rumen fluid associated with nutritionally induced polioencephalomalacia in calves. *American Journal of Veterinary Research* 52, 1164–1169.

Grace, N.D. (1988) Recent developments in trace elements in animal production. *Proceedings of the Australian Society of Animal Production* 17, 42–46.

Grace, N.D. and Wilson, P.R. (2002) Trace element metabolism, dietary requirements, diagnosis and prevention of deficiencies in deer. *New Zealand Veterinary Journal* 50, 252–259.

Hambidge, M., Cousins, R.J. and Costello, R.B. (2000) Introduction. *Journal of Nutrition* 130, 1341S–1343S.

Harper, A.F., Kornegay, E.T. and Schell, T.C. (1997) Phytase supplementation of low-phosphorus growing–finishing pig diets improves performance, phosphorus digestibility, and bone mineralization and reduces phosphorus excretion. *Journal of Animal Science* 75, 3174–3186.

Haydon, K.D., West, J.W. and McCarter, M.N. (1990) Effect of dietary electrolyte balance on performance and blood parameters of growing-finishing swine fed in high ambient temperatures. *Journal of Animal Science* 68, 2400–2406.

Henry, P.R. (1995a) Cobalt bioavailability. In: Ammerman, C.B., Baker, D.H. and Lewis, A.J. (eds) *Bioavailability of Nutrients for Animals*. Academic Press, San Diego, California, pp. 95–118.

Henry, P.R. (1995b) Manganese bioavailability. In: Ammerman, C.B., Baker, D.H. and Lewis, A.J. (eds) *Bioavailability of Nutrients for Animals*. Academic Press, San Diego, California, pp. 239–256.

Henry, P.R. and Miller, E.R. (1995) Iron bioavailability. In: Ammerman, C.B., Baker, D.H. and Lewis, A.J. (eds) *Bioavailability of Nutrients for Animals*. Academic Press, San Diego, California, pp. 169–199.

Hentze, M.W., Muckenthaler, M.U. and Andrews, N.C. (2004) Balancing acts: molecular control of mammalian iron metabolism. *Cell* 117, 285–297.

Hill, G.M. and Spears, J.W. (2001) Trace and ultratrace elements in swine nutrition. In: Lewis, A.J. and Southern, L.L. (eds) *Swine Nutrition*, 2nd edn. CRC Press, Boca Raton, Florida, pp. 229–261.

Holm, A.McR., Payne, A.L., Morgan, P.D. and Speijers, E.J. (1981) The response of weaner cattle grazing natural pastures in north western Australia to phosphoric acid, non protein nitrogen and sulphur in their drinking water. *Australian Rangeland Journal* 3, 133–141.

Horst, R.L., Goff, J.P., Reinhardt, T.A. and Buxton, D.R. (1997) Strategies for preventing milk fever in dairy cattle. *Journal of Dairy Science* 80, 1269–1280.

Hostetler, C.E., Kincaid, R.L. and Mirando, M.A. (2003) The role of essential trace elements in embryonic and fetal development in livestock. *Veterinary Journal* 166, 125–139.

Igbasan, F.A., Simon, O., Miksch, G. and Männer, K. (2001) The effectiveness of an *Escherichia coli* phytase in improving phosphorus and calcium bioavailabilities in poultry and young pigs. *Archiv für Tierernährung* 54, 117–126.

Jackson, J.A., Hopkins, D.M., Xin, Z. and Hemken, R.W. (1992) Influence of cation–anion balance on feed intake, body weight gain, and humoral response of dairy calves. *Journal of Dairy Science* 75, 1281–1286.

Jansen, B.C. (1969) Past, current and future control of epizootic diseases in South Africa. *Tropical Animal Health and Production* 1, 96–102.

Jarridge, R. (1989) *Ruminant Nutrition. Recommended Allowances and Feed Tables.* INRA and John Libbey Eurotext, Paris.

Jordan, R.M., Meyers, V.S., Yoho, B. and Spurrell, F.A. (1975) Effect of calcium and phosphorus levels on growth, reproduction, and bone development of ponies. *Journal of Animal Science* 40, 78–85.

Kegley, E.B., Spears, J.W., Flowers, W.L. and Schoenherr, W.D. (2002) Iron methionine as a source of iron for the neonatal pig. *Nutrition Research* 22, 1209–1217.

Kincaid, R.L., Garikipati, D.K., Nennich, T.D. and Harrison, J.H. (2005) Effect of grain source and exogenous phytase on phosphorus digestibility in dairy cows. *Journal of Dairy Science* 88, 2893–2902.

King, R.A., Markov, D., Sen, R., Severinov, K. and Weisberg, R.A. (2004) A conserved zinc binding domain in the largest subunit of DNA-dependent RNA polymerase modulates intrinsic transcription termination and antitermination but does not stabilize the elongation complex. *Journal of Molecular Biology* 342, 1143–1154.

Klein, G.L. (2005) Aluminum: new recognition of an old problem. *Current Opinion in Pharmacology* 5, 637–640.

Knights, G.I.,O'Rourke, P.K., Hopkins, P.S. (1979) Effects of iodine supplementation of pregnant and lactating ewes on the growth and maturation of their offspring. *Australian Journal of Experimental Agriculture and Animal Husbandry* 19, 19–22.

Kronfeld, D.S. (2001) Body fluids and exercise: replacement strategies. *Journal of Equine Veterinary Science* 21, 368–375.

Kuzniar, A., Mitura, P., Kurys, P., Szymonik-Lesiuk, S., Florianczyk, B. and Stryjecka-Zimmer, M. (2003) The influence of hypomagnesemia on erythrocyte antioxidant enzyme defence system in mice. *Biometals* 16, 349–357.

Langlands, J.P. (1987) Recent advances in copper and selenium supplementation of grazing ruminants. *Recent Advances in Animal Nutrition in Australia* 9, 144–151.

Lean, I.J., DeGaris, P.J., McNeil, D.M. and Block, E. (2006) Hypocalcemia in dairy cows: meta-analysis and dietary cation anion difference theory revisited. *Journal of Dairy Science* 89, 669–684.

Leonardi, C., Stevenson, M. and Armentano, L.E. (2003) Effect of two levels of crude protein and methionine supplementation on performance of dairy cows. *Journal of Dairy Science* 86, 4033–4042.

Levanger, O.A. and Whanger, P.D. (1996) Deliberations and evaluations of the approaches, endpoints and paradigms for selenium and iodine dietary recommendations. *Journal of Nutrition* 126S, 2427S–2440S.

Linder, M.C. and Hazegh-Azam, M. (1996) Copper biochemistry and molecular biology. *American Journal of Clinical Nutrition* 63, 797S–811S.

Liu, G. (2007) Reconfiguration of synaptic network and enhancement of memory by magnesium ion. *Neuroscience Research* 58, S18 (Abstr).

Lönnerdal, B. (2000) Dietary factors influencing zinc absorption. *Journal of Nutrition* 130, 1378S–1383S.

Lloyd Davies, H. and Chandrasekaran, M. (1980) Investigations on the effect of copper, cobalt and phosphorus on the growth of grazing Hereford cattle in coastal New South Wales. *Proceedings of the Australian Society of Animal Production* 13, 217–220.

Magos, L. and Clarkson, T.W. (2006) Overview of the clinical toxicity of mercury. *Annals of Clinical Biochemistry* 43, 257–268.

Mahan, D.C. (2001) Selenium and vitamin E in swine nutrition. In: Lewis, A.J. and Southern, L.L. (eds) *Swine Nutrition*, 2nd edn. CRC Press, Boca Raton, Florida, pp. 281–314.

Markov, D., Naryshkina, T., Mustaev, A. and Severinov, K. (1999) A zinc-binding site in the largest subunit of DNA-dependent RNA polymerase is involved in enzyme assembly. *Genes and Development*, September 15 13(18), 2439–2448.

Martin, G.B., White, C.L., Markey, C.M. and Blackberry, M.A. (1994) Effects of dietary zinc-deficiency on the reproductive-system of young male sheep – testicular growth and the secretion of inhibin and testosterone. *Journal of Reproduction and Fertility* 101, 87–96.

Martz, F.A., Belo, A.T., Weiss, M.F., Belyea, R.L. and Goff, J.P. (1990) True absorption of calcium and phosphorus from alfalfa and corn silage when fed to lactating cows. *Journal of Dairy Science* 73, 1288–1295.

McCosker, T. and Winks, L. (1994) *Phosphorus Nutrition of Beef Cattle in Northern Australia.* Queensland Department of Primary Industries, Brisbane, Australia.

McFarlane, J.D., Judson, G.J., Turnbull, R.K. and Kempe, B.R. (1991) An evaluation of copper-containing soluble glass pellets, copper oxide particles and injectable copper as supplements for cattle and sheep. *Australian Journal of Experimental Agriculture* 31, 165–174.

McFarlane, J.D., Judson, G.J., Woonton, T., Good, A.H. and Mitsioulis, A. (1992) The response in sheep of plasma and liver vitamin B12 concentrations to cobalt pellets containing various amounts of cobalt oxide. *Proceedings of the Australian Society of Animal Production* 19, 390 (Abstr).

McGavin, M.D., Ranby, P.D., and Tammemegai, L. (1962) Demyelination associated with low liver copper levels in pigs. *Australian Veterinary Journal* 38, 8–14.

McKenzie, R.A., Blaney, B.J. and Gartner, R.J.W. (1981) The effect of dietary oxalate on calcium, phosphorus and magnesium balances in horses. *Journal of Agricultural Science, Cambridge* 97, 69–74.

McLachlan, B.P. and Ternouth, J.H. (1985) Effects of N and P deficiencies on intake and liveweight of lambs. *Proceedings of the Nutrition Society of Australia* 10, 148 (Abstr).

McNeill, D.M., Roche, J.R., McLachlan, B.P. and Stockdale, C.R. (2002) Nutritional strategies for the prevention of hypocalcaemia at calving for dairy cows in pasture-based systems. *Australian Journal of Agricultural Research* 53, 755–770.

Meharg, A.A. and French, M.C. (1995) Heavy metals as markers for assessing environmental pollution from chemical warehouse and plastics fires. *Chemosphere* 30, 1987–1994.

Mertz, W. (ed.) (1986) *Trace Elements in Human and Animal Nutrition*, 5th edn. Academic Press, San Diego, California.

Meschy, F. and Ramirez-Perez, A.-H. (2005) Evolutions récentes des recommandations d'apport en phosphore pour les ruminants. *INRA Productions Animales* 18, 175–182.

Miller, S.M. and Thompson, R.P. (2007) Seasonal patterns of diet composition, herbage intake and digestibility limit the performance of cattle grazing native pasture in the Falkland Islands. *Grass and Forage Science* 62, 135–144.

Minson, D.J. and Norton, B.W. (1982) The possible cause of the absence of hypomagnesaemia in cattle grazing tropical pastures – a review. *Proceedings of the Australian Society of Animal Production* 14, 357–360.

Mongin, P. (1981) Recent advances in dietary anion–cation balance: applications in poultry. *Proceedings of the Nutrition Society* 40, 285–294.

Moran, T.A. and Viele, C.S. (2005) Normal clotting. *Seminars in Oncology Nursing* 21, 1–11.

Murray, R.C., Birch, H.L., Lakhan, K. and Goodship, A.E. (2001) Biochemical composition of equine carpal articular cartilage is influenced by short-term exercise in a site-specific manner. *Osteoarthritis and Cartilage* 9, 625–632.

Myers, B.J. and Ross, D.A. (1959) The effects of iodine and thyroxine administration on the Romney crossbred ewe. *New Zealand Journal of Agricultural Research* 2, 552–574.

National Research Council (NRC) (1981) *Nutrient Requirements of Goats: Angora, Dairy and Meat Goats in Temperate and Tropical Countries*. National Academies Press, Washington, DC.

National Research Council (NRC) (1985a) *Nutrient Requirements of Sheep*, 6th rev. edn. National Academy Press, Washington, DC.

National Research Council (NRC) (1985b) *Nutrient Requirements of Dogs*, rev. 1985. National Academies Press, Washington, DC.

National Research Council (NRC) (1986) *Nutrient Requirements of Cats*, rev. edn. National Academies Press, Washington, DC.

National Research Council (NRC) (1998) *Nutrient Requirements of Swine*, 10th rev. edn. National Academies Press, Washington, DC.

National Research Council (NRC) (2000) *Nutrient Requirements of Beef Cattle*, 7th rev. edn. Update. National Academies Press, Washington, DC.

National Research Council (NRC) (2001) *Nutrient Requirements of Dairy Cattle*, 7th rev. edn. National Academies Press, Washington, DC.

National Research Council (NRC) (2007) *Nutrient Requirements of Horses*, 6th rev. edn. National Academies Press, Washington, DC.

Neathery, M.W. and Miller, W.J. (1975) Metabolism and toxicity of cadmium, mercury, and lead in animals: a review. *Journal of Dairy Science* 58, 1767–1781.

Nicol, A.M., Keeley, M.J., Guild, C.D.H., Isherwood, P. and Sykes, A.R. (2003) Liveweight gain and copper status of young deer treated or untreated with copper oxide wire particles on ten deer farms in Canterbury. *New Zealand Veterinary Journal* 51, 14–20.

Nielsen, F.H. (1996) How should dietary guidance be given for mineral elements with beneficial actions or suspected of being essential? *Journal of Nutrition* 126S, 2377S–2385S.

Nielsen, F.H. (2000) The emergence of boron as nutritionally important throughout the life cycle. *Nutrition* 16, 512–514.

Nielsen, F.H., Hunt, C.D. and Uthus, E.O. (1980) Interactions between essential trace and ultratrace elements. *Annals of the New York Academy of Science* 355, 152–164.

Nisbet, D.I., Butler, E.J., Robertson, J.M. and Bannatyne, C.C. (1970) Osteodystrophic diseases of sheep. IV. Osteomalacia and osteoporosis in lactating ewes on west Scotland hill farms. *Journal of Comparative Pathology* 80, 535–542.

Norman, M.J.T. (1963) The pattern of dry matter and nutrient content changes in native pastures at Katherine, N.T. *Australian Journal of Experimental Agriculture and Animal Husbandry* 3, 119–124.

Opitz von Boberfeld, W. and Banzhaf, K. (2006) The effect of sward management on the mineral content of winter grazed herbage. *Journal of Agronomy and Crop Science* 192, 1–9.

Panda, N. and Sahu, B.K. (2002) Effect of dietary levels of oxalic acid on calcium and phosphorus assimilation in crossbred bulls. *Indian Journal of Animal Nutrition* 19, 215–220.

Partridge, I.G. (1978) Studies on digestion and absorption in the intestines of growing pigs. 3. Net movements of mineral nutrients in the digestive tract. *British Journal of Nutrition* 39, 527–537.

Patience, J.F. and Zijlstra, R.T. (2001) Sodium, potassium, chloride, magnesium and sulfur in swine nutrition. In: Lewis, A.J. and Southern, L.L. (eds) *Swine*

Nutrition, 2nd edn. CRC Press, Boca Raton, Florida, pp. 213–228.

Phillips, C.J.C., Youssef, M.Y.I., Chiy, P.C. and Arney, D.R. (1999) Sodium chloride supplements increase the salt appetite and reduce stereotypies in confined cattle. *Animal Science* 68, 741–748.

Qi, K., Owens, F.N. and Lu, C.D. (1994) Effects of sulfur deficiency on performance of fiber-producing sheep and goats: a review. *Small Ruminant Research* 14, 115–126.

Quirk, M.F. and Norton, B.W. (1982) The effects of cobalt supplementation of pregnant heifers on lactation and calf growth. *Proceedings of the Australian Society of Animal Production* 14, 293–296.

Raun, A., Cheng, E. and Burroughs, W. (1956) Phytate phosphorus hydrolysis and availability to rumen microorganisms. *Journal of Agricultural and Food Chemistry* 4, 869–871.

Reid, R.L. and Horvath, D.J. (1980) Soil chemistry and mineral problems in farm livestock. A review. *Animal Feed Science and Technology* 5, 95–167.

Reilly, C. (1996) *Selenium in Food and Health*. Blackie Academic and Professional, London.

Reis, P.J., Tunks, D.A. and Munro, S.G. (1990) Effects of the infusion of amino acids into the abomasum of sheep, with emphasis on the relative value of methionine, cystine and homocysteine for wool growth. *Journal of Agricultural Science, Cambridge* 114, 59–67.

Rogers, P.A.M. and Poole, D.B.R. (1988) Copper oxide needles for cattle: a comparison with parenteral treatment. *Veterinary Record* 123, 147–151.

Ross, J.G., Spears, J.W. and Garlich, J.D. (1994) Dietary electrolyte balance effects on performance and metabolic characteristics in growing steers. *Journal of Animal Science* 72, 1842–1848.

Ross, R.D., Cromwell, C.L. and Stahly, T.S. (1983) Biological availability of the phosphorus in high-moisture and pelleted corn. *Journal of Animal Science* 57(Suppl. 1), 96 (Abstr).

Ross, R.D., Cromwell, G.L. and Stahly, T.S. (1984) Effects of source and particle size on the biological availability of calcium in calcium supplements for growing pigs. *Journal of Animal Science* 59, 125–134.

Saris, N.-E.L., Mervaala, E., Karppanen, H., Khawaja, J.A. and Lewenstam, A. (2000) Magnesium. An update on physiological, clinical and analytical aspects. *Clinica Chimica Acta* 294, 1–26.

Schrauzer, G.N. (2002) Lithium: occurrence, dietary intakes, nutritional essentiality. *Journal of the American College of Nutrition* 21, 14–21.

Schröder, B. and Breves, G. (2007) Mechanisms and regulation of calcium absorption from the gastrointestinal tract in pigs and ruminants: comparative aspects with special emphasis on hypocalcemia in dairy cows. *Animal Health Research Reviews* 7, 31–41.

Schwab, C.G. (1996) Rumen-protected amino acids for dairy cattle: progress towards determining lysine and methionine requirements. *Animal Feed Science and Technology* 59, 87–101.

Shechter, Y., Goldwaser, I., Mironchik, M., Fridkin, M. and Gefel, D. (2003) Historic perspective and recent developments on the insulin-like actions of vanadium: toward developing vanadium-based drugs for diabetes. *Coordination Chemistry Reviews* 237, 3–11.

Sherlock, R.G., Harris, P.M., Lee, J., Wickham, G.A., Woods, J.L. and McCutcheon, S.N. (2001) Intake and long-term cysteine supplementation change wool characteristics of Romney sheep. *Australian Journal of Agricultural Research* 52, 29–36.

Shukla, R., Barve, V., Padhye, S. and Bhonde, R. (2004) Synthesis, structural properties and insulin-enhancing potential of bis(quercetinato)oxovanadium(IV) conjugate. *Bioorganic and Medicinal Chemistry Letters* 14, 4961–4965.

Smart, M.E., Cymbaluk, N.F. and Christensen, D.A. (1992) A review of copper status of cattle in Canada and recommendations for supplementation. *Canadian Veterinary Journal* 33, 163–170.

Soares, J.H. Jr (1995a) Calcium bioavailability. In: Ammerman, C.B., Baker, D.H. and Lewis, A.J. (eds) *Bioavailability of Nutrients for Animals*. Academic Press, San Diego, California, pp. 95–118.

Soares, J.H. Jr (1995b) Phosphorus bioavailability. In: Ammerman, C.B., Baker, D.H. and Lewis, A.J. (eds) *Bioavailability of Nutrients for Animals*. Academic Press, San Diego, California, pp. 257–294.

Spears, J.W. (1999) Re-evaluation of the metabolic essentiality of the minerals – Review. *Asian-Australasian Journal of Animal Sciences* 12, 1002–1008.

Spears, J.W. (2003) Trace mineral bioavailability in ruminants. *Journal of Nutrition* 133, 1506S–1509S.

Staaland, H., Garmo, T.H., Hove, K. and Pedersen, Ø. (1995) Feed selection and radiocaesium intake by reindeer, sheep and goats grazing alpine summer habitats in Southern Norway. *Journal of Environmental Radioactivity* 29, 39–56.

Standing Committee on Agriculture (SCA) (1987) *Feeding Standards for Australian Livestock. Pigs*. CSIRO Publications, Melbourne, Australia.

Standing Committee on Agriculture (SCA) (1990) *Feeding Standards for Australian Livestock. Ruminants*. CSIRO Publications, Melbourne, Australia.

Stapleton, S.R. (2000) Selenium: an insulin mimetic. *Cellular and Molecular Life Sciences* 57, 1874–1879.

Stewart, G.A. (1976) Mineral nutrients in Australian molasses for fattening cattle. *Proceedings of the Australian Society of Animal Production* 11, 373–376.

Suttle, N.F. (1974) A technique for measuring the biological availability of copper to sheep, using hypocupraemic ewes. *British Journal of Nutrition* 32, 395–405.

Suttle, N.F. (1975) Changes in the availability of dietary copper to young lambs associated with age and weaning. *Journal of Agricultural Science, Cambridge* 84, 255–261.

Suttle, N.F. and Jones, D.G. (1989) Trace-elements, disease resistance and immune responsiveness in ruminants. *Journal of Nutrition* 119, 1055–1061.

Svoboda, M. and Drabek, J. (2003) Efficiency of voluntary consumption of amino acid-chelated iron in preventing anaemia of suckling piglets. *Acta Veterinaria Brno* 72, 499–507.

Thilsing-Hansen, T., Jørgensen, R.J. and Østergaard, S. (2002) Milk fever control principles: a review. *Acta Veterinaria Scandinavica* 43, 1–19.

Thompson, K.H., Chiles, J., Yuen, V.G., Tse, J., McNeill, J.H. and Orvig, C. (2004) Comparison of anti-hyperglycemic effect amongst vanadium, molybdenum and other metal maltol complexes. *Journal of Inorganic Biochemistry* 98, 683–690.

Timlin, J.A., Carden, A., Morris, M.D., Rajachar, R.M. and Kohn, D.H. (2000) Raman spectroscopic imaging markers for fatigue-related microdamage in bovine bone. *Analytical Chemistry* 72, 2229–2236.

Traylor, S.L., Cromwell, G.L. and Lindemann, M.D. (2005a) Bioavailability of phosphorus in meat and bone meal for swine. *Journal of Animal Science* 83, 1054–1061.

Traylor, S.L., Cromwell, G.L. and Lindemann, M.D. (2005b) Effects of particle size, ash content, and processing pressure on the bioavailability of phosphorus in meat and bone meal for swine. *Journal of Animal Science* 83, 2554–2563.

Tremblay, G.F., Brassard, H., Belanger, G., Seguin, P., Drapeau, R., Bregard, A., Michaud, R. and Allard, G. (2006) Dietary cation anion difference of five cool-season grasses. *Agronomy Journal* 98, 339–348.

Truong-Tran, A.Q., Ho, L.H., Chai, F. and Zalewski, P.D. (2000) Cellular zinc fluxes and the regulation of apoptosis/gene-directed cell death. *Journal of Nutrition* 130(5S Suppl.), 1459–1466.

Underwood, E.J., and Somers, M. (1969) Studies of zinc nutrition in sheep.I. Relation of zinc to growth testicular development, and spermatogenesis in young rams. *Australian Journal of Agricultural Research* 20, 889–897.

Underwood, E.J. and Suttle, N.F. (1999) *The Mineral Nutrition of Livestock*, 3rd edn. CAB International, Wallingford, UK.

Uthus, E.O. and Seaborn, C.D. (1996) Deliberations and evaluations of the approaches, endpoints and paradigms for dietary recommendations of the other trace elements. *Journal of Nutrition* 126S, 2452S–2459S.

Van der Harst, M.R., Brama, P.A., van de Lest, C.H., Kiers, G.H., DeGroot, J. and van Weeren, P.R. (2004) An integral biochemical analysis of the main constituents of articular cartilage, subchondral and trabecular bone. *Osteoarthritis Cartilage* 12, 752–761.

Van Doorn, D.A., Everts, H., Wouterse, H. and Beynen, A.C. (2004) The apparent digestibility of phytate phosphorus and the influence of supplemental phytase in horses. *Horse Health Nutrition*. Second European Equine Health and Nutrition Congress, Equine Research Centre, Waiboerhoeve, Lelystad, The Netherlands, pp. 84–86.

Vanniekerk, F.E., Cloete, S.W.P., Vandermerwe, G.D., Heine, E.W.P. and Scholtz, A.J. (1995) Parenteral copper and selenium supplementation of sheep on legume-grass pastures – biochemical and production responses in lambs to maternal treatment. *Journal of the South African Veterinary Association* 66, 11–17.

Vincent, J.B. (2000) The biochemistry of chromium. *Journal of Nutrition* 130, 715–718.

Vincent, J.B. (2004) Recent advances in the nutritional biochemistry of trivalent chromium. *Proceedings of the Nutrition Society* 63, 41–47.

Wadsworth, J.C., Schlink, A.C., Miller, C.P. and Hendricksen, R.E. (1988) Deleterious effect of dietary phosphorus supplement on grazing cattle: a phosphorus copper interaction? *Proceedings of the Australian Society of Animal Production* 17, 346–349.

Warren, B.E., Bunny, C.J. and Bryant, E.R. (1990) A preliminary examination of the nutritive value of four saltbush (*Atriplex*) species. *Proceedings of the Australian Society of Animal Production* 18, 424–427.

Webster, W.R., Dimmock, C.K. and Moore, M.J. (2005) *Piglet Anaemia*. Available at: http://www2.dpi.qld.gov.au/pigs/4454.html

Wesley-Smith, R.N. and Schlink, A.C. (1990) Mineral nutrition studies of rangeland cattle in the north west coastal region of the Northern Territory. *Proceedings of the Australian Society of Animal Production* 18, 428–431.

West, J.W., Mullinix, B.G. and Sandifer, T.G. (1991) Changing dietary electrolyte balance for dairy cows in cool and hot environments. *Journal of Dairy Science* 74, 1662–1674.

Wilkinson, J.M., Hill, J. and Phillips, C.J.C. (2003) The accumulation of potentially-toxic metals by grazing ruminants. *Proceedings of the Nutrition Society* 62, 267–277.

Williams, R.J.P. (1996) Aluminium and biological systems: an introduction. *Coordination Chemistry Reviews* 149, 1–9.

Windisch, W., Wher, U., Rambeck, W. and Erben, R. (2002) Effect of Zn deficiency and subsequent Zn repletion on bone mineral composition and markers of bone tissue metabolism in ^{65}Zn-labelled, young-adult rats. *Journal of Animal Physiology and Animal Nutrition* 86, 214–221.

Yakovlev, A.G., Wang, G., Stoica, B.A., Boulares, H.A., Spoonde, A.Y., Yoshihara, K. and Smulson, M.E. (2000) A Role of the Ca^{2+}/Mg^{2+}-dependent endonucleave in apoptosis and its inhibition by poly(ADP-ribose) polymerase. *Journal of Biological Chemistry* 275, 21302–21308.

Yanke, L.J., Bae, H.D., Selinger, L.B., Cheng, K.J. (1998) Phytase activity of anaerobic ruminal bacteria. *Microbiology-UK* 144, 1565–1573.

Yokoi, K., Kimura, M. and Itokawa, Y. (1996) Effect of low dietary rubidium on plasma biochemical parameters and mineral levels in rats. *Biology Trace Element Research* 51, 199–208.

Yuan, P., Gould, T.D., Gray, N.A., Bachmann, R.F., Schloesser, R.J., Lan, M.J.K., Du, J., Moore, G.J. and Manji, H.K. (2004) Neurotrophic signaling cascades are major long-term targets for lithium: clinical implications. *Clinical Neuroscience Research* 4, 137–153.

Zhao, F.J., Adams, M.L., Dumont, C., McGrath, S.P., Chaudri, A.M., Nicholson, F.A., Chambers, B.J. and Sinclair, A.H. (2004) Factors affecting the concentrations of lead in British wheat and barley grain. *Environmental Pollution* 131, 461–468.

Chapter 9

10 Vitamins

History, Nomenclature, Classification and Definition

Vitamins are organic substances which are involved in the regulation of gene expression (vitamins A and D, thiamin, vitamin B12); in a structural role in the visual pigment, rhodopsin (vitamin A as retinol); and function as antioxidants (vitamins A, E and C); and as enzyme cofactors (the B group vitamins and vitamin K). Each vitamin has a unique role in animal metabolism and none can be substituted completely by another nutrient. They are essential nutrients and must be provided in the food or harvested (possibly by caecotrophy) from the digestive tract after having been synthesized by microorganisms. Vitamins are needed in very small amounts (µg/day) for a typical animal.

The first identified vitamin was isolated by Funk (1911, 1913). He characterized it as an amine and coined the term 'vital amine' or 'vitamine' to reflect its nitrogenous character (Funk, 1921). The word was later applied to all members of this class of nutrients.

It was realized in the latter part of the 19th century that animals needed some other 'accessory factors' in their diet in addition to the then-recognized carbohydrates, proteins and fats (Funk, 1914; McCollum, 1918; Hopkins, 1919). Credit for this observation has been given variously to Dumas and Lunin, who reported their findings in the 1870s and 1880s (McCollum, 1953). Funk (1913, 1914) found that beriberi, a neurological disorder which occurred particularly when people or animals ate diets rich in polished rice, could be cured by giving an aqueous extract of wheat grain, yeast or even rice bran. McCollum and Davis (1914) showed that a substance in butterfat was necessary to prevent ill health. By 1919, at least four accessory factors, or 'vitamines' were postulated: the 'fat-soluble A' found in animal fats, the 'water-soluble B' extracted from plant foods, an 'antiscorbutic factor' later called 'vitamin C' which was effective against scurvy and a fat-soluble antirachitic factor later named 'vitamin D' which prevented rickets (Hopkins, 1919; Mellanby, 1920; McCollum et al., 1922).

Vitamins are divided into the 'fat-soluble' and the 'water-soluble' types. This classification evolved early in the study of vitamins (McCollum et al., 1916) when a water-soluble substance in rice hulls was identified as an 'anti-beriberi' factor (called vitamin B1, thiamine or aneurin), and a fat-soluble factor (later identified as vitamin A) was found in milk.

Presently, 13 water-soluble vitamins are recognized (those in Table 10.1 plus myo-inositol and p-aminobenzoic acid; AINCN, 1987) and the four fat-soluble vitamins, A, D, E and K. Many other substances have been nominated as vitamins. Examples include adenine ('vitamin B4') and adenosine monophosphate ('vitamin B8') which have important biochemical roles but are not vitamins because they are synthesized in animal tissues; and also substances for which there is no evidence of biological effectiveness or requirement, e.g. pangamic acid ('vitamin B15'). The essential fatty acids have been referred to as 'vitamin F' but this usage has disappeared now that their chemistry and functions are well known. The same is happening now for the flavonoids ('vitamin P') but the essentiality and functions of these substances are still being worked out. There is some inconsistency in the naming of recognized vitamins. Folic acid is called 'vitamin B9' in France and 'vitamin B11' in the Netherlands (Van Dusseldorp, 1997). p-Aminobenzoic acid (PABA or vitamin B10) is a constituent of folate. Vitamin D1 was shown to be a mixture of vitamin D2 (ergocalciferol) and vitamin D3 (cholecaliferol) (Fraser, 1995).

Units of measurement

We use conventional mass units, i.e. µg or mg/kg of food or food dry matter for the B group vitamins and ascorbic acid, as the active forms of these vitamins occur naturally in foods. Vitamins A and D occur in both provitamin and active forms, and vitamin E occurs in several different forms which have different potencies. International Units (IU) are used to describe the potencies of different vitamin forms or

Table 10.1. Classification and chemistry of vitamins. (Names are according to the recommendations of AINCN, 1987).

Vitamin (chemical form)	Other names
Fat-soluble	
Vitamin A (retinoic acid; retinol; retinaldehyde)	
(provitamin: β-carotene; carotenoids)	
Vitamin D (cholecalciferol, ergocalciferol)	Vitamin D2, vitamin D3, calciferol
(active forms: 25-hydroxycholecalciferol;	
1,25-dihydroxycholecalciferol)	
Vitamin E (α-, β-, γ- and δ-tocopherols)	
Vitamin K (menadione, phylloquinone, menaquinone)	Vitamin K1, vitamin K2
Water-soluble[a]	
Thiamin	Vitamin B1, thiamine, aneurin
Riboflavin	Vitamin B2
Niacin (nicotinamide, nicotinic acid)	Vitamin B3, vitamin PP
Pantothenic acid	Vitamin B5
Vitamin B6 (pyridoxine, pyridoxal, pyridoxamine)	
Biotin	Vitamin B7, vitamin H, coenzyme R
Folate (folic acid, folacin, tetrahydrofolic acid)	Vitamin B9, vitamin B11, vitamin Bc, vitamin M
Vitamin B12 (cyanocobalamin, aquacobalamin, nitritocobalamin)	Vitamin B12a, vitamin B12b, vitamin B12c
Choline	
Lipoic acid	Thioctic acid
Vitamin C (ascorbic acid, dehydroascorbic acid)	

[a]All these vitamins except vitamin C comprise the B group of vitamins.

the potential activities of vitamin precursors. Despite being superseded by other units, the IU is still commonly used. The retinol activity equivalent (RAE) measures the potency of the active (or activated) form of vitamin A precursors, and vitamin E activity is quantified by reference to amounts (mg or μg) of *RRR*-α-tocopherol (AINCN, 1987; NRC, 2000; Blomhoff and Blomhoff, 2006):

$$1 \text{ IU of vitamin A} = \text{activity of } 0.300 \text{ μg of}$$
$$\text{all-}trans \text{ retinol} = 0.55 \text{ μg retinol palmitate} \tag{10.1a}$$

$$1 \text{ RAE} = 1 \text{ μg of retinol} = 3.3 \text{ IU}$$
$$\text{vitamin A} \tag{10.1b}$$

$$1 \text{ IU of vitamin D3} = 0.025 \text{ μg vitamin D3} \tag{10.2}$$

$$1 \text{ IU of vitamin E h} =$$
$$0.67 \text{ mg } RRR\text{-α-tocopherol} \tag{10.3}$$

The Fat-soluble Vitamins

Vitamin A

All-*trans* retinol (Fig. 10.1) is a pale yellow, crystalline solid, insoluble in water but soluble in fat and organic solvents. It is destroyed by UV radiation and readily oxidized in air. Dietary sources of vitamin A include that stored in animal tissues, and vitamin A precursors (e.g. α- and β-carotene, β-cryptoxanthin and other carotenoids; Fig. 10.1) which occur in yellow- or orange-coloured plant materials or in ruminant fats. Some 10% of the 600 carotenoids found in nature can be converted into retinol (Silveira and Moreno, 1998). Theoretically 1 μg of β-carotene should give 1 μg of retinol but the conversion in practice is not as efficient as this (Table 10.2). Note that cats, which are obligate carnivores, cannot activate β-carotene (Lakshmanan *et al.*, 1972); they have to obtain their vitamin A from animal livers, although they can absorb dietary β-carotene (Chew *et al.*, 2000). Carotenoids are absorbed by cattle, leading to the yellowing of milk and body fat. Other animal species generally convert carotenoids to retinol before they are absorbed (Handelman, 2001). Active vitamin A is stored in the stellate cells of the liver, as fatty acid esters, mainly retinyl palmitate (Silveira and Moreno, 1998). This can protect grazing animals against seasonal shortages for 2–4 months (NRC, 2000) or several months (SCA, 1990).

The first function of vitamin A to be discovered was its ability to prevent xerophthalmia, and its

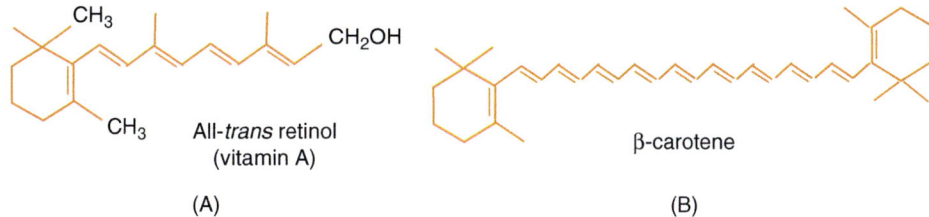

Fig. 10.1. Structural formulae of retinol and β-carotene.

Table 10.2. Activation of β-carotene. (From Lakshmanan *et al.*, 1972 (cat); Ullrey, 1972; NRC, 1998, 2007 (horse and pig).)

Animal species	Vitamin A/β-carotene equivalence (IU/mg)	Vitamin A/β-carotene equivalence (RAE/mg)
Rats[a]	72–1,667	22–505
Horses[b]	333–555	101–168
Ruminants[c]	300 or 400	90–127
Cattle	400	127
Sheep	250–680	76–206
Goats	400	127
Pigs	267	81
Cats	0	0

[a]The equivalence depends on the dose of β-carotene and approaches 100% (1,667 IU/mg) at 0.3 μmol/kg live weight (NRC, 1995).
[b]For growing and lactating horses, respectively.
[c]300 and 400 IU vitamin A activity/mg β-carotene are widely used conversions for ruminants (ARC, 1980; NRC, 1981, 1985, 2000, 2001; SCA, 1990).

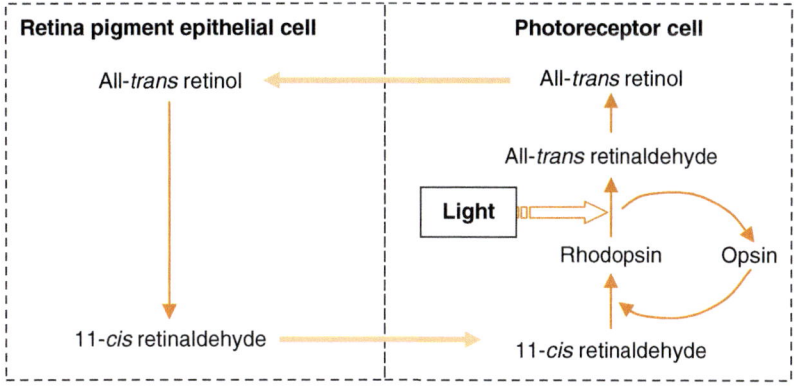

Fig. 10.2. The role of vitamin A in the synthesis of the visual pigment, rhodopsin (11-*cis* retinaldehyde-opsin).

role in visual pigment was elucidated soon afterwards (Wald, 1968). Retinol has a structural (rather than a catalytic) function in vision. 11-*cis* retinal binds to opsin forming rhodopsin (Fig. 10.2). When bleached by light the 11-*cis* retinaldehyde in rhodopsin changes to the *trans* form and the change in shape of the molecule generates a nerve impulse (Blomhoff and Blomhoff, 2006).

The active form of vitamin A is retinoic acid. Retinoic acid binds to nuclear proteins and controls gene expression (Liau *et al.*, 1981) and has a key role in the early development of embryos (Darroch, 2001). It is involved in the regulation of cell growth and intermediary glucose metabolism (Darroch, 2001). Retinoic acid is involved in glycoprotein synthesis (King and Pope, 1986) and the keratinization of skin, and controls the synthesis of surfactants and protects epithelial tissue and mucous membranes (Biesalski and Nohr, 2003). Rowland (1970) suggested that low vitamin A status is associated with the rumenitis/liver abscess syndrome, which is often found when cattle are fed low-fibre foods, as in typical cereal grain-based feedlot rations. A caveat though: any protective effect of vitamin A has not been proved and hyperkeratosis (a related disorder) has, in other animal species, been associated with vitamin A excesses. Carotenoids also act as antioxidants in addition to their role as retinol precursors (Handelman, 2001).

There are several symptoms of vitamin A deficiency, reflecting the multiplicity of vitamin A functions. Night blindness, the inability to see in conditions of low light intensity, is a symptom which is related to the role of retinol in the formation of visual pigment. Vitamin A deficiency reduces sperm production – retinoic acid controls the differentiation of sperm-forming cells and the release of sperm into the seminiferous tubules (Ghyselinck et al., 2006). Indeed, the gene coding for retinol-binding protein has been suggested as a marker of fertility in pigs (Rothschild et al., 2000). Other deficiency effects include rough hair, oedema, reduced growth rate, abnormal bone growth, blindness, abortion and stillbirths in cattle (NRC, 2000), and these symptoms plus hind limb paralysis, uncoordination, respiratory dysfunction and increased susceptibility to disease in pigs (Darroch, 2001).

Excessive consumption of retinol can overload the retinol-binding capacity of the intestinal epithelial cells (Suzuki et al., 1995). Excess free retinol is toxic and in the pig can cause fetal abnormalities, thickening of skin and fragile and overgrown bones

as well as other disorders (Darroch, 2001). Vitamin A toxicity in dogs causes reduced appetite, lethargy, vomiting, skin peeling, hair loss and gingivitis (Fitzgerald et al., 2006). Cats are particularly sensitive to vitamin A toxicity and bone deformity is a common symptom (Polizopoulou et al., 2005).

Vitamin D

'Vitamin D' is a generic term for all biological substances with vitamin D activity. The main substance is cholecalciferol (Fig. 10.3) although very small amounts of ergocalciferol, made by the UV irradiation of ergosterol, are found in some plant foods. The vitamin D3 provitamin is cholesterol, which is made by animals from endogenous fatty acids, and is also consumed in animal foods. Cholesterol is irradiated by UV light in the skin, giving cholecalciferol, which is later converted to the active substance 1,25-dihydroxy cholecalciferol or 'calcitriol' (Fig. 10.3).

Vitamin D3 is a steroid pro-hormone; the active form is 1,25-dihydroxy cholecalciferol. This substance mediates expression of the gene which codes for the synthesis of Ca-binding protein, and so regulates Ca absorption from the small intestines. Vitamin D also inhibits uncontrolled cell proliferation, promotes cell differentiation, enhances innate immunity and inhibits autoimmunity (Lips, 2006). The functions of vitamin D are reviewed by Fraser (1995) and Lal et al. (1999).

Vitamin D deficiency will occur if the diet supplies inadequate cholecalciferol and the animal is not exposed to sufficient sunlight. There appears to be no information on the minimum exposure to UV_B radiation needed to ensure that animals have

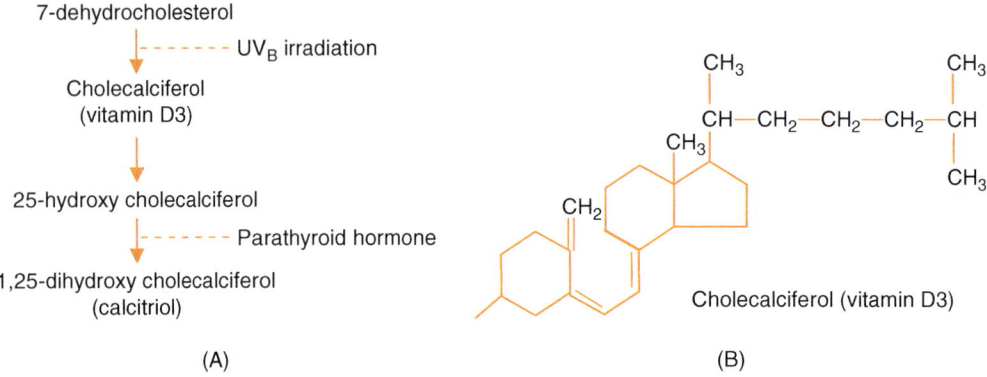

Fig. 10.3. Vitamin D3 (cholecalciferol): (A) synthesis and activation; (B) structural formula.

enough vitamin D. Vitamin D deficiencies may occur in housed pigs (Thompson and Robinson, 1989) and possibly in household pets kept indoors, but deficiencies are rare or non-existent in animals with access to sunlight. Clements *et al.* (1987) have reported an increased rate of vitamin D turnover and loss when rats were fed a low-Ca diet. It is possible, then, that diets based on cereal grains and other Ca-deficient foods may set up conditions which increase the vitamin D requirement.

Vitamin D deficiency may cause new bone to fail to mineralize (osteomalacia in adults, rickets in young animals) and osteoporosis (the withdrawal of mineral from mature bone) (Lips, 2006). In these conditions, the bone is weakened and prone to breakage, and in the case of rickets long bones may be bent and the young animal may grow more slowly than normal. In other cases, the deficiency leads to fetal deformities or stillbirths (NRC, 2000).

Excesses of vitamin D are toxic, causing elevated blood Ca, calcification of soft tissues (e.g. heart and blood vessels), extensive kidney damage, brittle and deformed bones and excessive bone mineral resorption in pigs (NRC, 1998). Similar symptoms occur in companion animals (Fitzgerald *et al.*, 2006). Animal managers should check the vitamin D content of multivitamin supplements, as vitamin A supplements for grazing animals may contain quite high levels of vitamin D.

Vitamin E

This is the generic name for a group of tocopherols which have a similar vitamin activity. α-tocopherol is the most common form (Fig. 10.4) and also the most active. Ullrey (1972) suggests that the different tocopherols have these activities, relative to α-tocopherol (100%): the β and ζ forms have activities of 33%, and the others (γ, δ, ε, η) have activities of less than 1%. Vitamin E occurs in green plants and seeds.

Vitamin E is an antioxidant. It protects against the oxidation of cell membrane fatty acids, especially those in mitochondria and the endoplasmic

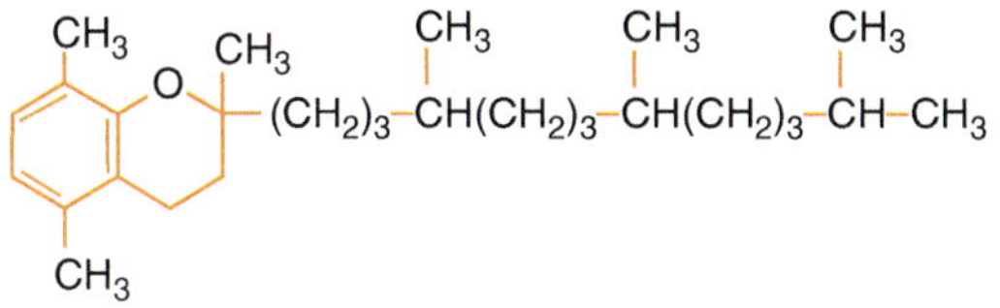

Fig. 10.4. Structural formula of α-tocopherol (vitamin E).

reticulum. A deficiency results in anaemia, muscle degeneration, digestive tract lesions and liver necrosis (NRC, 1998). Vitamin E-deficient sows may have higher incidences of mastitis, metritis and agalactia, and fewer pigs born, and pork quality is poorer due to greater lipid oxidation and reduced shelf life (Mahan, 1994, 2001).

Vitamin K

This is a group of three related compounds which are active in promoting blood clotting (Fig. 10.5): vitamin K1 (phylloquinone), vitamin K2 (menaquinone) and vitamin K3 (menadione). Phylloquinone occurs in foods derived from green plants and plant oils, but an important source of vitamin K for animals is the synthesis of menaquinone by intestinal bacteria. Although phylloquinone and menaquinone are fat-soluble substances, there are water-soluble analogues of menadione which have similar functions and which are used in animal feeds (CVM/FDA, 1990). Vitamin K cofactors enzymes which carboxylate glutamic acid to γ-carboxyglutamate; proteins with this amino acid are involved in Ca metabolism, including blood clotting and the mineralization of bone (Shearer, 1995). Animals deficient in vitamin K have impaired blood clotting and possibly poor bone mineralization but a deficiency is uncommon in grazing animals. Pigs may be given a dietary supplement. Vitamin K toxicity is rare.

Water-soluble Vitamins

These are the B group (or B complex) vitamins and ascorbic acid (vitamin C). The B group vitamins are found in plant foods and are synthesized by some intestinal bacteria because they are required for bacterial cell metabolism. When these plant or bacterial cells are digested, the vitamins become available for absorption by the animal. However, there is some evidence from pure culture work (Hungate, 1966) that some rumen bacteria require dietary sources of thiamin, niacin, *p*-aminobenzoic acid, folate, biotin, vitamin B6, pantothenic acid and/or vitamin B12.

Most B group vitamins are coenzymes involved in the reactions of cellular intermediary metabolism such as transamination, the electron transport pathway and the tricarboxylic acid cycle (Table 10.3; Fenech, 2001; FAO/WHO, 2002; Sybesma *et al.*, 2003; Depeint *et al.*, 2006).

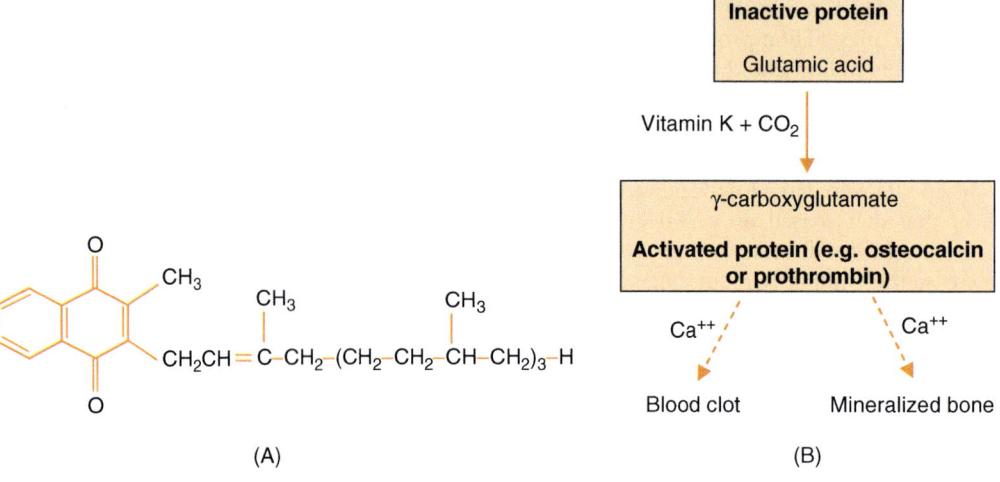

Fig. 10.5. Vitamin K: (A) structure of phylloquinone (vitamin K1); (B) carboxylation of glutamic acid and activation of glutamate-containing proteins.

Table 10.3. Functions of the B group vitamins.

Vitamin	Form as coenzyme or prosthetic group	Function
Thiamin	Thiamin pyrophosphate	Oxidative decarboxylation
Riboflavin	Flavin mononucleotide, flavin adenine dinucleotide	Oxidation–reduction reactions
Niacin	Nicotinamide adenine dinucleotide, nicotinamide adenine dinucleotide phosphate	H transfer reactions in fatty acid metabolism, glycolysis and the tricarboxylic acid cycle
Pantothenic acid	Coenzyme A, acyl carrier protein	Acyl group activation and transfer
Vitamin B6	Pyridoxal phosphate (occurs as pyridoxine, pyridoxal and pyridoxamine and the phosphorylated forms of these molecules)	Amino acid transamination, decarboxylation and racemization reactions; a cofactor in the formation of tetrahydrofolate and in glutathione biosynthesis
Folate	Tetrahydrofolic acid (also occurs as folic acid)	One-C group transfer, in the synthesis of amino acids, purines and pyrimidines; the methylation of DNA and prevention of chromosome breakage
Biotin	Biotinyl-protein enzymes	Carboxylation reactions
Lipoic acid	Lipoyl-protein enzymes	Dehydrogenation reactions
Vitamin B12	Adenosylcobalamin, methylcobalamin; (cyanocobalamin is the pharmaceutical form; the cyanide radicle is a contaminant resulting from a purification step)	Cofactor to methylmalonyl CoA mutase (conversion of propionate to succinyl-coA in gluconeogenesis) and methionine synthase (methionine synthesis)
Choline		Methyl group donor, in the synthesis of phospholipids, acetylcholine and methionine, and mitochondrial protein and nucleic acid synthesis

Sources of B group vitamins

All except vitamin B12 are present in foods of plant origin (Dove and Cook, 2001). For example, thiamin occurs in unmilled cereal grains (i.e. those with the seedcoat retained), in actively growing plant tissues, and yeast and animal tissues are rich sources. Folate is widely distributed in green plant material, although in amounts which may not meet animal requirements. Niacin similarly occurs widely in plant foods but its availability can be low for pigs. If foods are heated, e.g. in pelleting or extrusion, it is likely that a substantial proportion of thiamin (up to 77%) and vitamin B6 (65%) and a lesser proportion of pantothenic acid will be destroyed (Athar *et al.*, 2006).

B group vitamins are usually synthesized by rumen bacteria in amounts sufficient to meet the host animal's requirements, provided that sufficient bacterial nutrients like P, NH_3 and Co are present. Animals are entirely reliant on microbial synthesis of vitamin B12, as this vitamin does not exist in the higher plants. Cobalamin is synthesized by digestive tract bacteria (Martens *et al.*, 2002) to meet their need for this vitamin in substrate fermentation, methylation, acetogenesis and other metabolic pathways. This observation refers to adults with a functioning rumen and there are reports of deficiencies of almost all the B group vitamins in young calves (NRC, 2000).

Choline is synthesized by the animal using methionine and serine as starting compounds. Vitamin B6 is needed as a cofactor. This synthesis does not always supply the animal's requirements and so dietary choline is sometimes beneficial.

There is only limited absorption of B group vitamins from the caecum and colon of hindgut fermenters. Lindemann (1993) has suggested that caecotrophy may be necessary for optimum intake and absorption of folate in non-ruminants. Biotin and vitamin B12 are synthesized by large intestinal bacteria in the pig, but they are not absorbed from this part of the digestive tract (Mosenthin *et al.*, 1990; Dove and Cook, 2001). Reports summarized in NRC (2007) suggest that horses generally have adequate supplies of B group vitamins. However, there is some evidence that they may respond to thiamin and folate supplements.

B vitamin deficiencies

Mild deficiencies of these vitamins usually cause poor appetite and growth; more severe deficiencies give a wider range of symptoms (Kornegay, 1986; Lindemann, 1993; NRC, 1998; Dove and Cook, 2001). Symptoms of severe deficiencies in pigs include epileptic-like fits (in vitamin B6 deficiency), muscular uncoordination or other forms of lameness (biotin, choline, pantothenic acid, riboflavin, vitamin B6, vitamin B12), dermatitis and/or hair loss or roughness (biotin, choline, niacin, pantothenic acid, riboflavin, vitamin B12), and vomiting or diarrhoea (biotin, niacin, pantothenic acid,

riboflavin, thiamin, vitamin B6). Other effects of B group vitamin deficiencies include hypersensitivity (vitamin B12), impaired sight (riboflavin, vitamin B6) or immune function (pantothenic acid, riboflavin, vitamin B6), anaemia (choline, folate, vitamin B6, vitamin B12), digestive tract lesions (biotin, niacin), fatty liver (choline, pantothenic acid, riboflavin, vitamin B6) and sudden death (thiamin, vitamin B6).

Biotin deficiency has a characteristic effect on hoof and hair growth. Hoof lesions, e.g. splits in the hoof wall, are a commonly reported deficiency symptom in horses and pigs, and biotin supplements increase hoof growth and hardness (Comben *et al.*, 1984; Kornegay, 1986; Buffa *et al.*, 1992).

A thiamin deficiency can be induced by the action of thiaminase, an enzyme elaborated by some rumen bacteria which proliferate in the rumens of animals given concentrate-rich diets (Ramos *et al.*, 2003), and which is also found in some plants (e.g. bracken fern; Somogyi, 1973). Deficiencies in pigs, cattle and horses produce reduced weight gain, muscular uncoordination and periodic hypothermia in ears, hooves and the muzzle. Polioencephalomalacia (also called 'cerebrocortical necrosis', CCN) is associated with feeding concentrate-rich feedlot diets to cattle or goats. It causes degeneration of brain tissue resulting in listlessness, muscular tremors, blindness, convulsions and death.

Vitamin B12 deficiency causes progressive loss of appetite and of weight proceeding to gross emaciation in adult animals, anaemia, degeneration of nerves, cardiovascular lesions, hepatic dysfunction ('ovine white liver disease') and impaired immunity (Sutherland *et al.*, 1979; Mohammed and Lamand, 1986; Vellema *et al.*, 1996; Mullally *et al.*, 2004; Scalabrino, 2005). The deficiency in grazing animals is caused by a lack of Co in the pasture. While Co is a required nutrient for animals (as it is a central part of the corrin nucleus of the cobalamin molecule) it is not essential for plant metabolism, so apparently healthy and productive pastures may be quite unsuitable for animals. Co deficiency was the cause of widespread ill-thrift syndromes in New Zealand, southern and western Australia, Kenya, parts of North America, the Netherlands and Germany, Denmark, Ireland and Great Britain (Reid and Horvath, 1980). Vitamin B12 deficiency can occur in milk-fed young even when few symptoms are seen in the adult (Quirk and Norton, 1987). This is because milk has little vitamin B12 and if the dam is only marginally supplied with Co, and therefore with vitamin B12, the developing fetus will store little B12 in its liver prior to birth (Grace *et al.*, 1986).

B vitamin supplementation

Research consistently indicates that there is no benefit of folate supplementation to grower pigs fed conventional diets, but that supplementation to breeding sows increases the number of piglets born (Lindemann, 1993). A dietary source of folate is needed by monkeys, guinea pigs, foxes and mink (Siddons, 1978). While animals can synthesize choline, supplemental choline has sometimes improved pig performance (Dove and Cook, 2001). There is little or no effect of supplementary biotin given to sows (Lewis *et al.*, 1991; Watkins *et al.*, 1991) although an increased number of piglets born and a reduced parturition to oestrus interval (Bryant *et al.*, 1985) suggest that there may be positive effects of biotin supplementation on breeding sows which are distinct from the effect of biotin on hoof health. Hot environmental conditions (e.g. 35°C) may increase the thiamin requirement (Peng and Heitman, 1974).

Vitamin C

This vitamin is so called because it was the third vitamin to be identified. It is synthesized in most animals from glucose:

$$\text{glucose} + O_2 \longrightarrow \text{L-gulonolactone} \xrightarrow{A}$$
$$\text{L-ascorbate} + H_2O_2$$

Animals lacking the *od* gene (the guinea pig, some bats and primates) are unable to synthesize the enzyme gulonolactone oxidase (required for step A) and so need a dietary source of vitamin C (Hasan *et al.*, 2004).

The main function of vitamin C is as an antioxidant (Halliwell, 2001). It is a cofactor to a number of metalloenzymes where its role is to keep the metal ion in a reduced state (Padh, 1991). Some of the enzymes cofactored by vitamin C are involved in the hydroxylation of proline and lysine during collagen formation (Kivirikko and Myllylä, 1982). The symptoms of scurvy and other disorders can be associated with impaired collagen formation, e.g. 'leaky' blood vessels and haemorrhage, swollen gums, and loose teeth (the symptoms of scurvy), as well as navel bleeding in newborn pigs (Sandholm *et al.*, 1979). Other roles of vitamin C are in the syntheses of carnitine (Rebouche, 1991) and norepinephrine (Levine *et al.*, 1985). Carnitine is required for the oxidation of fatty acids (Rebouche, 1991); this function may be the cause of lack of energy seen in the onset of scurvy (Johnston *et al.*, 2006).

There is little evidence that vitamin C supplements have any positive effect on domestic animal performance or welfare. The observations that blood vitamin C levels of pigs in hot climates are lower than normal (Riker *et al.*, 1967) or that vitamin C status is reduced by overcrowding have not been repeated although in both cases we would expect lower than normal food intakes and the possibility of reduced vitamin C synthesis (Brown *et al.*, 1970). There is no evidence that dietary vitamin C supplements improve the health or productivity of young or mature pigs or that excesses are harmful (Dove and Cook, 2001).

References

Agricultural Research Council (ARC) (1980) *The Nutrient Requirements of Ruminant Livestock.* CAB, Farnham Royal, UK.

American Institute of Nutrition Committee on Nomenclature (AINCN) (1987) Nomenclature policy: Generic descriptors and trivial names for vitamins and related compounds. *Journal of Nutrition* 117, 7–14.

Athar, N., Hardacre, A., Taylor, G., Clark, S., Harding, R. and McLaughlin, J. (2006) Vitamin retention in extruded food products. *Journal of Food Composition and Analysis* 19, 379–383.

Biesalski, H.K. and Nohr, D. (2003) Importance of vitamin-A for lung function and development. *Molecular Aspects of Medicine* 24, 431–440.

Blomhoff, R. and Blomhoff, H.K. (2006) Overview of retinoid metabolism and function. *Journal of Neurobiology* 66, 606–630.

Brown, R.G., Sharma, V.D. and Young, L.G. (1970) Ascorbic acid metabolism in swine. Interrelationships between the level of energy intake and serum ascorbate levels. *Canadian Journal of Animal Science* 50, 605–609.

Bryant, K.L., Kornegay, E.T., Knight, J.W., Webb, K.E. Jr and Notter, D.R. (1985) Supplemental biotin for swine: II. Influence of supplementation to corn- and wheat-based diets on reproductive performance and various biochemical criteria of sows during four parities. *Journal of Animal Science* 60, 145–153.

Buffa, E.A., Van Den Berg, S.S., Verstraete, F.J. and Swart, N.G. (1992) Effect of dietary biotin supplement on equine hoof horn growth rate and hardness. *Equine Veterinary Journal* 24, 472–474.

Chew, B.P., Park, J.S., Weng, B.C., Wong, T.S., Hayek, M.G. and Reinhart, G.A. (2000) Dietary beta-carotene absorption by blood plasma and leukocytes in domestic cats. *Journal of Nutrition* 130, 2322–2325.

Clements, M.R., Johnson, L. and Fraser, D.R. (1987) A new mechanism for induced vitamin D deficiency in calcium deprivation. *Nature* 325, 62–65.

Comben, N., Clark, R.J. and Sutherland, D.J. (1984) Clinical observations on the response of equine hoof defects to dietary supplementation with biotin. *Veterinary Record* 115, 642–645.

CVM/FDA (1990) *Finding of No Significant Impact and Environmental Assessment for Vitamin K Active Substances (VKAS)*. Available at: http://www.fda.gov/cvm/foi/vkas_ea.pdf

Darroch, C.S. (2001) Vitamin A in swine nutrition. In: Lewis, A.J. and Southern, L.L. (eds) *Swine Nutrition*, 2nd edn. CRC Press, Boca Raton, Florida, pp. 263–280.

Depeint, F., Bruce, W.R., Shangari, N., Mehta, R. and O'Brien, P.J. (2006) Mitochondrial function and toxicity: role of B vitamins on the one-carbon transfer pathways. *Chemico-Biological Interactions* 163, 113–132.

Dove, C.R. and Cook, D.A. (2001) Water-soluble vitamins in swine nutrition. In: Lewis, A.J. and Southern, L.L. (eds) *Swine Nutrition*, 2nd edn. CRC Press, Boca Raton, Florida, pp. 315–355.

FAO/WHO (2002) *Human Vitamin and Mineral Requirements*. Report of a joint FAO/WHO Expert Consultation, Bangkok, Thailand. World Health Organization/Food and Agriculture Organization of the United Nations, Rome. Available at: http://www.fao.org/DOCREP/004/Y2809E/y2809e00.htm#Contents

Fenech, M. (2001) The role of folic acid and vitamin B12 in genomic stability of human cells. *Mutation Research* 475, 57–67.

Fitzgerald, K.T., Bronstein, A.C. and Flood, A.A. (2006) 'Over-the-counter' drug toxicities in companion animals. *Clinical Techniques in Small Animal Practice* 21, 215–226.

Fraser, D.R. (1995) Vitamin D. *The Lancet* 345, 104–107.

Funk, C. (1911) On the chemical nature of the substance which cures polyneuritis in birds induced by a diet of polished rice. *Journal of Physiology* 43, 395–400.

Funk, C. (1913) Studies on beri-beri: VII. Chemistry of the vitamine-fraction from yeast and rice-polishings. *Journal of Physiology* 46, 173–179.

Funk, C. (1914) Is polished rice plus vitamine a complete food? *Journal of Physiology* 48, 228–232.

Funk, C. (1921) The antiberiberi vitamine. *Journal of Industrial and Engineering Chemistry* 13, 1110–1111.

Ghyselinck, N.B., Vernet, N., Dennefeld, C., Giese, N., Nau, H., Chambon, P., Viville, S. and Mark, M. (2006) Retinoids and spermatogenesis: lessons from mutant mice lacking the plasma retinol binding protein. *Developmental Dynamics* 235, 1608–1622.

Grace, N.D., Clark, R.G. and Mortleman, L. (1986) Hepatic storage of vitamin B_{12} by pregnant ewes and the foetus during the third trimester. *New Zealand Journal of Agricultural Research* 29, 231–232.

Halliwell, B. (2001) Vitamin C and genomic stability. *Mutation Research/Fundamental and Molecular Mechanisms of Mutagenesis* 475, 29–35.

Handelman, G.J. (2001) The evolving role of carotenoids in human biochemistry. *Nutrition* 17, 818–822.

Hasan, L., Vogeli, P., Stoll, P., Kramer, S.S., Stranzinger, G. and Neuenschwander, S. (2004) Intragenic deletion in the gene encoding L-gulonolactone oxidase causes vitamin C deficiency in pigs. *Mammalian Genome* 15, 323–333.

Hopkins, F.G. (1919) The practical importance of vitamins. *British Medical Journal* April 26, 507–510.

Hungate, R.E. (1966) *The Rumen and Its Microbes*. Academic Press, New York.

Johnston, C.S., Corte, C. and Swan, P.D. (2006) Marginal vitamin C status is associated with reduced fat oxidation during submaximal exercise in young adults. *Nutrition & Metabolism* 3, 35.

King, I.A. and Pope, F.M. (1986) Synthesis of cellular and extracellular glycoproteins by cultured human keratinocytes and their response to retinoids. *Biochimica et Biophysica Acta* 887, 263–274.

Kivirikko, K.I. and Myllylä, R. (1982) Posttranslational enzymes in the biosynthesis of collagen: intracellular enzymes. *Methods in Enzymology* 82, 245–304.

Kornegay, E.T. (1986) Biotin in swine production: a review. *Livestock Production Science* 14, 65–89.

Lakshmanan, M.R., Chansang, H. and Olson, J.A. (1972) Purification and properties of carotene 15,15'-dioxygenase of rabbit intestine. *Journal of Lipid Research* 13, 477–482.

Lal, H., Pandey, R. and Aggarwal, S.K. (1999) Vitamin D: non-skeletal actions and effects on growth. *Nutrition Research* 19, 1683–1718.

Levine, M., Morita, K., Heldman, E. and Pollard, H.B. (1985) Ascorbic acid regulation of norepinephrine biosynthesis in isolated chromaffin granules from bovine adrenal medulla. *Journal of Biological Chemistry* 260, 15598–15603.

Lewis, A.J., Cromwell, G.L. and Pettigrew, J.E. (1991) Effects of supplemental biotin during gestation and lactation on reproductive performance of sows: a cooperative study. *Journal of Animal Science* 69, 207–214.

Liau, G., Ong, D.E. and Chytil, F. (1981) Interaction of the retinol/cellular retinol-binding protein complex with isolated nuclei and nuclear components. *Journal of Cell Biology* 91, 63–68.

Lindemann, M.D. (1993) Supplemental folic acid: a requirement for optimizing swine reproduction. *Journal of Animal Science* 71, 239–246.

Lips, P. (2006) Vitamin D physiology. *Progress in Biophysics and Molecular Biology* 92, 4–8.

Martens, J.-H., Barg, H., Warren, M.J. and Jahn, D. (2002) Microbial production of vitamin B12. *Applied Microbiology and Biotechnology* 58, 275–285.

McCollum, E.V. (1918) The 'vitamin' hypothesis and the diseases referable to faulty diet. *Journal of the American Medical Association* 71, 937–941.

McCollum, E.V. (1953) Who discovered vitamins? *Science* 118, 632.

McCollum, E.V. and Davis, M. (1914) Observations on the isolation of the substance in butter fat which exerts a stimulating influence on growth. *Journal of Biological Chemistry* 19, 245–250.

McCollum, E.V., Simmonds, N. and Pitz, W. (1916) The relation of the unidentified dietary factors, the fat-soluble *A*, and water-soluble *B*, of the diet to the growth-promoting properties of milk. *Journal of Biological Chemistry* 27, 33–43.

McCollum, E.V., Simmonds, N., Shipley, P.G. and Park, E.A. (1922) Studies on experimental rickets. XII. Is there a substance other than fat-soluble A associated with certain fats which plays an important role in bone development? *Journal of Biological Chemistry* 49, 5–31.

Mahan, D.C. (1994) Effects of dietary vitamin E on sow reproductive performance over a five-parity period. *Journal of Animal Science* 72, 2870–2879.

Mahan, D.C. (2001) Selenium and vitamin E in swine nutrition. In: Lewis, A.J. and Southern, L.L. (eds) *Swine Nutrition*, 2nd edn. CRC Press, Boca Raton, Florida, pp. 281–314.

Mellanby, E. (1920) Accessory food factors (vitamines) in the feeding of infants. *Lancet* April 17, 856–862.

Mohammed, R. and Lamand, M. (1986) Cardiovascular lesions in cobalt-vitamin B12 deficient sheep. *Annales de Recherches Veterinaire* 17, 447–450.

Mosenthin, R., Sauer, W.C., Völker, L. and Frigg, M. (1990) Synthesis and absorption of biotin in the large intestine of pigs. *Livestock Production Science* 25, 95–103.

Mullally, A.M., Vogelsang, G.B. and Moliterno, A.R. (2004) Wasted sheep and premature infants: the role of trace metals in hematopoiesis. *Blood Reviews* 18, 227–234.

National Research Council (NRC) (1981) *Nutrient Requirements of Goats: Angora, Dairy, and Meat Goats in Temperate and Tropical Countries*. National Academy Press, Washington, DC.

National Research Council (NRC) (1985) *Nutrient Requirements of Sheep*, 6th rev. edn. National Academy Press, Washington, DC.

National Research Council (NRC) (1995) *Nutrient Requirements of Laboratory Animals*, 4th rev. edn. National Academy Press, Washington, DC.

National Research Council (NRC) (1998) *Nutrient Requirements of Swine*, 10th rev. edn. National Academy Press, Washington, DC.

National Research Council (NRC) (2000) *Nutrient Requirements of Beef Cattle*, 7th rev. edn. Update. National Academy Press, Washington, DC.

National Research Council (NRC) (2001) *Nutrient Requirements of Dairy Cattle*, 7th rev. edn. National Academy Press, Washington, DC.

National Research Council (NRC) (2007) *Nutrient Requirements of Horses*, 6th rev. edn. National Academy Press, Washington, DC.

Padh, H. (1991) Vitamin C: newer insights into its biochemical functions. *Nutrition Reviews* 49, 65–70.

Peng, C.-L. and Heitman, H. (1974) The effect of ambient temperature on the thiamine requirement of growing-finishing pigs. *British Journal of Nutrition* 32, 1–9.

Polizopoulou, Z.S., Kazakos, G., Patsikas, M.N. and Roubies, N. (2005) Hypervitaminosis A in the cat: a case report and review of the literature. *Journal of Feline Medicine and Surgery* 7, 363–368.

Quirk, M.F. and Norton, B.W. (1987) The relationship between the cobalt nutrition of ewes and the vitamin B_{12} status of ewes and their lambs. *Australian Journal of Agricultural Research* 38, 1071–1082.

Ramos, J.J., Marca, C., Loste, A., García de Jalón, J.A., Fernández, A. and Cubel, T. (2003) Biochemical changes in apparently normal sheep from flocks affected by polioencephalomalacia. *Veterinary Research Communications* 27, 111–124.

Rebouche, C.J. (1991) Ascorbic acid and carnitine biosynthesis. *American Journal of Clinical Nutrition* 54, 1147S–1152S.

Reid, R.L. and Horvath, D.J. (1980) Soil chemistry and mineral problems in farm livestock. A review. *Animal Feed Science and Technology* 5, 95–167.

Riker, J.T., Perry, T.W., Pickett, R.A. and Heidenreich, C.J. (1967) Influence of controlled temperatures on growth rate and plasma ascorbic acid values in swine. *Journal of Nutrition* 92, 99–103.

Rothschild, M.F., Messer, L., Day, A., Wales, R., Short, T., Southwood, O. and Plastow, G. (2000) Investigation of the retinol-binding protein 4 (*RBP4*) gene as a candidate gene for increased litter size in pigs. *Mammalian Genome* 11, 75–77.

Rowland, A.C. (1970) Rumenitis and liver abscess complex and vitamin-A status in beef cattle. *Animal Production* 12, 291–298.

Sandholm, M., Honkanen-Buzalski, T. and Rasi, V. (1979) Prevention of navel bleeding in piglets by preparturient administration of ascorbic acid. *Veterinary Record* 104, 337–338.

Scalabrino, G. (2005) Cobalamin (vitamin B_{12}) in sub-acute combined degeneration and beyond: traditional interpretations and novel theories. *Experimental Neurology* 192, 463–479.

Shearer, M.J. (1995) Vitamin K. *Lancet* 345, 229–234.

Siddons, R.C. (1978) Nutrient deficiencies in animals – folic acid. In: Rechcigl, M. (ed.) *CRC Handbook Series in Nutrition and Food. Vol. 11. Nutritional Disorders*. CRC Press, West Palm Beach, Florida.

Silveira, E.R. and Moreno, F.S. (1998) Natural retinoids and β-carotene: from food to their actions on gene expression. *Journal of Nutritional Biochemistry* 9, 446–456.

Somogyi, J.C. (1973) Antivitamins. In: *Toxicants Occurring Naturally in Foods.* National Academies Press, Washington, DC, pp. 254–275.

Standing Committee on Agriculture (SCA) (1990) *Feeding Standards for Australian Livestock. Ruminants.* CSIRO, Canberra.

Sutherland, R.J., Cordes, D.O. and Carthew, G.C. (1979) Ovine white liver disease – an hepatic dysfunction associated with vitamin B12 deficiency. *New Zealand Veterinary Journal* 27, 227–232.

Suzuki, R., Goda, T. and Takase, S. (1995) Consumption of excess vitamin A, but not excess β-carotene, causes accumulation of retinol that exceeds the binding capacity of cellular retinol-binding protein, Type II in rat intestine. *Journal of Nutrition* 125, 2074–2082.

Sybesma, W., Starrenburg, M., Kleerebezem, M., Mierau, I., de Vos, W.M. and Hugenholtz, J. (2003) Increased production of folate by metabolic engineering of *Lactococcus lactis. Applied and Environmental Microbiology* 69, 3069–3076.

Thompson, K.G. and Robinson, B.M. (1989) An osteodystrophy apparently caused by vitamin D deficiency in growing pigs. *New Zealand Veterinary Journal* 37, 155–157.

Ullrey, D.E. (1972) Biological availability of fat-soluble vitamins – vitamin-A and carotene. *Journal of Animal Science* 35, 648–657.

Van Dusseldorp, M. (1997) Classification of folic acid. *Lancet* 349, 289.

Vellema, P., Rutten, V.P.M.G., Hoek, A., Moll, L. and Wentink, G.H. (1996) The effect of cobalt supplementation on the immune response in vitamin B_{12} deficient Texel lambs. *Veterinary Immunology and Immunopathology* 55, 151–161.

Wald, G. (1968) The molecular basis of visual excitation. *Nature* 219, 800–807.

Watkins, K.L., Southern, L.L. and Miller, J.E. (1991) Effect of dietary biotin supplementation on sow reproductive performance and soundness and pig growth and mortality. *Journal of Animal Science* 69, 201–206.

11 Voluntary Food Intake

Introduction: Food Consumption Is the Result of a Large Number of Interacting Factors

The amount of food eaten by an animal in a given meal is influenced by a large number of factors (Fig. 11.1). The factors which influence intake can be grouped into classes: those which act at the level of the digestive tract; those which are social or psychological in nature; those which are characteristics of the animal's physical environment (including the type of food available); and those which are related to the animal's physiology.

Food intake by all animals is subject to the physiological integration and social and psychological influences shown in Fig. 11.1. Those which are related to pasture composition (within the physical environment) of course apply only to herbivores, and the 'mechanical' factors which operate in the digestive tract apply more to ruminants than other herbivores. Figures 11.2 and 11.3 illustrate how food and environmental characteristics can influence food consumption in animal production contexts.

Rees *et al.* (1972) investigated the reasons for low milk production in a district in subtropical Australia. Among several other factors (e.g. grazing area provided and use of fertilizer), milk yields were constrained by the proportion of grazing provided from a farm's night paddocks – paddocks near the milking shed used to facilitate mustering cows in the morning. The sequence of events is outlined in Fig. 11.2.

The second example comes from commercial Australian pig production. Vajrubukka *et al.* (1983) found that pigs grew at 594 g/day in winter but growth declined to 545 g/day in summer. The digestible energy (DE) intake of the weaned pig decreases approximately linearly as the ambient temperature increases from 5°C to 30°C (NRC, 1987):

$$\Delta DEI = 126.3 - 1.65 \times T$$

where
ΔDEI = % change in DE intake (intake at 15°C = 100%)
T = ambient temperature (°C)

In general, growing pigs eat about 10% less food in summer than in winter in subtropical Australia. The reasons for the lower summer growth rate of Australian pigs are summarized in Fig. 11.3, based on the ideas discussed in SCA (1987).

Beef cattle in feedlots have varying intakes depending on the food roughage/concentrate ratio (Costa *et al.*, 2005). Intakes increase as the roughage content is reduced from 100% to about 30%, and then plateaus and may decline slightly as more roughage is removed. This can happen even while growth rates are unaffected (Loerch and Fluharty, 1998). The interplay and relative importance of intake regulation mechanisms operating at the digestive tract and neuro-hormonally are described in Fig. 11.4. Consumption of the concentrate-rich diet is less influenced by rumen fill than for the roughage-rich diet.

Regulation of Food Consumption

An overview of the signals and neuro-endocrine pathways involved in the control of food intake is given in Fig. 11.5. Also indicated are the interrelationships between food intake, energy stores and losses, signals to the hypothalamus, and the influence of 'non-nutritional' and external factors such as social influences and food palatability.

Motivation or compulsion?

It is easy to see that any mechanism which *enforces* eating behaviour could have disastrous results, e.g. through the animal eating something poisonous, or continuing to eat when its safety was threatened. Activity of the hypothalamus *motivates* eating but does not enforce it, and this motivation can be

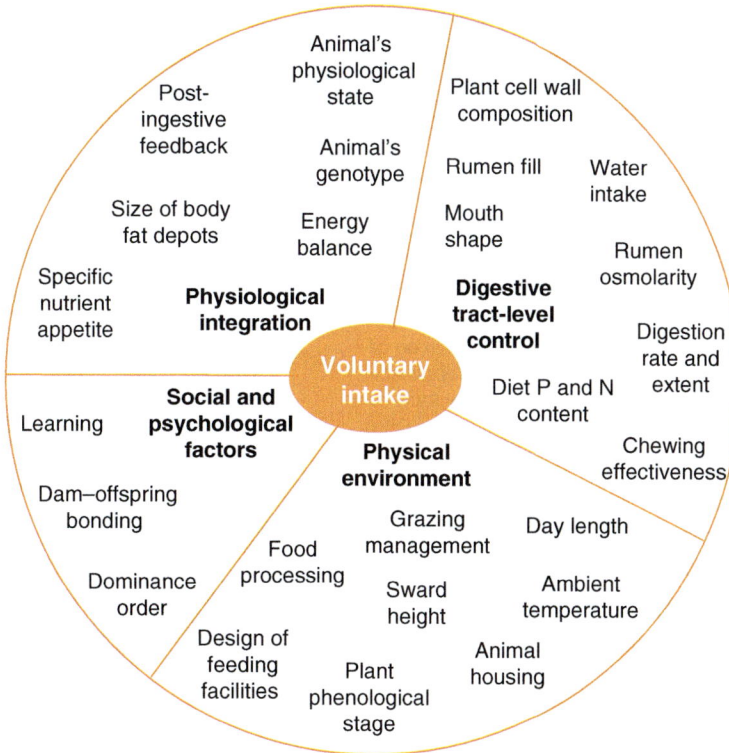

Fig. 11.1. Some factors which influence the consumption of food by animals.

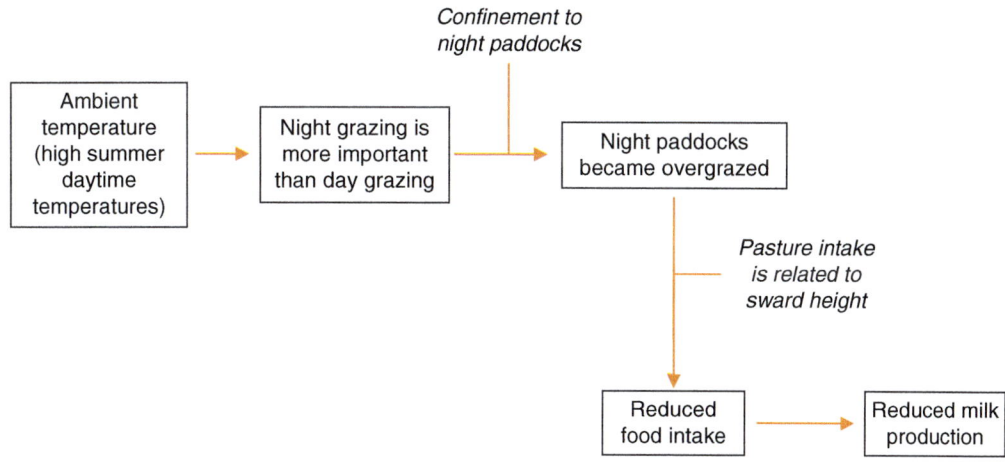

Fig. 11.2. Events in dairy cattle management which constrained food intake and thus milk production.

overridden by other stimuli. For example, if we give an animal a very palatable food it may eat to excess (Baumont *et al.*, 1997), and the reverse effect occurs when the food offered is unpalatable (Greenhalgh and Reid, 1971; Karda *et al.*, 1998). It is unlikely that a hard-wired system would enable an animal to learn about its nutritional environment and it has been shown that post-ingestive feedback has a strong influence on short-term feeding behaviour (Provenza, 1996).

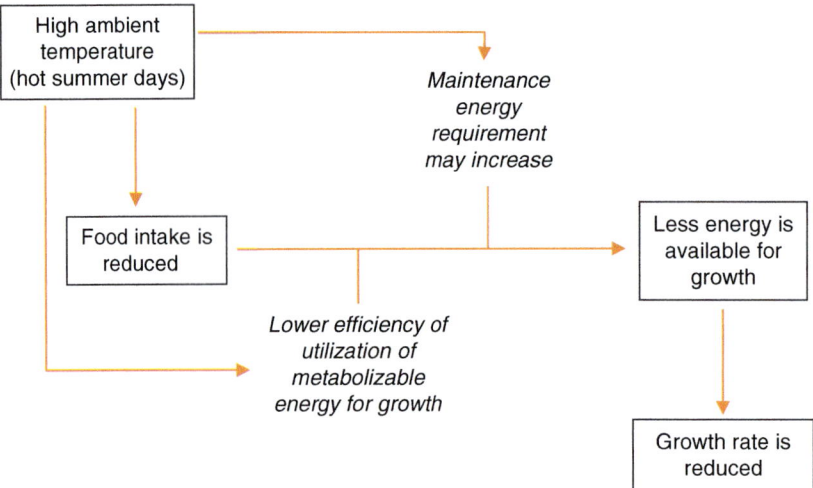

Fig. 11.3. The sequence of events which leads to slow growth by pigs in hot climates.

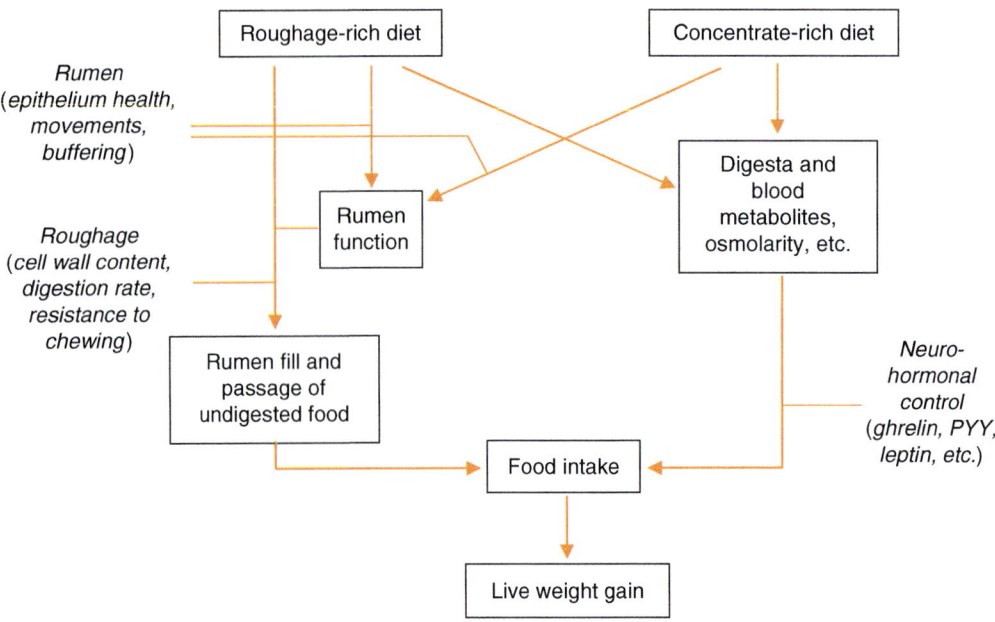

Fig. 11.4. The relative importance of digestive tract and neuro-hormonal factors in the control of food intake by lot-fed beef cattle.

Short- and long-term control

Short-term control of intake, i.e. control of the start and finish of bouts of eating, is initiated by signals from the digestive tract. These include the absence of digesta, rumen distension, the presence of digestion products (glucose in monogastrics, and osmolarity and acidity in ruminants), signals from receptors for glucose or propionic acid in the liver and insulin production by the pancreas (Forbes, 1996). These signals are integrated in the hypothalamus and the animal responds to signals of hunger or satiety by beginning to eat or by stopping eating. Animals given a meal after a period of fasting eat quickly at first, but the rate of intake declines

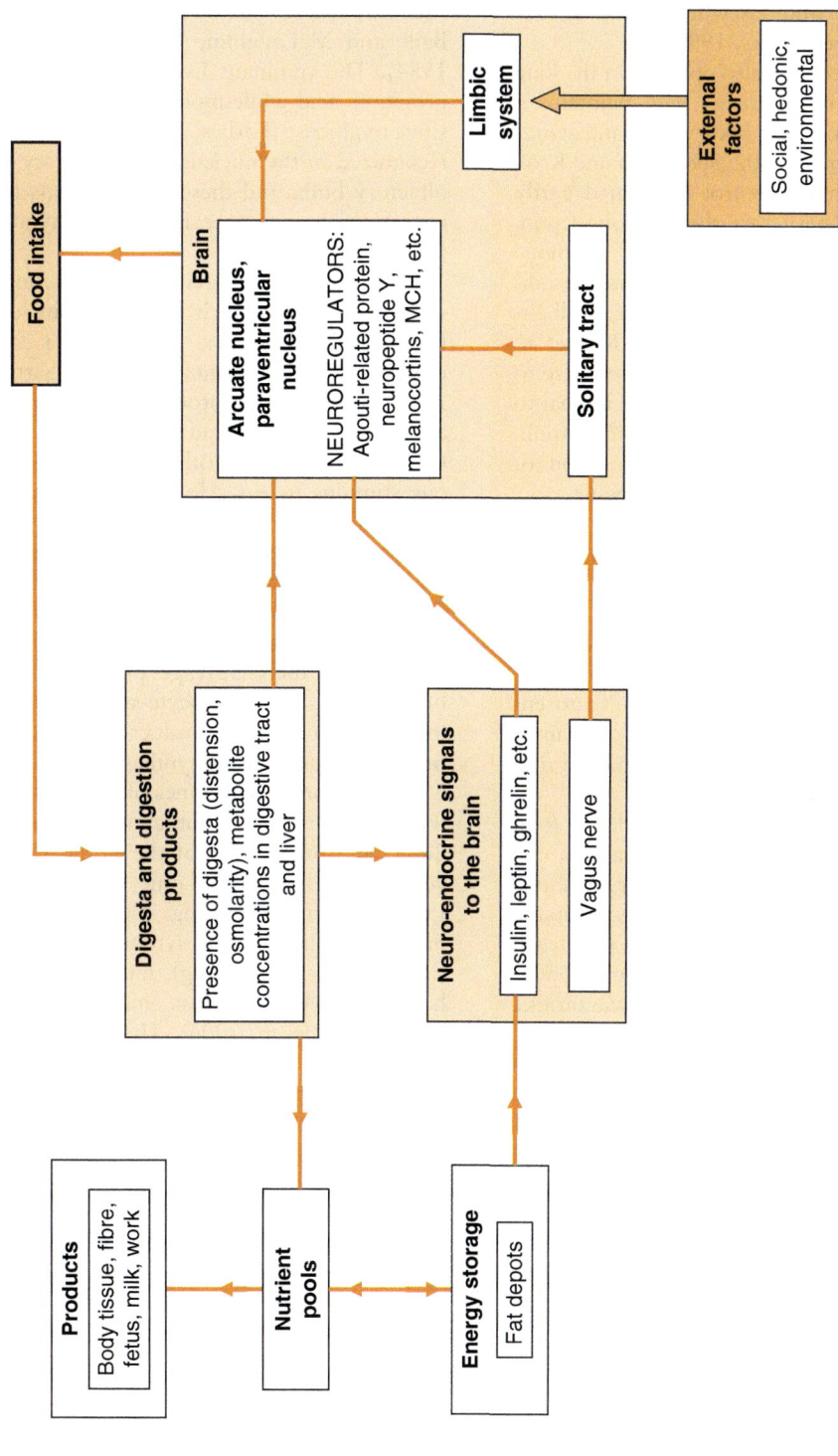

Fig. 11.5. Information sources, signals and brain centres involved in food intake control. (From Harding and Leek, 1972; Weston and Poppi, 1987; Horvath *et al.*, 2001; Holst and Schwartz, 2004; Sartin *et al.*, 2005; Huda *et al.*, 2006; Yeomans, 2006.)

(Faverdin, 1985, cited by Baumont *et al.*, 1989, 2000), possibly as a variety of satiety signals from the digestive tract and elsewhere build up in strength and number (Forbes, 1996).

Voluntary intake control also operates in the long term so that the animal reaches and maintains a more or less constant mature body weight and avoids obesity (Speakman *et al.*, 2002; Speakman and Król, 2005). Long-term intake control is initiated partly from the adipose tissue where the hormone leptin, which is an 'anorectic' or 'anorexigenic' hormone (i e. acts to stop eating), is synthesized in the adipocytes. This mechanism does not operate in all circumstances without appropriate control. Król *et al.* (2006) have shown that voles become insensitive to leptin during the summer. This allows the animal to accumulate body fat in preparation for hibernation, without being subject to the usual repression of intake by leptin produced in the adipose tissue.

The Neuro-endocrine Control of Eating Behaviour

Feeding behaviour involves a complex neuro-endocrine system (Fig. 11.6) which has a set of messengers to relay information about the animal's nutrient status to its brain, and nuclei in the hypothalamus which receive and integrate this information and stimulate other brain centres.

Peptide hormones have a central role in food intake control. They provide the signals to the hypothalamus about the animal's energy status. Important among these are leptin (synthesized in the adipose tissue), insulin (synthesized in the pancreas), and ghrelin, peptide YY and several other peptides (oxyntomodulin, cholecystokinin, glucagon-like peptide 1, etc.; Huda *et al.*, 2006) which are synthesized in the digestive tract. Except for ghrelin, which is a hunger-signalling hormone, all of these give satiety signals. Some act directly on receptors in the hypothalamus but most also act via the vagus nerve (Huda *et al.*, 2006). Short-term intake control may be mediated, in part, by these peptide hormones and signals from digestive tract and liver receptors.

Information about the presence of digesta and digestion products in the reticulo-rumen, stomach and intestines is also carried by the vagus nerve to the solitary tract nucleus in the brain stem. The vagus transmits information from distension receptors in the reticulo-rumen (Harding and Leek, 1972) and is also stimulated by ghrelin and cholecystokinin (Holst and Schwartz, 2004). Receptors

have been demonstrated in the digestive tract for glucose, amino acids and pH (in monogastrics; Mei, 1985) and volatile fatty acids (in ruminants; Baile and McLaughlin, 1970; Cottrell and Iggo, 1984). The ruminant liver can apparently detect propionic acid while monogastric animals are sensitive to glucose (Forbes, 1988). Taste and smell are recognized in the nucleus of the solitary tract and olfactory bulb, and these brain regions have neurones which innervate the lateral hypothalamus (Laing, 1985).

The hypothalamic arcuate nucleus contains two sets of neurones which have opposite effects on feeding behaviour. One type exerts a continuous repression of the stimulus to eat (Sartin *et al.*, 2005). These are the proopiomelanocortin/cocaine and amphetamine-regulated transcript (POMC/CART) neurones. The other set provides a continuous stimulus to eat (Holst and Schwartz, 2004). These are the neuropeptide Y/agouti-related protein (NPY/AGRP) neurones. The balance between the activities of these neurones determines whether the animal is motivated to eat or to stop eating. The neuroregulators (NPY, AGRP, POMC and CART) and their cleavage products affect other brain centres. α-melanocyte-stimulating hormone (α-MSH) is a cleavage product of POMC and, with other melanocortins, is synthesized by the POMC/CART neurones. Both types of neurones project to the paraventricular hypothalamic nucleus (Bloom, 2003) and other areas of the brain including the ventromedial hypothalamic nucleus (Haskell-Luevano *et al.*, 1999; King, 2006). Neurones in the paraventricular nucleus synthesize the orexigenic (i.e. appetite-stimulating) melanin-concentrating hormone (MCH) (Holst and Schwartz, 2004; Druce and Bloom, 2006). These neurones project to the cortex (Tritos *et al.*, 1998). There are also 'reward' stimuli as a response to eating (Druce and Bloom, 2006) which involve opioids, endocannabinoids and serotonin.

Control of food intake is complex and not fully understood. However, the essential basis has been established. Aspects of neuro-hormonal control and the roles of motivation and reward in eating behaviour are discussed in more detail by Berridge (2004) and Broberger (2005).

Ghrelin is an important orexigenic hormone

When the stomach is empty, ghrelin is synthesized in endocrine cells in the fundus and released into the

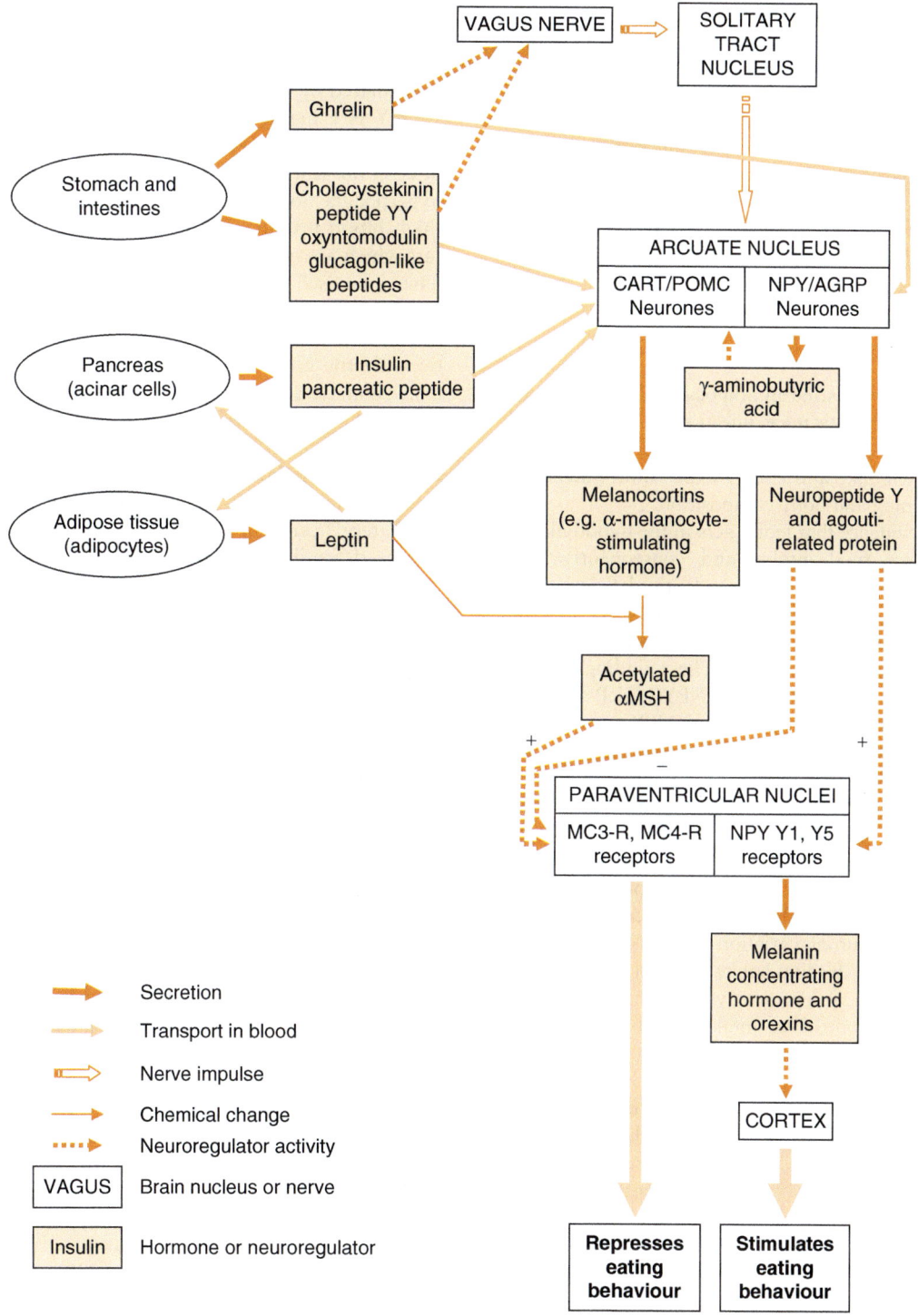

Fig. 11.6. Hormones and neuroregulators which are involved in the hypothalamic regulation of eating behaviour.

blood (Holst and Schwartz, 2004). It travels to the brain, crosses the blood–brain barrier and contacts receptors in the NPY/AGRP neurones in the arcuate nuclei where it promotes the synthesis of NPY and AGRP. These neuroregulators stimulate the synthesis of MCH and orexins in the paraventricular nuclei, which in turn promotes the eating of food. The ghrelin stimulus also causes the NPY/AGRP neurones to secrete γ-aminobutyric acid. This reduces the synthesis of melanocortins by the POMC/CART neurones (Holst and Schwartz, 2004).

Leptin is a satiety hormone

Leptin is synthesized in adipocytes. In both normal and obese/diabetic mice, injections of leptin decrease food intake. Leptin regulates body weight by affecting food intake and energy expenditure, metabolic partitioning and body composition.

The evidence that leptin acts in the hypothalamus to reduce food intake, and consequently body weight, comes primarily from leptin-deficient *ob/ob* mice (the homozygous *ob/ob* mouse is a strain which is genetically obese), with supporting evidence from human studies. These findings suggest that leptin action inhibits the NPY/AGRP neurones, thus decreasing the amount of NPY secreted and inhibiting the action of γ-aminobutyric acid on the POMC/CART neurones (Rahmouni and Haynes, 2001). Leptin also promotes the acetylation (i.e. activation) of α-MSH (Guo *et al.*, 2004). Leptin synthesis appears to be sensitively influenced by eating as its concentration in the blood rises within 2–8 h of feeding in sheep (Marie *et al.*, 2001), and there is a diurnal variation in rats which reflects these animals' feeding habits (Saladin *et al.*, 1995).

The action of leptin may allow both long- and short-term control of intake. In the short term, meal-eating increases leptin mRNA expression (Harris *et al.*, 1996). There is a further increase in leptin mRNA expression once significant obesity is established, but this is varied according to the animal's physiological status. The control of leptin synthesis thus allows long-term control of intake while allowing receptiveness to the animal's varying energy demands (Chilliard *et al.*, 2005).

Neuropeptide Y and agouti-related protein stimulate eating

NPY and AGRP are neuroregulators which stimulate feeding behaviour and which appear to act independently (Horvath *et al.*, 2001). AGRP blocks the melanocortin receptors in the paraventricular nuclei (Stutz *et al.*, 2005). AGRP synthesis is upregulated by fasting.

NPY is a highly potent appetite promoter. For example, infusion of NPY into the hypothalamus or cerebral ventricles (Marks *et al.*, 1996) produces hyperphagia (overeating) and hyperinsulinaemia (high blood insulin). More convincingly, lactating rats which are not allowed to satisfy their appetites have increased amounts of NPY mRNA in the arcuate nuclei and the paraventricular nuclei (Pickavance *et al.*, 1996). In ruminants, NPY can override the effects of satiety signals such as gut distension. Miner *et al.* (1990) gave a complete pelleted diet ad lib to sheep fitted with lateral cerebral ventricular guide cannulae and rumen cannulae. Water-filled balloons placed into the rumen decreased food intake. NPY was then injected into the lateral cerebral ventricle and food intake increased. Similarly, intraruminal infusion of propionate decreased food intake. This effect was also reversed by NPY injection. Increases in brain NPY synthesis can be achieved by lactation, fasting, running, and by helminth infestation, all conditions which impose an added energy demand on the animal (Lewis *et al.*, 1993; Smith, 1993; Horbury *et al.*, 1995). These findings show that a negative energy balance, whether caused by reduced energy intake or by increased energy expenditure, increases NPY synthesis in the hypothalamus and results in increased food intake.

Insulin has an integrating role by responding to the body's energy status

Glucose is a key energy source, and blood glucose levels reflect the animal's 'energy balance'. In other words, the blood glucose pool reflects the rates at which glucose is provided from the diet (or by metabolism of body tissue) and is used in the animal's metabolism. Animals which have low blood glucose levels also have low circulating insulin levels. Insulin is secreted from the pancreas when blood glucose levels rise after a meal.

Leptin release is stimulated by increased levels of insulin in the blood (Sonnenberg *et al.*, 2001) and this appears to be the result of a direct influence on leptin gene expression in adipocytes (Saladin *et al.*, 1995). There is also evidence that increased levels of leptin downregulate the synthesis of insulin (Seufert *et al.*, 1999). These interactions are illustrated in Fig. 11.7.

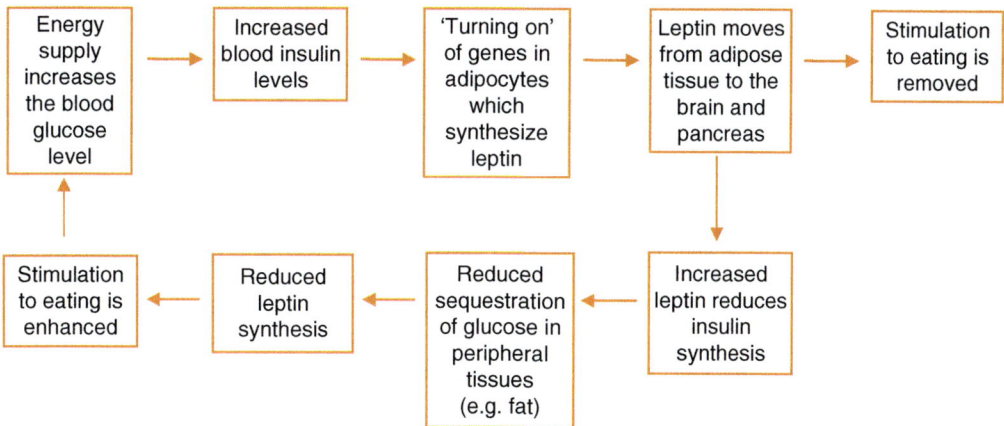

Fig. 11.7. Interactions between leptin and insulin which regulate food intake in response to changes in blood glucose.

Food Intake and Digesta Flow Through the Ruminant Digestive Tract

Forage consumption by ruminants is constrained by the amount of digesta in the reticulo-rumen, i.e. by the 'rumen fill'. For rumen fill to influence food intake the animal must have some way of detecting it. This is done by vagus- and splanchnic-innervated receptors which sense the amount of 'stretch' in the rumen wall (Harding and Leek, 1972), the osmotic pressure of the rumen fluid (Grovum, 1995), the concentration of metabolites (e.g. volatile fatty acids) and the temperature. These different stimuli may operate in conjunction with each other, rather than as single effectors (Forbes and Barrio, 1992). Non-ruminants (e.g. the cat and ferret; Mei and Garnier, 1986; Blackshaw and Grundy, 1989) also have receptors which are sensitive to osmotic pressure and distension, which are probably involved in the movement of digesta in the stomach and small intestine, and thus may influence food intake.

While there is evidence that rumen fill regulates food intake, there are some caveats. First, descriptions of the role of rumen fill often imply that the reticulo-rumen is a constant-volume tank, but this is not so. Seasonal changes in rumen capacity have been reported in cattle, sheep, goats and camels, and in red deer (Lechner-Doll *et al.*, 1990; Freudenberger *et al.*, 1994); when animals eat different foods (e.g. in sheep, Weyreter and von Engelhardt, 1984); and during pregnancy and lactation in the sheep and dairy cow (Coffey *et al.*, 1989; Kaske and Groth, 1997). Second, distension may not be effective alone. Mbanya *et al.* (1993) could

not reduce intake by simple distension of the rumen, but intake was depressed by distension in combination with volatile fatty acid infusions. Third, as Forbes (1996) has pointed out, it is unlikely that the ruminant uses information from only one source in controlling food intake, and that stimuli from stretch, acidity and other receptors in the digestive tract and liver are probably all integrated. Fourth, we have to acknowledge that other intake controls may override signals from the digestive tract. Ketelaars and Tolkamp (1992) review the shortcomings of a strict adherence to digestive tract-level intake control theory and imply that the animal's energy demand has more influence on intake than signals from the digestive tract.

A Digesta Flow/Food Intake Model

Simplistically, we can say:

Food DM intake = (DM removal from the reticulo-rumen)   (11.1)

= (passage of undigested DM to the abomasum) + (absorption of digested DM through the rumen wall)

This relationship implies that there are some characteristics of the way food is handled in the rumen which constrain food intake. Food dry matter (DM) can leave the rumen either as products of microbial digestion or as particles of undigested food. Digestion products may be absorbed through the rumen wall, or incorporated into microbial cells or flow from the rumen in the rumen liquor. Undigested food particles will pass into the abomasum and lower digestive tract (Fig. 11.8). Consistent with this, a variety of digestion attributes have been shown to influence forage intake in ruminants:

1. Changes in particulate digesta flow rates, which influence the time which digesta remains in the rumen and in the whole digestive tract (Mertens, 1973, cited by Van Soest, 1975; Hyer *et al.*, 1991; Bruining *et al.*, 1998; Doreau and Diawara, 2003);
2. The rate at which the slowly digesting DM fraction is digested and the lag time before digestion starts (Mertens, 1973; Hyer *et al.*, 1991);
3. The rate at which food particles are comminuted (reduced in size) to smaller particles (Bruining *et al.*, 1998), and the effects of food-processing methods such as milling which reduce the food particle size (Laredo and Minson, 1975; Udén, 1988).

Digestion in the rumen

Digestion occurs at rates which are characteristic of the forage type, i.e. its anatomical structure and chemical composition. The mesophyll, vascular bundles (bundle sheaths and vascular vessels), axial and abaxial epidermis, and schlerenchyma are digested at different rates – the phloem and mesophyll most rapidly, the bundle sheath and epidermis less quickly, and the xylem vessels and the schlerenchyma little if at all (Akin, 1979). Using another approach, we can distinguish between cell contents and cell walls. Cell contents are about 98% digestible (Van Soest, 1967) while cell wall digestibility varies according to the degree of lignification and the time this material stays in the rumen. Thus, it is useful to distinguish rapidly digesting, slowly digesting and indigestible forage

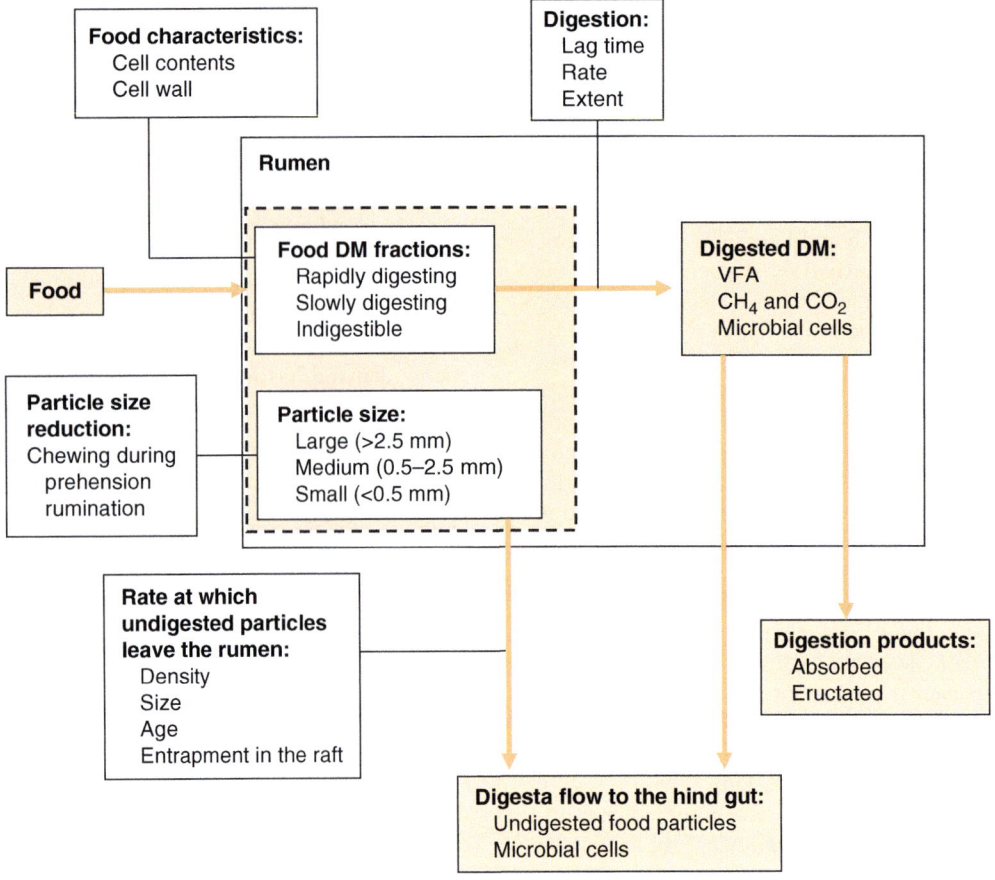

Fig. 11.8. The influences of food digestion, comminution and passage on the loss of food dry matter from the rumen. (From Welch and Smith, 1969; Mertens and Ely, 1979; Poppi *et al.*, 1980; Dryden *et al.*, 1995; Weston, 1996; Kaske and Midasch, 1997; Bruining *et al.*, 1998; Doreau and Diawara, 2003; Lund *et al.*, 2006.)

fractions (Mertens and Ely, 1979). Measurements of the rate of digestion in the rumen often suggest that there is a lag period between the entry of food particles into the rumen and the start of digestion (e.g. Waldo *et al.*, 1972; Dryden and Kempton, 1983). Both digestion rate and the lag time may influence the time needed to remove DM from the rumen, and thus influence food intake.

Flow of digesta from the rumen

The undigested remains of all the food eaten by a ruminant have to pass through the reticulo-omasal orifice. The probability of undigested food leaving the reticulo-rumen is maximized when it is comminuted to small particles (Poppi *et al.*, 1980; Fig. 11.9). There is no absolute upper limit to the size of food particles which leave the rumen, but small ruminants (sheep, deer and goats) need to comminute their food to about 1–1.2 mm (Poppi *et al.*, 1980; Domingue *et al.*, 1991) to maximize the likelihood that a particle will escape, while large ruminants, e.g. cattle, moose (*Alces alces*) and wapiti (*Cervus elaphus canadensis*), chew theirs to 3–7 mm particles (Renecker and Hudson, 1990). Rumination is an important part of ruminant digestive physiology (Welch, 1982).

Microbial digestion does not contribute much to reduction in particle size. Food particles are mostly comminuted by chewing, either during ingestion or rumination (Van Soest, 1975; Fig. 11.10). Those factors which influence the ease with which a food is chewed also influence the amount of it which is characteristically eaten. However, some easily chewed foods (e.g. the small-leafed legume, *Lotus*

corniculatus) may be swallowed almost whole, while larger and tougher leaves, e.g. from ryegrass, may be chewed more because the animal cannot swallow them until they are formed into a bolus (Dryden *et al.*, 1995). Notwithstanding this, forages with more cell wall will need more chewing before they leave the rumen than those with less cell wall (Welch and Smith, 1969).

Muscular contractions of the walls move food particles around the reticulo-rumen and bring them near the reticulo-omasal orifice (Evans *et al.*, 1973; Waghorn and Reid, 1977). Digesta flow is related to these movements (Ulyatt *et al.*, 1984) and intake falls if rumen motility is impaired (Kaske and Midasch, 1997).

Newly swallowed forage pieces form a mat of long particles (the 'raft') which lies in the dorsal rumen at the top of the rumen contents. As a result of rumination, hydration and microbial attack, undigested food particles become smaller and denser and move from the raft to the ventral rumen. Modelling studies (Wylie *et al.*, 2000; Poppi *et al.*, 2001) suggest that food particles leave the raft relatively slowly, but once detached most leave the rumen very quickly.

Size is only one of several attributes which influence the probability of a food particle leaving the rumen. Particle shape may be important as cuboidal particles may be more likely to leave the rumen than long fibres (Troelsen and Campbell, 1968). Density is also important, and particles with densities of about 1.5 g/cc are more likely to leave the rumen. Food particle density is influenced by the amount of gas and liquid associated with them. Ensalivation during chewing, and exposure to the rumen liquor, hydrates the fibrous part of these

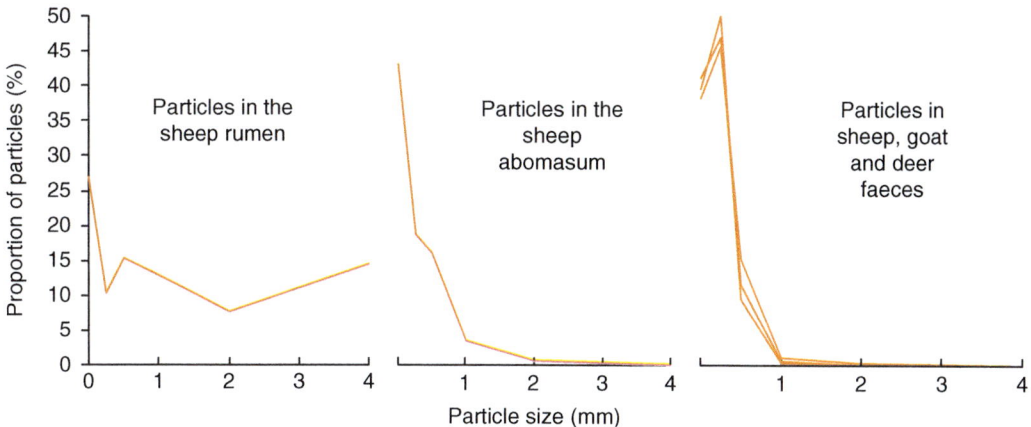

Fig. 11.9. Size of undigested food particles in small ruminants. (Drawn from data in Waghorn *et al.*, 1986; Domingue *et al.*, 1991.)

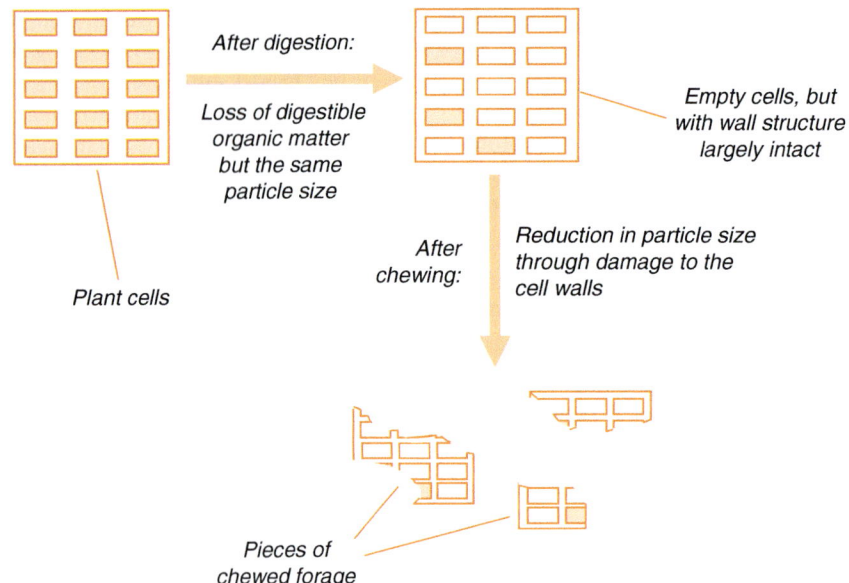

After digestion:

Loss of digestible organic matter but the same particle size

Empty cells, but with wall structure largely intact

Plant cells

After chewing:

Reduction in particle size through damage to the cell walls

Pieces of chewed forage

Fig. 11.10. The effect of chewing on forage particle size. (Based on a concept of Van Soest, 1975.)

particles (Wattiaux *et al.*, 1992) and increases their density. Higher cell wall contents, or lower cell wall digestibilities, are associated with reduced forage intakes; and cattle eat more legume than grass forage possibly because legume cell walls are more fragile (Allen, 2000).

Food Palatability, Diet Selection and Intake

Foods differ in their 'palatability', i.e. animals display different degrees of willingness to eat different foods. One of the earliest illustrations of this, and of the effect it can have on food intake, was given by Greenhalgh and Reid (1971), who fed sheep either grass hay or oaten straw and equalized intakes by introducing the other forage directly into the rumen. They demonstrated that the lower consumption of straw was probably because the sheep did not like its taste or because of some metabolic consequence of eating it. Other examples include the reluctance of rusa deer (*Cervus timorensis*) to eat rhodes grass (*Chloris guyana* cv. Pioneer) to the extent that they will lose weight rather than eat large amounts of it (Hmeidan and Dryden, 1998), and the refusal of goats and sheep to eat a common cultivar of the fodder tree gliricidia (*Gliricidia sepium*) (Karda *et al.*, 1998).

There have been many attempts to describe in mechanistic terms why animals choose to eat what they do. Diet selection is a topic of major impor-

tance in animal nutrition, for all animal species, because it determines what part of the available diet an animal consumes and thus greatly influences the animal's nutrient intake. Factors which may influence diet selection include:

1. Food nutrient content: pigs avoid foods which have an unbalanced amino acid content, and may select for threonine and isoleucine but not for lysine (Forbes, 1995). Na-deficient cattle will drink saline water and eat sodium bicarbonate, which is usually unpalatable (Underwood and Suttle, 1999). Ruminants which are very deficient in P may eat bones. This could be either an example of 'euphagia' or eating to maximize nutrient status, or it could be a dysfunctional habit. Evidence that it is an example of post-ingestive feedback comes from work by Villalba *et al.* (2006a). In this experiment, P-deficient sheep chose foods which had flavours the sheep associated with a P supplement, but not when fed adequate P.

2. The suite of other available foods: seasonal changes in the diet composition of sheep, cattle and goats grazing tropical grass/legume pastures in northern Australia (Norton *et al.*, 1990) and of the Australian bilby (*Macrotis lagotis*) (Gibson, 2001) are associated with seasonal changes in pasture species composition or the availability of different food types.

3. Social factors: eating behaviour is influenced by social pressures within a group (e.g. dominance and

familiarity) and the group's feeding experience. Food preferences may be learnt from more experienced animals in the herd, often the animal's dam (Green *et al.*, 1984). Dominance alters food preferences and intake. Whitney and Olson (2006) found that lambs ate more of a novel food when alone than when with their mothers. Competition for limited, and palatable, food can lead to considerable intra-group antagonism, as described in deer fed a grain supplement (Sydenham and Dryden, 2007).

4. Taste: molasses is often used to mask the taste of unpalatable foods (although sweet tastes are not preferred by all animals).

5. The animal's ability to find and prehend the preferred food component: diet selection, at least by grazing animals, is influenced by their ability to prehend the preferred food or food part (Black *et al.*, 1987). An example is young leaf which is 'protected' by dry mature stems.

6. Competition: there is competition for preferred foods between sympatric species; e.g. between kangaroos and sheep in Australian rangelands (Edwards *et al.*, 1995).

Learning, diet selection and food intake

Post-ingestive feedback is one of the more important mechanisms which animals use to learn about their nutritional environment. Animals learn to identify suitable (i.e. nutritious and non-poisonous) foods by associating a 'good' feeling after eating with the food which is eaten (Provenza, 1996). This effect can be very strong and can override other influences. For example, Scott *et al.* (1995) showed that lambs continued to prefer the cereal grain they had been previously conditioned to eat, even when their herd-mates (conditioned to prefer another grain) ate a different grain.

The theory of post-ingestive feedback is that animals seek those consequences of eating which increase their comfort by satisfying their nutrient requirements or by minimizing the consumption of toxins. Animals are thought to behave in an action/feedback/learning loop. They choose foods which maximize a feeling of comfort after the food has been eaten and associate food tastes, appearance and odour with nutritional or health outcomes (Provenza *et al.*, 1992; Villalba and Provenza, 2000). These associations remain for several months in sheep (Provenza, 1996; Villalba *et al.*, 2006a). How animals find out about the different foods available, what type of metabolic changes are needed to cause

a change in diet and what amount of perturbation in their metabolisms are needed to cause learning are discussed by Kyriazakis *et al.* (1999).

One of the more interesting observations about post-ingestive feedback is that animals learn to ameliorate their diet in positive ways, as well as simply avoiding harmful foods. Villalba *et al.* (2006b) have described how sheep learn to 'self-medicate' by selecting sodium bentonite, polyethylene glycol (PEG) or dicalcium phosphate supplements to offset the effects of eating diets rich in starch (predisposing to rumen acidosis), tannins or oxalate (which complexes with Ca). Tannins have contrasting effects, depending on the concentration in the food. They are highly aversive at concentrations over about 7% (Kimball and Nolte, 2005) because they are astringent, reduce nutrient availability and cause intestinal injury (Dawson *et al.*, 1999). On the other hand, at lower dietary concentrations (e.g. 2–5%) tannins may improve animal performance by increasing the dietary bypass protein content and through an anthelmintic effect (Barry and McNabb, 1999; Hoskin *et al.*, 2000). Douglas *et al.* (1995) offered sheep either lotus (*Lotus corniculatus*; 3.2% condensed tannins) or lucerne (*Medicago sativa*; 0.2% condensed tannins). The sheep given lotus selected a diet which was significantly richer in tannins (5.7%) than that on offer, grew faster (228 versus 183 g/day) and produced more wool. Provenza *et al.* (2000) fed sheep on tannin-rich foods with a barley grain/PEG mixture (PEG complexes with, and inactivates, tannins). When the barley was removed, the sheep continued to eat PEG while they were fed the tannin-rich food.

Sheep, goats and cattle are able to recognize food nutrient contents. For example, growing sheep selected diets of different protein contents in amounts which met their protein requirements (Kyriazakis and Oldham, 1993), and sheep first given a low-energy food subsequently ate more of a high-energy food than sheep initially given the high-energy food (James and Kyriazakis, 2002). Arsenos and Kyriazakis (2001) fed sheep either low- or high-protein diets, and then offered them protein supplements. The low-protein group initially exhibited neophobia, but after a week began to eat more of the protein supplements than those initially given the high-protein diet. Cattle have shown similar abilities to recognize and choose diets on the basis of their rumen degradable protein contents (Tolkamp *et al.*, 1998).

Animals avoid novel foods unless, through repeated experience of them, they learn that the food is 'safe' to eat, i.e. neophobia helps to minimize the risk of poisoning (Kimball and Nolte, 2005). In contrast, the sampling of a wide range of foods is also a common behaviour of grazing animals. Provenza *et al.* (2003) suggest that, because many foods contain toxins, this also reduces the risk of excessive nutrient or toxin consumption. Satiety is influenced by sensory cues as well as post-ingestive ones. Neurones responsive to taste, smell and sight of monkeys sated with a particular food ceased to respond when that food was presented, but continued to respond to other foods (Critchley and Rolls, 1996). Thus, extensive sampling of the available foods may help to maintain food consumption and nutrient intake.

Diet selection involves choosing a grazing area, as well as plants or plant parts, and dealing with spatial and temporal changes in the food resource. The interrelationships between diet selection, food intake and grazing behaviour involve concepts like optimal foraging, dietary breadth and the effects of pasture structure on intake. These are outside the scope of this discussion. Some of the relevant reviews are by Hodgson and Illius (1996), Duncan and Gordon (1999), Baumont *et al.* (2000) and Provenza *et al.* (2003). The review of Illius and Gordon (1999) is especially interesting as it bridges the gap between dietary preferences, digestive tract anatomy and feeding classification.

Final Thoughts on Food Intake Control

Food consumption is essential to maintain life and the capacity to reproduce. Given this, it is quite reasonable that the system which has evolved to control intake, i.e. to allow the animal to recognize its nutritional status and rectify under-nutrition, is able to sense information from a variety of different sources and is 'redundant', in the sense of having several different ways of sending that information to the brain and a variety of ways in which the hypothalamic and other nuclei process that information (Fig. 11.11).

The issue of obesity in domestic animals, especially pets, is not addressed here. If customary levels of food intake are the result of balances between

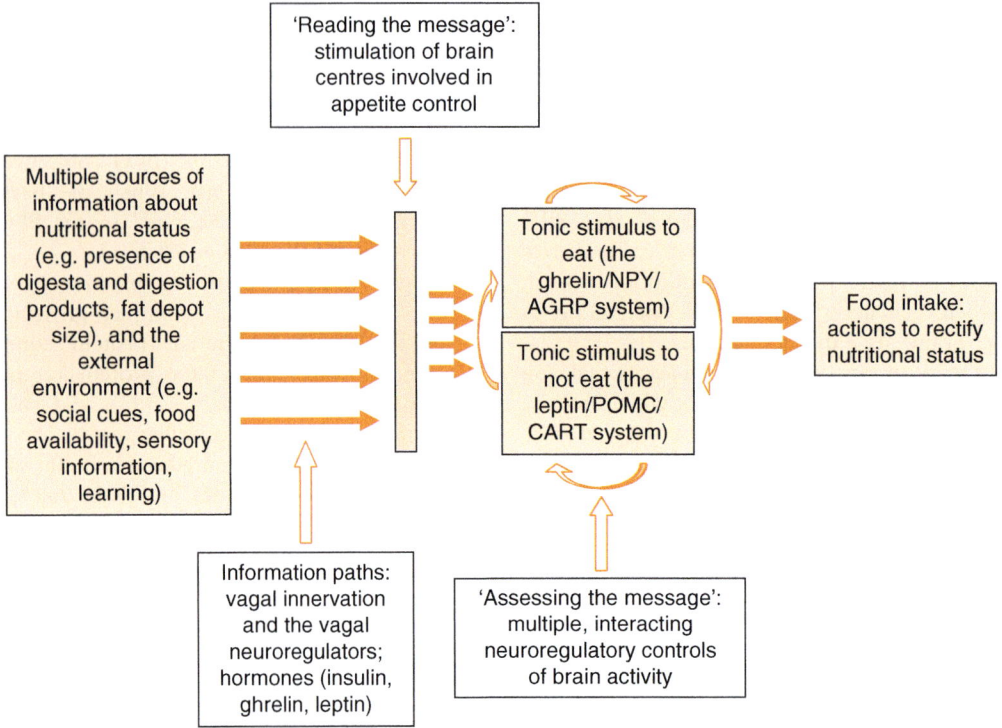

Fig. 11.11. The multifaceted, 'redundant' system which controls food intake.

two continuously acting, opposing systems (the NPY/AGRP and POMC/CART systems), then a conventional view is that this balance defines a 'set point'. Alternatively, food intake might be determined by the animal's energy status varying from a 'settling point' (Bolles, 1980, cited by Berridge, 2004). The suggestion (from Berridge, 2004) is that the appetite settling point changes according to the animal's current nutritional environment, rising when food is plentiful (e.g. easy access to palatable foods) and perhaps declining when it is scarce.

Finally, we must acknowledge that nothing in this chapter answers the question of how intake controls are 'set'. How does an animal 'know' what its mature body size or composition should be? How does it 'know' what amount of food is the 'right' amount? There are no good answers to these questions. Speakman *et al*. (2002) discuss the idea that energy intake responds to daily deficits and excesses in energy intakes, Ketelaars and Tolkamp (1996) introduce the interesting idea that food intake might be related to the optimal use of O by the animal and Forbes (2001) suggests that animals 'experiment with the amount eaten per day, . . . until the total of signals generated from excesses or deficiencies . . . is minimized'.

References

Akin, D.E. (1979) Microscopic evaluation of forage digestion by rumen microorganisms – a review. *Journal of Animal Science* 48, 701–710.

Allen, M.S. (2000) Effects of diet on short-term regulation of feed intake by lactating dairy cattle. *Journal of Dairy Science* 83, 1598–1624.

Arsenos, G. and Kyriazakis, I. (2001) Does previous protein feeding affect the response of sheep towards foods that differ in their rumen availability, but not content, of nitrogen? *Physiology and Behaviour* 72, 533–541.

Baile, C.A. and McLaughlin, C.L. (1970) Feed intake of goats during volatile fatty acid injections into four gastric areas. *Journal of Dairy Science* 53, 1058–1063.

Barry, T.N. and McNabb, W.C. (1999) The implications of condensed tannins on the nutritive value of temperate forages fed to ruminants. *British Journal of Nutrition* 81, 263–272.

Baumont, R., Brun, J.P. and Dulphy, J.P. (1989) Influence of the nature of hay on its ingestibility and the kinetics of intake during large meals in sheep and cow. In: Jarrige, R. (ed.) *Proceedings of the XVI International Grassland Congress, Nice, France*. French Grassland Society, Nice, France, pp. 787–788.

Baumont, R., Dulphy, J.P. and Jailler, M. (1997) Dynamic of voluntary intake, feeding behaviour and rumen function in sheep fed three contrasting types of hay. *Annales de Zootechnie* 46, 231–244.

Baumont, R., Prache, S., Meuret, M. and Mohrand-Fehr, P. (2000) How forage characteristics influence behaviour and intake in small ruminants: a review. *Livestock Production Science* 64, 15–28.

Berridge, K.C. (2004) Motivation concepts in behavioural neuroscience. *Physiology and Behaviour* 81, 179–209.

Black, J.L., Kenney, P.A. and Colebrook, W.F. (1987) Diet selection by sheep. In: Wheeler, J.L., Pearson, C.J. and Robards, G.E. (eds) *Temperate Pastures: Their Production, Use and Management*. CSIRO, Melbourne, Australia, pp. 331–334.

Blackshaw, L.A. and Grundy, D. (1989) Responses of vagal efferent fibres to stimulation of gastric mechano- and chemoreceptors in the anaesthetized ferret. *Journal of the Autonomic Nervous System* 27, 39–45.

Bloom, S.R. (2003) How the brain controls appetite. *Endocrine Abstracts* 5, S2.

Bolles, R.W. (1980) Some functionalistic thoughts about regulation. In: Toates, T.W. and Halliday, T.W. (eds) *Analysis of Motivational Processes*. Academic Press, New York, pp. 63–75.

Broberger, C. (2005) Brain regulation of food intake and appetite: molecules and networks. *Journal of Internal Medicine* 258, 301–327.

Bruining, M., Bakker, R., Van Bruchem, J. and Tamminga, S. (1998) Rumen digesta kinetics in dairy cows fed grass, maize and alfalfa silage. 1. Comparison of conventional, steady-state and dynamic methods to estimate microbial degradation, comminution and passage of particles. *Animal Feed Science and Technology* 73, 37–58.

Chilliard, Y., Delavaud, C. and Bonnet, M. (2005) Leptin expression in ruminants: nutritional and physiological regulations in relation with energy metabolism. *Domestic Animal Endocrinology* 29, 3–22.

Coffey, K.P., Paterson, J.A., Saul, C.S., Coffey, L.S., Turner, K.E. and Bowman, J.G. (1989) The influence of pregnancy and source of supplemental protein on intake, digestive kinetics and amino acid absorption by ewes. *Journal of Animal Science* 67, 1805–1814.

Costa, M.A.L., Filho, S.deC.V., Paulino, M.F., Valadares, R.F.D., Cecon, P.R., Paulino, P.V.R., de Moraes, E.H.B.K. and Magalhães, K.A. (2005) Productive performance, digestibility and carcass characteristics of zebu steers fed diets with different concentrate levels. *Revista Brasileira de Zootecnia* 34, 268–279.

Cottrell, D.F. and Iggo, A. (1984) Mucosal enteroceptors with vagal afferent fibers in the proximal jejunum of sheep. *Journal of Physiology* 354, 497–522.

Critchley, H.D. and Rolls, E.T. (1996) Hunger and satiety modify the responses of olfactory and visual neurons in the primate orbitofrontal cortex. *Journal of Neurophysiology* 75, 1673–1686.

Dawson, J.M., Buttery, P.J., Jenkins, D., Wood, C.D. and Gill, M. (1999) Effects of dietary quebracho tannin on nutrient utilization and tissue metabolism in sheep and rats. *Journal of the Science of Food and Agriculture* 79, 1423–1430.

Domingue, B.M.F., Dellow, D.W., Wilson, P.R. and Barry, T.N. (1991) Comparative digestion in deer, sheep and goats. *New Zealand Journal of Agricultural Research* 34, 45–53.

Doreau, M. and Diawara, A. (2003) Effect of level of intake on digestion in cows: influence of animal genotype and nature of hay. *Livestock Production Science* 81, 35–45.

Douglas, G.B., Wang, Y., Waghorn, G.C., Barry, T.N., Purchas, R.W., Foote, A.G. and Wilson, G.F. (1995) Liveweight gain and wool production of sheep grazing *Lotus corniculatus* and lucerne (*Medicago sativa*). *New Zealand Journal of Agricultural Research* 38, 95–104.

Druce, M. and Bloom, S.R. (2006) The regulation of appetite. *Archives of Disease in Childhood* 91, 183–187.

Dryden, G.McL. and Kempton, T.J. (1983) Digestion of organic matter and nitrogen in ammoniated barley straw. *Animal Feed Science and Technology* 10, 65–75.

Dryden, G.McL., Stafford, K.J., Waghorn, G.C. and Barry, T.N. (1995) Comminution of roughages by red deer (*Cervus elaphus*) during the prehension of feed. *Journal of Agricultural Science, Cambridge* 125, 407–414.

Duncan, A.J. and Gordon, I.J. (1999) Habitat selection according to the ability of animals to eat, digest and detoxify foods. *Proceedings of the Nutrition Society* 58, 799–805.

Edwards, G.P., Dawson, T.J. and Croft, D.B. (1995) The dietary overlap between red kangaroos (*Macropus rufus*) and sheep (*Ovis aries*) in the arid rangelands of Australia. *Australian Journal of Ecology* 20, 324–334.

Evans, E.W., Pearce, G.R., Burnett, J. and Pillinger, S.L. (1973) Changes in some physical characteristics of the digesta in the reticulo-rumen of cows fed once daily. *British Journal of Nutrition* 29, 357–376.

Faverdin, P. (1985) Regulation de l'ingestion des vaches laitières en début de lactation. Thèse de Doctorat INA Paris-Grignon, France.

Forbes, J.M. (1988) Metabolic aspects of the regulation of voluntary food intake and appetite. *Nutrition Research Reviews* 1, 145–168.

Forbes, J.M. (1995) *Voluntary Food Intake and Diet Selection in Farm Animals*. CAB International, Wallingford, UK.

Forbes, J.M. (1996) Integration of regulatory signals controlling forage intake in ruminants. *Journal of Animal Science* 74, 3029–3035.

Forbes, J.M. (2001) Consequences of feeding for future feeding. *Comparative Biochemistry and Physiology – Part A: Molecular & Integrative Physiology* 128, 461–468.

Forbes, J.M. and Barrio, J.P. (1992) Abdominal chemo- and mechanosensitivity in ruminants and its role in the control of food intake. *Experimental Physiology* 77, 27–50.

Freudenberger, D.O., Toyokawa, K., Barry, T.N., Ball, A.J. and Suttie, J.M. (1994) Seasonality in digestion and rumen metabolism in red deer (*Cervus elaphus*) fed on a forage diet. *British Journal of Nutrition* 71, 489–499.

Gibson, L.A. (2001) Seasonal changes in the diet, food availability and food preference of the greater bilby (*Macrotis lagotis*) in south-western Queensland. *Wildlife Research* 28, 121–134.

Green, C., Elwin, R.L., Mottershead, B.E., Keogh, R.G. and Lynch, J.J. (1984) Long-term effects of early experience to supplementary feeding in sheep. *Animal Production in Australia* 15, 373–380.

Greenhalgh, J.F.D. and Reid, G.W. (1971) Relative palatability to sheep of straw, hay and dried grass. *British Journal of Nutrition* 26, 107–116.

Grovum, W.L. (1995) Mechanisms explaining the effects of short chain fatty acids on feed intake in ruminants – osmotic pressure, insulin and glucagons. In: von Engelhardt, W., Leonhard-Marek, S., Breves, G. and Giesecke, D. (eds) *Ruminant Physiology: Digestion, Metabolism, Growth and Reproduction*. Ferdinand Enke Verlag, Stuttgart, Germany, pp. 173–197.

Guo, L., Munzberg, H., Stuart, R.C., Nillni, E.A. and Bjorbaek, C. (2004) N-acetylation of hypothalamic α-melanocyte-stimulating hormone and regulation by leptin. *PNAS* 101, 11797–11802.

Harding, R. and Leek, B.F. (1972) Rapidly adapting mechanoreceptors in the reticulo-rumen which also respond to chemicals. *Journal of Physiology* 223, 32P–33P.

Harris, R.B.S., Ramsay, T.G., Smith, S.R. and Bruch, R.C. (1996) Early and late stimulation of *ob* mRNA expression in meal-fed and overfed rats. *Journal of Clinical Investigation* 97, 2020–2026.

Haskell-Luevano, C., Chen, P., Li, C., Chang, K., Smith, M.S., Cameron, J.L. and Cone, R.D. (1999) Characterization of the neuroanatomical distribution of agouti-related protein immunoreactivity in the rhesus monkey and the rat. *Endocrinology* 140, 1408–1415.

Hmeidan, M.C. and Dryden, G.McL. (1998) Effect of hay quality and grain supplementation on feed intake, liveweight and digestibility in young rusa deer (*Cervus timorensis*) stags. *Animal Production in Australia* 22, 383.

Hodgson, J. and Illius, A.W. (eds) (1996) *The Ecology and Management of Grazing Systems*. CAB International, Wallingford, UK.

Holst, B. and Schwartz, T.W. (2004) Constitutive ghrelin receptor activity as a signalling set point in appetite regulation. *TRENDS in Pharmacological Sciences* 25, 113–117.

Horbury, S.R., Mercer, J.G. and Chappell, L.H. (1995) Anorexia induced by the parasitic nematode, *Nippostrongylus brasiliensis:* effects on NPY and CRF gene expression in the rat hypothalamus. *Journal of Neuroendocrinology* 7, 867–873.

Horvath, T.L., Diano, S., Sotonyi, P., Heiman, M. and Tschop, M. (2001) Minireview: ghrelin and the regulation of energy balance – a hypothalamic perspective. *Endocrinology* 142, 4163–4169.

Hoskin, S.O., Wilson, P.R., Barry, T.N., Charleston, W. A.G. and Waghorn, G.C. (2000) Effect of forage legumes containing condensed tannins on lungworm (*Dictyocaulus* sp.) and gastrointestinal parasitism in young red deer (*Cervus elaphus*). *Research in Veterinary Science* 68, 223–230.

Huda, M.S.B., Wilding, J.P.H. and Pinkney, J.H. (2006) Gut peptides and the regulation of appetite. *Obesity Reviews* 7, 163–182.

Hyer, J.C., Oltjen, J.W. and Galyean, M. (1991) Development of a model to predict forage intake by grazing cattle. *Journal of Animal Science* 69, 827–835.

Illius, A.W. and Gordon, I.J. (1999) The physiological ecology of mammalian herbivory. In: Jung, H.-J.G. and Fahey, G.C. (eds) *Nutritional Ecology of Herbivores*. American Association of Animal Science, Savoy, Illinois, pp. 71–96.

James, S.M. and Kyriazakis, I. (2002) The effect of consumption of foods that differ in energy density and/or sodium bicarbonate supplementation on subsequent diet selection in sheep. *British Journal of Nutrition* 88, 81–90.

Kaske, M. and Groth, A. (1997) Changes in factors affecting the rate of digesta passage during pregnancy and lactation in sheep fed on hay. *Reproduction, Nutrition and Development* 37, 573–588.

Kaske, M. and Midasch, A. (1997) Effects of experimentally impaired reticular contractions on digesta passage in sheep. *British Journal of Nutrition* 78, 97–110.

Karda, W., Dryden, G.McL. and Gutteridge, R.C. (1998) Oven-dried gliricidia leaf (*Gliricidia sepium*) is unpalatable to sheep and goats. *Animal Production in Australia* 22, 366.

Ketelaars, J.J.M.H. and Tolkamp, B.J. (1992) Towards a new theory of feed intake regulation in ruminants 1. Causes of differences in voluntary feed intake: critique of current views. *Livestock Production Science* 30, 269–296.

Ketelaars, J.J.M.H. and Tolkamp, B.J. (1996) Oxygen efficiency and the control of energy flow in animals and humans. *Journal of Animal Science* 74, 3036–3051.

Kimball, B.A. and Nolte, D.L. (2005) Herbivore experience with plant defense compounds influences acquisition of new flavor aversions. *Applied Animal Behaviour Science* 91, 17–34.

King, B.M. (2006) The rise, fall, and resurrection of the ventromedial hypothalamus in the regulation of feeding behaviour and body weight. *Physiology and Behaviour* 87, 221–244.

Król, E., Duncan, J.S., Redman, P., Morgan, P.J., Mercer, J.G. and Speakman, J.R. (2006) Photoperiod regulates leptin sensitivity in field voles, *Microtus agrestis*. *Journal of Comparative Physiology B* 176, 153–163.

Kyriazakis, I. and Oldham, J.D. (1993) Diet selection in sheep: the ability of growing lambs to select a diet that meets their crude protein (nitrogen × 6.25) requirements. *British Journal of Nutrition* 69, 617–629.

Kyriazakis, I., Tolkamp, B.J. and Emmans, G. (1999) Diet selection and animal state: an integrative framework. *Proceedings of the Nutrition Society* 58, 765–772.

Laing, D.G. (1985) The role of smell and taste in the regulation of food intake in humans and animals. *Proceedings of the Nutrition Society of Australia* 10, 70–77.

Laredo, M.A. and Minson, D.J. (1975) The effect of pelleting on the voluntary intake and digestibility of leaf and stem fractions of three grasses. *British Journal of Nutrition* 33, 159–170.

Lechner-Doll, M., Rutagwenda, T., Schwartz, H.J., Schultka, W. and von Engelhardt, W. (1990) Seasonal changes of ingesta mean retention time and forestomach volume in indigenous grazing camels, cattle, sheep and goats on a thornbush savanna pasture. *Journal of Agricultural Science, Cambridge* 115, 409–420.

Lewis, D.E., Shellard, L., Koeslag, D.G., Boer, D.E., McCarthy, H.D., McKibbin, P.E., Russell, J.C. and Williams, G. (1993) Intense exercise and food restriction cause similar hypothalamic neuropeptide Y increases in rats. *American Journal of Physiology* 264, E279–E284.

Loerch, S.C. and Fluharty, F.L. (1998) Effects of corn processing, dietary roughage level, and timing of roughage inclusion on performance of feedlot steers. *Journal of Animal Science* 76, 681–685.

Lund, P., Weisbjerg, M.R. and Hvelplund, T. (2006) Passage kinetics of fibre in dairy cows obtained from duodenal and faecal ytterbium excretion. Effect of forage type. *Animal Feed Science and Technology* 128, 229–252.

Marie, M., Findlay, P.A., Thomas, L. and Adam, C.L. (2001) Daily patterns of plasma leptin in sheep: effects of photoperiod and food intake. *Journal of Endocrinology* 170, 277–286.

Marks, J.L., Waite, K. and Davies, L. (1996) Intracerebroventricular neuropeptide Y produces hyperinsulinemia in the presence and absence of food. *Physiology and Behavior* 60, 685–692.

Mbanya, J.N., Anil, M.H. and Forbes, J.M. (1993) The voluntary intake of hay and silage by lactating cows in response to ruminal infusion of acetate or propionate, or both, with or without distension of the rumen by a balloon. *British Journal of Nutrition* 69, 713–720.

Mei, N. (1985) Intestinal chemosensitivity. *Physiological Reviews* 65, 211–237.

Mei, N. and Garnier, L. (1986) Osmosensitive vagal receptors in the small intestine of the cat. *Journal of the Autonomic Nervous System* 16, 159–170.

Mertens, D.R. (1973) *Application of Theoretical Mathematical Models to Cell Wall Digestion and Forage Intake in Ruminants.* PhD Thesis, Cornell University, Ithaca, New York.

Mertens, D.R. and Ely, L.O. (1979) A dynamic model of fiber digestion and passage in the ruminant for evaluating forage quality. *Journal of Animal Science* 49, 1085–1095.

Miner, J.L., Della-Fera, M.A. and Paterson, J.A. (1990) Blockade of satiety factors by central injection of neuropeptide Y in sheep. *Journal of Animal Science* 68, 3805–3811.

National Research Council (NRC) (1987) *Predicting Feed Intake of Food-Producing Animals.* National Academy Press, Washington, DC.

Norton, B.W., Kennedy, P.J. and Hales, J.W. (1990) Grazing management studies with Australian cashmere goats. 3. Effect of season on the selection of diet by cattle, sheep and goats from two tropical grass-legume pastures. *Australian Journal of Experimental Agriculture* 30, 783–788.

Pickavance, L., Dryden, S., Hopkins, D., Chen, B., Frankish, H., Qiong, W., Vernon, R.G. and Williams, G. (1996) Relationships between hypothalamic neuropeptide Y and food intake in the lactating rat. *Peptides* 17, 577–582.

Poppi, D.P., Norton, B.W., Minson, D.J. and Hendrickson, R.E. (1980) The validity of the critical size theory for particles leaving the rumen. *Journal of Agricultural Science, Cambridge* 94, 275–280.

Poppi, D.P., Ellis, W.C., Matis, J.H. and Lascano, C.E. (2001) Marker concentration patterns of labelled leaf and stem particles in the rumen of cattle grazing bermuda grass (*Cynodon dactylon*) analysed by reference to a raft model. *British Journal of Nutrition* 85, 553–563.

Provenza, F.D. (1996) Familiarity and novelty in animal diets. *Proceedings of the Australian Society of Animal Production* 21, 12–16.

Provenza, F.D., Pfister, J.A. and Cheney, C.D. (1992) Mechanisms of learning in diet selection with reference to phytotoxicosis in herbivores. *Journal of Range Management* 45, 136–145.

Provenza, F.D., Burritt, E.A., Perevolotsky, A. and Silanikove, N. (2000) Self-regulation of intake of polyethylene glycol by sheep fed diets varying in tannin concentrations. *Journal of Animal Science* 78, 1206–1212.

Provenza, F.D., Villalba, J.J., Dziba, L.E., Atwood, S.B. and Banner, R.E. (2003) Linking herbivore experience, varied diets, and plant biochemical diversity. *Small Ruminant Research* 49, 257–274.

Rahmouni, K. and Haynes, W.G. (2001) Leptin signaling pathways in the central nervous system: interactions between neuropeptide Y and melanocortins. *BioEssays* 23, 1095–1099.

Rees, M.C., Minson, D.J. and Kerr, J.D. (1972) Relation of dairy productivity of feed supply in the Gympie district of south-eastern Queensland. *Australian Journal of Experimental Agriculture and Animal Husbandry* 12, 553–560.

Renecker, L.A. and Hudson, R.J. (1990) Digestive kinetics of moose (*Alces alces*), wapiti (*Cervus elaphus*) and cattle. *Animal Production* 50, 51–61.

Saladin, R., de Vos, P., Guerre-Millo, M., Leturque, A., Girard, J., Staels, B. and Auwerx, J. (1995) Transient increase in obese gene expression after food intake or insulin administration. *Nature* 377, 527–529.

Sartin, J.L., Wagner, C.G., Marks, D.L., Daniel, J.A., McMahon, C.D., Obese, F.Y. and Partridge, C. (2005) Melanocortin-4 receptor in sheep: a potential site for therapeutic intervention in disease models. *Domestic Animal Endocrinology* 29, 446–455.

Scott, C.B., Provenza, F.D. and Banner, R.E. (1995) Dietary habits and social interactions affect choice of feeding location by sheep. *Applied Animal Behaviour Science* 45, 225–237.

Seufert, J., Kieffer, T.J. and Habener, J.F. (1999) Leptin inhibits insulin gene transcription and reverses hyperinsulinemia in leptin-deficient *ob/ob* mice. *Proceedings of the National Academy of Sciences of the USA* 96, 674–679.

Smith, M.S. (1993) Lactation alters neuropeptide-Y and proopiomelanocortin gene expression in the arcuate nucleus of the rat. *Endocrinology Philadelphia* 133, 1258–1283.

Sonnenberg, G.E., Krakower, G.R., Hoffmann, R.G., Maas, D.L., Hennes, M.M. and Kissebah, A.H. (2001) Plasma leptin concentrations during extended fasting and graded glucose infusions: relationships with changes in glucose, insulin, and FFA. *Journal of Clinical Endocrinology and Metabolism* 86, 4895–4900.

Speakman, J.R. and Król, E. (2005) Limits to sustained energy intake. IX: a review of hypotheses. *Journal of Comparative Physiology B* 175, 375–394.

Speakman, J.R., Stubbs, R.J. and Mercer, J.G. (2002) Does body mass play a role in the regulation of food intake? *Proceedings of the Nutrition Society* 61, 473–487.

Standing Committee on Agriculture and Resource Management (SCA) (1987) *Feeding Standards for Australian Livestock. Pigs.* CSIRO, Melbourne, Australia.

Stutz, A.M., Morrison, C.D. and Argyropoulous, G. (2005) The agouti-related protein and its role in energy homeostasis. *Peptides* 26, 1771–1781.

Sydenham, C. and Dryden, G.McL. (2007) Agonistic behaviour of red deer hinds given a supplement. *Recent Advances in Animal Nutrition in Australia* 16, 280 (Abstr).

Tolkamp, B.J., Kyriazakis, I., Oldham, J.D., Lewis, M., Dewhurst, R.J. and Newbold, J.R. (1998) Diet choice by dairy cows. 2. Selection for metabolizable protein or for ruminally degradable protein? *Journal of Dairy Science* 81, 2670–2680.

Tritos, N.A., Vicent, D., Gillette, J., Ludwig, D.S., Flier, E. and Maratos-Flier, E. (1998) Functional interactions between melanin-concentrating hormone, neuropeptide Y, and anorectic neuropeptides in the rat hypothalamus. *Diabetes* 47, 1687–1692.

Troelsen, J.E. and Campbell, J.B. (1968) Voluntary consumption of forage by sheep and its relation to the size and shape of particles in the digestive tract. *Animal Production* 10, 289–296.

Udén, P. (1988) The effect of grinding and pelleting hay on digestibility, fermentation rate, digesta passage and rumen and faecal particle size in cows. *Animal Feed Science and Technology* 19, 145–157.

Ulyatt, M.J., Waghorn, G.C., John, A., Reid, C.S.W. and Munro, J. (1984) Effect of intake and feeding frequency on feeding-behavior and quantitative aspects of digestion in sheep fed chaffed lucerne hay. *Journal of Agricultural Science* 102, 645–657.

Underwood, E.J. and Suttle, N.F. (1999) *The Mineral Nutrition of Livestock*, 3rd edn. CAB International, Wallingford, UK.

Vajrabukka, C., Thwaites, C.J. and Farrell, D.J. (1983) A field survey and experiments to determine the effect of high temperature on the biological performance of pigs. *Recent Advances in Animal Nutrition in Australia* 7, 192–194.

Van Soest, P.J. (1967) Development of a comprehensive system of feed analyses and its application to forages. *Journal of Animal Science* 26, 119–128.

Van Soest, P.J. (1975) Physico-chemical aspects of fibre digestion. In: McDonald, I.W. and Warner, A.C.I. (eds) *Digestion and Metabolism in the Ruminant*. University of New England Publishing Unit, Armidale, New South Wales, Australia, pp. 351–365.

Villalba, J.J. and Provenza, F.D. (2000) Roles of novelty, generalization, and postingestive feedback in the recognition of foods by lambs. *Journal of Animal Science* 78, 3060–3069.

Villalba, J.J., Provenza, F.D., Hall, J.O. and Peterson, C. (2006a) Phosphorus appetite in sheep: dissociating taste from postingestive effects. *Journal of Animal Science* 84, 2213–2223.

Villalba, J.J., Provenza, F.D. and Shaw, R. (2006b) Sheep self-medicate when challenged with illness-inducing foods. *Animal Behaviour* 71, 1131–1139.

Waghorn, G.C. and Reid, C.S.W. (1977) Rumen motility in sheep and cattle as affected by feeds and feeding. *Proceedings of the New Zealand Society of Animal Production* 37, 176–181.

Waghorn, G.C., Reid, C.S.W., Ulyatt, M.J. and John, A. (1986) Feed comminution, particle composition and distribution between the 4 compartments of the stomach in sheep fed chaffed lucerne hay at 2 feeding frequencies and intake levels. *Journal of Agricultural Science, Cambridge* 106, 287–296.

Waldo, D.R., Smith, L.W. and Cox, E.L. (1972) Model of cellulose disappearance from the rumen. *Journal of Dairy Science* 55, 125–129.

Wattiaux, M.A., Mertens, D.R. and Satter, L.D. (1992) Kinetics of hydration and effect of liquid uptake on specific gravity of small hay and silage particles. *Journal of Animal Science* 70, 3597–3606.

Welch, J.G. (1982) Rumination, particle-size and passage from the rumen. *Journal of Animal Science* 54, 885–894.

Welch, J.G. and Smith, A.M. (1969) Influence of forage quality on rumination time in sheep. *Journal of Animal Science* 28, 813–818.

Weston, R.H. (1996) Some aspects of constraint to forage consumption by ruminants. *Australian Journal of Agricultural Research* 47, 175–197.

Weston, R.H. and Poppi, D.P. (1987) Comparative aspects of food intake. In: Hacker, J.B. and Ternouth, J.H. (eds) *The Nutrition of Herbivores*. Academic Press, Sydney, Australia, pp. 133–162.

Weyreter, H. and von Engelhardt, W. (1984) Adaptation of Heidschnucken, Merino and Blackhead sheep to a fibrous roughage diet of poor quality. *Canadian Journal of Animal Science* 64(Suppl.), 152–153.

Whitney, T.R. and Olson, B.E. (2006) Conditioning ewes and lambs to increase consumption of spotted knapweed. *Applied Animal Behaviour Science* 100, 193–206.

Wylie, M.J., Ellis, W.C., Matis, J.H., Bailey, E.M., James, W.D. and Beever, D.E. (2000) The flow of forage particles and solutes through segments of the digestive tracts of cattle. *British Journal of Nutrition* 83, 295–306.

Yeomans, M.R. (2006) Olfactory influences on appetite and satiety in humans. *Physiology and Behavior* 87, 800–804.

12 Quantitative Nutrition: Requirements for Energy and Protein

Approaches to Quantifying Nutrient Requirements

A factorial approach is often used to quantify animals' requirements for energy and amino acids, and for some major minerals. To do this, we identify all the physiological processes that use the nutrient of interest and determine the amount of that nutrient used in each process. For example, a lactating and pregnant dairy cow uses energy for body tissue maintenance, milk production and fetal growth. The energy required for milk synthesis depends on the amount and chemical composition of milk produced, and we can measure (or predict) these and calculate the energy required for milk production. Similarly, we can calculate the energy needed for maintenance and pregnancy. Although these are all separate functions, they do not have to be treated as independent of each other because we can take into account nutrient flows between the animal's different physiological 'compartments' (e.g. adipose tissue, the fetus and the mammary gland). For example, the amount of dietary energy we need to feed to support milk production early in the lactation can be reduced by the amount supplied from the breakdown of body tissue.

A factorial approach is not always possible. We may not have a detailed enough understanding of the animal's physiology (e.g. the metabolism of some vitamins and trace minerals), or we may understand the physiological concepts but not have enough quantitative information (e.g. the interrelationships between lean growth genotype and reproduction in the pig). In these cases, we have no option but to use an empirical, or whole-animal, basis for stating nutrient requirements.

The factorial method has several advantages over an empirical approach. Because each physiological process is identified and quantified separately, we can fine-tune the conceptual and quantitative basis of each individual process, and of the interrelationships between the various pro-

cesses, as our understanding about these develops. We can thus modify the nutrient requirements for a particular process without necessarily affecting our estimations of other requirements. The factorial approach suits mathematical modelling which is widely used in quantitative nutrition and nutritional management.

'Maintenance' and 'Production' Requirements for Nutrients

Animals have a variety of bodily functions. Some of these are basic to the maintenance of life, such as breathing, blood circulation, etc. Others involve the synthesis of products, e.g. wool or cashmere fibres, milk, meat and fetuses. In some cases muscular activity is used for productive purposes such as draught work or competition in sporting events. If we keep animals as production units, then it is useful to distinguish between nutrient requirements for 'maintenance' and those for 'production'. If we are working with wildlife or companion animals, then we may not need to separate these, as we will be more interested in the animal's total nutrient demand.

Rates of Energy Use

Basal metabolic rate

Basal metabolism (also called the 'minimal metabolism') is the animal's basic life-sustaining functions, e.g. breathing, maintaining adequate blood circulation, synthesizing molecules to replace catabolized tissue, translocating substrates between cells or organelles, contracting muscle fibres, passing electrical impulses along nerve fibres and maintaining internal body temperature (in homeotherms). According to estimations by Rolfe and Brown (1997) about 70% of oxygen use (i.e. energy substrate oxidation) is for the synthesis of ATP. The major energy-using processes are protein synthesis

Table 12.1. The energy cost of some activities of wild and companion dogs.

Species and activity	Energy requirement (MJ NE/kgW$^{0.75}$/h)	Source
African wild dog, hunting	0.281	Gorman *et al.*, 1998
African wild dog, total daily activity	0.057	Gorman *et al.*, 1998
African wild dog, resting	0.019	Gorman *et al.*, 1998
Sled dogs, 490 km Arctic race	1.960	Hinchcliff *et al.*, 1997
Labrador pups, suckling	0.023	Scantlebury *et al.*, 2001
Schnauzer pups, suckling	0.022	Scantlebury *et al.*, 2001

(25–30% of the total ATP use), regulation of intracellular and extracellular concentrations of Na$^+$, K$^+$ and Ca^{2+} (23–36%), gluconeogenesis (7–10%), with urea and nucleic acid syntheses, and muscle movement, making other requirements (Rolfe and Brown, 1997). The basal metabolic rate (BMR) is the amount of energy used each day for these purposes and is the lowest rate at which the normally nourished animal uses energy.

The BMR declines with increasing age (a difference of 20–25% between sheep aged 1 and 4–6 years; Blaxter, 1962; Toutain *et al.*, 1977) and if the animal is asleep. The BMR falls by about 0.5% per day if a sheep is starved (Blaxter, 1967). The BMR can be elevated in temperamental animals, such as some horses, or if an animal is stressed by confinement or competition.

Standard metabolic rate

The standard metabolic rate (SMR) is the rate of energy use measured under 'standard conditions'. These are imposed in an attempt to put the animal into a condition of basal metabolism. Another name for the SMR is 'fasting heat production'. This reflects one of the ways in which the SMR is measured, i.e. by confining a fasted animal in a chamber and measuring its heat production or oxygen consumption. If the SMR is measured when the animal is allowed to eat, the rate of energy use will increase by about 25% over the rate measured in a fasting animal (Gessaman and Nagy, 1988; Rolfe and Brown, 1997). The rate of energy use when the animal is digesting food is also called the 'resting metabolic rate' (Rolfe and Brown, 1997).

Field metabolic rate and maximum metabolic rate

The field metabolic rate (FMR) is the animal's total rate of energy use. This concept is useful for work with wildlife or companion animals where there is no good reason to separate the energy requirements for maintenance and production. Nagy (1987) has collated a large amount of data on the FMR of eutherians, metatherians and cold-blooded animals. The FMR can be substantially greater than the SMR; illustrative values for FMR of dogs are summarized in Table 12.1.

The maximum metabolic rate (MMR) is reached when additional activity or cold stress does not further increase the animal's rate of oxygen use. This is the highest rate at which an animal's aerobic metabolism can function (called the 'summit metabolism') and is eight to ten times the SMR (SCA, 1990; Rolfe and Brown, 1997). However, animals can synthesize ATP anaerobically for short periods, incurring an oxygen debt at the expense of lactic acid accumulation. Barrey (1993) gives examples of how this can be incorporated into assessments of the energy expenditures of horses working at high rates.

Relationships Between the Rate of Energy Use and Body Measurements

Several workers in the 19th century developed the theory that an animal's basal heat production was related to its surface area, or to the 2/3 power of its body weight. The basis of the relationship between surface area and heat production under standard conditions is:

1. Heat loss is proportional to the animal's free body surface ('free' means 'exposed to the environment'); Kleiber (1975) discusses the difficulty of measuring an animal's surface area.
2. If body temperature is to remain constant, as it does in homeotherms, then heat loss from the animal must be equalled by its heat production.
3. Therefore, heat production is proportional to the animal's free body surface, which in turn is proportional to the square of the linear dimensions of the animal.

For a sphere:

surface area = $4\pi a^2$ (where a = the radius) (12.1)
volume = $(4/3)\pi a^3$
mass = volume × density
so mass $\propto a^3$
and $a \propto \text{mass}^{1/3}$
thus, surface area $\propto (\text{mass}^{1/3})^2$
$\propto \text{mass}^{2/3}$

This 'surface area law' was demonstrated experimentally, more or less rigorously, by Rubner (1883), Voit (1901) and others (cited by Kleiber, 1975).

Metabolic size

For practical use, we need an easily measured attribute of the animal which is closely related to its rate of energy use. The allometric relationship ('allometry' is the relationship between body measurements and body metabolism) between mass (M, kg) and heat production or the SMR (MJ NE/day) is, in general terms:

$$\text{SMR} = a\text{M}^b \qquad (12.2)$$

where
M^b = the animal's 'metabolic size'
a = a factor which relates SMR to metabolic size

There is currently a robust debate about the size and meaning of the exponent b in defining metabolic size. Rubner's surface area law inferred that metabolic size should be $\text{M}^{2/3}$. $\text{M}^{0.73}$ was adopted at a conference in Maryland in 1935, based on the evidence available at that time. An EAAP conference in Scotland in 1964 adopted the value 3/4 (note, not 0.75) to facilitate interspecies comparisons and avoid the perception of 'spurious accuracy' (Kleiber and Flatt, 1965). This decision was based on a consideration of data published by Kleiber (1932, 1947), Brody and Proctor (1932), Brody et al. (1934) and Benedict (1938). It is customary now to define metabolic size as $\text{M}^{0.75}$, e.g. ARC (1980; for sheep but not cattle), SCA (1990), NRC (1998, 2000, 2001). Nevertheless, opinion about the size of the exponent still oscillates between 2/3 (Dodds et al., 2001; White and Seymour, 2003) and 0.75 or 3/4 (Darveau et al., 2002) and other values. Da Silva et al. (2006) have summarized the various exponents which have been used to define metabolic size and discussed the logical flaws which may exist in the derivation

of some of these. Painter (2005) reanalysed two large data sets and concluded that the exponent is not constant. For animals of body mass less than 0.2 kg, he obtained b of about 0.62, and for animals weighing more than 10 kg, b was about 0.88. Interestingly, MMR scales at about $\text{W}^{0.87}$ (Burger and Johnson, 1991; Darveau et al., 2002; Weibel et al., 2004) and the FMR of eutherians scales as $\text{W}^{0.81}$ (Nagy, 1987).

Of perhaps greater fundamental importance is the debate about the biological meaning of metabolic size. Some proponents argue that this represents a biological concept. For example, West et al. (1999) have argued that the 'fractal-like networks' of blood vessels and other body organs is 'the origin of quarter-power scaling that is so pervasive in biology'. Darveau et al. (2002) arrived at an exponent of 0.75 by considering the ways that a number of different physiological processes contributed to the organism's overall metabolism. Rolfe and Brown (1997) also cite several tissue and organelle allometries which scale with b equal to approximately 0.75.

Alternatively, we can be empirical, and look for an exponent for M which gives a good statistical fit to the regression of energy requirement on body mass. It is an advantage if the exponent is a constant for all of the animal types we deal with. For the moment, $\text{M}^{0.75}$ is satisfactory, and we will use it to define metabolic size in this text.

It was first thought that the value for a (Equation 12.2) would be constant for all animals. It is not, but there are surprising similarities, at least within eutherians. For adults, a (MJ net energy (NE)/ $\text{kg}^{0.75}$/day) is approximately 0.31 for cattle (Flatt et al., 1965; Jarrige, 1989), 0.26 for goats (Jarrige, 1989), 0.25 for llamas (Carmean et al., 1991), 0.36 for horses (Vermorel et al., 1997, after a 20% correction for the effects of eating in the calorimeter) and 0.37 for dogs (Walters et al., 1993).

Some animals have metabolic rates (values for a) which differ from the usual domestic eutherians' rates. The sloth bear (Ursus ursinus) has a low SMR (about 0.17 MJ/$\text{kg}^{0.75}$/day; McNab, 1992). Lovegrove (2000) concluded that tropical and subtropical mammals have lower SMR than mammals which originated in temperate regions. The SMR of metatherians is about 70% of that of eutherians (Dawson and Hulbert, 1970). Values for SMR (MJ NE/$\text{W}^{0.75}$/day) for marsupials compiled by Hume (1999) include 0.21 for the red kangaroo, 0.19 for the brushtailed possum and 0.16 for the koala. Judging by their FMR (which is influenced by the BMR), cold-blooded animals such as the Iguanidae (Nagy, 1987) have a SMR which is about 6% of that of eutherians.

Measuring the Basal Metabolic Rate

Calorimetry (from the Greek *calor*, 'heat', and *metor*, 'to measure') is the measurement of heat flow between the animal and its environment. All of the food energy used for basal metabolism is ultimately released from the animal's body as heat (Blaxter, 1989; Rolfe and Brown, 1997 discuss the reasons for this). The laws of thermodynamics assure us that there is an exact equivalence between the amount of chemical energy accumulated in ATP which is used for basal metabolism and the amount of heat produced by a homeothermic animal under conditions which mimic the basal use of energy (i.e. 'standard conditions'). This assertion is supported by some empirical evidence (e.g. Rubner, 1894; Atwater and Benedict, 1903, cited by Webb, 1991).

Although we have defined the BMR as the amount of energy needed to fuel the animal's life processes, we can also think of it as being the rate at which heat is lost from the body when an animal is kept in conditions in which energy is used only for supporting life. With animals, it is very difficult to achieve this state, and the animal is allowed to stand, lie and move in the calorimeter. Accordingly,

we measure the SMR, rather than the BMR, by measuring the heat produced when the animal is confined under these standard conditions:

1. At rest, but not asleep. Sheep use about 1.5 kJ NE/kgW$^{0.75}$/h, or 20–25% more energy (Toutain *et al.*, 1977) when awake than when they are asleep (in slow-wave sleep; see Ruckebusch, 1975 for a discussion of sleep in ruminants).

2. In a thermoneutral environment. This is the range of environmental temperatures at which the animal does not need extra food energy to keep warm or cool. As defined by the NRC (1981), the thermoneutral zone is where body temperature remains normal and the animal does not pant or sweat, and its heat production remains at a basal level. The boundaries of the thermoneutral zone (Fig. 12.1) are influenced by coat characteristics, body condition, food intake, wind, rainfall and humidity. Lower critical temperatures (NRC, 1981, 2007; SCA, 1990) are about –30°C for lactating dairy cattle, –20°C for beef cattle at maintenance, 10°C for newborn calves, 0–30°C for adult, woolled sheep depending on the fleece length, 12°C for goats, 14–30°C for pigs of various ages and –15°C to 5°C for mature horses. See Silanikove (2000) for a discussion of the upper critical temperature.

3. In a post-absorptive state. With ruminants, this is taken to be when methane production has declined to

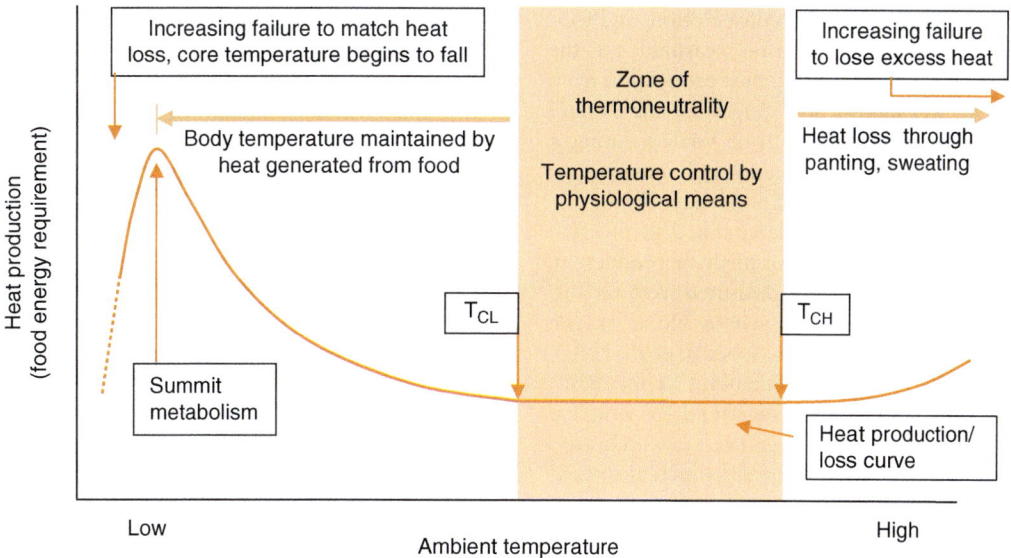

Fig. 12.1. Concepts of heat production and loss from animals. T_{CL} = lower critical temperature; T_{CH} = higher critical temperature; see text for definitions.

500 or 2000 ml/day (normal rates are 30 and 200 l/day in sheep and cattle, respectively); or alternatively when the respiratory quotient (RQ) is less than 1.0, i.e. indicating that carbohydrate (which must be of dietary origin) is not the main energy source. An RQ of 0.82 is often used; this indicates that a mixture of nutrients is being oxidized. It usually takes 2–3 days of fasting to achieve this in ruminants (Blaxter, 1967). Monogastric animals take less time although it may take 4 days for a pig to reach a post-absorptive state. Metabolic rates measured when animals are not post-absorptive are 20–30% greater than the SMR (McNab, 1992; Rolfe and Brown, 1997).

4. In good health and nutritional condition immediately before these measurements are made. Animals should have been given a maintenance intake of all nutrients immediately before the fasting period.

Direct and indirect calorimetry

When we set out to measure the energy used by an animal we can either measure heat flow, i.e. we use 'direct' calorimetry, or we measure the amount of oxygen the animal uses, i.e. 'indirect' calorimetry. Indirect calorimetry can be either 'closed-circuit', in which the animal is enclosed without a continually replenished oxygen supply; or 'open-circuit', in which there is a continual stream of air passing through the chamber. The construction and use of calorimeters are described by Blaxter (1967) and Matarese (1997). Most modern calorimeters are open-circuit, indirect calorimeters in which oxygen consumption, and CO_2 and methane production, can be measured. For the greatest accuracy in estimating heat production from oxygen use we should also determine the N and bicarbonate contents of the urine voided during a measurement. Ignoring N excretion will introduce an error of about 3% (Gessaman and Nagy, 1988).

Animals do not have to be confined to measure oxygen consumption or CO_2 or methane production. Oxygen consumption can be measured from the difference between arterial and venous blood oxygen concentrations and blood flow (Gooden *et al.*, 1991), or the rates of ^2H and ^{18}O excretion (Webb, 1991). Methane production can be measured by sampling the exhaled air. The animal is given a rumen-controlled release device charged with SF_6 which escapes at a known rate. The amount of methane in the exhaled breath is calculated from the methane and SF_6 concentrations (Grainger *et al.*, 2007).

Oxygen consumption is related to energy use in a straightforward way. Most energy-producing reactions

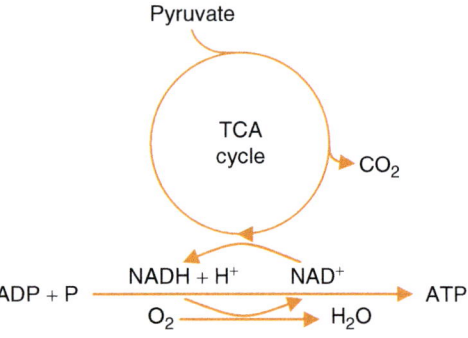

Fig. 12.2. A simplified illustration of how energy production (as ATP in this example) and oxygen consumption are related. TCA cycle = tricarboxylic acid cycle; NAD = nicotinamide adenine dinucleotide; ATP = adenosine triphosphate; ADP = adenosine diphosphate.

involve the stripping of H atoms from the substrate molecule. An example is the oxidation (dehydrogenation) of pyruvate through the TCA cycle (Fig. 12.2). About 80% of the body's oxygen use is related to ATP production. The rest is used in extra-mitochondrial oxidation reactions (Rolfe and Brown, 1997).

From these gas consumption and production data we can calculate:

1. The RQ:

$$RQ = (moles\ CO_2\ produced)/(moles\ O_2\ used)$$
$$(12.3a)$$
$$= (CO_2\ produced,\ l)/(O_2\ used,\ l) \qquad (12.3b)$$

2. The heat production (Brouwer, 1965):

$$
\begin{aligned}
Heat\ (kJ/day) = \ & 16.175\ O_2\ (l/day) \\
& + 5.021\ CO_2\ (l/day) \\
& - 2.167\ CH_4\ (l/day) \\
& - 5.987\ urinary\ N(g/day)\ (12.4)
\end{aligned}
$$

Equation 12.4 gives the heat equivalences in Table 12.2.

3. The type of substrate that is being used as the energy source (Table 12.2). The RQ is greater than 1 when animals are synthesizing fat or using bicarbonate to buffer body fluids (Matarese, 1997).

The Maintenance Energy Requirement

If it is not fed, an animal will get the energy it needs to keep itself alive by breaking down its own tissues. Obviously, this will cause a loss of body weight (Fig. 12.3). If we give enough food energy to fuel the animal's basic life processes, we will prevent this loss of live

Table 12.2. Heat equivalence of respired oxygen (calculated from the Brouwer equation applied to pigs).

Respiratory quotient	Heat equivalence of oxygen used (kJ/l)	Substrate oxidized
0.70	19.7	Fatty acids
0.75	19.9	
0.80	20.2	A mixed food
0.82	20.3	
0.85	20.4	Amino acids
0.90	20.7	
0.95	20.9	
1.00	21.2	Monosaccharides

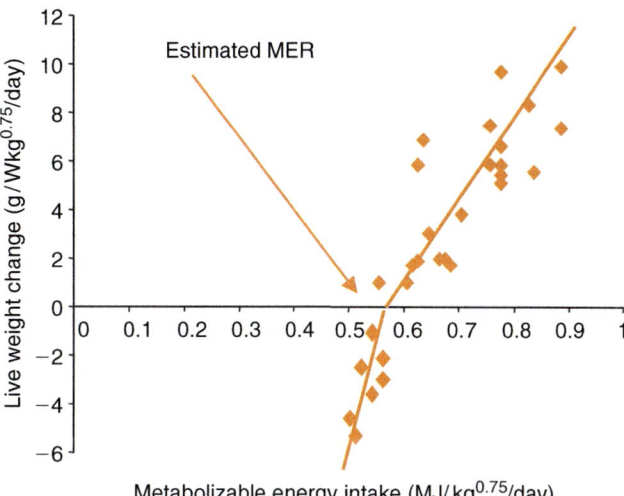

Fig. 12.3. Change in body mass in rusa (*Cervus timorensis*) stags fed above- and below-maintenance energy intakes. (From M.C. Hmeidan and G.McL. Dryden, 2000, Gatton, Australia.)

weight, and 'maintain' the animal. More formally, the maintenance energy requirement (MER) is the amount of food energy needed to keep the animal at a constant body weight and composition (i.e. zero energy retention; ARC, 1980), when it is in its normal environment.

The MER has three components. The first, and quantitatively the most important, is the basal metabolism, or SMR. The others are the amounts of energy needed for voluntary activity, and to maintain the animal's core body temperature (of particular importance to homeotherms in cold environments). If the animal is housed in a small pen or a loose box, or if it is kept indoors and inactive for most of the time, its MER can be similar to the SMR, but the effects of the environment can increase the MER to double this amount.

Values for the MER of animals confined in stalls or small yards are (MJ ME/kg[0.75]/day) 0.46 for cattle (NRC, 2001), 0.5 for rusa stags (Dryden *et al.*, 2002), 0.44 for pigs (NRC, 1998) and 0.52 for horses (Vermorel *et al.*, 1997). MER for the red kangaroo, brushtailed possum and the koala, respectively, are (MJ ME/kg[0.75]/day) 0.45, 0.37 and 0.39 (Hume, 1999).

Measuring the Maintenance Energy Requirement

Change in live weight

Because the animal will gain or lose body mass according to whether its energy intake is greater or less than its MER, we can give animals varying amounts of dietary energy, measure the response and regress these data to find the energy intake at which energy balance (i.e. weight change) is zero. Data for yearling rusa (*Cervus timorensis*) stags are illustrated in Fig. 12.3. These studies generally last for several weeks as it is difficult to get a good estimate of live weight change in only a few days. Accordingly, these estimates of MER include the effects of food consumption and digestion, and environmental effects.

Comparative slaughter

If an animal is fed a known energy intake and the amount of energy retained in the body is calculated from its live weight gain and change in body composition (estimated from chemical analysis of body tissues obtained after slaughter), then the difference between energy absorbed and that retained in tissues estimates the amount of energy used for maintenance. Lofgreen (1965) describes the calculations used in this method. The procedures used in a comparative slaughter experiment are described by Thompson (1965) and Noblet *et al.* (1987). Like the change in live weight method, comparative slaughter gives an estimate of MER which includes the effects of eating and the environment.

Factors Which Influence the Maintenance Energy Requirement

Temperature regulation

Grazing animals' MER is sometimes nearly double the SMR (Dryden, 1981; Lachica and Aguilera, 2003). Animals use extra energy to keep warm in cold conditions. Heat is lost in three ways: by sensible loss (radiation, convection and conduction), by vaporization of water (in exhaled air and vaporized sweat), and by the heat used to warm ingested feed and water or which is lost with the elimination of urine and faeces. The relative importance of these routes of heat loss varies with the environmental temperature. Sensible losses are important at low temperatures, while evaporation is the only route possible when the ambient temperature rises above 39°C, or the animal's core temperature. Body heat is lost quickly in cold weather, or when wind and rain reduce the insulation provided by the animal's coat. Newborn animals (which have a large surface area relative to their body mass), recently shorn sheep (which have less coat cover) and thin animals (which have less insulation pro-

vided by subcutaneous fat) have an increased MER (Blaxter, 1989) and are at greater risk of dying from exposure than older, fatter or fully fleeced animals. When an animal needs to produce extra heat to keep its body warm it does this by eating more food, and using this extra food energy to fuel muscular movements (shivering) or energy-spilling metabolism (e.g. the uncoupled electron transport in brown fat; Rolfe and Brown, 1997; Alberts *et al.*, 2002). Animals may use energy to reduce their body heat load in hot weather, e.g. by seeking shade or wallowing.

Environmental factors such as ambient temperature, rainfall, wind speed, depth of mud in yards and coat insulative properties alter the MER and their effects are included in several nutritional models (e.g. SCA, 1990; NRC, 1998, 2000).

Energy use for activity

Free-ranging animals seek and consume food, play, fight, and move to and from water and shade. They may display temperament or stress through increased movement, including stereotypical movements if they are confined inappropriately. This energy use is included in the MER because it is incidental to an animal's normal life. We may also include some 'forced' activity in the MER. For example, dairy cows have to walk from the paddock to the milking shed. This is a necessary adjunct to production, but the energy used is not harvested in milk, and so is best assigned to the MER.

Racehorses and greyhounds 'produce' by racing, riding hacks by trotting, cantering, etc.; draught animals by pulling carts; and companion animals play at the instigation of their owners. If we wish, we can call this productive activity, but it is probably better to use the FMR in these situations.

The NE requirement for movement (Table 12.3) is influenced by the distance covered and the terrain (i.e. whether the animal is walking uphill or downhill) and the speed at which it moves. Lachica and Aguilera (2005) have summarized data on the energy cost of walking: ruminants use about 2.75 J/horizontal metre (range 1.6–4.7 J/m); equines are more economical, using about 1.2 J/m (range 1–1.5 J/m); pigs may use between 1.6 and 6.7 J/m; while the kangaroo is very economical as it uses only about 0.75 J/m. The cost of walking up a gradient is about 27 J/vertical metre (range 13–40 J/m) and may be similar for all domestic species, although pigs are at the upper end of the scale.

Table 12.3. The net energy costs of activity in cattle, sheep, pigs and horses.

Animal type	Energy expenditure	Source
Cattle		
Standing	14.5 kJ/kg$^{0.75}$/h	Vercoe, 1973
Lying	12.2 kJ/kg$^{0.75}$/h	Vercoe, 1973
Getting up then lying down	0.45 kJ/kg$^{0.75}$ per event	Vercoe, 1973
Eating long lucerne hay	363 kJ/kg DM	Adam *et al.*, 1984
Eating chopped dry grass	274 kJ/kg DM	Adam *et al.*, 1984
Sheep		
Standing	15.9 kJ/kg$^{0.75}$/h	Toutain *et al.*, 1977
Lying	12.9 kJ/kg$^{0.75}$/h	Toutain *et al.*, 1977
Getting up then lying down[a]	0.53 J/kg$^{0.75}$ per event	Toutain *et al.*, 1977
Eating fresh grass	346 kJ/kg DM	Osuji *et al.*, 1975
Eating dry, pelleted grass	23.5 kJ/kg DM	Osuji *et al.*, 1975
Ruminating chopped grass	0.22–0.35 kJ/min	Osuji *et al.*, 1975; Toutain *et al.*, 1977
Pigs		
Standing	14.1 kJ/kg$^{0.75}$/h	Susenbeth and Menke, 1991
Standing	15.6 kJ/kg$^{0.75}$/h	Noblet *et al.*, 1993a
Standing	19.2 kJ/kg$^{0.75}$/h	Schiemann *et al.*, 1971
Eating concentrate diet	100 kJ/kg DM	Noblet *et al.*, 1993a
Horses		
Eating long hay	493 kJ/kg DM	Vermorel and Mormède, 1991
Eating pelleted food	93 kJ/kg DM	Vermorel and Mormède, 1991

[a]Twice the estimated energy cost of getting up, assuming that this act takes 1 min.

The amount of energy used in eating increases with the time spent eating, and so is greatest (kJ/kg) for those foods which are eaten slowly, such as long hay and pasture, and least for foods like pelleted forages and concentrate mixtures. NE expenditures for eating are summarized in Table 12.3. Other data about the energy cost of eating by herbivores are summarized by Osuji (1974) and Vermorel and Mormède (1991). Grazing is energetically more expensive than eating from troughs or bins. The NRC (2001) suggests that dairy cattle use about 8.5 kJ NE/kg live weight daily (about 5.25 MJ NE/day for a Friesian cow) more in grazing than they would if their food was supplied in stalls. It seems that eating is a much more energetically expensive activity for pigs than for herbivores, with pigs spending up to five times more NE in eating comparable foods.

Energy Transactions

Most of the modern energy systems are factorial systems in which the animal's NE requirements are determined for each function. NE requirements are calculated from the amount of heat lost from the body at maintenance (NE$_m$) or when the animal is working, and the amounts of energy accumulated in the body's physical products (milk, tissue growth and the products of conception). In metabolizable energy (ME)-based systems of expressing energy requirements (e.g. AFRC, 1990b, 1993), these NE requirements (MJ/day) are converted to the equivalent amounts of ME by applying appropriate values for the efficiency of use of ME:

$$ME = NE/k \qquad (12.5)$$

where

k = the efficiency of utilization of ME

The estimates of ME for each physiological function are then summed to give a total ME requirement. Because each physiological function has a different k, we have to calculate each of the NE requirements separately (Fig. 12.4). In this example the maintenance requirement is calculated from the SMR with necessary allowances for activity and regulation of core temperature. Requirements for pregnancy, lactation and growth are calculated from the amounts of energy retained in the growing fetus and body tissue, and excreted as milk. Each of these functions uses ME with a particular efficiency (k) and these efficiencies are usually different.

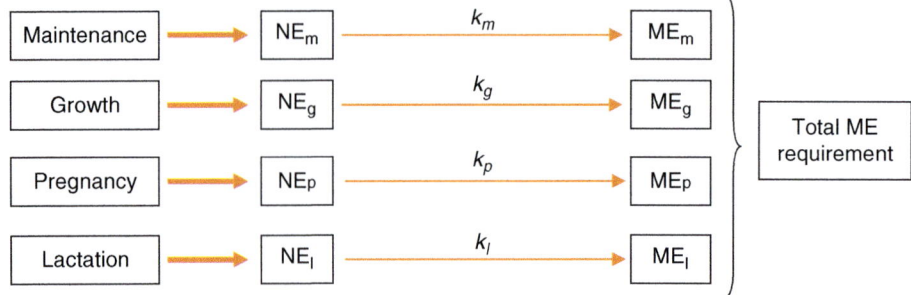

Fig. 12.4. Calculating the ME requirement of a dairy cow.

In the NE-based systems used by INRA and the NRC (Jarrige, 1989; NRC, 2000, 2001) there is a slightly different approach. Instead of expressing energy requirements in ME and using food ME content data, the k values are used to calculate food NE contents and these are applied to the animal's NE requirements.

Why does k change with different foods?

The efficiency of utilization of ME (k) also changes with food type, particularly for ruminants. Energy from foods with lower ME content (MJ/kg DM; M/D) is used less efficiently, especially for growth. Low M/D foods are usually more fibrous and are fermented to an acetate-rich volatile fatty acids (VFA) mixture. It was proposed by McClymont (1952) that most of the absorbed acetate was rapidly oxidized by animal tissues and this resulted in 'extremely inefficient utilisation of the energy liberated'. This idea was incorporated into early concepts of ruminant energy use (ARC, 1965; Blaxter, 1967). However, later work (reviewed by Ørskov and MacLeod, 1990) showed that when mixtures of the three important VFA (acetate, propionate and *n*-butyrate), similar to those obtained when conventional foods are fermented, were administered into the rumen, they were used with similar efficiencies, irrespective of the mixture composition (Ørskov *et al.*, 1991). Several reasons for the relationship between M/D and the efficiency of use of ME have been suggested. Blaxter (1989) carefully reviewed the possibilities and concluded that there were multiple causes. The major sources of inefficiency are probably:

1. Energy expenditure during food prehension; this includes chewing during prehension and bolus formation, rumination and muscular activity when food is given (e.g. standing, anticipation of eating and competition for food).
2. Heat of fermentation in the rumen, especially when microorganisms digest food rich in fermentable fibre.
3. Other inefficiencies are associated with the synthesis of ATP, nutrient absorption and transport, excretion of CO_2, and protein synthesis and degradation.

Reconciling food energy content and animal energy requirement data

If we want to use linear programming methods to formulate rations we have to express the animal's energy requirement in the same units as the food energy contents. To do this the total ME requirement (MJ/day) is divided by an estimate of daily DM intake (kg/day) to give an ME requirement which is expressed in concentration units (MJ/kg DM):

M/D = total ME requirement/daily DM intake

(12.6)

But the total ME requirement depends on k, which in turn depends on the various proportions of the ME intake which are used for maintenance, lactation, growth, etc., and also the ration M/D. We have a circular argument (Fig. 12.5). This problem is greatest with growing animals because k_g is more influenced by food type than k_m or k_l.

We encounter the same problem with NE systems. The NE requirement is easily calculated, as it is simply the sum of the requirements for maintenance, lactation, work, etc., and the required diet NE concentration is obtained by dividing the total NE requirement by the expected DM intake. However, we cannot exactly calculate the NE

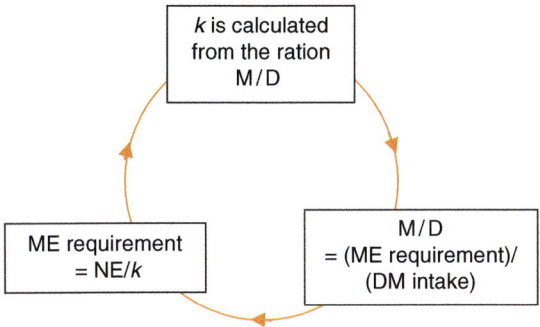

Fig. 12.5. To know how much ME we need, we have to know how much ME we need.

content of a food, because the overall efficiency with which its ME is converted to NE varies with the mix of physiological processes (maintenance plus the production uses) that it will be used for.

Several solutions to the problem of varying k have been proposed:

1. Using a 'variable net energy' system in which each food has several NE_{m+p} contents which are calculated from the ratio of $(NE_g + NE_m)/NE_m$ (this ratio is called the Animal Production Level). The NE_{m+p} values are a composite NE content which takes into account the proportions of the food's energy which will be used for maintenance and production. This approach works if a single equation for calculating k_g is used for all foods, as in MAFF (1984).
2. Making assumptions about the nature of the finished ration. An example is the Takeaway ration formulation programme (Barber, 1990) where the expected M/D is estimated from the ME requirement of the specified animal and an estimated maximum DM intake. This approach depends on reliable prediction of food intake and a consistent ration concentrate/roughage composition so that k_g does not vary widely.
3. Developing multiple regression equations to predict M/D from inputs such as live weight, growth rate, etc. (Dryden, 1995).
4. Using a single value for k. Because their k_m and k_l are, in practice, very similar (0.62 and 0.64, respectively for conventional diets) the NRC (2001) uses a single NE_l value to establish the energy content of foods for lactating and/or pregnant adult dairy cattle. This value does not vary with the type of food. This approach was originally devised by Van Es (1975). We can then calculate ME_m and ME_l, and thus the total ME requirement, without

any need to guess the M/D of the final ration. Similarly, NRC (2000) uses k_m to describe the energy requirements for pregnancy and lactation of beef cattle. INRA uses k_m for all physiological processes in the equine. They justify this (Vermorel and Martin-Rosset, 1997) from the proportion of the total NE requirement which is used for maintenance – 50% in lactating mares, 90% in pregnant mares, 60–90% in growing horses and 70–90% in working horses (although the last assumption needs validation).

Energy Requirements of Ruminants

There are many sets of energy feeding standards for ruminants, including the Australian (SCA, 1990), British (AFRC, 1990b, 1993; Sutton and Alderman, 2000), French (Jarrige, 1989), Norwegian (Ekern, 1991) and US (Fox *et al.*, 1992; NRC, 2000, 2001). Some of these systems have been compared by Agnew and Yan (2000), Agnew and Newbold (2002) and Yan *et al.* (2003). The description of ruminant energy requirements in this text uses information taken from several of these systems, selected to illustrate particular points. Where necessary, the original units used for energy have been converted to MJ.

Metabolizable energy requirement for maintenance

We attempt to identify all the factors which influence the MER and to quantify their effects, as in the following equation (SCA, 1990):

$$NE_m = K \times S \times M \times [0.28W^{0.75} \times e^{(-0.03A)}] + 0.1NEI + EGRAZE + ECOLD \quad (12.7)$$

where
NE_m = NE for maintenance (MJ/day)
K, S, M = factors which modify the calculated value of NE_m to take account of animal species and breed, sex (either entire male or not) and the proportion of milk in the diet (these values vary between 1 and 1.4, 1 and 1.15, and 1 and 1.23)
$W^{0.75}$ = metabolic size (kg)
A = age (years, to a maximum of 6)
NEI = net energy intake (MJ/day)
EGRAZE = net energy used in activity associated with grazing (MJ/day)
ECOLD = net energy needed to maintain internal body temperature in cold conditions (MJ/day)

The SCA (1990) equation to predict the efficiency of utilization of ME for maintenance (k_m) is:

$$k_m = 0.02M/D + 0.5 \tag{12.8}$$

where
M/D = ME content of the diet, and assuming that the diet total energy content is 18.4 MJ/kg DM.

Equation 12.8 does not apply to milk as a food; k_m for milk is assumed to be 0.85. The equation gives values of 0.77 for a typical concentrate-rich ration, 0.68 for a medium-quality forage and 0.62 for a low-quality cereal straw.

Metabolizable energy requirement for growth

The SCA (1990) equation for predicting the NE requirement for a particular growth rate is given below. The requirement is obtained by estimating the energy content of the tissue accumulated during growth and multiplying this by the growth rate. This equation applies to sheep and all cattle except large European breeds:

$$NE_g = LWG \times \{(6.7+R)+(20.3-R)/ [1+e^{(-6(P-0.4))}]\} \tag{12.9}$$

where
LWG = growth rate (kg/day)
R = adjustment for weight gain or loss
= $[(EBC)/(4\ SRW^{0.75})] - 1$
EBC = empty body weight change (g/day)
= 0.92 LWG
SRW = standard reference weight (kg); the weight of a mature animal of that type which has a body condition score in the middle of the normal range SCA (1990)
P = (current W)/SRW

The energy content of body tissue increases with the animal's age because of its increasing proportion of fat and decreasing proportion of protein (ARC, 1980). Although age does not appear in this equation as such, the ratio (P) of the animal's current live weight relative to SRW is a surrogate for age.

The equations for k_g vary with the type of food:

For concentrates and very digestible early-growth temperate forages:

$$k_g = 0.042M/D + 0.006 \tag{12.10a}$$

For all other forages:

$$k_g = 0.063M/D - 0.308 \tag{12.10b}$$

Metabolizable energy requirement for milk production

The NE requirement for milk production is calculated from the milk energy content and the amount of milk produced. The SCA (1990) suggests three alternative equations to predict the energy content from fat, protein, lactose and solids-not-fat (SNF):

Tyrell and Reid (1965):
$$NE_1 = MY \times (0.0386F + 0.0205SNF - 0.236) \tag{12.11a}$$

Perrin (1958):
$$NE_1 = MY \times (0.0381F + 0.0165L + 0.1563N) \tag{12.11b}$$

Gaines and Davidson (1923):
$$NE_1 = MY \times 3.054(0.4 + 0.015F) \tag{12.11c}$$

where
F = fat content (g/kg milk)
SNF = solids-not-fat content (g/kg milk)
L = lactose content (g/kg milk)
N = N content (g/kg milk)

When applied to Friesian milk (typically 3.7% fat, 8.5% SNF, 4% lactose and 3.6% protein), Equations 12.11a, b and c predict energy contents of 2.93, 2.3 and 2.92 MJ/kg, respectively.

The value for k_l is influenced by the diet ME content:

$$k_l = 0.02M/D + 0.4 \tag{12.12}$$

k_l is 0.63 and 0.59 for foods containing 11.5 and 9.5 MJ ME/kg DM, i.e. a typical dairy total mixed ration and a good-quality subtropical pasture.

The changes in dairy cows' milk composition and daily milk production throughout the lactation are well known (Holmes and Wilson, 1984; SCA, 1990). Milk yield rises to a peak early in the lactation, then falls gradually (at about 10% per month). Milk fat and protein contents are initially high, fall to their lowest point at the time of peak yield, then gradually rise throughout the rest of the lactation. Lactose content remains more or less constant throughout the lactation. Generally, we do not try to match food intake with these changes in milk composition. We do try to match feeding methods to changes in milk yield, but it is important that allowance is made for the changes in the partitioning of nutrients between milk production and live weight gain which occur as the lactation proceeds (SCA, 1990; NRC, 2001), and that food allocations do not 'lead' the downward slope of milk production.

Metabolizable energy requirement for pregnancy

The well-known exponential growth of the fetus is reflected in the equations which predict the amount of energy accumulated in the conceptus (i.e. fetus, uterine and placental tissues; NE_c, MJ/day). The NRC (2000) equation is:

$$NE_c = BW \times (1.1811) \times e^{((0.03233 - 0.0000275t) \times t)}$$

$$(12.13)$$

where
BW = birth weight (kg)
t = time after conception (days)

The efficiency of use of ME for pregnancy (k_p) is about 0.13 (ARC, 1980). The total amount of NE used by the placental and uterine tissues, and for the maintenance of the fetus, is greater than, or equivalent to, the amount of NE accumulated in the fetus each day depending on the size of the fetus (SCA, 1990; NRC, 2000). When we compare the increase in total energy use by the pregnant animal with the amount of energy accumulated in the developing fetus we get an apparently very low efficiency of energy use. It is the inclusion of these other energy requirements in the calculation of k_p that leads to the apparently low efficiency with which ME is used for fetal growth.

It is well recognized that the later part of the pregnancy imposes a high energy demand, but we have to reconsider the idea that the first half of the pregnancy has only low nutrient needs. Fibre initiation and maturation in the ovine fetus begin about halfway through gestation (Hardy and Lyne, 1956), and the nutrient supply early in gestation may influence the offspring's ability to withstand diseases in adult life (Robinson et al., 1999). We should also take note of the essentiality of glucose for fetal growth (Hay, 2006).

Energy Requirements of Non-ruminants

There are several sets of feeding systems for pigs and horses, e.g. Australian (SCA, 1987), British (ARC, 1981), French (Blum, 1984; Martin-Rosset, 1990) and US (NRC, 1998, 2007). As in the discussion of ruminant energy requirements, this description of the requirements of pigs and horses uses information taken from each of these systems, selected to illus-

trate the point under discussion. Where necessary, the original units used for energy have been converted to MJ.

Efficiencies of energy use by the pig

The efficiency of conversion of digestible energy (DE) to ME for conventional, concentrate-rich pig foods is approximately 0.9–0.96 (AFRC, 1990a; Noblet et al., 1993b; NRC, 1998) and is likely to be at the higher end of this range if most of the energy substrates are absorbed from the small, rather than the large, intestine (Noblet et al., 1993b). Efficiencies of conversion of ME to NE by growing pigs are assumed to be 0.6–0.7 and 0.75–0.8, for protein and fat deposition, respectively (AFRC, 1990a; NRC, 1998). There was no difference between diets rich in fermentable non-starch polysaccharides and conventional concentrate-rich diets.

Everts and Dekker (1991) reported the efficiency of use of ME for growth in pregnant sows (k_g) of 0.73 and lactation (k_l) of 0.72. The value for k_l adopted by NRC (1998, after Noblet and Etienne, 1987) is 0.72. Longland et al. (1991) obtained an average k of 0.75 for maintenance and production in growing pigs and pregnant sows. This value has been used in this section to convert DE and ME values to NE.

Maintenance energy requirements of pigs

Equation 12.14 predicts the NE requirement (MJ/day) for maintenance (NE_m), adopted by NRC (1998):

$$NE_m = 0.333W^{0.75} + [(0.00098W) + 0.071] \times (T_{CL} - T_A)$$

$$(12.14)$$

where
W = live weight (kg)
$(T_{CL} - T_A)$ = difference between the ambient and lower critical temperatures (°C); if ambient temperature > lower critical temperature then $(T_{CL} - T_A) = 0$

The lower critical temperature of adult pigs is 18–23°C (AFRC, 1990a; NRC, 1998). When it is greater than the ambient temperature, pigs require extra food energy to maintain their core body temperature. Like other animals, pigs use energy for voluntary movement, but this cost is relatively low – about 10–25% of the total MER of housed pigs (Van Milgen et al., 1998; Van Milgen and Noblet, 2000). The MER is higher for modern, lean-genotype pigs than for older genotypes (Whittemore et al., 2001b), and declines with age (Everts and Dekker, 1991).

Whittemore (1983) and Tess *et al.* (1984a) have suggested that there is a better fit between NE_m and body protein content than total live weight. Body fat mass has either zero, or negative, effects on NE_m. An example equation to predict NE_m from body composition is given by Van Milgen *et al.* (1998) for pigs of 20–60 kg:

$$NE_m = 0.457 \times muscle^{0.81} + 1.969 \times viscera^{0.81} - 0.644 \times fat^{0.81} \quad (12.15)$$

where
NE_m = net energy for maintenance (MJ/day)
muscle = muscle mass (kg)
viscera = viscera mass (kg)
fat = fat mass (kg)

Energy requirements for growth in the pig

The relationship between growth rate and energy requirement varies with the pig's age, sex and genotype (Campbell and Taverner, 1988; Rao and McCracken, 1990; Quiniou *et al.*, 1995; Möhn *et al.*, 2000; Szabó *et al.*, 2001; Moughan *et al.*, 2006). Tess *et al.* (1984b) summarized the relationships between protein deposition (PD, g/day) and ME requirement (MJ/g PD) which were published between 1965 and 1980. The NE requirements (MJ/day) for protein retention and fat deposition (NE_{pd}, NE_{fd}) in Equations 12.16a and b are averages of those in Tess *et al.* (1984b) and other values published more recently.

$$NE_{pd} = 0.0425 \times PD \quad (12.16a)$$

$$NE_{fd} = 0.0418 \times FD \quad (12.16b)$$

where
PD = whole-body PD rate (g/day)
FD = whole-body fat deposition (g/day)

The NE_{pd} requirement is larger, i.e. PD is less energy-efficient, for pigs of inferior PD genotype, and for castrates and females compared to entire boars (Campbell *et al.*, 1985; Campbell and Taverner, 1988).

We cannot consider the NE requirement separately from other nutrients. In particular, the energy requirement is closely related to the pig's amino acid intake. The PD rate increases with increasing energy intake, up to a maximum as long as sufficient amino acids are available (Fig. 12.6). Sex and genotype effects alter the level at which the relationship plateaus. There is no single NE 'requirement' for growth because the optimum energy intake is determined by the pig's genotype,

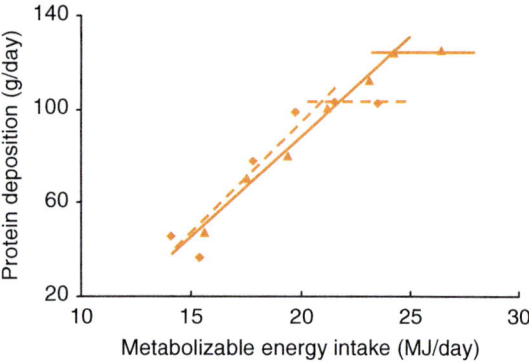

Fig. 12.6. Relationship between protein deposition and metabolizable energy intake at two levels of lysine supply. (Drawn from data in Möhn *et al.*, 2000.)

as well as market-related factors such as consumer preferences for pig meat composition (i.e. the fat/lean ratio) and the relative costs of the limiting amino acids compared to product prices.

Energy requirements of the breeding sow

AFRC (1990a) suggests the following relationships to predict the energy requirement of the pregnant sow. In using these, we should note that NE_m increases by about $0.75\,MJ/kgW^{0.75}$/day after the 40th day of pregnancy (Geuyen *et al.*, 1984).

$$NE_m = 0.323W^{0.75} \quad (12.17a)$$

$$NE_{pd} = 0.149NR \quad (12.17b)$$

where
NE_m = net energy for maintenance (MJ/day)
NE_{pd} = net energy required for PD (MJ/day)
W = live weight (kg)
NR = N retention (g/day)
  = (0.37NI – 2.47), with a maximum of 16 g/day
NI = N intake (g/day)

The sow's milk production is influenced by the size of the litter, so the NE for lactation (NE_l, MJ/day) is (Noblet and Etienne, 1989):

$$NE_l = [(0.02058 \times ADG) - 0.377)] \times N \quad (12.18)$$

where
ADG = growth rate (g/day) averaged over the litter
N = number of piglets

Breeding sows are generally not fed strictly according to these recommendations because the manager must feed to optimize the sow's body condition. Food consumption by lactating sows is reduced if they farrow in fat condition, or vice versa (Danielsen and Vestergaard, 2001). On the other hand, sows must not lose true live weight over the breeding cycle, as this can precipitate a situation where she loses live weight over successive pregnancies and lactations with adverse welfare and production outcomes. Verstegen *et al.* (1987) suggest a gain of 40–45 kg during pregnancy, consisting of about 20 kg of conceptus gain (lost at farrowing) and 20–25 kg of true live weight gain, followed by a weight loss of 5–10 kg during the lactation. Although there is no strong evidence that pregnant sows accumulate body tissue (other than that in the gravid uterus) more efficiently than non-pregnant sows, i.e. that they exhibit 'pregnancy anabolism' (Close *et al.*, 1984; Walach-Janiak St Raj and Fandrejewski, 1986b), the body weight gain by pregnant sows is certainly influenced by the amount of food they eat (Close *et al.*, 1984; Walach-Janiak St Raj and Fandrejewski, 1986a). To achieve the recommended live weight changes we have to limit the food intake of the pregnant sow to about 125% of maintenance (Verstegen *et al.*, 1987). Food restrictions sufficient to cause the lactating sow to lose 10% or more of her live weight at weaning will adversely affect performance in the next pregnancy (Thaker and Bilkei, 2005). Young sows (Dourmad, 1988; Noblet *et al.*, 1990) and sows of high lean growth genotype (Smith *et al.*, 1991; Kerr and Cameron, 1996) have relatively poor appetites, especially immediately after farrowing. Feeding programmes for modern sows need careful attention to avoid excessive weight loss during lactation. Managers may use a 'step-up' feeding plan in which the newly farrowed sow is given about 2.5 kg of food, with the allowance being increased by 250–500 g daily throughout the early lactation (J. Gaughan, 2007, Gatton, Australia, personal communication).

Energy requirements of horses for maintenance and growth

The NRC (1989, 2007) system uses DE, while the French horse standards are based on an NE unit, 'l'unité fourragère cheval' (UFC; Martin-Rosset *et al.*, 1994):

$$1 \text{ UFC} = NE_m \text{ of 1 kg barley grain}$$
$$= 9.414 \text{ MJ NE} \qquad (12.19)$$

We can compare the two sets of recommendations if we assume that, for the NRC equations, the efficien-

cies of converting DE to ME and of ME to NE are as reported by Pagan and Hintz (1986a,b). The NRC (1989) and INRA (Martin-Rosset *et al.*, 1994) estimations of equine energy requirements for maintenance (MJ/day) are Equations 12.20a, b and Equation 12.20c, respectively:

$$DE_m = 5.86 + 0.126W \qquad (12.20a)$$

$$NE_m = 3.46 + 0.076W \qquad (12.20b)$$

$$NE_m = 0.351W^{0.75} \qquad (12.20c)$$

where
W = live weight (kg)

These give somewhat similar estimates: for a 500 kg horse, the INRA and NRC equations predict NE_m of 42 and 37 MJ/day, respectively. The French NE_m is varied according to breed, sex (an increase of 10–20% for stallions) and work (an increase of 5–15%) (Martin-Rosset *et al.*, 1994). NRC (2007) applies a 10% impost over the 'general' NE_m estimate for especially active or temperamental horses (temperamental horses generally express this through additional movement). Both sets of recommendations allow for a reduced NE_m with age. Martin-Rosset *et al.* (1994) reduce the value by 1% between yearlings and 2-year-olds. NRC (2007) suggests a curvilinear relationship between age and NE_m for growing yearlings, although this relationship requires further development.

Equations 12.21a and b to predict k_m are from Vermorel and Martin-Rosset (1997). They include an allowance for the energy cost of eating (see Table 12.3).

Forages:

$$k_m = 0.5756 - 0.00011CF + 0.000105CP$$
$$+ 0.00027CC + 0.00015DOM \qquad (12.21a)$$

Cereals and legume seeds:

$$k_m = 0.7234 + 0.000119CF - 0.000081CP$$
$$+ 0.000112CC \qquad (12.21b)$$

where
CF = crude fibre (g/kg DM)
CP = crude protein (g/kg DM)
CC = cell content carbohydrates (g/kg DM)
DOM = digestible organic matter (g/kg DM)

The NE requirement for growth (NE_g, MJ/day) of horses between 4 and 24 months is influenced by the horse's age in a manner similar to that for ruminants, and for the same reasons. If we assume that

ME is 87% of DE and k_g = 0.59 (Pagan and Hintz, 1986a,b), then the NRC (2007) equation to predict NE_g (MJ/day) is:

$$NE_g = LWG \times (4.27 + 2.58age - 0.045age^2) \tag{12.22}$$

where
LWG = live weight gain (kg/day)
age = horse's age (months)

Energy requirements for work in horses

The horse's energy requirement for work may be measured by indirect calorimetry, i.e. by measuring oxygen consumption, often while on a treadmill. However, its muscles go into oxygen debt during intense activity, and we have to also measure the amount of lactic acid which is produced when ATP is synthesized anaerobically. Relationships between the amount of NE used for activity, and an animal's oxygen consumption and lactic acid production (Seeherman *et al.*, 1981), are:

$$1 \text{ l } O_2/kg/s = 183.3 \text{ mmol lactate/kg/s} \tag{12.23}$$

At an RQ of 0.82 these are equivalent to an NE expenditure of 20.3 kJ/kg/s.

The NE requirement for work (NE_w, J/kg/min) is influenced by the horse's live weight plus the weight of rider and tack (saddle, etc.), the speed or gait (Pagan and Hintz, 1986b) and the nature of the terrain, especially its slope. The Pagan and Hintz (1986b) prediction Equation (12.24) was generated from observations on horses moving at no more than 400 m/min, and so should not be applied to speeds greater than this.

$$NE_w = 4.184[e^{(3.02 + 0.0065X)}] - 58.2 \tag{12.24}$$

where
X = horse's speed (m/min)

Estimates of the energy expenditure of horses for activity are given in Table 12.4.

The NRC (2007) energy requirements for work are expressed as multiples of the daily MER. If we make some assumptions about the durations of particular activities in a horse's day, then, using the same approach, we can compare the relative intensities of the energy expenditures of horses engaged in these activities (Table 12.5).

Essential Amino Acids

About 12 amino acids cannot be synthesized by animals *de novo* or from other amino acids. The list changes with species, age and physiological function, but the list for pigs (Table 12.6) gives a good indication of those amino acids which are required in the diets of monogastric animals. These amino acids are referred to as being 'essential' or 'indispensable'. Because of their digestive physiology, ruminants are partially protected from essential amino acid (EAA) deficiencies, but they still have the same requirements for EAA at a tissue level as other animals.

Lysine and sometimes other EAA such as isoleucine and valine are limiting amino acids in pig diets (Kracht *et al.*, 1976; Figueroa *et al.*, 2002). The first-limiting nature of lysine is often recognized by expressing the requirements for other EAA as amino acid/lysine ratios, and using a lysine/energy ratio to ensure that pigs have sufficient amino acids for good lean growth. EAAs should be at least 45% of the total dietary amino acid for the growing pig (Wang and Fuller, 1989).

Lysine is also expected to be the first-limiting amino acid in the metabolizable protein (MP; see Table 12.9) flowing into the ruminant small intestine, although methionine and cystine are first-limiting for wool growth. Leucine may be the first-limiting amino acid in the undegraded protein of some protein meals (Boisen *et al.*, 2000).

Table 12.4. Estimates of the net energy required for work by horses.

Gait	Speed (m/min)	NE for work (J/kg/min)	
		Martin-Rosset *et al.*, 1994	Pagan and Hintz, 1986b
Walk	110	317	119
Slow trot	200	697	260
Normal trot	300	1014	550
Canter	500	2219	2175
Gallop	600	2663	–

Table 12.5. Daily energy requirements of working horses (multiples of the maintenance energy requirement).

Daily activity	DE requirement (multiple of the maintenance requirement)	Source
Walking (8 h/day)	1.6	Guerouali *et al.*, 2003
Walking (8 h/day)	1.62	Pagan and Hintz, 1986b
Walking with load (8 h/day)	1.8	Guerouali *et al.*, 2003
Light work (hacking, equitation)[a]	1.2	NRC, 2007
Moderate work (stock work, jumps, gymkhana)[a]	1.4	NRC, 2007
Heavy work (race training, show jumping)[a]	1.6	NRC, 2007
Very heavy work (racing, polo, 3-day events)[a]	1.9	NRC, 2007
Polocrosse (4 games in a 2-day carnival)[b]	1.9	Buzas, 2007
Ploughing (6 h/day)	2.24	Perez *et al.*, 1996
3-day event (cross-country day)[c]	2.2–2.8[d]	Jones and Carlson, 1995
Endurance ride[e]	4.5–10[f]	Barrey, 1993

[a]Generalized from the NRC (2007) descriptions.
[b]Energy expenditure estimated from heart rate (Coenen, 2005).
[c]Based on a CCI 4-star event: 23 min roads and tracks, 4.5 min steeplechase, 50 min roads and tracks, 13 min jumps; 9.8 kJ/kg/vertical m allowed for jumping (Harris, 1997).
[d]Estimates of NE for work by Jones and Carlson (1995; the lower estimate) and Martin-Rosset *et al.* (1994; the upper estimate).
[e]100 km in 6.2 h (56 min walking, 241 min trotting, 75 min galloping; rider weighs 70 kg).
[f]Calculated from NE_w provided by Martin-Rosset *et al.* (1994; the lower estimate) and Barrey (1993; the upper estimate).

Table 12.6. Amino acids which must be provided in the diet of pigs.

Essential amino acids		Non-essential amino acids	
Arginine[a]	Methionine[c]	Alanine	Glutamine
Histidine	Phenylalanine[c]	Asparagine	Hydroxyproline
Cyst(e)ine[b,c]	Threonine	Aspartic acid	Proline
Isoleucine	Tryptophan	Glycine	Taurine[d]
Leucine	Tyrosine[c]	Glutamic acid	
Lysine	Valine		

[a]Synthesized in the urea cycle at rates sufficient for maintenance in the pig.
[b]Includes cystine and cysteine.
[c]Cystine can provide 50–80% of the requirement for sulfur amino acids, and tyrosine can provide 50% of the tyrosine + phenylalanine requirement of pigs (Fuller *et al.*, 1989).
[d]Taurine is not essential for the pig, but is essential for felines.

The Maintenance Requirement for Protein

Animals need a constant supply of protein, even when they are not growing or otherwise producing. This is because body tissue is constantly being catabolized and replaced (Ratner *et al.*, 1940; Hershko *et al.*, 2000; Duggleby and Waterlow, 2005; Table 12.7), and because some nitrogenous material is lost as secretions into the digestive tract, or as detritus. The amino acids needed to rebuild this tissue, and the purine and pyrimidine bases used to synthesize the nucleic acids needed to code for the required enzymes, make up the animal's maintenance requirement for protein. Much of this degraded protein and nucleic acid is recycled, but there is also a steady excretion of waste nitrogenous products which cannot be reused. This material must be replaced, and it can only come from the protein in the animal's diet.

The endogenous loss of nitrogen occurs in three ways

The loss of tissue N from the body is called the endogenous N loss. Endogenous N is excreted in urine (the endogenous urinary N, EUN), in faeces (the metabolic faecal N, MFN) and detritus. The total of endogenous N lost in these ways represents the maintenance protein requirement.

Table 12.7. Turnover rates of mammalian tissues and enzymes. (From Conn and Stumpf, 1976; Abdul-Razzaq and Bickerstaffe, 1989; Doherty and Mayer, 1992.)

Substrate	Organelle	Turnover rate (half-life) (h)
Enzymes		
Acetyl coA carboxylase	Cytosol	48
Alanine aminotransferase	Mitochondrion	18
Ornithine decarboxylase	Cytosol	0.17–0.5
Tissues		
Liver	Whole organ	96–120
	Mitochondria	145–170
	Endoplasmic reticulum	48
Skin	Whole organ	61
Muscle	Whole organ	389
Brain	Whole organ	118
Heart	Whole organ	171
Rumen	Whole organ	38
Abomasum	Whole organ	48
Proximal small intestine	Whole organ	11
Kidney	Whole organ	28

EUN contains the N of nucleic acid degradation products (allantoin, xanthine, hypoxanthine and uric acid), urea produced from the deamination of those amino acids which were derived from catabolized tissue protein, and creatinine (the dehydration product of creatine). It is difficult to measure. Chemical analysis of the nitrogenous substances in urine is not feasible because a great deal of the endogenous N is excreted in urea, and urea is also made in large amounts as a result of the deamination of surplus dietary amino acids. One approach is to feed the animal an N-free or very low-N diet and to assume that the daily urinary N excretion equals the EUN (Majumdar, 1960; Tullis *et al.*, 1986; Luo *et al.*, 2004). This works reasonably well, but it takes about 10 days to reach a minimum level of urinary N in ruminants, and it may underestimate the EUN excretion. Pigs which had been fed a low-N diet had a period of greater than normal N retention when they were subsequently given an N-rich diet (Tullis *et al.*, 1986). An alternative method is to regress urinary N excretion at several levels of N intake against diet N intake and extrapolate this back to zero N intake (Luo *et al.*, 2004).

MFN can be estimated by regression in the same away as EUN (Bosshardt and Barnes, 1946; Arman *et al.*, 1975) or by feeding an N-free diet (Majumdar, 1960; Tullis *et al.*, 1986). We may not be able to estimate MFN excretion accurately with low-N diets because they are associated with reduced food intake and diarrhoea (Blaxter and Wood, 1951; Majumdar,

1960). These difficulties might be avoided by feeding a diet with a completely digestible N source (Blaxter and Wood, 1951) but the method is not appropriate for ruminant animals because residues of the rumen bacteria which grow on these N sources may appear in the faeces. Mason (1969) showed that the neutral detergent-soluble N of ruminant faeces is a good estimator of MFN. The reasons for this are described by Dryden (1982). First, most of the microbial cells leaving the rumen are digested in the small intestine. Endogenous N is secreted into the small intestine in the form of bile salts and pigments, mucus, epithelial cells and enzymes. Some of this, especially the mucus and some enzymes, reaches the caecum where, together with more mucus and cells originating from the large intestine, it provides N for large intestinal microorganisms. Mammalian and microbial cells, mucus and enzyme and plasma proteins are soluble in neutral detergent.

Net Protein Requirement

This is the amount of amino acid needed for maintenance and each productive process, and is analogous to net energy requirement. The maintenance protein requirement (Pr_M, g/day) is calculated by measuring or estimating the total endogenous N loss (the basal endogenous excretion) and expressing this as the equivalent amount of protein:

$$Pr_M = 6.25(EUN + MFN + detritis N) \quad (12.25)$$

where
EUN = endogenous urinary N (g/day)
MFN = metabolic faecal N (g/day)

Net protein requirements for production are calculated by multiplying the protein content (often estimated as $6.25 \times N$) of the product of interest by the rate at which it is synthesized.

Efficiency of use of absorbed amino acids

There are inescapable inefficiencies in protein use because, in general, dietary and animal tissue proteins have dissimilar amino acid profiles, and animals do not store amino acids against a later need. Accordingly, to use net protein data we have to convert them to the amount of protein which will supply the required amino acids. For non-ruminant (or pre-ruminant) animals we refer to the amino acid composition of the diet; for ruminants it is the amino acid composition of MP.

Protein efficiency coefficients (i.e. the efficiency with which the N in the dietary protein or MP is used for a given physiological function) are composites of the similarity between the amino acid compositions of absorbed protein and the product to be synthesized, together with the expected efficiency with which an ideal amino acid mixture would be used for that purpose. It may seem unnecessary to introduce the latter idea. If the amino acids absorbed from the intestine are an 'ideal' mixture, i.e. one which exactly matches the amino acid composition of the proteins to be synthesized, then we would expect it to be used with an efficiency of 100%. However, this is not so. The efficiency of retention of an absorbed ideal amino acid mixture is probably no more than 85%, for both pigs (Whittemore et al., 2001c) and ruminants (Oldham, 1987).

There are several ways of obtaining protein efficiency coefficients:

1. Empirical estimates: estimates of the efficiency of use of dietary protein by horses are 60% for pregnancy, 55% for lactation (NRC, 1989), and 45% for growth (Martin-Rosset et al., 1994).
2. Calculated from 'relative values': these are assessments, for each physiological process, of the similarities between body tissue amino acids and those in MP. Relative values are multiplied by the efficiency of use of an ideal amino acid mixture to derive k_{NX}, i.e. the efficiency with which absorbed protein is used for function 'X' (AFRC, 1992).
3. Biological value (BV) of dietary protein: this is the ratio of the amount of dietary amino acid N retained to the amount absorbed. In effect, BV is the same as k_{NX}, as it integrates the concepts of relative value and the efficiency of use of an ideal amino acid mixture. BVs for some foods are reported in Table 12.8. The BV of MP entering the ruminant duodenum is about 66% (NRC, 2000), but this is influenced by the

Table 12.8. Estimates of the biological values of selected animal food ingredients.

Foodstuff	Biological value (%)	Reference (and comments)
Pigs		
Soybean meal	61	Kracht et al., 1976
Horse bean meal	59	Kracht et al., 1976
Fish meal	54	Kracht et al., 1976
Maize gluten	58	Kracht et al., 1976
Wheat	40–42	Rundgren, 1988
Triticale	45–48	Rundgren, 1988
Rye	58–61	Rundgren, 1988
Casein	82	Richert et al., 1994
Wheat gluten	74–76	Richert et al., 1994
Soybean meal	74	Richert et al., 1994
Calves		
Skimmed milk powder	92	Blaxter and Wood, 1952 (sole protein source)
Casein	79	Blaxter and Wood, 1952 (sole protein source)
Gelatin	30	Blaxter and Wood, 1952 (sole protein source)
Whey powder	66	Babella et al., 1988 (sole protein source)
Sheep		
Casein	87	Asplund, 1987 (given intragastrically)
Gelatin	7	Asplund, 1987 (given intragastrically)

proportion of food protein degraded in the rumen, and the amino acid composition of the undegraded food protein. BVs are not constant; they vary according to the physiological process the absorbed amino acids are used for, and there is evidence from calf-feeding studies that BV declines as protein intake increases (Donnelly and Hutton, 1976; Blome *et al.*, 2003).

4. Efficiency of use of ideal protein (e_p): this is the efficiency with which dietary protein is retained in the pig's lean tissue. e_p is calculated by regressing protein retention against the amount of ideal protein absorbed (Sandberg *et al.*, 2005b). Note that k_{NG} is the efficiency of use of a particular food protein, while e_p is the efficiency of use of ideal protein. The intake of ideal protein is calculated as described in Equation 12.38 (Moughan, 2003).

Nitrogen Transactions in the Ruminant

Ruminant protein systems distinguish between the N requirements of the rumen microorganisms and the animal's amino acid requirements. Both of these have to be met. If the rumen microorganisms are not properly fed, their activity, including the fermentation of plant cell walls, will be diminished and this will reduce the amount of food eaten and the animal's supply of energy.

Many microorganisms, especially the fibre-digesters, require NH_3. This is obtained from that fraction of the food protein which is degraded in the rumen. Not all food protein is so degraded; the proportion which is degraded is called the 'ruminally degradable protein' (RDP; SCA, 1990) or the 'degradable intake protein' (DIP; NRC, 2000). Most of the transactions in the rumen are based on N, not protein, and the expression 'ruminally degradable N' is sometimes used instead of RDP. However, this usage implies that microorganisms can somehow degrade N, which is impossible. A better term is 'ruminally available N' (RAN). The proportion of dietary protein which escapes microbial attack is called the 'undegradable dietary protein' (UDP), the 'undegradable intake protein' (UIP), the 'bypass protein', or in the French system the 'protein truly digestible in the intestine, fraction A' (PDIA; Jarrige, 1989).

The extent to which food protein is digested by rumen microorganisms depends on the chemical and physical composition of the food, the nature of the microbial population and the rate of passage of undigested food through the rumen. The protein of rumen bacteria, together with undigested food protein, passes to the abomasum and small intestine where it is exposed to digestion by abomasal, pancreatic and intestinal enzymes. Because the animal absorbs a mixture of food and microbial amino acids, the amino acids absorbed from the small intestine may bear little similarity to the amino acid composition of the food eaten. Also, there is a loss of food protein N through its absorption as NH_3 from the rumen. NH_3, coming largely from the digestion of endogenous secretions and some remnants of rumen bacterial cells, may also be absorbed from the hind gut. These processes are illustrated in Fig. 12.7.

The N economy of the ruminant

The concepts used in ruminant protein systems are defined in Table 12.9. MP is the concept which links amino acid supply to amino acid requirements. The amount of amino acid (expressed as MP) required by the animal is compared with the amount of MP

Table 12.9. Concepts relating to the supply of amino acid N in the ruminant.

Concept	Symbol	Definition
Food protein degradability	dg	The proportion of food protein degraded in the rumen; 'effective degradability' (Edg) is dg corrected for the rate of passage of undigested food through the rumen
Ruminally degradable protein	RDP	That food protein which is (able to be) degraded by microbial action in the rumen; ruminally available N (RAN) = RDP/6.25
Undegradable dietary protein (bypass protein)	UDP	That food protein which is not degraded in the rumen
Microbial crude protein	MCP	The total N of rumen microorganisms, expressed as crude protein (i.e. N × 6.25)
Metabolizable protein	MP	The sum of microbial (principally bacterial), endogenous and undegraded food protein which leaves the reticulo-rumen and is capable of being digested in the abomasum and small intestine; it excludes any non-protein N

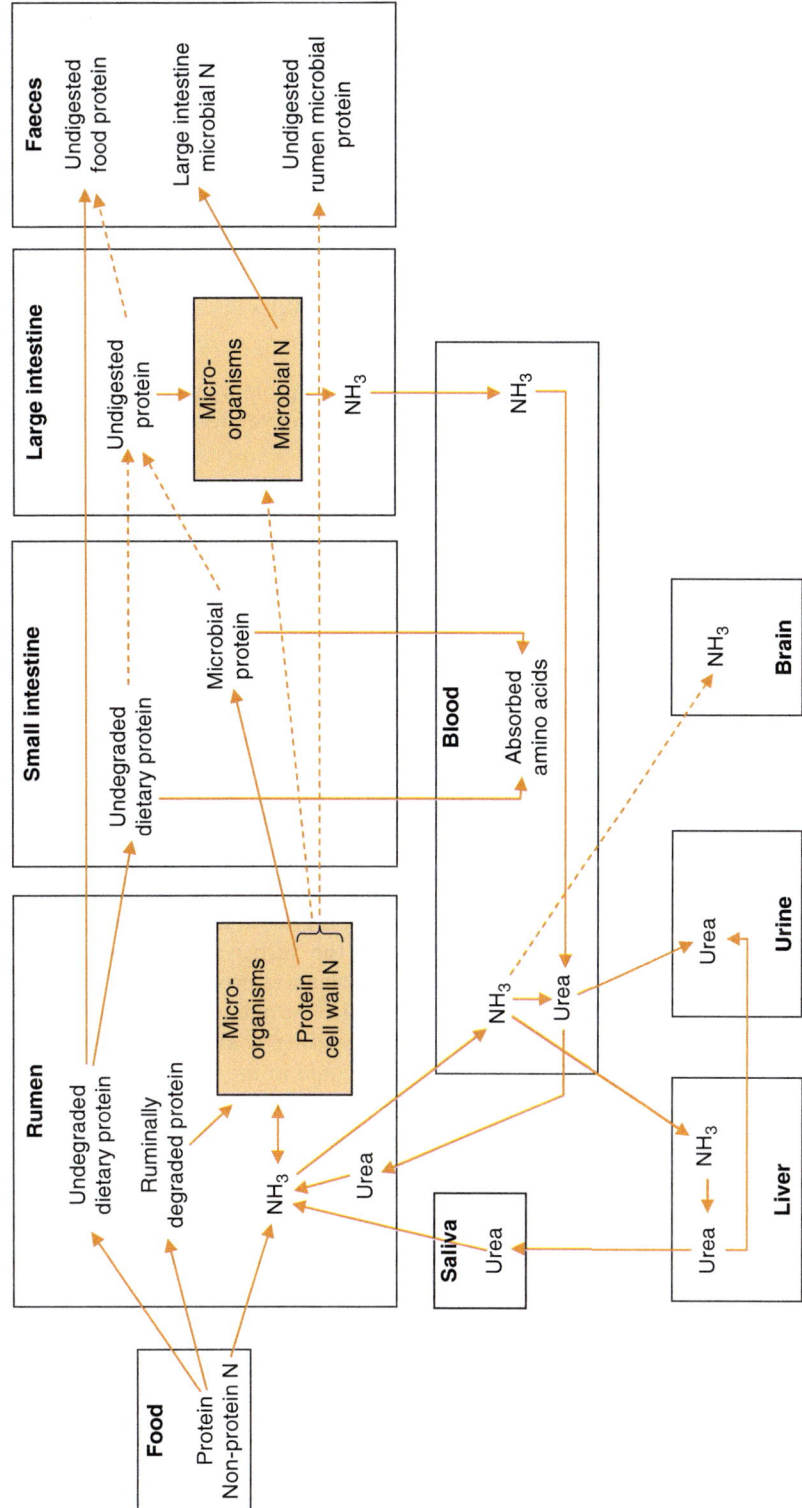

Fig. 12.7. Nitrogen flow through the ruminant digestive tract (endogenous losses into the digestive tract are not shown).

obtained from the diet, as a mixture of microbial protein (MCP) and UDP. If there is a deficiency and less MCP is produced than required, the difference must be made up with extra UDP.

Quantifying the ruminally degradable and undegradable protein

The total protein of the animal's diet can be divided into these components:

1. Quickly degradable protein (QDP) or 'fraction a' protein: this is defined either as the non-protein N content (NRC, 2001) or as the water-soluble protein (AFRC, 1992).
2. Slowly degraded protein (SDP) or 'fraction b' protein: this is the potentially degradable water-insoluble protein; it can be calculated as the difference between the total and QDP protein, corrected for the fraction c protein. SDP = total protein − (QDP + ADIN).
3. Indigestible protein or 'fraction c' protein: this is the protein which cannot be digested by either rumen microorganisms or intestinal enzymes − it should not be confused with UDP, which is potentially digestible food protein that escapes digestion in the rumen. Fraction c protein may be estimated from the acid detergent-insoluble N (ADIN), although some of this is digestible.

If we have information for the concentrations of fractions a and b in the food and the rate of protein degradability in the rumen, then we can calculate the protein rumen degradability (dg) as described by Ørskov and McDonald (1979). Suggested values for the ruminal protein degradation rate, and for food protein fractions a and b, are given in SCA (1990), AFRC (1992) and NRC (2000, 2001) for a variety of foods. For forages, dg can be estimated from a relationship with forage protein and crude fibre contents (SCA, 1990):

$$dg = (Pr − 0.125CF)/Pr \qquad (12.26)$$

where
Pr = crude protein content (g/kg DM)
CF = crude fibre content (g/kg DM)

If we also know the rate of passage of undigested food from the rumen we can calculate the 'effective degradability' (Edg). If measured passage rates (p, /h) are not available, they can be predicted (AFRC, 1992):

$$p = −0.024 + 0.179[1−e^{(−0.278L)}] \qquad (12.27)$$

where
L = food intake (multiples of the amount needed to give a maintenance ME intake)

Example values for Edg (calculated from the NRC, 2001 data for fractions a, b and c) are 0.68 (temperate species grass hays) to 0.8 (legume hays) at $p = 0.05$; 0.39 (sorghum grain) to 0.91 (triticale grain) at $p = 0.07$; and 0.46 (sunflower meal) to 0.64 (coconut meal) at $p = 0.07$.

We also need to know the amount of undegraded food protein leaving the rumen (UDP, g/day). This is a function of dietary protein intake and the Edg in the rumen of this protein:

$$UDP = PrI \times (1−Edg) \qquad (12.28)$$

where
PrI = diet protein intake (g/day)
Edg = effective protein degradability

Predicting the metabolizable protein supply

When we calculate MP we assume that the diet provides adequate amounts of RAN, minerals, vitamins and other microbial nutrients so that the only factor limiting microbial growth is energy. This is not always the case. Although the amount of RAN which can be converted into MCP depends on the amount of fermentable dietary carbohydrate the animal eats (Erdman et al., 1986; Kolade and Ternouth, 1996), microbial protein synthesis is likely to be reduced when the rumen NH_3 content is below 5 mg/100 ml (Satter and Slyter, 1974). Minson (1990) noted that many tropical grasses do not have enough RDP to provide for sufficient MCP synthesis to meet the maintenance protein requirements of cattle. In practice the maximum yield of MCP when RAN is not limiting is about 360 g/kg of organic matter fermented in the rumen (SCA, 1990). The French system recognizes that energy does not always limit microbial protein synthesis and calculates one measure of rumen microbial protein where energy is limiting and another where RAN is limiting (Jarrige, 1989). The NRC (2001) adopts a similar approach, allowing either ruminally fermentable organic matter intake or RDP supply to limit MCP synthesis.

Some digestible food constituents are not fermented in the rumen. In particular, dietary fat and the organic acids in silage are not digested by rumen microorganisms (AFRC, 1992). The energy content of the diet, corrected for these constituents, is called the 'fermentable metabolizable energy' (FME, MJ/kg DM) by AFRC (1992) and is:

FME = diet total ME content − (diet fat
ME + diet organic acids ME)   (12.29)

In addition, not all of the digestible starch is digested in the rumen, although Equation 12.29 implies that it is. The FME content of diets with appreciable amounts of ruminally undegradable but subsequently digestible starch, e.g. feedlot rations based on sorghum grain, should be corrected to allow for the proportion of bypass starch as, for example, in the Dutch system (Tamminga *et al.*, 1994). The French system also excludes the energy of UDP (Jarrige, 1989).

MP (g/day) flow from the rumen can be modelled as in Equation 12.30. The expression is based on an SCA (1990) equation and includes concepts and coefficients from AFRC (1992) and NRC (2001). It is emphasized that this equation is not intended to accurately predict the MP supply but to illustrate the kinds of concepts which are used to calculate it.

$$MP = \{b_1 \times [b_2 \times (b_3 \times FMEI)]\} + (b_4 \times UDP) + (b_5 \times EP)$$ (12.30)

where
b_1 = digestibility of microbial protein produced in the rumen
b_2 = proportion of MCP which is true protein
b_3 = the ratio of MCP synthesis to dietary ME intake (g/MJ)
    $= 7 + 6\,[1 - e^{(-0.35L)}]$
L = food intake (multiples of the amount needed to give a maintenance ME intake)
b_4 = digestibility of undegraded food protein entering the intestines
b_5 = digestibility of endogenous protein secreted into the digestive tract
FMEI = fermentable ME intake (MJ/day)
UDP = amount of undegraded food protein leaving the rumen (g/day)
EP = endogenous protein secreted into the rumen (g/day)
    = 0.5 (11.9DMI)
DMI = DM intake (kg/day)

MP is the amount of *digestible* true protein which enters the duodenum (Table 12.9) and it is influenced by factors which vary the efficiency of MCP synthesis and the amount of non-microbial true protein flowing into the duodenum:

1. The first term in the equation estimates the amount of digestible microbial protein leaving the rumen. Total MCP synthesis is estimated from

$b_3 \times$ FMEI. Energy is assumed to be the limiting nutrient. b_3 can take different values, depending on the level of food intake (AFRC, 1992) as suggested above. Another approach is to use different values of b_3 for different types of food. The SCA (1990) recommended values are 11.0, 8.4 and 6.1 g/MJ ME for fresh temperate legumes and temperate grasses where digestibility is more than 70%, all other forages and mixed diets including concentrates, and silages, respectively. NRC (2001) recognizes that the efficiency of microbial growth (g MCP/kg fermentable organic matter) declines as the amount of surplus RAN increases, and warns that the relationship between MCP production and energy availability is not linear. b_2 is the proportion of MCP which is true protein (i.e. excluding the nucleic acid-N and other non-protein-N of these organisms) and is estimated to be 75% (AFRC, 1992) or 80% (NRC, 2001). b_1, the digestibility of microbial true protein, is assumed to be 85% (AFRC, 1992) or 80% (NRC, 2001). Thus, in both systems, the yield of MP from MCP is 64%.

2. UDP is calculated by multiplying the dietary total protein (N × 6.25) by its Edg. The value for UDP digestibility (b_4) also varies. Cited ranges are 26–65% for forages and 70% for concentrate-rich foods (SCA, 1990), 55–95% (Jarrige, 1989), 50–100% (NRC, 2001) and 90% after allowing for the amount of acid detergent lignin-insoluble N in the diet (AFRC, 1992).

3. The amount of endogenous protein secreted into the rumen is estimated from the DM intake (NRC, 2001), using a relationship which assumes that 50% of endogenous crude protein is true protein. It is assumed that EP is 80% digestible (b_5) in the small intestine (NRC, 2001).

While there is general agreement about the components of MP, there is little unanimity between authorities on how it should be calculated. Part of the problem is that different authorities have used different data sets, but we also have much to learn about the details of microbial growth in the rumen and the factors which regulate it.

N requirements of rumen microorganisms

If the uptake of RAN (either from RDP or non-protein N) and its conversion into MCP were completely efficient, then the RAN or RDP requirements could be calculated directly from MCP production.

However, there are significant inefficiencies, so the amounts of RAN or RDP (g/day) required are:

$$RAN = MCP/(6.25k_{RAN}) \qquad (12.31a)$$

$$RDP = MCP/k_{RAN} \qquad (12.31b)$$

where
MCP = amount of microbial protein synthesized (g/day)
k_{RAN} = efficiency of uptake of RAN (or RDP-N) by rumen microorganisms

k_{RAN} can take different values. These vary according to the relative concentrations of peptide- and NH_3-N in the rumen, the nature of the carbohydrates being fermented and the bacterial species present. The topic is thoroughly discussed in NRC (2001). AFRC (1992) uses k_{RAN} of 0.8 for QDP and 1.0 for SDP, and NRC (2001) adopts a maximum value of 0.85. If a large amount of urea is given in a single meal (assuming no toxicity), k_{RAN} for urea-N will approach zero, while k_{RAN} for an SDP will approach unity. The values for k_{RAN} in Table 12.10 should be used as a guide only.

Table 12.10. Efficiency of incorporation of urea-N into microbial protein. (From SCA, 1990.)

N source and method of feeding	k_{RAN}
Protein from concentrates and (mature) forages	1.0
Non-protein N from silages (equivalent to QDP)	0.8
Water-soluble protein and non-protein N in immature forages	0.8
Urea-N fed from roller-drum dispensers	0.2

Net protein requirements

The net protein requirement for maintenance (Pr_M, g/day) is the endogenous N loss expressed as an equivalent amount of protein. The basal endogenous N excretion (BEN, g/day) can be calculated (AFRC, 1992) as:

$$BEN = 0.368W^{0.75} \qquad (12.32)$$

This gives a similar estimate to the INRA prediction equation (Jarrige, 1989). BEN can also be estimated factorially from the SCA (1990) equations which include separate terms for EUN, MFN and detritus N. These are the first, second and third terms, respectively, in Equation 12.33:

$$BEN = A \times [37.0 \times \log_{10}(W) - 42.2] + 15.2DMI + 0.11W^{0.75} \qquad (12.33)$$

where
A = 1.0 for *Bos taurus* cattle or 0.8 for *Bos indicus* cattle
W = live weight (kg)
DMI = dry matter intake (kg/day)

Equations 12.32 and 12.33 include the amounts of N lost as detritus but not the requirement for growing wool. The SCA (1990) estimates the amount of net protein deposited in the fleece (Pr_W, g/day) from a relationship between ME intake and SRW. Equation 12.34 predicts the 'clean' (i.e. scoured) wool yield of adult sheep, allowing for shearing losses:

$$Pr_W = [MEI \times (8SFW/SRW)] \qquad (12.34)$$

where
MEI = ME intake (MJ/day)
SFW = standard fleece weight (kg), i.e. the amount of clean wool produced by a mature non-lactating, non-pregnant sheep (of a given breed) under good pastoral conditions
SRW = standard reference weight (kg)

The net protein requirement for milk production (Pr_L, g/day) is obtained by multiplying the milk protein content by the milk yield. Milk protein content is usually measured directly, but if necessary it can be estimated from the milk fat content (ARC, 1980). Milk yield changes with time after calving. We can predict a cow's yield through the lactation from a Gompertz function (AFRC, 1998), from the simpler Wood (1979) curve or from herd production records. Bypass amino acids (e.g. methionine; Pisulewski *et al.*, 1996) may improve the milk protein content. This aspect of MP supply has not yet been fully incorporated into feeding standards for dairy cattle, but see NRC (2001), Rulquin *et al.* (2001) and Thomas (2004).

The net protein for growth (Pr_G) is calculated from the expected protein content of gain. This is influenced by the animal's maturity. The AFRC (1992) predicts this with a quadratic function of current live weight, while the SCA (1990) uses the ratio of current live weight to the expected mature weight. The AFRC (1992) equation for Pr_G (g/day) is:

$$Pr_G = K \times S \times [LWG \times (168.07 - 0.16869W + 0.0001633W^2) \times (1.12 - 0.1223LWG)] \qquad (12.35)$$

where

K = factor to correct for breed (1.1 for large breeds, and 0.9 for small breeds)

S = factor to correct for sex (1.1 for males, and 0.9 for females)

LWG = growth rate (kg/day)

W = initial (or current) live weight (kg)

The net protein requirement for pregnancy (Pr_P, g/day) is predicted from an estimate of the growth rate of the gravid uterus (fetus plus uterine tissues) and its expected protein content (AFRC, 1992). This example is for cattle:

$$Pr_p = \{10^{(3.707 - 5.698 \times e^{(-0.00262t)})}\} \times [34.37 \times e^{(-0.00262t)}] \quad (12.36)$$

where

t = days since conception

Metabolizable protein requirements

Net protein requirements are converted to the equivalent amounts of MP by applying k_{NX} (Table 12.11) which recognize the amino acid profiles of the protein in body lean tissue, milk, the products of conception and wool. Most values of k_{NX} are 0.6 or more, although that for wool production is substantially less, reflecting the high cystine content of wool compared to the lower cystine and methionine contents of many plant proteins.

The total MP requirement is calculated (Equation 12.37) in a way that is similar to calculating an ME requirement. We cannot simply sum the amounts

of net protein and apply a single k_{NX} value, because these are all different.

$$\text{Total MP requirement} = Pr_M/k_{NM} + Pr_G/k_{NG} + Pr_L/k_{NL} + Pr_P/k_{NP} + Pr_W/k_{NW} \quad (12.37)$$

Nitrogen Transactions in the Non-ruminant

Amino acid requirements can be expressed in several different ways: as minimum required concentrations of total or available amino acid in the food; as amounts of amino acid relative to lysine, because lysine is usually the first-limiting amino acid in pig foods (Fuller et al., 1979); or relative to energy, which helps to avoid energy limitations to PD, and over-fat carcasses.

Because lysine is generally the first-limiting amino acid there is a tendency to think that once the lysine requirement has been met, requirements for the other amino acids will be satisfied as well. This is not necessarily so, and the supplies of all EAA, and the total protein content, should be monitored. Requirements for the other EAA, relative to lysine, are illustrated in Table 12.12. The data are for growing pigs but can be applied to

Table 12.11. Relative values and efficiencies of use of metabolizable protein for different physiological functions. (From AFRC, 1992.)

Function	Relative value[a]	Assumed efficiency[b]	k_{NX}[c]
Maintenance	1.0	1.00	1.00
Growth	0.7	0.85	0.59
Lactation	0.8	0.85	0.68
Pregnancy	1.0	0.85	0.85
Wool production[d]	0.3	0.85	0.26

[a]The similarity between the amino acid compositions of MP and the product to be synthesized.
[b]The efficiency with which an 'ideal' amino acid mixture would be used for this purpose.
[c]k_{NX} = (relative value) × (assumed efficiency).
[d]SCA (1990) uses k_{NW} = 0.6 for wool production on the grounds that some of the amino acid which cannot be used for wool growth is available (and used) for other protein syntheses.

Table 12.12. Essential amino acid requirements of the growing pig. (From Wang and Fuller, 1989; Fuller et al., 1989.)

	Requirements for growth[a]	
Amino acid	Relative to lysine	Required for tissue protein synthesis (mg/g protein)
Lysine	100	68
Threonine	72	47
Valine	75	53
Isoleucine	60	43
Leucine	110	78
Methionine + cystine[b]	63	36
Phenylalanine + tyrosine[c]	120	84
Tryptophane	18	12

[a]Requirements for arginine and histidine were not determined (in vivo synthesis of arginine can meet the pig's requirements for maintenance and some growth (NRC, 1998)).
[b]Cystine can provide 80% of the methionine + cystine requirement for maintenance and 50% for growth (Fuller et al., 1989).
[c]Tyrosine can supply 50% of the tyrosine + phenylalanine requirement (Fuller et al., 1989).

maintenance as well. Other authorities have different recommendations; in particular, NRC (1998) suggests higher ratios of arginine and leucine for milk production and growth than for maintenance. The amino acid pattern for a particular physiological process does not change; e.g. the same pattern holds over a range of milk yields (Close and Cole, 2000).

Amino acid requirements for maintenance in pigs

We can use BEN excretion to estimate the pig's Pr_M in the same way as we have discussed for ruminants. Whittemore et al. (2001c) have reviewed the limited data on this topic. Alternatively, we can determine the lysine requirement for maintenance and apply the ideal protein concept to calculate the Pr_M. The maintenance requirement for total lysine is 36–39 mg/kgW$^{0.75}$/day (Fuller et al., 1989; NRC, 1998). As these numbers indicate, the maintenance requirement is a small proportion (less than 5%) of the growing pig's total lysine requirement.

Amino acid requirements for growth in pigs

We should ensure that its amino acid and energy intakes allow the pig to express its full lean growth potential. Pig genotypes have changed markedly in the last 20 years, and the lean growth potential in superior genotypes now exceeds 500 g/day (Whittemore et al., 2001c). Consistent with this, the NRC (1998) recommendations for lysine are higher than both NRC (1988) and SCA (1987) values. Feeding standards must be appropriate to the animals' genotype.

As protein intake increases from a low level, with energy intake held constant, the PD rate first rises, and then plateaus when the capacity to retain more dietary N reaches a ceiling imposed by the energy available for this process (Campbell et al., 1984; Bikker et al., 1994). This is illustrated in Fig. 12.8.

According to Sandberg et al. (2005b) there is no effect of age (live weight), sex or genotype on the *efficiency* of the retention of dietary amino acids, i.e. the slope of the ascending line in Fig. 12.8. This applies both when amino acid intakes increase when energy intake is not limiting (Fig. 12.8), and when energy intake increases when amino acid intake is limiting (Fig. 12.6). The efficiency of retention of an ideal protein (e_p), when energy is not limiting, is about 0.76 g/g ideal digestible pro-

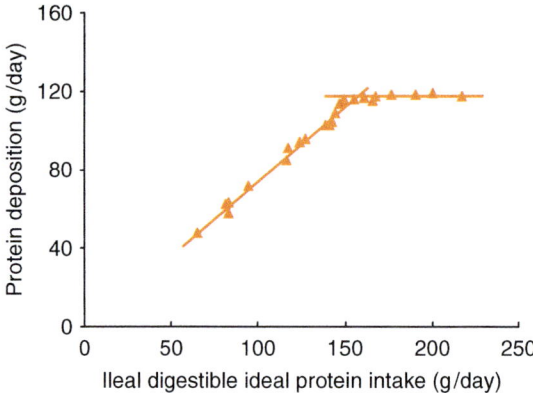

Fig. 12.8. Protein retention by growing pigs given a diet with adequate energy (mean, 109 MJ ME/kg digestible protein) and increasing amounts of protein. (Redrawn from Ferguson and Gous, 1997.)

tein (IDP). It may seem unusual that e_p is not 1.0, but we have previously noted that pigs do not use ideal proteins with complete efficiency (Whittemore et al., 2001c). According to data examined by Sandberg et al. (2005b), energy ceases to limit PD when the ME/IDP ratio is more than about 70 MJ/kg. At ME/IDP ratios less than this, e_p decreases by 0.0112/MJ/kg (Kyriazakis and Emmans, 1992; Sandberg et al., 2005b). The shape of the e_p versus ME/IDP response is a line/plateau graph, similar to Figs 12.6 and 12.8.

The pig's *daily* amino acid requirements for growth are influenced by its age, sex, ability to deposit protein (i.e. its 'lean growth' genotype) and energy intake (Whittemore et al., 1988; Dunshea et al., 1993; Sandberg et al., 2005a). There are several equations to predict PD, e.g. Black et al. (1986) as modified by NRC (1998). A simple one, which satisfies the theory as it is presently understood (Sandberg et al., 2005b), is based on Kyriazakis and Emmans (1992) where PD (kg/day) is calculated as:

$$PD = e_p \times (IP - IP_m) \qquad (12.38)$$

where

e_p = efficiency of retention of absorbed ideal protein

= 0.0112[MEC/(PrC × d$_i$)]

MEC = food energy content (MJ ME/kg)

PrC = food protein content (kg/kg)

IP = intake of ileal digestible ideal protein (kg/day)

= (FI × PrC × v × d$_i$)

FI = food intake (kg/day)

$$v = \frac{\text{(concentration of the first-limiting amino acid in digested protein)}}{\text{(concentration of the first-limiting amino acid in ideal protein)}}$$

d_i = ileal digestibility of food protein
IP_m = maintenance requirement for ileal digestible ideal protein (kg/day)
    = $(0.00064\,W)$
W = live weight (kg)

Although this expression calculates PD for a given ME/IDP and IP intake, it does not help to identify the PD plateau for any particular sex or genotype. Animals of high lean growth genotype, and males (compared to females and castrates), have higher plateaux (Batterham *et al.*, 1990) and so they need more absorbable amino acids (and energy) if they are to reach their genetic potential. NRC (1998) deals with this by multiplying the expected PD by MPAR/125 (in which MPAR is the animal's actual mean protein accretion rate, g/day). In effect, this increases e_p as the genotype improves, but this approach is not supported by the available literature. A better method is to allow PD to increase until it reaches the maximum value allowable for that genotype. NRC (1998) suggests several ways of estimating the lean growth potential for commercial pig herds.

The rate of daily PD reaches a maximum when the pig is about 37% of its mature size (Whittemore *et al.*, 2001c). This is reflected in the NRC (1998) polynomial Equation (12.39a) for the total lysine requirement (Lys, % diet) for starter pigs (live weight range of 5–20 kg). Alternatively, we can estimate the expected PD and apply to this the lysine content of body protein (7.2% in protein retained during growth; Susenbeth, 1995) and the expected efficiency of retention of ingested lysine (0.55; Susenbeth, 1995), as in Equation 12.39b:

$$\text{Lys} = 1.793 - 0.0873W + 0.00429W^2 - 0.000089W^3 \tag{12.39a}$$

$$\text{Lys} = (PD \times 0.072/0.55)/(10FI) \tag{12.39b}$$

where
W = live weight (kg)
PD = protein deposition (g/day)
FI = food intake (kg/day)

The lysine requirements (g/MJ ME) suggested by SCA (1987) for growing pigs are in Table 12.13. These recommend lower lysine levels for finishing females (gilts) than entire males. This approach is based on data published by Batterham *et al.* (1985),

Table 12.13. Total lysine requirements, relative to metabolizable energy, of ad lib-fed growing pigs. (From SCA, 1987.)

Pig description	Total lysine requirement (g/MJ ME)	
	Entire males	Females
Birth to 5 kg live weight	0.69	0.69
5–20 kg live weight	0.75	0.75
20–50 kg live weight	0.71	0.64
50–90 kg live weight	0.57	0.57

Campbell *et al.* (1985), Giles *et al.* (1986), Ettle *et al.* (2003) and others. Barrows (castrated males) are less responsive to high lysine intakes than gilts (Ettle *et al.*, 2003). Whittemore *et al.* (2001c) summarize data which similarly distinguish between the sexes and lean growth genotypes. Between-sex differences appear only at about 50 kg live weight and are not apparent in young growers or baby pigs (Hill *et al.*, 2007).

If food intake is reduced for some reason, perhaps due to hot conditions (Lopez *et al.*, 1991; Ferguson and Gous, 1997; Collin *et al.*, 2001) or to an inherently lower appetite (e.g. in Piétrain cross pigs; Warnants *et al.*, 2003), dietary amino acid contents must be increased by an amount equivalent to the expected reduction in food intake.

Amino acid requirements of the breeding sow

Older standards used an empirical approach to deriving protein requirements. ARC (1981) and SCA (1987) recommended 180 g/day protein for a pregnant sow, as long as the diet provided the appropriate amounts of EAA.

The more recent standards use a factorial approach. If a sow carries 10 fetuses and is pregnant for 115 days, then she will deposit on average 2.7 g/day N (AFRC, 1990a) or 3.4 g/day N (NRC, 1998) in the gravid uterus and mammary gland. Data reviewed by AFRC (1990a) suggest that dietary N is retained with an efficiency of 37%, so the sow requires 46 g (AFRC, 1990a) or 57 g (NRC, 1998) of protein daily for this function. Young sows grow while pregnant, increasing their true live weights by 20–40 kg or more per parity until they reach their mature size (SCA, 1987; AFRC, 1990a; King, 1990; NRC, 1998). AFRC (1990a) uses a 'line/plateau' model for total protein retention during pregnancy and adopts a maximum retention of 100 g/day.

These guidelines for total protein assume that the diet has an appropriate EAA composition.

Recommendations for available lysine are 5.6–7 g/day (ARC, 1981), 7 g/day (SCA, 1987) or 5.6 g/day (NRC, 1998). The relationships in Table 12.12 can be used as a guide to the requirements for other EAA.

Lactating sows require at least 620 g/day protein, providing about 25 g/day of available lysine (SCA, 1987). The NRC (1998) recommendation is for 22 g of ileal digested lysine per kilogramme of piglet growth, or 44 g/day if the sow has a litter of 10 piglets and thus an average total litter growth rate of 2 kg/day between birth and 21 days. Lactating sows lose body protein during the lactation and this contributes to the lysine in milk, at about 6 g/day (NRC, 1998).

Sows are often given a restricted diet during the pregnancy and fed ad lib during the lactation. Even so, the lactating sow usually loses live weight. The NRC (1998) model allows the manager, for a given sow live weight, litter size and expected litter growth rate, to specify either the intended energy intake or the allowable rate of live weight loss. These inputs are used to predict either: (i) the live weight loss rate and the amounts of amino acids to be supplied from the diet; or (ii) the required energy and amino acid intakes.

Protein requirements of horses

Most horse feeding standards use a conventional whole-tract apparent digestibility to quantify the available protein in foods (Julliand and Martin-Rosset, 2004). In contrast, the French MADC (matières azotées digestibles cheval) system (Martin-Rosset et al., 1994) sums the amount of food amino acids absorbed from the small intestine and microbial amino acids absorbed from the large intestine. This is questionable because it is doubtful that nitrogenous materials absorbed from the large intestine can be used in the horse's amino acid metabolism (see Potter, 2004 for a discussion of this). NRC (2007) introduces the concept of 'available protein' (AP, g/kg DM), calculated as:

$$AP = CP - 6.25(ADIN + NPN) \qquad (12.40)$$

where
CP = total food protein (g/kg DM) (i.e. total N × 6.25)
ADIN = acid detergent-insoluble N (g/kg DM)
NPN = non-protein N (g/kg DM)

AP identifies and discounts those forms of dietary N which do not contribute to the horse's pool of absorbed amino acid. However, we could go further and recognize that dietary amino acids flowing to the large intestine do not contribute to the horse's amino acid economy (although they do nourish the bacteria which are responsible for digestion of plant cell wall polysaccharides). We could also discount the dietary nucleic acid-N absorbed from the small intestine. The concept of ideal protein is introduced by NRC (2007) but there is yet only limited information on the possible composition of equine ideal protein. In the future we may see equine protein feeding standards incorporating a mixture of pig and ruminant approaches. Indeed, if we apply a digestibility coefficient to AP, we have a concept which closely resembles the MP of ruminant feeding standards.

Horse protein requirements can be derived empirically or factorially. The NRC (1989) standards estimated the total protein requirement from the horse's predicted energy requirement. For example, after assuming a digestibility of 46% for the protein in a forage-based diet the protein required for maintenance was 9.56 g/MJ DE. The factorial approach outlined in Equations 12.41 and 12.42a and b partitions the horse's protein requirement into the net protein requirements for maintenance (Pr_M, g/day), pregnancy (Pr_P, g/day, assuming exponential growth of the fetus plus uterine tissues) and lactation (Pr_L, g/day). These equations are derived, using a value of 0.55 for k_{NX}, from the MADC data reported by Martin-Rosset et al. (1994).

$$Pr_M = 1.54W^{0.75} \qquad (12.41)$$

De Almeida et al. (1998) and Meyer (1985, cited by NRC, 2007) suggest values for Pr_M of 2.47 and 1.68 g/kgW$^{0.75}$/day, for yearlings and mature horses, respectively. The higher value for yearlings is unexplained, but might reflect the higher metabolic rate of young animals.

Most feeding standards assume that the pregnant mare requires little extra protein until the eighth or ninth month, and that milk yield declines steadily from foaling. Equations 12.42a and b reflect this:

$$Pr_P = 0.2356 \times e^{(0.562t)} \qquad (12.42a)$$

$$Pr_L = 405 - 45t \qquad (12.42b)$$

where
W = live weight (kg)
t = time after conception or foaling (months)

Growth requirements are influenced by the horse's age, growth rate and exercise regime (NRC, 2007), and vary between breeds (Hintz, 1982; Martin-Rosset *et al.*, 1994). Light breeds reach about 66% of mature live weight at 12 months, and warmbloods reach about 55% at this age. The suggested net protein requirements for growth (Martin-Rosset *et al.*, 1994, after assuming k_{NG} of 45%) are 200 and 120 g/kg LWG for weaners and 2- to 3-year-olds, respectively. NRC (2007) recommends that the dietary protein contains 4.3% lysine, and warns that lower lysine contents will require the growing horse's protein requirement to be increased.

The effect of muscular activity (work) on protein requirements is unclear. There is some loss of N through sweat (Ott, 2005) and perhaps wear-and-tear on joint cartilage (Bird *et al.*, 2000). Horses increase muscle protein synthesis during training, especially if given extra amino acids (Matsui *et al.*, 2006). These observations support the recommendations of Martin-Rosset *et al.* (1994) and NRC (2007) to increase the horse's protein intake during training. However, excess protein consumption leads to more urea excretion, and so increases an already stressed animal's water requirement (Ott, 2005). It seems safest to adopt the NRC (1989) approach, i.e. to not increase the protein content of diets for working horses, but to rely on their increased voluntary intake to provide any extra protein which may be needed.

Predicting Food Intake

This is one of the most fundamental aspects of quantitative nutrition and it is important that we make these predictions accurately. Estimates of food intake are used to predict the minimum nutrient concentrations needed in animal foods to sustain particular levels of production and for planning food supplies and storages.

There are many published relationships for predicting food intake. Common predictor variables are live weight, age, current live weight in relation to mature weight, body fat content, stage of lactation, NE requirement, food chemical composition and ambient temperature. Examples of some dry matter intake (DMI, kg/day) or food intake (FI, kg/day) prediction equations are given below.

The NRC (1987) predictions of food intake by weaner pigs (Equation 12.43a), growers (Equation 12.43b) and lactating sows (Equation 12.43c) eating diets with 13.4, 14.2 and 14.2 MJ DE/kg, respectively, are:

$$FI = -0.4784 + (0.1423W) - (0.00296W^2) \quad (12.43a)$$

$$FI = 3.87 [1 - e^{(-0.176W)}] \quad (12.43b)$$

$$FI = 3.8235 + 0.1753t - 0.0051t^2 \quad (12.43c)$$

where
W = live weight (kg)
t = time since farrowing (days)

These equations apply to diets with the nominated DE contents. Note that pigs eat more of diets with lower energy contents (Ettle *et al.*, 2003).

The NRC (1989) suggests that the expected daily DMI of an adult horse at maintenance is about 1.5–2 kg/100 kg live weight, with intakes rising to 3 kg/100 kg for horses in intense work or by growing yearlings. Other authorities relate intake to food quality, e.g. Lawrence *et al.* (2001):

$$DMI = W/100 \times (124.55 - 2.5742NDF + 0.0155NDF^2) \quad (12.44)$$

where
W = live weight (kg)
NDF = diet neutral detergent fibre content (%)

There are many equations for growing beef cattle. They include (Equation 12.45a) the very simple but fairly reliable estimate of an upper limit to DMI based on data in Church (1977). The SCA (1990) Equation (12.45b) relates intake to the animal's physiological maturity as indicated by its current live weight. Equation 12.45c predicts the FI of cattle eating concentrate-rich feedlot diets (NRC, 2000), and Equation 12.45d predicts silage intake (AFRC, 1991b).

$$DMI = 0.27W \quad (12.45a)$$

$$DMI = a \times SRW \times Z \times (1.7 - Z) \quad (12.45b)$$

$$DMI = [SBW^{0.75} \times (0.2435CNE_m - 0.0466CNE_m^2 - 0.1128)]/CNE_m \quad (12.45c)$$

$$DMI = (W^{0.75}/1000) \times (9.92 + 0.1196TDM + 0.0607DOMD - 0.5798CDMI) \quad (12.45d)$$

where
W = live weight (kg)
a = 0.04 for sheep and 0.024 for cattle
SRW = standard reference weight (kg)
Z = W/SRW, with a maximum value = 1.0
SBW = shrunk body weight (kg) (i.e. live weight after a fast, typically = 0.96 W)
CNE_m = dietary concentration of NE for maintenance (Mcal/kg DM)

TDM = silage DM content (measured by distillation under toluene, g/kg)

DOMD = digestible organic matter content (g/kg TDM)

CDMI = concentrate DMI ($g/W^{0.75}$)

The two example lactating dairy cow equations include (Equation 12.46a) a very simple model (MAFF, 1984) more appropriate for cattle in mid-lactation and another (Equation 12.46b) which is more complex (NRC, 2001) and which attempts to describe FI over the whole-lactation, including the depression in FI which occurs after calving:

$$DMI = 0.025W + 0.1MY \qquad (12.46a)$$

$$DMI = (0.372FCM + 0.0968W^{0.75}) \times [1 - e^{(-0.192(t + 3.67))}] \qquad (12.46b)$$

where
W = animal live weight (kg)
MY = milk yield (kg/day)
FCM = fat-corrected milk
= MY × (0.4 + 0.15 milk fat%)
t = time since calving (weeks)

These prediction equations forecast the mean intake of groups of animals over time. They do this with quite variable accuracy. The NRC claims that their beef cattle prediction equation explains 70% of the variation in food intake (NRC, 2000), and the weaner pig equation explains 92% (NRC, 1998). However, most prediction equations are less reliable. Pittroff and Kothmann (2001a,b,c) concluded that most existing ruminant models are inadequate. Whittemore et al. (2001a) arrived at a similar conclusion for growing pig models.

If we can predict it acceptably well, average food intake gives useful information for planning food supplies and storages, or to forecast average animal performance. But it has some limitations. Although we expect day-to-day variations in the amounts of food an animal will eat, we have very little ability to predict the magnitude of these, to explain why these variations occur or to predict meal-eating behaviour. It may be impossible to model short-term variations in intake, as examination of some sets of intake data indicates that there is a 'random' variation around the mean. Little is known about the structure of this variation. It 'could be a result of the interaction of an unknown number of deterministic factors, as it seems counter-intuitive that animal metabolism should act in a random fashion' (Dryden, 2006).

Quantitative Nutrition in Practice

Feeding standards: requirements and allowances

'Feeding standards' are sets of recommendations about the amounts of energy, or other nutrients, needed to obtain certain production outcomes. They are based on the concepts of nutrient use which we have discussed above. In practice, estimates of nutrient requirements are translated into 'allowances'. These allow for between-animal variations in nutrient use and ingredient nutrient contents (AFRC, 1992). Nutrient requirements are calculated for an 'average' animal, and their use without safety margins would see 50% of animals underfed. In the case of the AFRC (1990b) energy standards, the margins needed to reduce the proportion of underfed animals to 20% are +10%, +27% and +32%, for lactating cows, growing female and growing male cattle, respectively. Additionally, it should be acknowledged that nutrient requirements may be derived from quite limited databases, and that environmental effects (see NRC, 2001 for a discussion in relation to dairy cow feeding) will often increase an animal's energy requirements, at least.

Nutritional models

Nutrition models incorporate the types of equations which we have discussed in this chapter. These models are generally static, mechanistic and deterministic. Types of model are discussed by AFRC (1991a), Black et al. (1993) and Baldwin (1995). A 'static' model is one which describes the outcome of a given set of inputs. In comparison, a 'dynamic' model describes the rates at which outputs occur. Dynamic models are 'potentially capable of tracing impacts of previous and current management decisions on subsequent performance' (Baldwin, 1995). A 'mechanistic' model is based on a theoretical understanding of animal metabolism, and quantifies this sufficiently to allow production outcomes to be predicted from inputs like the amount and quality of the animal's food. 'Empirical' models simply use relationships which have been established from cause-and-effect observations. They work at the whole-animal, or even whole-farm, level and predict outputs from inputs provided at those levels. Nutrition models often are situated somewhere between empirical and mechanistic. Finally, 'deterministic' models apply to the 'average animal'. No allowance is made for between-animal variation. This leads to the observations made by the AFRC (1990b) about the safety margins that are needed to ensure that a predetermined proportion of animals are not underfed. 'Stochastic' models give some idea of the range of responses which we can expect from a group of animals given the same nutritional treatment.

Examples (but certainly not a complete list) of nutrient prediction models are the NRC (1998, 2000, 2007) pig, beef cattle and horse nutrient requirement equations; the Cornell Net Carbohydrate and Protein system (see references in Fox *et al.*, 1995); 'Molly' (Baldwin, 1995; named after the cow which Baldwin was given to learn to milk); 'RUMNUT' (Chamberlain, 2007); 'Karoline', a Scandinavian nutrient system (Danfær *et al.*, 2006); the SCA (1987, 1990) equations for pigs and ruminants; and the pig nutrient prediction models of Black *et al.* (1986) and Danfær (2000).

Nutrient requirements models are used in decision support systems (DSS). These are mathematical simulations of farming enterprises which are used to make decisions about levels of feeding, nutrient supply, etc. often in an economic or environmental context. Examples of nutrition DSS are:

1. For ruminants: 'CAMDAIRY' (Hulme *et al.*, 1986); the CPM-Dairy model (Boston *et al.*, 2000); 'NorFor-plan' (Danfær *et al.*, 2006), which uses the Karoline nutrient system; 'Grazfeed' (Freer *et al.*, 1997), a DSS for grazing ruminants; and 'Feed into Milk' (FiM) (Thomas, 2004).
2. For pigs: AUSPIG (Black *et al.*, 1988), which includes a module to predict nutrient requirements, and algorithms for ration formulation and enterprise profit maximization; and the IMS pig model (Green and Whittemore, 2005), which links estimations of pig growth obtained from real-time visual image analysis with automated logging of food intake data.

The value of models, compared to tables of nutrients requirements, is that the user can describe the animal of interest, and its environment, and possibly couple this information with financial and other information. Models will answer 'what if?' questions. Users can apply techniques such as sensitivity analysis and parametric ranging to investigate the ways in which altering nutrient supply can influence enterprise profitability, or nutrient wastage, or product quality.

References

Abdul-Razzaq, H.A. and Bickerstaffe, R. (1989) The influence of rumen volatile fatty acids on protein metabolism in growing lambs. *British Journal of Nutrition* 62, 297–310.

Adam, I., Young, B.A., Nicol, A.M. and Degen, A.A. (1984) Energy cost of eating in cattle given diets of different form. *Animal Production* 38, 53–56.

Agriculture and Food Research Council (AFRC) (1990a) AFRC Technical Committee on Responses to Nutrients, Report No. 4: Nutrient requirements of sows and boars. *Nutrition Abstracts and Reviews (Series B)* 60, 383–406.

Agricultural and Food Research Council (AFRC) (1990b) AFRC Technical Committee on Responses to Nutrients. Report No. 5. Nutritive requirements of ruminant animals: energy. *Nutrition Abstracts and Reviews (Series B)* 60, 729–804.

Agricultural and Food Research Council (AFRC) (1991a) AFRC Technical Committee on Responses to Nutrients. Report No. 7. Theory of response to nutrients by farm animals. *Nutrition Abstracts and Reviews (Series B)* 61, 681–722.

Agricultural and Food Research Council (AFRC) (1991b) AFRC Technical Committee on Responses to Nutrients. Report No. 8. Voluntary intake of cattle. *Nutrition Abstracts and Reviews (Series B)* 61, 815–823.

Agricultural and Food Research Council (AFRC) (1992) AFRC Technical Committee on Responses to Nutrients. Report No. 9. Nutritive requirements of ruminant animals: protein. *Nutrition Abstracts and Reviews (Series B)* 62, 787–835.

Agricultural and Food Research Council (AFRC) (1993) *Energy and Protein Requirements of Ruminants.* AFRC Technical Committee on Responses to Nutrients. CAB International, Wallingford, UK.

Agricultural and Food Research Council (AFRC) (1998) *Response in the Yield of Milk Constituents to the Intake of Nutrients by Dairy Cows.* AFRC Technical Committee on Responses to Nutrients, Report No. 11. CAB International, Wallingford, UK.

Agricultural Research Council (ARC) (1965) *The Nutrient Requirements of Farm Livestock, No. 2 Ruminants, Technical Reviews.* Agricultural Research Council, London.

Agricultural Research Council (ARC) (1980) *The Nutrient Requirements of Ruminant Livestock.* CAB, Farnham Royal, UK.

Agricultural Research Council (ARC) (1981) *The Nutrient Requirements of Pigs.* CAB, Farnham Royal, UK.

Agnew, R.E. and Newbold, J.R. (2002) *Nutritional Standards for Dairy Cattle.* Report of the British Society of Animal Science Nutritional Standards Working Group: Dairy Cows, March 2002. Available at: http://www.bsas.org.uk/downloads/reports/FinalDairy.pdf

Agnew, R.E. and Yan, T. (2000) Impact of recent research on energy feeding systems for dairy cattle. *Livestock Production Science* 66, 197–215.

Alberts, B., Johnson, A., Lewis, J., Raff, M., Roberts, K. and Walter, P. (2002) *Molecular Biology of the Cell*, 4th edn. Garland Science, New York.

Arman, P., Hopcraft, D. and McDonald, I. (1975) Nutritional studies on East African herbivores. 2. Losses of nitrogen in the faeces. *British Journal of Nutrition* 33, 265–276.

Asplund, J.M. (1987) Amino acid requirements and biological value of proteins for sheep. *Journal of Nutrition* 117, 1207–1211.

Atwater, W.O. and Benedict, F.G. (1903) *Experiments on the Metabolism of Matter and Energy in the Human Body, 1900–1902*. US Department of Agriculture, Office of Experiment Stations, Bull. No. 136, Washington, DC.

Babella, G., Novak, A., Schmidt, J. and Kaszas, I. (1988) Influence of changing the casein/whey protein ratio on the feeding value of calf milk replacers. *Milchwissenschaft* 43, 551–554.

Baldwin, R.L. (1995) *Modeling Ruminant Digestion and Metabolism*. Chapman & Hall, London, pp. 11–18, 469–518.

Barber, A.A. (1990) TAKE-AWAY: a ruminant nutrition software package. *Proceedings of the Australian Society of Animal Production* 18, 136–139.

Barrey, E. (1993) Simulation des dépenses énergétiques chez le cheval en compétition. *Science and Sports* 8, 109–115.

Batterham, E.S., Giles, L.R. and Dettmann, E.B. (1985) Amino acid and energy interactions in growing pigs. 1. Effects of food intake, sex and live weight on the responses of growing pigs to lysine concentration. *Animal Production* 40, 331–343.

Batterham, E.S., Andersen, L.M., Braigent, D.R. and White, E. (1990) Utilisation of ileal digestible amino acids by growing pigs: effect of dietary lysine concentration on efficiency of lysine retention. *British Journal of Nutrition* 64, 81–94.

Benedict, F.G. (1938) Vital energetics, a study in comparative metabolism. *Publications of the Carnegie Institute of Washington* 503, Washington, DC.

Bikker, P., Verstegen, M.W.A., Campbell, R.G. and Kemp, B. (1994) Digestible lysine requirement of gilts with high genetic potential for lean gain, in relation to the level of energy intake. *Journal of Animal Science* 72, 1744–1753.

Bird, J.L.E., Platt, D., Wells, T., May, S.A. and Bayliss, M.T. (2000) Exercise-induced changes in proteoglycan metabolism of equine articular cartilage. *Equine Veterinary Journal* 32, 161–163.

Black, J.L., Campbell, R.G., Williams, I.H., James, K.J. and Davies, G.T. (1986) Simulation of energy and amino acid utilization in the pig. *Research and Development in Agriculture* 3, 121–145.

Black, J.L., Fleming, J.F. and Davies, G.T. (1988) AUS-PIG: a computer program for the optimal management of pigs. *Animal Production in Australia* 17, 366.

Black, J.L., Davies, G.T. and Fleming, J.F. (1993) Role of computer simulation in the application of knowledge to animal industries. *Australian Journal of Agricultural Research* 44, 541–555.

Blaxter, K.L. (1962) The fasting metabolism of adult wether sheep. *British Journal of Nutrition* 16, 615–626.

Blaxter, K.L. (1967) *The Energy Metabolism of Ruminants*, rev. edn. Hutchinson, London.

Blaxter, K.L. (1989) *Energy Metabolism in Animals and Man*. Cambridge University Press, Cambridge, UK.

Blaxter, K.L. and Wood, W.A. (1951) Nutrition of the young Ayrshire calf. 1. The endogenous nitrogen and basal energy metabolism of the calf. *British Journal of Nutrition* 5, 11–25.

Blaxter, K.L. and Wood, W.A. (1952) The nutrition of the young Ayrshire calf. 7. The biological value of gelatin and of casein when given as the sole source of protein. *British Journal of Nutrition* 6, 56–71.

Blome, R.M., Drackley, J.K., McKeith, F.K., Hutjens, M.F. and McCoy, G.C. (2003) Growth, nutrient utilization, and body composition of dairy calves fed milk replacers containing different amounts of protein. *Journal of Animal Science* 81, 1641–1655.

Blum, J.-C. (ed.) (1984) *L' Alimentation des Animaux Monogastriques: Porc, Lapin, Volailles*. INRA, Paris.

Boisen, S., Hvelplund, T. and Weisbjerg, M.R. (2000) Ideal amino acid profiles as a basis for feed protein evaluation. *Livestock Production Science* 64, 239–151.

Bosshardt, D.K. and Barnes, R.H. (1946) The determination of metabolic fecal nitrogen and protein digestibility. *Journal of Nutrition* 31, 13–21.

Boston, R.C., Fox, D.G., Sniffen, C., Janczewski, E., Munson, R. and Chalupa, W. (2000) The conversion of a scientific model describing dairy cow nutrition and production to an industry tool: the CPM Dairy project. In: McNamara, J.P., France, J. and Beever, D.E. (eds) *Modelling Nutrient Utilization in Farm Animals*. CAB International, Wallingford, UK, pp. 361–378.

Brody, S. and Proctor, R.C. (1932) *Growth and Development with Special Reference to Domestic Animals: Further Investigations of Surface Area in Energy Metabolism*. University of Missouri Agricultural Experiment Station, Research Bulletin No. 14., Missouri.

Brody, S., Proctor, R.C. and Ashworth, U.S. (1934) Growth and development. XXXIV. Basal metabolism endogenous nitrogen, creatinine and neutral sulphur excretions as functions of body weight. *University of Missouri Agricultural Experiment Station, Research Bulletin* 220, 1–40.

Brouwer, E. (1965) Report of subcommittee on constants and factors. In: Blaxter, K.L. (ed.) *Energy*

Metabolism. EAAP Publication No.11. Academic Press, London, pp. 441–443.

Burger, I.H. and Johnson, J.V. (1991) Dogs large and small: the allometry of energy requirements within a single species. *Journal of Nutrition* 121, S18–S21.

Buzas, A.M. (2007) *Physiological Variables and Energy Expenditure of Competing Polocrosse Horses Using Field Studies of Heart Rate and Velocity with a Global Positioning System*. BAppSc (Hons) thesis, University of Queensland, Queensland, Australia.

Campbell, R.G. and Taverner, M.R. (1988) Genotype and sex effects on the relationship between energy intake and protein deposition in growing pigs. *Journal of Animal Science* 66, 676–686.

Campbell, R.G., Taverner, M.R. and Curic, D.M. (1984) The effect of feeding level and dietary protein content on the growth, body composition anf rate of protein deposition in pigs growing from 45 to 90 kg. *Animal Production* 38, 233–240.

Campbell, R.G., Taverner, M.R. and Curic, D.M. (1985) Effects of sex and energy intake between 48 and 90 kg live weight on protein deposition in growing pigs. *Animal Production* 40, 497–503.

Carmean, B.R., Johnson, K.A., Johnson, D.E. and Johnson, L.W. (1991) Maintenance energy requirement of the llama. In: Wenk, C. and Boessinger, M. (eds) *Energy Metabolism of Farm Animals*. Proceedings 12th Symposium, Switzerland. 1–7 September, 1991. EAAP Publication No. 58, Institut fur Nutztierwissenschaften, Zurich, pp. 454–457.

Chamberlain, A.T. (2007) *RUMNUT 5.2 The Ruminant Nutrition Program*. Available at: http://www.rumnut.com/abrum.htm

Church, D.C. (1977) *Livestock Feeds and Feeding*. O&B Books, Corvallis, Oregon, pp. 136–141.

Close, W.H. and Cole, D.J.A. (2000) *Nutrition of Sows and Boars*. Nottingham University Press, Nottingham, UK.

Close, W.H., Noblet, J. and Heavens, R.P. (1984) The partition of body-weight gain in the pregnant sow. *Livestock Production Science* 11, 517–527.

Collin, A., van Milgen, J., Dubois, S. and Noblet, J. (2001) Effect of high temperature on feeding behaviour and heat production in group-housed young pigs. *British Journal of Nutrition* 86, 63–70.

Coenen, M. (2005) About the predictability of oxygen consumption and energy expenditure in the exercising horse. *Proceedings of the 19th Equine Science Society Conference*, Tuscon Arizona, p. 123.

Conn, E.E. and Stumpf, P.K. (1976) *Outlines of Bichemistry*, 4th edn. Wiley, New York.

Danfær, A. (2000) A pig model for feed evaluation. In: McNamara, J.P., France, J. and Beever, D.E. (eds) *Modelling Nutrient Utilization in Farm Animals*. CAB International, Wallingford, UK, pp. 393–408.

Danfær, A., Huhtanen, P., Udén, P., Sveinbjörnsson, J. and Volden, H. (2006) The Nordic dairy cow model, Karoline – description. In: Kebreab, E., Dijkstra, J., Bannink, A., Gerrits, W.J.J. and France, J. (eds) *Nutrient Digestion and Utilization in Farm Animals: Modelling Approaches*, CAB International, Wallingford, UK, pp. 383–406.

Darveau, C.-A., Suarez, R.K., Andrews, R.D. and Hochachka, P.W. (2002) Allometric cascade as a unifying principle of body mass effects on metabolism. *Nature* 417, 166–170.

Da Silva, J.K.L., Garcia, G.J.M. and Barbosa, L.A. (2006) Allometric scaling laws of metabolism. *Physics of Life Reviews* 3, 229–261.

Danielsen, V. and Vestergaard, E.-M. (2001) Dietary fibre for pregnant sows: effect on performance and behaviour. *Animal Feed Science and Technology* 90, 71–80.

Dawson, T.J. and Hulbert, A.J. (1970) Standard metabolism, body temperature, and surface areas of Australian marsupials. *American Journal of Physiology* 218, 1233–1238.

de Almeida, F.Q., Valadares, S.D.C., Cecon, P.R., Leao, M.I., Donzele, J.L., de Silva, J.F.C. and de Queiroz, A.C. (1998) Prececal, postileal, fecal, and urinary endogenous nitrogen compounds in equines. *Revista Brasileira de Zootecnia* 27, 538–545.

Dodds, P.S., Rothman, D.H. and Weitz, J.S. (2001) Re-examination of the '3/4-law' of metabolism. *Journal of Theoretical Biology* 209, 9–27.

Doherty, F.J. and Mayer, R.J. (1992) *Intracellular Protein Degradation*. IRL Press/Oxford University Press, Oxford, UK.

Donnelly, P.E. and Hutton, J.B. (1976) Effects of dietary protein and energy on the growth of Friesian bull calves. I. Food uptake, growth and protein requirements. *New Zealand Journal of Agricultural Research* 19, 289–297.

Dourmad, J.Y. (1988) Voluntary feed intake in lactating sows: numerous factors of variation. *INRA Productions Animales* 1, 141–146.

Dryden, G.McL. (1981) Simulation of growth rate in grazing sheep – a comparison of the NRC and ARC feeding standards. *Agricultural Systems* 7, 189–197.

Dryden, G.McL. (1982) Endogenous nitrogen in ruminant faeces. *Proceedings of the Nutrition Society of Australia* 7, 132–135.

Dryden, G.McL. (1995) ME requirements for Australian beef cattle. In: Sauvant, D. (ed.) *Modelling Nutrient Responses of Herbivores*. INRA, Paris.

Dryden, G.McL. (2006) Investigating daily changes in food intake by ruminants. In: Kebreab, E., Dijkstra, J., Bannink, A., Gerrits, W.J.J. and France, J. (eds) *Nutrient Digestion and Utilization in Farm Animals: Modelling Approaches*, CAB International, Wallingford, UK, pp. 314–327.

Dryden, G.McL., Hmeidan, M.C., Puttoo, K. and Yape Kii, W. (2002) Energy, protein and water requirements of rusa (*Cervus timorensis*) stags. *Proceedings of the 5th International Deer Biology Congress*, Quebec City, Canada, August, 2002, 49 (Abstr).

Duggleby, S.L. and Waterlow, J.C. (2005) The end-product method of measuring whole-body protein turnover: a review of published results and a comparison with those obtained by leucine infusion. *British Journal of Nutrition* 94, 141–153.

Dunshea, F.R., King, R.H., Campbell, R.G., Sainz, R.D. and Kim, Y.S. (1993) Interrelationships between sex and ractopamine on protein and lipid deposition in rapidly growing pigs. *Journal of Animal Science* 71, 2919–2930.

Ekern, A. (1991) A new system of energy evaluation of food for ruminants. *Norsk Landbruksforskning* 5, 273–277.

Erdman, R.A., Proctor, G.H. and Vandersall, J.H. (1986) Effect of rumen ammonia concentration on *in situ* rate and extent of digestion of feedstuffs. *Journal of Dairy Science* 69, 2312–2320.

Ettle, T., Roth-Maier, D.A. and Roth, F.X. (2003) Effect of apparent ileal digestible lysine to energy ratio on performance of finishing pigs at different dietary metabolisable energy levels. *Journal of Animal Physiology and Animal Nutrition* 87, 269–279.

Everts, H. and Dekker, R.A. (1991) Effect of protein/lysine supply and of parity number on energy metabolism during pregnancy and lactation in sows. In: Wenk, C. and Boessinger, M. (eds) *Energy Metabolism of Farm Animals.* Proceedings 12th Symposium, Zurich, Switzerland. 1–7 September, 1991. EAAP Publication No. 58, Institut fur Nutztierwissenschaften, Zurich, pp. 317–320.

Ferguson, N.S. and Gous, R.M. (1997) The influence of heat production on voluntary food intake in growing pigs given protein-deficient diets. *Animal Science* 64, 365–378.

Figueroa, J.L., Lewis, A.J., Miller, P.S., Fischer, R.L., Gómez, R.S. and Diedrichsen, R.M. (2002) Nitrogen metabolism and growth performance of gilts fed standard corn-soybean meal diets or low-crude protein, amino acid-supplemented diets. *Journal of Animal Science* 80, 2911–2919.

Flatt, W.P., Coppock, C.E. and Moore, L.A. (1965) Energy balance studies with lactating, non-pregnant dairy cows consuming rations with varying hay to grain ratios. In: Blaxter, K.L. (ed.) *Energy Metabolism.* EAAP Publication No. 11. Academic Press, London, pp. 121–130.

Fox, D.G., Sniffen, C.J., O'Connor, J.D., Russell, J.B. and Van Soest, P.J. (1992) A net carbohydrate and protein system for evaluating cattle diets: 3. Cattle requirements and diet adequacy. *Journal of Animal Science* 70, 3578–3596.

Fox, D.G., Barry, M.C., Pitt, R.E., Roseler, D.K. and Stone, W.C. (1995) Application of the Cornell net carbohydrate and protein model for cattle consuming forages. *Journal of Animal Science* 73, 267–277.

Freer, M., Moore, A.D. and Donnelly, J.R. (1997) GRAZPLAN: decision support systems for Australian grazing enterprises: II. The animal biology model for feed intake, production and reproduction and the GrazFeed DSS. *Agricultural Systems* 54, 77–126.

Fuller, M.F., Livingstone, R.M., Baird, B.A. and Atkinson, T. (1979) The optimal amino acid supplementation of barley for the growing pig. 1. Response of nitrogen metabolism to progressive supplementation. *British Journal of Nutrition* 41, 321–331.

Fuller, M.F., McWilliam, R., Wang, T.C. and Giles, L.R. (1989) The optimum dietary amino acid pattern for growing pigs. 2. Requirements for maintenance and for tissue protein accretion. *British Journal of Nutrition* 62, 255–267.

Gaines, W.L. and Davidson, F.A. (1923) *Bulletin of the Illinois Agricultural Experiment Station,* No. 245. Urbana, Illinois.

Gessaman, J.A. and Nagy, K.A. (1988) Energy metabolism: errors in gas-exchange conversion factors. *Physiological Zoology* 61, 507–513.

Geuyen, T.P.A., Verhagen, J.M.F. and Verstegen, M.W.A. (1984) Effect of housing and temperature on metabolic rate of pregnant sows. *Animal Production* 38, 477–485.

Giles, L.R., Batterham, E.S. and Dettmann, E.B. (1986) Amino acid and energy interactions in growing pigs. 2. Effects of food intake, sex and live weight on responses to lysine concentration in barley-based diets. *Animal Production* 42, 133–144.

Gooden, J.M., Huang, M.D., McCredie, F.C., Sommer, J.L. and Annison, E.F. (1991) Continuous measurement of changes in energy expenditure in sheep. In: Wenk, C. and Boessinger, M. (eds) *Energy Metabolism of Farm Animals.* Proceedings 12th Symposium, Switzerland. 1–7 September, 1991. EAAP Publication No. 58, Institut fur Nutztierwissenschaften, Zurich, pp. 24–27.

Gorman, M.L., Mills, M.G., Raath, J.P. and Speakman, J.R. (1998) High hunting costs make African wild dogs vulnerable to kleptoparasitism by hyaenas. *Nature* 391, 479–481.

Grainger, C., Clarke, T., McGinn, S.M., Auldist, M.J., Beauchemin, K.A., Hannah, M.C., Waghorn, G.C., Clark, H. and Eckard, R.J. (2007) Methane emissions from dairy cows measured using the sulfur hexafluoride (SF_6) tracer and chamber techniques. *Journal of Dairy Science* 90, 2755–2766.

Green, D.M. and Whittemore, C.T. (2005) Calibration and sensitivity analysis of a model of the growing pig for weight gain and composition. *Agricultural Systems* 84, 279–295.

Guerouali, A., Bouayad, H. and Taouil, M. (2003) Estimation of energy expenditures in horses and donkeys at rest and when carrying a load. In: Pearson, R.A., Lhoste, P., Saatamoinen, M. and Martin-Rosset, W. (eds) *Working Animals in Agriculture and Transport. A Collection of Some Current Research and Development Observations.* Wageningen Academic Publishers, Wageningen, The Netherlands, pp. 75–78.

Hardy, M.H. and Lyne, A.G. (1956) The prenatal development of wool follicles in Merino sheep. *Australian Journal of Biological Sciences* 9, 423–441.

Harris, P. (1997) Energy sources and requirements of the exercising horse. *Annual Reviews of Nutrition* 17, 185–210.

Hay, W.W. Jr (2006) Recent observations on the regulation of fetal metabolism by glucose. *Journal of Physiology* 572, 17–24.

Hershko, A., Ciechanover, A. and Varshavsky, A. (2000) The ubiquitin system. *Nature Medicine* 6, 1073–1081.

Hill, G.M., Baido, S.K., Cromwell, G.L., Mahan, D.C., Nelssen, J.L. and Stein, H.H. (2007) Evaluation of sex and lysine during the nursery period. *Journal of Animal Science* 85, 1453–1458.

Hinchcliff, K.W., Reinhart, G.A., Burr, J.R., Schreier, C.J. and Swenson, R.A. (1997) Metabolizable energy intake and sustained energy expenditure of Alaskan sled dogs during heavy exertion in the cold. *American Journal of Veterinary Research* 58, 1457–1462.

Hintz, H.F. (1982) Growth rate of horses. In: *Equine Nutrition*, Proceedings of the Fourth Bain-Fallon Memorial Symposium. Australian Equine Veterinary Association, Sydney, Australia, pp. 97–103.

Holmes, C.W. and Wilson, G.F. (1984) *Milk Production from Pasture*. Butterworths of New Zealand, Wellington, New Zealand.

Hulme, D.J., Kellaway, R.C., Booth, P.J. and Bennett, L. (1986) The CAMDAIRY model for formulating and analysing dairy cow rations. *Agricultural Systems* 22, 81–108.

Hume, I.D. (1999) *Marsupial Nutrition*. Cambridge University Press, Cambridge, UK.

Jarrige, R. (ed.) (1989) *Ruminant Nutrition. Recommended Allowances and Feed Tables*. INRA, John Libbey Eurotext, Paris.

Jones, J.H. and Carlson, G. (1995) Estimation of metabolic energy cost and heat production during a 3-day-event. *Equine Veterinary Journal, Suppl.* 20, 23–31.

Julliand, V. and Martin-Rosset, W. (eds) (2004) *Nutrition of the Performance Horse*. EAAP Publication No. 111. Wageningen Academic Publishers, Wageningen, The Netherlands.

Kerr, J.C. and Cameron, N.D. (1996) Responses in gilt post-farrowing traits and pre-weaning piglet growth to divergent selection for components of efficient lean growth rate. *Animal Science* 63, 523–531.

King, R.H. (1990) Feeding breeding stock. In: Gardner, J.A.A., Dunkin, A.C. and Lloyd, L.C. (eds) *Pig Production in Australia*. Butterworths, Sydney, Australia, pp. 72–77.

Kleiber, M. (1932) Body size and metabolism. *Hilgardia* 6, 315–353.

Kleiber, M. (1947) Body size and metabolic rate. *Physiological Reviews* 27, 511–541.

Kleiber, M. (1975) *Fire of Life: An Introduction to Animal Energetics*, rev. edn. Robert E. Krieger, New York.

Kleiber, M. and Flatt, W.P. (1965) General discussion. In: Blaxter, K.L. (ed.) *Energy Metabolism*. EAAP Publication No. 11. Academic Press, London, pp. 432–435.

Kolade, M.M. and Ternouth, J.H. (1996) Effect of dietary protein levels on microbial nitrogen flow from the rumen. *Animal Production in Australia* 21, 448.

Kracht, W., Hennig, A. and Gruhn, K. (1976) Protein utilization of mixed feed rations in lactating pigs with reference to the essential amino acid content of the feed proteins. 2. Report. Utilization of the feed proteins in the use of soy bean extraction residue, waste liquor yeast, horse bean meal, fish meal, and maize gluten for a basic ration. *Archive der Tierernahrung* 26, 267–274.

Kyriazakis, I. and Emmans, G.C. (1992) The effects of varying protein and energy intakes on the growth and body-composition of pigs. 2. The effects of varying both energy and protein-intake. *British Journal of Nutrition* 68, 615–625.

Lachica, M. and Aguilera, J.F. (2003) Estimation of energy needs in the free-ranging goat with particular reference to the assessment of its energy expenditure by the ^{13}C-bicarbonate method. *Small Ruminant Research* 49, 303–318.

Lachica, M. and Aguilera, J.F. (2005) Energy expenditure of walk in grassland for small ruminants. *Small Ruminant Research* 59, 105–121.

Lawrence, A.St.C., Lawrence, L.M. and Coleman, R.J. (2001) Using an empirical equation to predict voluntary intake of grass hays by mature equids. *Proceedings of the 17th Equine Nutrition and Physiology Symposium*, Kentucky, pp. 99–100.

Lofgreen, G.P. (1965) A comparative slaughter technique for determining net energy values with beef cattle. In: Blaxter, K.L. (ed.) *Energy Metabolism*. EAAP Publication No.11. Academic Press, London, pp. 309–317.

Longland, A.C., Close, W.H., Sharpe, C.E. and Low, A.G. (1991) The efficiency of energy utilization by pigs fed diets containing varying proportions of non-starch polysaccharides. In: Wenk, C. and Boessinger, M. (eds) *Energy Metabolism of Farm Animals*. Proceedings 12th Symposium, Switzerland. 1–7 September, 1991. EAAP Publication No. 58, Institut fur Nutztierwissenschaften, Zurich, pp. 154–157.

Lopez, J., Jesse, G.W., Becker, B.A. and Ellersieck, M.R. (1991) Effects of temperature on the performance of finishing swine: I. Effects of a hot, diurnal temperature on average daily gain, feed intake, and feed efficiency. *Journal of Animal Science* 69, 1843–1849.

Lovegrove, B.G. (2000) The zoogeography of mammalian basal metabolic rate. *American Naturalist* 156, 201–220.

Luo, J., Goetsch, A.L., Moore, J.E., Johnson, Z.B., Sahlu, T., Ferrell, C.L., Galyean, M.L. and Owens, F.N. (2004) Prediction of endogenous urinary nitrogen of goats. *Small Ruminant Research* 53, 293–308.

McClymont, G.L. (1952) Specific dynamic actions of acetic acid and heat increment of feeding in ruminants. *Australian Journal of Scientific Research, Series B* 5, 374–383.

McNab, B.K. (1992) Rate of metabolism in the termite-eating sloth bear (*Ursus ursinus*). *Journal of Mammallogy* 73, 168–172.

Majumdar, B.N. (1960) Studies on goat nutrition. Part I. Minimum protein requirement of goats for maintenance – endogenous urinary nitrogen and metabolic faecal nitrogen excretion studies. *Journal of Agricultural Science, Cambridge* 54, 329–334.

Martin-Rosset, W. (ed.) (1990) *L'Alimentation des Chevaux*. INRA Publications, Versailles, France.

Martin-Rosset, W., Vermorel, M., Doreau, M., Tisserand, J.L. and Andrieu, J. (1994) The French horse feed evaluation systems and recommended allowances for energy and protein. *Livestock Production Science* 40, 37–56.

Mason, V.C. (1969) Some observations on the distribution and origin of nitrogen in sheep faeces. *Journal of Agricultural Science, Cambridge* 73, 99–111.

Matarese, L.E. (1997) Indirect calorimetry technical aspects. *Journal of the American Dietetic Association* 97 Suppl.1, S154–S160.

Matsui, A., Ohmura, H., Asai, Y., Takahashi, T., Hiraga, A., Okamura, K., Tokimura, H., Sugino, T., Obitsu, T. and Taniguchi, K. (2006) Effect of amino acid and glucose administration following exercise on the turnover of muscle protein in the hindlimb femoral region of thoroughbreds. *Equine Veterinary Journal* 36 Suppl., 611–616.

Meyer, H. (1985) Investigations to determine endogenous faecal and renal N losses in horses. In: *Proceedings of the 9th Equine Nutrition and Physiology Society Symposium*, East Lansing, Michigan, p. 68.

Ministry of Agriculture Fisheries and Food (MAFF) (1984) *Energy Allowances and Feeding Systems for Ruminants*, 2nd edn. HMSO, London.

Minson, D.J. (1990) *Forage in Ruminant Nutrition*. Academic Press, San Diego, California.

Möhn, S., Gillis, A.M., Moughan, P.J. and de Lange, C.F.M. (2000) Influence of dietary lysine and energy intakes on body protein deposition and lysine utilization in the growing pig. *Journal of Animal Science* 78, 1510–1519.

Moughan, P.J. (2003) Amino acid availability: aspects of chemical analysis and bioassay methodology. *Nutrition Research Reviews* 16, 127–141.

Moughan, P.J., Jacobson, L.H. and Morel, P.C.H. (2006) A genetic upper limit to whole-body protein deposition in a strain of growing pigs. *Journal of Animal Science* 84, 3301–3309.

Nagy, K.A. (1987) Field metabolic rate and food requirement scaling in mammals and birds. *Ecological Monographs* 57, 111–128.

National Research Council (NRC) (1981) *Effect of Environment on Nutrient Requirements of Domestic Animals*. National Academy Press, Washington, DC.

National Research Council (NRC) (1987) *Predicting Feed Intake of Food-Producing Animals*. National Academy Press, Washington, DC.

National Research Council (NRC) (1988) *Nutrient Requirements of Swine*, 9th rev. edn. National Academy Press, Washington, DC.

National Research Council (NRC) (1989) *Nutrient Requirements of Horses*, 5th rev. edn. National Academy Press, Washington, DC.

National Research Council (NRC) (1998) *Nutrient Requirements of Swine*, 10th rev. edn. National Academy Press, Washington, DC.

National Research Council (NRC) (2000) *Nutrient Requirements of Beef Cattle*, 7th rev. edn. update 2000. National Academy Press, Washington, DC.

National Research Council (NRC) (2001) *Nutrient Requirements of Dairy Cattle*, 7th rev. edn. National Academy Press, Washington, DC.

National Research Council (NRC) (2007) *Nutrient Requirements of Horses*, 6th rev. edn. National Academy Press, Washington, DC.

Noblet, J. and Etienne, M. (1987) Metabolic utilization of energy and maintenance requirements in lactating sows. *Journal of Animal Science* 64, 774–781.

Noblet, J. and Etienne, M. (1989) Estimation of sow milk nutrient output. *Journal of Animal Science* 67, 3352–3359.

Noblet, J., Henry, Y. and Dubois, S. (1987) Effect of protein and lysine levels in the diet on body gain composition and energy utilization in growing pigs. *Journal of Animal Science* 65, 717–726.

Noblet, J., Dourmad, J.Y. and Etienne, M. (1990) Energy utilization in pregnant and lactating sows: modeling of energy requirements. *Journal of Animal Science* 68, 562–572.

Noblet, J., Shi, X.S. and Dubois, S. (1993a) Energy cost of standing activity in sows. *Livestock Production Science* 34, 127–136.

Noblet, J., Shi, X.S. and Dubois, S. (1993b) Metabolic utilization of dietary energy and nutrients for maintenance energy requirements in sows: basis for a net energy system. *British Journal of Nutrition* 70, 407–419.

Oldham, J.D. (1987) Efficiencies of amino acid utilisation. In: Jarrige, R. and Alderman, G. (eds) *Feed Evaluation and Protein Requirement Systems for Ruminants*. CEC, Luxembourg, pp. 171–186.

Ørskov, E.R. and McDonald, I. (1979) The estimation of protein degradability in the rumen from incubation measurements weighted according to rate of passage. *Journal of Agricultural Science, Cambridge* 92, 499–503.

Ørskov, E.R. and MacLeod, N.A. (1990) Dietary-induced thermogenesis and feed evaluation in ruminants. *Proceedings of the Nutrition Society* 49, 227–237.

Ørskov, E.R., MacLeod, N.A. and Nakashima, Y. (1991) Effect of different volatile fatty acids mixtures on energy metabolism in cattle. *Journal of Animal Science* 69, 3389–3397.

Osuji, P.O. (1974) The physiology of eating and the energy expenditure of the ruminant at pasture. *Journal of Rangeland Management* 27, 437–443.

Osuji, P.O., Gordon, J.G. and Webster, A.J.F. (1975) Energy exchanges associated with eating and rumination in sheep given grass diets of different physical forms. *British Journal of Nutrition* 34, 59–71.

Ott, E.A. (2005) Influence of temperature stress on the energy and protein metabolism and requirements of the working horse. *Livestock Production Science* 92, 123–130.

Pagan, J.D. and Hintz, H.F. (1986a) Equine energetics. I. Relationship between body weight and energy requirements in horses. *Journal of Animal Science* 63, 815–821.

Pagan, J.D. and Hintz, H.F. (1986b) Equine energetics. II. Energy expenditure in horses during submaximal exercise. *Journal of Animal Science* 63, 822–830.

Painter, P.R. (2005) Data from necropsy studies and *in vitro* tissue studies lead to a model for allometric scaling of basal metabolic rate. *Theoretical Biology and Medical Modelling* 2, 39.

Perez, R., Valenzuela, S., Merino, V., Cabezas, I., Garcia, M., Bou, R. and Ortiz, P. (1996) Energetic requirements and physiological adaptation of draught horses to ploughing work. *Animal Science* 63, 343–351.

Perrin, D.R. (1958) The calorific value of milks of different species. *Journal of Dairy Research* 25, 215–220.

Pittroff, W. and Kothmann, M.M. (2001a) Quantitative prediction of feed intake in ruminants. I. Conceptual and mathematical analysis of models for sheep. *Livestock Production Science* 71, 131–150.

Pittroff, W. and Kothmann, M.M. (2001b) Quantitative prediction of feed intake in ruminants. II. Conceptual and mathematical analysis of models for cattle. *Livestock Production Science* 71, 151–169.

Pittroff, W. and Kothmann, M.M. (2001c) Quantitative prediction of feed intake in ruminants. III. Comparative example calculations and discussion. *Livestock Production Science* 71, 171–181.

Pisulewski, P.M., Rulquin, H., Peyraud, J.L. and Verite, R. (1996) Lactational and systemic responses of dairy cows to postruminal infusions of increasing amounts of methionine. *Journal of Dairy Science* 79, 1781–1791.

Potter, G.D. (2004) Protein requirements of horses for maintenance and work. In: Julliand, V. and Martin-Rosset, W. (eds) *Nutrition of the Performance Horse*. EAAP Publication No. 111. Wageningen Academic Publishers, Wageningen, The Netherlands, pp. 149–156.

Quiniou, N., Noblet, J., van Milgen, J. and Dourmad, J.-Y. (1995) Effect of energy intake on performance, nutrient and tissue gain and protein and energy utilization in growing boars. *Animal Science* 61, 133–143.

Rao, D.S. and McCracken, K.J. (1990) Protein requirements of boars of high genetic potential for lean growth. *Animal Production* 51, 179–187.

Ratner, S., Rittenberg, D., Keston, A.S. and Schoenheimer, R. (1940) Studies in protein metabolism. XIV. The chem-

ical interaction of dietary glycine and body proteins in rats. *Journal of Biological Chemistry* 134, 665–676.

Richert, B.T., Hancock, J.D., and Morrill, J.L. (1994) Effects of replacing milk and soybean products with wheat glutens on digestibility of nutrients and growth performance in nursery pigs. *Journal of Animal Science* 72, 151–159.

Robinson, J.J., Sinclair, K.D., Randel, R.D. and Sykes, A.R. (1999) In: Jung, H.-J.G. and Fahey, G.C. (eds) *Nutritional Ecology of Herbivores*. ASAS, Savoy, Illinois, pp. 550–608.

Rolfe, D.F.S. and Brown, G.C. (1997) Cellular energy utilization and molecular origin of standard metabolic rate in mammals. *Physiological Reviews* 77, 731–758.

Rubner, M. (1883) Uber den Einfluss der Korpergrosse auf Stoff- und Kraftwechsel. *Zeitschrift für Biologie* 19, 535–562.

Rubner, M. (1894) Die Quelle der Thierischen Wärme. *Zeitschrift für Biologie* 30, 73–142.

Ruckebusch, Y. (1975) Motility of the ruminant stomach associated with states of sleep. In: McDonald, I.W. and Warner, A.C.I. (eds) *Digestion and Metabolism in the Ruminant*. UNE Publishing Unit, Armidale, New South Wales, Australia, pp. 77–90.

Rulquin, H., Verite, R. and Guinard-Flament, J. (2001) Acides aminés digestibles dans l'intestin. Le système AADI et les recommandations d'apport pour la vache laitière. *INRA Productions Animales* 14, 265–274.

Rundgren, M. (1988) Evaluation of triticale given to pigs, poultry and rats. *Animal Feed Science and Technology* 19, 359–375.

Sandberg, F.B., Emmans, G.C. and Kyriazakis, I. (2005a) Partitioning of limiting protein and energy in the growing pig: description of the problem, possible rules and their qualitative evaluation. *British Journal of Nutrition* 93, 205–212.

Sandberg, F.B., Emmans, G.C. and Kyriazakis, I. (2005b) Partitioning of limiting protein and energy in the growing pig: testing quantitative rules against experimental data. *British Journal of Nutrition* 93, 213–224.

Satter, L.C. and Slyter, L.L. (1974) Effect of ammonia concentration on rumen microbial protein production *in vitro*. *British Journal of Nutrition* 32, 199–208.

Scantlebury, M., Butterwick, R. and Speakman, J.R. (2001) Energetics and litter size variation in domestic dog *Canis familiaris* breeds of two sizes. *Comparative Biochemistry and Physiology – Part A* 129, 919–931.

Schiemann, R., Nehring, K., Hoffmann, L., Jetsch, W. and Chudy, A. (1971) *Energetische Futterbewertung und Energienormen*. Deutscher Landwirtschaftsverlag, Berlin.

Seeherman, H.J., Taylor, C.R., Maloiy, G.M.O. and Armstrong, R.B. (1981) Design of the mammalian respiratory system. II. Measuring maximum aerobic capacity. *Respiration Physiology* 44, 11–23.

Silanikove, N. (2000) Effects of heat stress on the welfare of extensively managed domestic ruminants. *Livestock Production Science* 67, 1–18.

Smith, W.C., Ellis, M., Chadwick, J.P. and Laird, R. (1991) The influence of index selection for improved growth and carcass characteristics on appetite in a population of Large White pigs. *Animal Production* 52, 193–199.

Standing Committee on Agriculture (SCA) (1987) *Feeding Standards for Australian Livestock. Pigs.* CSIRO, Canberra.

Standing Committee on Agriculture (SCA) (1990) *Feeding Standards for Australian Livestock. Ruminants.* CSIRO, Canberra.

Susenbeth, A. (1995) Factors affecting lysine utilization in growing pigs: an analysis of literature data. *Livestock Production Science* 43, 193–204.

Susenbeth, A. and Menke, K.H. (1991) Energy requirement for physical activity in pigs. In: Wenk, C. and Boessinger, M. (eds) *Energy Metabolism of Farm Animals.* Proceedings 12th Symposium, Switzerland. 1–7 September, 1991. EAAP Publication No. 58, Institut fur Nutztierwissenschaften, Zurich, pp. 416–419.

Sutton, J.D. and Alderman, G. (2000) The energy and protein requirements of pregnant and lactating dairy goats. The Agriculture and Food Research Council report. *Livestock Production Science* 64, 3–8.

Szabó, C., Jansman, A.J.M., Babinszky, L., Kanis, E. and Verstegen, M.W.A. (2001) Effect of dietary protein source and lysine:DE ratio on growth performance, meat quality, and body composition of growing–finishing pigs. *Journal of Animal Science* 79, 2857–2865.

Tamminga, S., Van Straalen, W.M., Subnel, A.P.J., Meijer, R.G.M., Steg, A., Wever, C.J.G. and Blok, M.C. (1994) The Dutch protein evaluation system: the DVE/OEB-system. *Livestock Production Science* 40, 139–155.

Tess, M.W., Dickerson, G.E., Nienaber, J.A. and Ferrell, C.L. (1984a) The effects of body composition on fasting heat production in pigs. *Journal of Animal Science* 58, 99–110.

Tess, M.W., Dickerson, G.E., Nienaber, J.A., Yen, J.T. and Ferrell, C.L. (1984b) Energy costs of protein and fat deposition in pigs fed ad libitum. *Journal of Animal Science* 58, 111–122.

Thaker, M.Y.C. and Bilkei, G. (2005) Lactation weight loss influences subsequent reproductive performance of sows. *Animal Reproduction Science* 88, 309–318.

Thomas, C. (2004) *Feed Into Milk: A New Applied Feeding System for Dairy Cows.* Nottingham University Press, Nottingham, UK.

Thompson, D.J. (1965) Energy retention in lambs as measured by the comparative slaughter technique. In: Blaxter, K.L. (ed.) *Energy Metabolism.* Academic Press, London, pp. 319–326.

Toutain, P.-L., Toutain, C., Webster, A.J.F. and McDonald, J.D. (1977) Sleep and activity, age and fatness, and the energy expenditure of confined sheep. *British Journal of Nutrition* 38, 445–453.

Tullis, J.B., Whittemore, C.T. and Phillips, P. (1986) Compensatory nitrogen retention in growing pigs following a period of N deprivation. *British Journal of Nutrition* 56, 259–261.

Tyrell, H.F. and Reid, J.T. (1965) Prediction of the energy value of cow's milk. *Journal of Dairy Science* 48, 1215–1223.

Van Es, A.J.H. (1975) Feed evaluation for dairy cows. *Livestock Production Science* 2, 95–107.

Van Milgen, J., Bernier, J.F., Lecozler, Y., Dubois, S. and Noblet, J. (1998) Major determinants of fasting heat production and energetic cost of activity in growing pigs of different body weight and breed/castration combination. *British Journal of Nutrition* 79, 509–517.

Van Milgen, J. and Noblet, J. (2000) Modelling energy expenditure in pigs. In: McNamara, J.P., France, J. and Beever, D.E. (eds) *Modelling Nutrient Utilization in Farm Animals.* CAB International, Wallingford, UK, pp. 103–114.

Vercoe, J.E. (1973) The energy cost of standing and lying in adult cattle. *British Journal of Nutrition* 30, 207–210.

Vermorel, M. and Martin-Rosset, W. (1997) Concepts, scientific bases, structure and validation of the French horse net energy system (UFC). *Livestock Production Science* 47, 261–275.

Vermorel, M., Martin-Rosset, W. and Vernet, J. (1997) Energy utilization of twelve forages or mixed diets for maintenance by sport horses. *Livestock Production Science* 47, 157–167.

Vermorel, M. and Mormède, P. (1991) Energy cost of eating in ponies. In: Wenk, C. and Boessinger, M. (eds) *Energy Metabolism of Farm Animals.* Proceedings 12th Symposium, Switzerland. 1–7 September, 1991. EAAP Publication No. 58, Institut fur Nutztierwissenschaften, Zurich, pp. 437–440.

Verstegen, M.W.A., Verhagen, J.M.F. and den Hartog, L.A. (1987) Energy requirements of pigs during pregnancy: a review. *Livestock Production Science* 16, 75–89.

Voit, E. (1901) Über die Grösse des Energiebedarfs der Tiere in Hungerzustande. *Zeitschrift für Biologie* 41, 113–154.

Walach-Janiak St Raj, M. and Fandrejewski, H. (1986a) Protein and energy balance in pregnant gilts. *Livestock Production Science* 15, 249–260.

Walach-Janiak St Raj, M. and Fandrejewski, H. (1986b) The effect of pregnancy on protein, water and fat deposition in the body of gilts. *Livestock Production Science* 15, 261–269.

Walters, L.M., Ogilvie, G.K., Salman, M.D., Joy, L., Fettman, M.J., Hand, M.S. and Wheeler, S.L. (1993) Repeatability of energy expenditure measures in clinically normal dogs by use of indirect calorimetry. *American Journal of Veterinary Research* 54, 1881–1885.

Wang, T.C. and Fuller, M.F. (1989) The optimum dietary amino acid pattern for growing pigs. 1. Experiments by amino acid deletion. *British Journal of Nutrition* 62, 77–89.

Warnants, N., Van Oeckel, M.J. and De Paepe, M. (2003) Response of growing pigs to different levels of ileal standardised digestible lysine using diets balanced in threonine, methionine and tryptophan. *Livestock Production Science* 82, 201–209.

Webb, P. (1991) The measurement of energy expenditure. *Journal of Nutrition* 121, 1897–1901.

Weibel, E.R., Bacigalupe, L.D., Schmitt, B. and Hoppeler, H. (2004) Allometric scaling of maximal metabolic rate in mammals: muscle aerobic capacity as determinant factor. *Respiratory Physiology & Neurobiology* 140, 115–132.

West, G.B., Brown, J.H. and Enquist, B.J. (1999) The fourth dimension of life: fractal geometry and allometric scaling of organisms. *Science* 284, 1677–1679.

White, C.R. and Seymour, R.S. (2003) Mammalian basal metabolic rate is proportional to body mass$^{2/3}$. *Proceedings of the National Academy of Sciences* 100, 4046–4049.

Whittemore, C.T. (1983) Development of recommended energy and protein allowances for young pigs. *Agricultural Systems* 11, 159–186.

Whittemore, C.T., Tullis, J.B. and Emmans, G.C. (1988) Protein growth in pigs. *Animal Production* 46, 437–445.

Whittemore, C.T., Green, D.M. and Knap, P.W. (2001a) Technical review of the energy and protein requirements of growing pigs: food intake. *Animal Science* 73, 3–17.

Whittemore, C.T., Green, D.M. and Knap, P.W. (2001b) Technical review of the energy and protein requirements of growing pigs: energy. *Animal Science* 73, 199–215.

Whittemore, C.T., Green, D.M. and Knap, P.W. (2001c) Technical review of the energy and protein requirements of growing pigs: protein. *Animal Science* 73, 363–373.

Wood, P.D.P. (1979) A simple model of lactation curves for milk yield, food requirement and body weight. *Animal Production* 28, 55–63.

Yan, T., Agnew, R.E., Murphy, J.J., Ferris, C.P. and Gordon, F.J. (2003) Evaluation of different energy feeding systems with production data from lactating dairy cows offered grass silage-based diets. *Journal of Dairy Science* 86, 1415–1428.

13 Ration Formulation

Ration formulation is a mathematical technique for devising a mixture of food ingredients which, when eaten in the expected amounts, will provide the animal with specified amounts of nutrients. Ration formulation allows the animal manager to:

1. Prepare a ration with the required nutrient content characteristics at least cost or optimized in some other way, e.g. maximum profit or minimum pollution.
2. Plan ingredient buying strategies based on 'price for inclusion' data provided for those ingredients which were excluded from the formulation.
3. Impose constraints on the use of ingredients, either by 'forcing in' nominated ingredients, i.e. requiring the formulation to contain certain ingredients at specified levels or by preventing the use of an ingredient at greater than, or less than, a specified level.
4. Specify that a particular nutrient is used in a set ratio to another nutrient, or to set upper or lower limits on the concentration of a nutrient.
5. Optimize the allocation of scarce ingredients across a number of rations; this is called 'multiblending'.
6. In some cases, use a sub-programme to calculate nutrient requirements from a description of the animal and its environment; this removes some of the imprecision inherent in using data from published feeding standards tables.

Most ration formulation programmes use a linear programming algorithm (Dent and Casey, 1967), a mathematical routine to calculate the proportions of ingredients which meet specified nutrient requirements. These formulations usually have a 'least-cost' solution. A least-cost ration is a mixture of ingredients which meet all the specified nutrient concentration requirements at the lowest possible cost. Linear programming is commonly used in ration formulation because, as pointed out by Lara and Stancu-Minasian (1999), it can readily be related to real problems, there is a practical algorithm available to solve the linear programming problem (the Simplex

algorithm; Greenwald, 1957) and there is now extensive practical experience that this approach works.

Basic Concepts

Methods of expressing food nutrient contents and animal nutrient requirements

All foods (even those which are 'air-dry') contain water. The nutrients other than water, i.e. amino acids, fatty acids, minerals, vitamins and energy, are found within the food dry matter (DM). Accordingly, it is possible to express nutrient contents (other than water) as either the concentration in the DM or the concentration in the whole food (DM plus water), as in Fig. 13.1. Changing the basis of expression can have a large effect on the numerical value, particularly if the food has a high water content. For example, the Ca content of maize silage (NRC, 2000) is 0.25% on a DM basis, but 0.09% on an as-fed basis.

Food nutrient contents can be converted between DM and as-fed bases by multiplying the value by the appropriate ratio of the food DM content. If the DM content is given as a percentage, then:

$$\Phi \ (g/kg \ DM) = \Phi \ (g/kg, \ as\text{-}fed \ basis) \times (100/DM\%) \tag{13.1}$$

$$\Phi \ (g/kg, \ as\text{-}fed \ basis) = \Phi \ (g/kg \ DM) \times (DM\%/100) \tag{13.2}$$

where
Φ = nutrient content
DM% = dry matter content (%)

Nutrient requirements can be expressed as either amounts per day or concentrations in the food. The most mathematically efficient ration formulation methods use nutrient requirements expressed as concentrations. This introduces a problem because the most accurate way of stating a nutrient requirement is to give the amount required each day. To convert this to a concentration means that we have to introduce an estimate of voluntary food consumption:

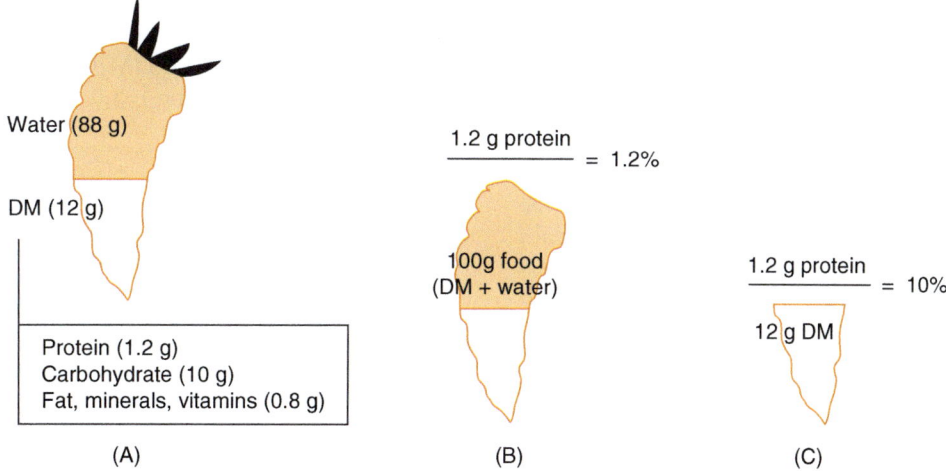

Fig. 13.1. Two ways of expressing the protein content of a carrot: (A) the proximate composition of the carrot; (B) protein content calculated on a whole-food or as-fed basis; (C) protein content calculated on a DM basis.

$$\Phi = \frac{\text{amount of nutrient required (g/day)}}{\text{amount of food DM eaten (kg/day)}} \quad (13.3)$$

where

Φ = required nutrient content (g/kg DM)

Estimates of expected food intake are not very accurate (Pittroff and Kothmann, 2001a,b), so feeding standards expressed in concentration units are less accurate than those expressed as daily amounts. Nevertheless, concentration units must be used because the nutrient contents and requirements must be given in the same units and there is no option but to use concentrations to express the nutrient content of foods. Further, both nutrient requirements and nutrient contents must be on the same basis, i.e. DM or as-fed.

Calculating the nutrient content of a ration

The general formula which describes the content of a particular nutrient (Φ) in a mixture is:

$$\Phi_n = \sum_{k}^{n=1} (\Phi_n \times \text{proportion}_n) \quad (13.4)$$

where

Φ_n = content (DM basis) of Φ in the nth ingredient ($n = 1$ to k)

proportion$_n$ = proportion (decimal) of the nth ingredient in the ration ($n = 1$ to k)

k = number of ingredients available to be included in the ration

Example

We can use this formula to calculate the metabolizable energy (ME) content of a mixture of 15 kg of fresh rhodes grass, 1 kg of rice bran and 0.1 kg of salt (using the nutrient content data, DM basis, suggested by NRC, 2000). The amount of grass DM is $(15 \times 23\%) = 3.45$ kg (the DM content of rhodes grass is 23%). The DM supplied by the other two ingredients is calculated similarly, giving a total DM of $(3.45 + 0.91 + 0.1) = 4.46$ kg. Thus, we have $(3.45/4.46 \times 100) = 78\%$ grass, and (using the same calculations) 20% bran and 2% salt on a DM basis. The ME contents of ryegrass, rice bran and salt are 12.7, 10.5 and 0 MJ/kg DM. The ME content (MJ/kg DM) of the mixture is:

$$ME = (12.7 \times 0.78) + (10.5 \times 0.20) + (0 \times 0.02)$$
$$= 12.0$$

Ration Formulation Methods

If we reverse Equation 13.4, we have the type of equation which we would solve to formulate a ration so that it has a required nutrient content, i.e. for ME in the above example:

$$(12.7 \times 0.78) + (10.5 \times 0.20) + (0 \times 0.02)$$
$$= 12.0 \ (\text{MJ/kg DM})$$

or generally:

$$\sum_{k}^{n=1} (\Phi_n \times \text{proportion}_n) \geq \Phi_r \quad (13.5)$$

where
Φ_r = the animal's requirement (DM basis) for nutrient Φ

This is called a 'nutrient constraint' (also called a 'restraint'). There is one constraint for each nutrient specification. Note that Equation 13.5 is an inequality; it allows the nutrient content of the formulated ration to be more than the animal's requirement. This is acceptable for most nutrients and it gives the programme the flexibility needed to solve practical formulation problems. In addition to the nutrient constraints we need another equation, the 'quantity constraint', to prevent unrealistic amounts of ingredients being calculated:

$$\sum_{k}^{n=1} (\text{proportion}_n) = 1.0 \qquad (13.6)$$

Equation 13.6 forces the nutrient proportions to sum to 1 and ensures that all the nutrient in each ingredient is considered in the calculations. Finally, we need an equation which describes the ration cost:

$$\sum_{k}^{n=1} (\text{cost}_n \times \text{proportion}_n) = \text{total cost} \qquad (13.7)$$

where
cost_n = the cost (e.g. $/t) of the nth ingredient

A least-cost formulation based on two ingredients and two nutrients

This is a very simple example, but it illustrates the mathematics which is used in the least-cost algorithm. Table 13.1 lists the data; the animal is a 150 kg weaned beef steer.

The equations are (using the symbols in Table 13.1):

$$(\text{ME}_L \times \text{proportion}_L) + (\text{ME}_S \times \text{proportion}_S)$$
$$\geq \text{ME}_R \text{ (MJ/kg DM)}$$
$$(\text{protein}_L \times \text{proportion}_L) + (\text{protein}_S \times \text{proportion}_S)$$
$$\geq \text{protein}_R \text{ (\%)}$$
$$(\text{proportion}_L) + (\text{proportion}_S) = 1.0$$

Table 13.1. Data (dry matter basis) for a least-cost ration formulation.

Constraint	ME content (MJ/kg)	Protein content (%)	Price ($/t)
Lucerne hay (L)	9.1	19.9	360
Sorghum grain (S)	12.4	12.6	250
Requirement (R)	11.3	17.0	Minimum

Substituting the data in Table 13.1, we have
$$(9.1 \times \text{proportion}_L) + (12.4 \times \text{proportion}_S)$$
$$\geq 11.3 \text{ (MJ/kg DM)}$$
$$(\text{proportion}_L) + (\text{proportion}_S) = 1.0$$

and for protein:

$$(19.1 \times \text{proportion}_L) + (12.6 \times \text{proportion}_S)$$
$$\geq 17.0 \text{ (MJ/kg DM)}$$
$$(\text{proportion}_L) + (\text{proportion}_S) = 1.0$$

These sets of equations can be solved simultaneously. For the ME equations, $\text{proportion}_L = 0.32$ and $\text{proportion}_S = 0.68$, and for protein, the values are 0.60 and 0.40. In this example, the optimum combination of ingredients for ME is different to that of protein, a common result in ration formulation problems. We choose between the two alternatives on the basis of which mixture of ingredients will meet or exceed the ME and protein requirements, and which is the cheapest or least-cost option. A graphical solution is illustrated in Fig. 13.2.

In Fig. 13.2, the protein constraint line represents all the possible combinations of lucerne hay and sorghum grain which contain 12% protein, and the ME constraint line has all the combinations of these ingredients which contain 11.3 MJ ME/kg. However, all but one of these possible combinations (for each nutrient) are unfeasible because they would require our mixture to have more or less than 100% of ingredient. For example, if a ration of 100% lucerne hay was fed, the animal would have to eat 124% of its maximum daily food intake to consume enough ME. The only feasible solutions are those where the ME and protein lines intersect the quantity constraint line.

Each of the two possible mixtures of hay and grain will contain exactly the required amount of one of the nutrients, but there is no guarantee that it will also contain enough of the other. It seems that the problem cannot be solved, i.e. it is a 'non-feasible' problem. However, a feasible solution can be found because animals will usually tolerate nutrient surpluses. Figure 13.2B shows the combinations of hay and grain which provide exactly the ME and protein requirements. Note that the mixture which gives the correct ME requirement will provide more protein than is needed. Although it is not shown in the figure, a similar situation exists for the combination of hay and grain calculated for protein. Both the ME and protein mixtures will provide more than the required concentration of the other nutrient. We calculate the cost of each mixture and choose the

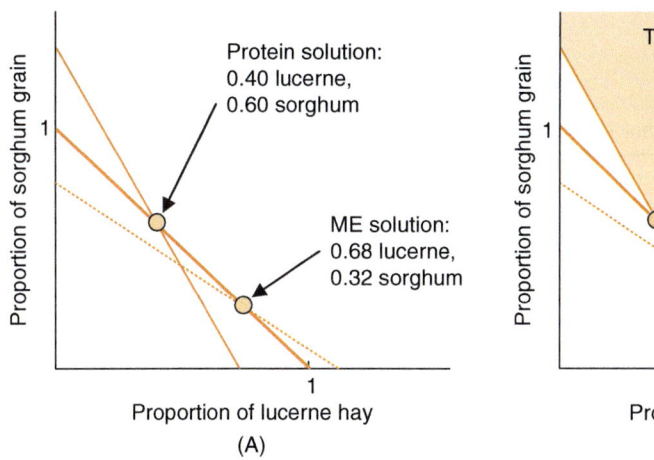
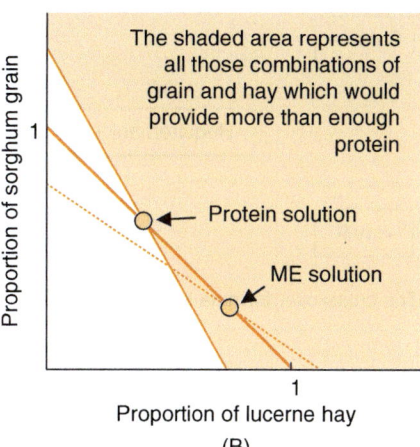

Fig. 13.2. Solutions to the ration formulation problem described in Equations 13.1 and 13.2, displayed graphically (— protein constraint line; ⋯⋯ ME constraint line; — quantity constraint line; shaded area in (B) represents solutions which have more than the required protein concentration).

least-cost formulation. Equation 13.7 gives $285/t for the ME formulation and $316/t for the protein formulation. The first of these is the least-cost ration.

The Pearson square

This method (also known as the 'cross-over formula' or the 'blender's formula') will find the mixture of two ingredients which will exactly meet a specified nutrient content. It is essentially a graphical method of solving a set of two equations in two variables. In this example, we will use oats grain (14 MJ digestible energy (DE)/kg DM) and timothy hay (9 MJ DE/kg DM) to formulate a mixture which contains 12 MJ DE/kg.

1. Choose two ingredients; one must have more than the required nutrient content, and the other less than the required content, the method gives a nonsensical answer otherwise.
2. Set up the calculation as shown in Fig. 13.3.
3. Calculate on the diagonal – always subtract the smaller value (irrespective of whether it is a food nutrient content or the requirement) from the larger one.
4. Read off horizontally the amounts of ingredients in the mixture.

The Pearson square can be used in an iterative way, with the first solution providing one of the ingredients for the second solution, and so on. This is cumbersome, and it becomes inaccurate with successive iterations.

The Pearson square is one of several manual formulation methods. All of these are much more restricted than methods based on linear programming, both in their ability to give a least-cost solution and in the number of ingredients and nutrients that can be considered.

Ration Formulation Approaches and Techniques

Choosing nutrients and ingredients and controlling the shape of the ration

The first step in any formulation is to decide which nutrients will be used. Many nutrients are present in conventional feed mixtures in adequate concentrations, so in practice we need to consider only those which might limit animal performance in each specific circumstance. For ruminants, these are usually energy (ME or DE), protein (either total protein or ruminally degradable protein and undegradable dietary protein), Ca, P and Na. Local circumstances will dictate whether other minerals (often Cu, Co and Mg) and vitamins (especially vitamin A) need to be considered. We need to consider more nutrients for monogastrics. In addition to those for ruminants we include a selection of the essential amino acids and fatty acids, and B group vitamins.

The second step is to choose a variety of ingredients. Typically, we would list 20 or so

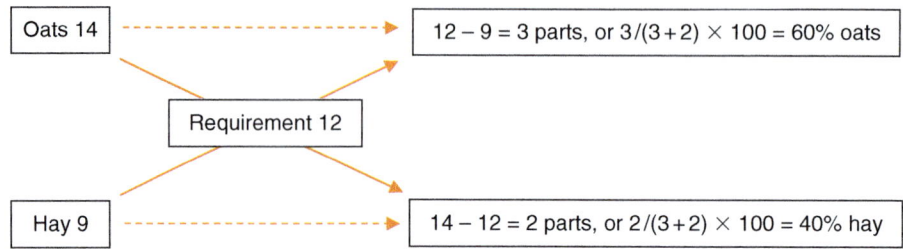

Fig. 13.3. The cross-over formula in use: calculate diagonally, read off the answers horizontally.

ingredients, although only a proportion of these will be incorporated into the ration. The choice of ingredients is influenced by cost, seasonal availability, reliability of supply, what is presently in storage (this is often a good guide to which ingredients are readily available, and it is likely that the necessary storage and processing equipment needed will be available), ingredient palatability, the possibility of nutritional disorders if a ration contains an excessive amount of a problem ingredient, and the need to ensure a source of all required nutrients.

A ration formulation should be changed when the animal's nutrient requirements change (e.g. it becomes pregnant, is given a different type of work or grows larger), ingredients become unavailable, the formulated ration turns out to be unsuitable or ingredient prices change.

Dealing with formulation problems

Quite often, a formulation problem cannot be solved (i.e. there is a 'non-feasible solution'). The most common reasons are:

1. Too many constraints in the problem – these include maximum or minimum ingredient limits, requirements to include ingredients at exact levels, specified nutrient maxima or nutrient ratios (e.g. ratios between lysine and DE, or between Ca and P). Delete or relax these constraints until the ration can be formulated. However, maximum limit constraints that protect the animal against toxicities (e.g. from nutrient toxicities or secondary substances in the feeds) must be retained.

2. A nutrient requirement which is higher than the concentration of that nutrient in all or most of the ingredients – some programmes report which nutrients are causing problems, so dealing with this can be relatively simple.

3. There are too few ingredients – in some cases this will prevent a solution, even though the nutrient content and requirement data would suggest that a solution is possible. Deal with this by including additional ingredients.

All formulations must be checked for suitability. Not uncommonly a ration will have an excessive amount of an ingredient which is unpalatable, or which makes the mixture difficult or impossible to mix or otherwise handle, or which in some other way makes it unwise to use. This type of problem is usually best handled by imposing constraints (often upper limits) to the inclusion of particular ingredients.

Using ration formulation programmes to design ingredient buying strategies

The technique is to assign a high cost (e.g. $30,000/t) to ingredients which are not in store, but which could be bought if the price was favourable. The high cost ensures that they are not used in the formulation, but the price at which they would enter (and should therefore be bought at) will be displayed as a 'price for inclusion'. It is wise to repeat the formulation using the new cost, as there may be consequential changes in the ration which cause unexpected changes in cost or general diet suitability. A least-cost result is not necessarily a global least cost because it is limited to the ingredients which have been made available for the formulation.

Parametric ranging

It is wise to carry out sensitivity tests to find out which nutrient or other constraint is determining the ration formulation and cost. The constraint which is associated with the largest shadow price will have the largest effect on the ration price or composition if that constraint is modified. Another technique is to alter nutrient specifications (either the requirements or the nutrient contents) by a series of standard amounts (e.g. 10%) and recording the resulting changes in specifications and costs. Once the constraint which has the greatest influence on the ration cost has been identified it may be possible to reduce the cost by relaxing this requirement (if this is feasible) or by finding alternative ingredients. It may be possible to make substantial savings in ration cost by making quite small changes in nutrient specifications, at the cost of only minor effects on animal performance.

Multiblending

This is formulating more than one ration at a time, so that scarce resources are allocated on an enterprise-wide basis, rather than in each ration considered separately. This more effectively minimizes the cost of feeding animals over the whole enterprise. Pesti and Miller (1988) describe multiblending as follows:

> Sometimes feed formulators are faced with the problem of having a limited amount of one or more ingredients. The decision must be made to use each ingredient in one or more diets. One approach . . . is to formulate each diet (*separately*) with and without the ingredient. The savings resulting from including the ingredient in each diet can then be compared. This is a bit tedious, but gives the correct answer to the problem. An alternative method is to make up a single formulation problem that includes all the separate problems. . . . The complete matrices for each problem. . . . are included in the larger matrix. The method is known as multiblending.

Linear Programming and the Problems of Variability and Multiple Optimization in Ration Formulation

Linear programming problems are formulated in terms of single, invariant values for the ingredient nutrient contents. This does not reflect real life as the nutrient contents of ingredients, with the exception of some mineral and vitamin sources, can vary quite widely. For example, the coefficient of variation for the protein content of cereal grains is about 18% (NRC, 2001). The protein content of any particular sample is likely to differ from the expected value and this may lead us to formulate rations with less than the required protein content.

There are essentially two ways in which this problem can be tackled:

1. The nutrient content values used in a formulation are decreased by an amount which will ensure that rations are formulated to desired standards a predetermined fraction of the time. If the protein content of barley grain is 12.4±2.1% (mean±SD), then by using a protein content of 8.2% in the formulation problem we should produce, in 95% of cases, a mixture which contains at least 12.4% protein. Torres-Rojo (2001) describes a stochastic programming approach to this problem which uses the expected mean nutrient contents of the ingredients and their variances modified by the probability of achieving a ration which contains the required nutrient content in a specified proportion of cases. This approach wastes resources and does not guarantee a ration which always meets declared or regulatory standards. Also, the apparent shortfall in protein content must be made up by including, generally expensive, protein meals.

2. The nutrient content of each batch of ingredient is measured as it is received into storage. This approach is expensive with regard to analytical facilities (although near infrared reflectance spectroscopy and other techniques can reduce time and cost substantially) and storage facilities, and requires good warehouse management. However, it avoids the problem of some batches having less than the required protein content and others having much more than is needed, and it is likely to be much less expensive than the alternative method.

Cost minimization is an important objective in ration formulation, but in modern animal nutrition there are several other objectives which should be optimized. These include minimizing nutrient excretion (particularly N and P), maximizing ration palatability or processing characteristics (pelletability or segregation during transport), etc. Linear programming allows us to optimize only one objective (usually cost) but we can include constraints which help us to shape a ration which meets other objectives as well. If we want to ensure that a pig produces minimum backfat we can set a minimum lysine/DE ratio; if we want the roughage content of a dairy cow ration to be at least 60% we can indicate whether ingredients are roughages or concentrates and set a minimum roughage constraint. However, with every additional constraint we increase the risk of being unable to formulate a feasible ration.

An alternative method is to consider objectives (e.g. cost, nutrient content) as goals to be approached but not necessarily met (Rehman and Romero, 1984). In other words, provided that the nutritionist is happy with the general outcome, 'near enough is good enough'. Programming techniques which can be used to optimize a group of objectives include weighted goal programming, lexicographic goal programming and multiple objective programming (Rehman and Romero, 1984). Tozer and Stokes (2001) describe an application of multiple objective programming to the formulation of dairy cattle rations. They attempted to simultaneously reduce N and P excretion and minimize the ration cost. They found these objectives to be incompatible, but they were able to reduce P excretion at the cost of a small increase in ration price. This may not be a universal result, although in most cases it seems that minimizing N wastage increases the cost of the ration, but it does illustrate the possibilities of multiple objective formulations. A different example is given by Castrodeza et al. (2005). These authors used an interactive form of multiple objective programming to formulate a pig ration in which four criteria (cost, P excretion, amino acid balances and the lysine/energy ratio) were optimized. This is an iterative method which involves decisions by a 'decision maker', a nutritionist who re-evaluates the outcomes after each iteration and decides when an acceptable result for the basket of objectives has been reached. Zhang and Roush (2002) illustrate the use of a weighted (or ranked) approach. In this method the nutritionist gives weights (or ranks) to the various objectives at the start of the formulation process so that more emphasis in the calculations is given to optimizing the heavily weighted objective(s).

Nutritional Modelling in Ration Formulation

As noted by Lara and Romero (1994), we cannot define nutrient requirements with complete accuracy, and it is certainly true that tabulated feeding standards fit specific animal management situations only approximately. This is recognized by most publishers of feeding standards and it is now commonplace for these publications to include the equations from which the tabular standards have been derived. Examples are in the SCA (1990) and NRC (1998, 2000, 2001) beef and dairy cattle and pig feeding standards. With these equations we could calculate a more closely fitted set of nutrient requirements than can be obtained from feeding standards tables.

Even though we may have calculated the 'ideal' nutrient requirements of our animal, it is beneficial to test combinations of nutrient levels which differ from this to see if we can find a ration which, while not supporting maximum productivity, may nevertheless maximize profit or some other desirable outcome. Sensitivity analysis (parametric ranging) will indicate what will happen to the ration cost if we relax a nutrient requirement specification. However, we cannot know how much that relaxation will reduce profit or animal welfare.

Nutritional models are mathematical descriptions of animal components (e.g. rumen, whole-animal) or of the animal in its environment. Many nutritional management models have a nutrient requirements module which allows the user to describe the animal (species, breed, age, sex, production status, coat type) and its environment (pasture conditions, land topography, weather, etc.) and from this information to calculate its nutrient requirements. These models can also allow for different efficiencies of nutrient utilization in different diets, e.g. the effect of diet ME content on the efficiency of utilization of ME for growth by ruminants (SCA, 1990). In contrast to straightforward ration formulation, nutritional models give solutions that are specific for the particular animal, its environment and the composition of the diet that is chosen to be used. They can be used to investigate different nutritional scenarios (e.g. relaxing nutritional constraints) or to identify areas of physiology in which our understanding needs to be strengthened (McNamara, 2004). Examples of nutritional management models are AUSPIG (Mania Software, 1993) for commercial pig production, Grazfeed (Horizon Agriculture, 1999) for modelling intake and nutrient use by grazing ruminants and CPM-Dairy (Boston et al., 2000) for dairy cattle.

References

Boston, R.C., Fox, D.G., Sniffen, C., Janczewksi, E., Munson, E.R. and Chalupa, W. (2000) The conversion of a scientific model describing dairy cow nutrition and production to an industry tool: the CPM dairy project.

In: McNamara, J.P., France, J. and Beever, D. (eds) *Modelling Nutrient Utilization in Farm Animals*. CAB International, Wallingford, UK, pp. 361–378.

Castrodeza, C., Lara, P. and Peña, T. (2005) Multicriteria fractional model for feed formulation: economic, nutritional and environmental criteria. *Agricultural Systems* 86, 76–96.

Dent, J.B. and Casey, H. (1967) *Linear Programming and Animal Nutrition*. Lippincott, Philadelphia, Pennsylvania.

Greenwald, D.U. (1957) *Linear Programming: An Explanation of the Simplex Algorithm*. Ronald Press, New York.

Horizon Agriculture (1999) *GRAZFEED – A Nutritional Management System for Grazing Animals*. CSIRO, Sydney, Australia.

Lara, P. and Romero, C. (1994) Relaxation of nutrient requirements in livestock rations through interactive multigoal programming. *Agricultural Systems* 45, 443–453.

Lara, P. and Stancu-Minasian, I. (1999) Fractional programming: a tool for the assessment of sustainability. *Agricultural Systems* 62, 131–141.

McNamara, J.P. (2004) Research, improvement and application of mechanistic, biochemical, dynamic models of metabolism in lactating dairy cattle. *Animal Feed Science and Technology* 112, 155–176.

Mania Software (1993) *FEEDMANIA*. ABRI/UNE, Armidale, New South Wales, Australia.

National Research Council (NRC) (1998) *Nutrient Requirements of Swine*, 10th rev. edn. National Academy Press, Washington, DC.

National Research Council (NRC) (2000) *Nutrient Requirements of Beef Cattle*, 6th rev. edn. Update. National Academy Press, Washington, DC.

National Research Council (NRC) (2001) *Nutrient Requirements of Dairy Cattle*, 7th rev. edn. National Academy Press, Washington, DC.

Pesti, G.M. and Miller, B.R. (1988) *Least-Cost Poultry Feed Formulation: Principles, Practices and a New Microcomputer Program*. Georgia Agricultural Experiment Stations, College of Agriculture, University of Georgia, Atlanta, Georgia.

Pittroff, W. and Kothmann, M.M. (2001a) Quantitative prediction of feed intake in ruminants. I. Conceptual and mathematical analysis of models for sheep. *Livestock Production Science* 71, 131–150.

Pittroff, W. and Kothmann, M.M. (2001b) Quantitative prediction of feed intake in ruminants. II. Conceptual and mathematical analysis of models for cattle. *Livestock Production Science* 71, 151–169.

Rehman, T. and Romero, C. (1984) Multiple-criteria decision-making techniques and their role in livestock ration formulation. *Agricultural Systems* 15, 23–49.

SCA (1990) *Feeding Standards for Australian Livestock – Ruminants*. CSIRO, Melbourne, Australia.

Torres-Rojo, J.M. (2001) Risk management in the design of a feeding ration: a portfolio theory approach. *Agricultural Systems* 68, 1–20.

Tozer, P.R. and Stokes, J.R. (2001) A multi-objective programming approach to feed ration balancing and nutrient management. *Agricultural Systems* 67, 201–215.

Zhang, F. and Roush, W.B. (2002) Multiple-objective (goal) programming model for feed formulation: an example for reducing nutrient variation. *Poultry Science* 81, 182–192.

14 Nutritional Investigations: Measures of Nutritional Status

General Investigative Methods

A request to identify and treat a nutritional problem may come with no warning, and with pressure to make a judgement based on little information and without an opportunity for reflection or checking the facts. Resist being pressured into offering advice. Take time to get an accurate and complete description of the problem, backed up by production records where these are relevant and available. Relevant information includes:

- How the animals are managed; animals, even pets, are 'managed' to the extent that the owner determines the size and type of housing (pens, stables, cages and paddocks), and if the animal is kept by itself or in a group.
- If the problem is general or is found only in a particular age group, sibling group, sex, stage of lactation or pregnancy, season of the year, etc.
- Ration composition, the amounts fed (amounts of food allocated by volume must be converted to mass units); the feeding methods including the type and size of food troughs, the frequency of meals, access to food and the extent of between-animal competition for food; effect of grazing management on food availability and quality; procedures used to mix foods if this is done by the owner; how the food is stored, and if live insects are present or the food is mould-affected; how probiotics and vitamin premixes are stored as these may lose potency with age or if exposed to heat and/or humidity.
- The source of water and how it is presented to the animal(s), water temperature, the amount of water given if it is not freely available.
- Any non-nutritional diseases which may be present, such as internal or external parasites.

Support from professionals in other disciplines and the willingness of the owner to pay for any needed tests and to apply any recommendations are important factors in the successful identification and treatment of a nutritional problem.

General Approaches to Assessing Nutritional Status

1. Appearance and behaviour. Useful information includes estimated live weight or body condition, occurrence of diarrhoea or constipation, presence of bloat or rumen impaction, lameness or other skeletal problems, hair loss or change in hair colour, appearance of eyes, presence of rumination and some assessment of appetite.

2. Symptoms of nutrient deficiencies. Compare the animal(s) under consideration with the appearance and behaviour of the 'normal' animal. Some vitamin and mineral deficiencies have more or less specific symptoms. However, other nutrient deficiencies, e.g. of energy or protein, cause a general loss of productivity but have no specific symptoms.

3. Response to nutrient supplementation. Nutrient deficiencies can be investigated by giving a nutrient supplement and measuring the response, but an animal which is suffering from multiple deficiencies (e.g. N plus S plus energy plus P in low-quality, mature pasture) may not respond to a single supplemented nutrient.

4. Comparison with performance norms. National or local production norms are benchmarks for animal condition and production levels. Some owners will have overly optimistic ideas about the levels of performance which they feel are achievable, while others may fail to recognize problems when they occur. Animals should achieve certain body condition scores (BCS) or live weights (target weights) prior to events such as mating or parturition.

5. Specific nutrient tests. In many cases, there are tests for specific nutrients. Choose among the alter-

natives on the basis of the test accuracy and the availability of reliable reference values.

Food Analysis

Measured or estimated diet nutrient contents can be compared with feeding standards for domestic animals. Feeding standards are not usually available for wild animals but data on standard or field metabolic rates (Nagy, 1987) can be used to estimate energy requirements. If a ration does not have the recommended nutrient contents or has more than the recommended (or permitted by law) levels of secondary substances or contaminants, it is unlikely that the animal will perform satisfactorily. Nutrient content data may support an assessment of nutrient deficiency symptoms. Analyses to determine secondary substances (e.g. tannins, alkaloids) and contaminants (e.g. aflatoxins, insecticides, heavy metals, poisonous seeds) can give useful information. Collect samples which accurately represent the whole mass of food being tested (European Community, 1976). However, if the problem seems to revolve around some contamination, then take separate samples of the more badly affected material. Record the sample origin, the number of bags or approximate weight of food from which the sample was taken, the number of samples which were collected, copies of any product labels, the date on which the food was purchased, the expiry date of any active ingredients (e.g. probiotics and vitamins) and the date of sampling. Process the sample according to the requirements of the analytical laboratory. This may involve low-temperature drying (at not more than 55°C) or refrigeration. Seal the sample so as to prevent changes in moisture content.

If we know the probable diet composition, we can calculate approximate values for diet nutrient content by reference to the nutritive value of the diet components. This approach is also useful when combined with the faecal alkane method for predicting the composition of grazing animals' diets or diet composition information obtained from the scats of wild carnivores.

Indices of Energy Status

Kidney fat index and bone marrow fat content

The kidney fat index was developed by Riney (1955) to describe the energy status of wild deer in New Zealand. It is measured by dissecting the perirenal fat and weighing it and the 'defatted' kidney:

$$\text{KFI} = [(\text{weight of perirenal fat})/(\text{weight of kidney})] \times 100 \qquad (14.1)$$

where
KFI = kidney fat index (%)

KFI values range from about 5% to more than 250%, and KFI is an acceptably accurate indicator of the amount of total body fat in red and mule deer (Riney, 1955; Anderson et al., 1972).

Bone marrow fat content is also an indicator of body fat content. Riney (1955) developed a visual index but this has been superseded by chemical determination of the fat content. The index is measured by removing the marrow from the femur bone, and either extracting the fat by a conventional hexane extraction or by simply drying the marrow (e.g. oven drying at 60°C for 4 days or until there is no further weight loss) and assuming that the residual material after drying is all fat (Neiland, 1970; Marquez and Coblentz, 1987). Bone marrow fat content may be a more sensitive indicator than KFI of energy status at low body fat contents (Torbit et al., 1988; Fig. 14.1).

Body condition scores

These are measures of the amount of soft tissue in an animal's body relative to the frame size and indicate recent nutritional history. BCS are made by assessing subcutaneous fat thickness and muscle mass especially along the back, over the ribs, the tailhead, brisket and flank. BCS are repeatable provided that the scorer is experienced (Grainger and McGowan, 1982; Ferguson et al., 1994) although there may be consistent differences between scorers.

There are condition scoring systems for pigs (Menzies, 1997; Fearon, 2005), sheep (Jeffries, 1961; Russel et al., 1969), dairy and beef cattle (Earle, 1976; East of Scotland Agricultural College, 1976; Wildman et al., 1982; Nicholson and Butterworth, 1986), horses (NRC, 2007) and deer (Audigé et al., 1998; Tuckwell, 2003). They are all similar, although they may use scales of different size and/or use whole or half-score intervals.

BCS can be used to predict animal performance. Ovulation rates in sheep increase by

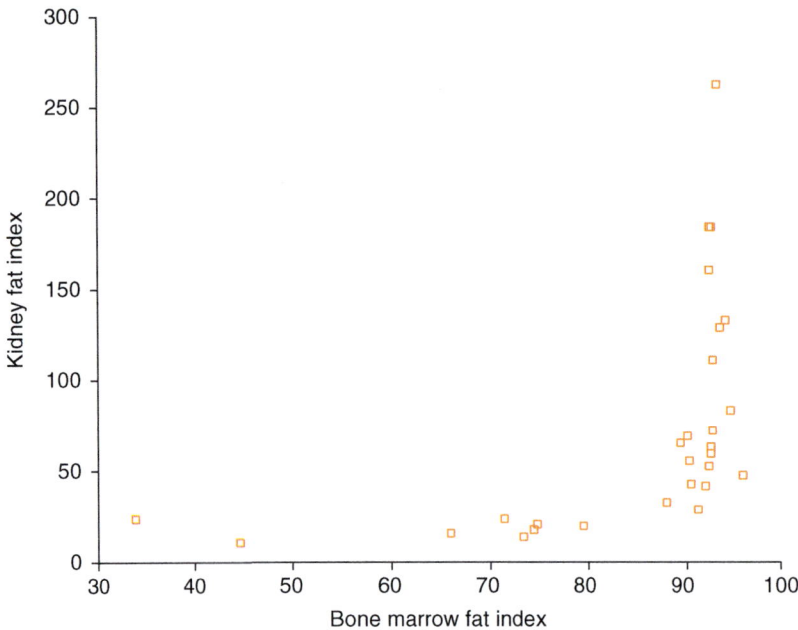

Fig. 14.1. The relationship between bone marrow fat content and the kidney fat index in red deer. (From Finch, 2000.)

0.13–0.56 eggs/BCS, and the post-partum anoestrus interval decreases in beef and dairy cattle as BCS increases (SCA, 1990). Grainger *et al.* (1982) showed that BCS was positively associated with improved milk yield in early lactation of dairy cattle.

SCA (1990) has summarized the live weight change represented by a unit change in BCS for a variety of animal types. These range (modified to apply to a 5-unit scale) from 7 kg/BCS for the Merino

sheep, 42–67 kg/BCS for dairy cattle, to 100 kg/BCS for British breed beef cattle (Table 14.1).

Tyler (1987) developed a muscle index for reindeer which also compares muscle mass with skeletal size, but uses a more objective approach:

$$\text{Muscle index} = M/f^3 \qquad (14.2)$$

where
M = dry weight of *M. gluteobiceps* (g)
f = length of the femur (cm)

Table 14.1. Live weight change and metabolizable energy required for an increase of one body condition score in sheep and cattle.[a] (From SCA, 1990.)

Type of animal	Live weight change (kg/BCS)	Energy change (MJ/kg W)	ME required (MJ/BCS)[3b]
Adult sheep (scale 0–5)			
Merino	7	21	340
Other breeds	12	21	585
Beef cows (scale 0–5)			
Hereford, Angus	100	21	4900
Dairy cows (scale 1–8)			
Jersey	26	20	1210
Jersey × Friesian	34	20	1580
Friesian	42	20	1950

[a]Approximate values only for adult sheep and beef cows in the range of BCS 2–4, and for dairy cows for ±1 unit from BCS 4.
[b]Diet ME content = 10 MJ/kg DM and kg = 0.43.

Indices of Protein Status

Faecal N content

The diet protein content of herbivorous animals can be estimated from their faecal N content. For beef cattle grazing tropical pastures, a faecal N content of less than 13 mg/g DM indicates a probable dietary protein deficiency (Winks *et al.*, 1979). Diet/faecal relationships have been reported for African zebu cattle eating *Chloris* and *Cynodon* pasture (Bredon *et al.*, 1963; Equation 14.3a, R^2 = 0.92), North American elk (*Cervus elaphus canadensis*) eating diets with low tannin contents (Mould and Robbins, 1981, Equation 14.3b) and African herbivores (Arman *et al.*, 1975, Equation 14.3c):

$$N_{diet} = 1.677 \times N_{faeces} - 1.11 \qquad (14.3a)$$

$$N_{diet} = 2.04 \times N_{faeces} - 1.57 \qquad (14.3b)$$

$$N_{diet} = 2.04 \times N_{faeces} - 15.2 \qquad (14.3c)$$

where

N_{diet} = dietary N content (% DM)
N_{faeces} = faecal N content (% DM)

This approach has been used with white-tailed deer (Brown *et al.*, 1995) and to predict the protein contents of diets eaten by snowshoe hares (Sinclair *et al.*, 1982) and colobus monkeys (Chapman *et al.*, 2005). The level of faecal N for the snowshoe hare which represents a diet protein content which will maintain live weight is 12 g/kg DM, a value which is surprisingly similar to the value for cattle.

Rumen ammonia concentration

Rumen microorganisms require in the rumen liquor at least 5 mg NH_3-N/100 ml for optimal microbial protein synthesis (Satter and Slyter, 1974; review by Miller, 1982), and fibrolytic activity is reduced at levels below 60–100 mg/100 ml (Mehrez *et al.*, 1977; Oosting *et al.*, 1989). Accordingly, we would expect that food intake, and less commonly digestibility, would be adversely affected if rumen NH_3 levels fall to low levels.

Stephenson (1985) developed a method of measuring the rumen NH_3 concentrations of grazing sheep. A sample (about 50 ml) of strained rumen liquor is collected by stomach tube from sheep mustered before they have finished their morning grazing; time since last eating and recent water consumption both influence the results (Stephenson and McGuigan, 1985). Reaction with sodium salicylate, Na nitroprusside and dichloroisocyanuric acid in alkaline solution gives a deep blue colour for a positive test. The colour is compared to NH_3 standards.

Milk urea content

It can be used to diagnose excessive protein intakes in dairy cattle (Eicher *et al.*, 1999). A meta-analysis by Broderick and Clayton (1997) on the relationships between milk urea N (MUN) and various measures of N economy gave R^2 of 0.84 for diet protein content, 0.77 for excess N intake, 0.62 for N efficiency and 0.57 for rumen NH_3 content. Sannes *et al.* (2002) reported even higher R^2 for similar indices, e.g. 0.98 for ruminal NH_3 content and 0.94 for N utilization efficiency. The test should be used on a herd basis, rather than for individual cows (Eicher *et al.*, 1999; Trevaskis and Fulkerson, 1999).

Non-nutritional factors influence the MUN content, including stage of lactation, breed of cow, number of previous pregnancies, mastitis, animal live weight, milk yield, housing and time of sample collection (Westwood *et al.*, 1998), and there may be consistent differences between herds (Eicher *et al.*, 1999).

Symptoms of Mineral Deficiencies

Deficiency symptoms can, with appropriate care, be used to make a tentative diagnosis of mineral deficiency. However, they should be confirmed with a more specific test, e.g. tissue mineral concentrations or enzyme activities, or the response to a supplement. Accessible lists of deficiency symptoms are in Underwood and Suttle (1999) and NRC (1998, 2000, 2001, 2007). A key for important mineral deficiencies is in Table 14.2.

Specific Tests for Mineral Deficiencies

There are several quite reliable and specific tests for mineral status. They are particularly useful for investigating the mineral status of grazing or wild animals, where we do not know the diet composition with any certainty.

Bone

Biopsies carried out on the live animal, or on samples collected post-mortem, give sensitive assays of

Table 14.2. Some symptoms of mineral deficiencies in cattle, horses and pigs. (From symptoms described in NRC, 1998, 2001, 2007; Crenshaw, 2001; Hill and Spears, 2001; Mahan, 2001.)

Deficient nutrient	Important deficiency symptoms	Other deficiencies which may cause these symptoms
Ca	Cattle	
	• Weak bones, osteomalacia, rickets	• P, vitamin D
	Horses	
	• Rickets, osteomalacia	• P, vitamin D
	• Developmental orthopaedic disease	• P, Cu, vitamin D
	• Osteodystrophia fibrosa	• P, oxalate in some tropical grasses
	Pigs	
	• Rickets, osteomalacia	• P, vitamin D
	• Hind leg paralysis (posterior paralysis) of sows	
P	Cattle	
	• Depraved and/or reduced appetite	• Protein, Na
	• Lameness, rickets, osteomalacia	• Ca, vitamin D
	• Pale coat	• Protein, Cu
	Horses	
	• As for Ca	
	Pigs	
	• As for Ca	
Cu	Cattle	
	• Pale coat colour, hair loss	• Protein, P
	• Diarrhoea	
	• Weak, possibly deformed, bones	• Ca, P, vitamin D
	Horses	
	• Developmental orthopaedic disease	• Ca, P, vitamin D, possibly excess energy
	Pigs	
	• Anaemia	• Fe
	• Leg deformities and spontaneous fractures	
	• Depigmentation of hair	
Na	Cattle and horses	
	• Abnormal appetite for salt, reduced and/or depraved appetite	• P, protein
	• Rough coat	• Cu (in cattle)
	• Uncoordination	
	Pigs	
	• Reduced rate and efficiency of growth	• Energy, protein, amino acids, Fe
Fe	Cattle and horses	
	• Anaemia (but rare in cattle)	
	Pigs	
	• Reduced growth	• Energy, protein, amino acids, Na, Cu
	• Anaemia	• Cu
	• Laboured breathing in baby pigs	
I	All species	
	• Lethargy, goitre (except in horses), birth of weak or dead offspring with goitre	
Se	Cattle	
	• Muscular dystrophy	
	• Low reproduction rate, retained placenta	
	Horses	
	• Muscular dystrophy	
	Pigs	
	• Sudden death	• Fe, possibly Mg

the animal's P and Ca status. The bone biopsy is done under local analgesia. The skin is incised to expose the rib, and a 1–1.5 cm diameter bone sample is taken by trephine. Samples are taken from the 12th rib, on either side of the body. The sample consists of compact bone down to the medulla (Little, 1972). The P content of rib bone varies with distance from the vertebral column (Poppi and Ternouth, 1978) but ribs on different sides of the animal have similar compositions.

The best measure of the animal's P status is to express the P content as either a proportion of the fresh bone mass, or in relation to the volume of bone removed (e.g. mg/kg bone or mg/cc bone), or compared to the bone N content (Little, 1972; Holst *et al.*, 2002). Holst *et al.* (2002) have reported reference values for cattle (% of fat-free bone, 95% confidence interval) for Ca (25.1, 24.14–26.06) and P (11.4, 10.82–11.98), but they found that Ca and P analyses were less sensitive than ash density (total mineral/bone volume, mg/ml) in identifying mineral-deficient cattle.

Bone densitometry (Murray *et al.*, 1994) has been used to investigate P and Ca status, but it is less accurate and more inconvenient than chemical analysis.

Faeces

These are easy to sample and can usually be obtained non-invasively. Even if direct rectal sampling is required it is not difficult to do and does not distress the animal if done correctly.

The content of inorganic P in faeces indicates the P content of the animal's diet (Wrench *et al.*, 1997; Wu *et al.*, 2001). As the P content of an animal's diet rises above its requirement, the proportion of water-soluble P in the faecal P increases (Dou *et al.*, 2002). Regressions between total faecal P and diet P have been published for different animals and diet types. Examples include the equation of Wrench *et al.* (1997) for grazing and browsing African herbivores (Equation 14.4a, $R^2 = 0.97$) and that of Cohen (1974) for cattle grazing Australian tropical grass pastures (Equation 14.4b, $R^2 = 0.79$):

Diet P = 0.393 × (faecal P) (14.4a)

Diet P = 3.401 × (faecal P) − 0.173 (14.4b)

where
Diet P = diet P content (g/kg DM)

Faecal P = faecal total P content (g/kg DM)

Care is needed when interpreting results from P-deficient ruminants because the total faecal P content may decrease as the amount of dry matter (DM) eaten increases, because less endogenous P is excreted (Ternouth, 1989). On the other hand, consistent with the data cited above, endogenous and total faecal P increases with increasing P intake in adequately nourished sheep (Grace, 1981).

The concentration of Na in faeces falls when low-Na foods are eaten, but the test is unreliable (Murphy and Gartner, 1974).

Blood

Mineral status can be tested by measuring total mineral levels in serum, plasma or whole blood; or blood constituents such as an enzyme (e.g. glutathione peroxidase, GSH_{px}) to indicate mineral levels. In many cases, blood tests are better used to indicate herd mineral status than the status of individual animals (Waldner *et al.*, 1998; Ortman *et al.*, 1999; Herdt *et al.*, 2000). To perform these tests, divide the whole herd into classes of animals, e.g. dry cows, low producers and high producers and take blood samples from five to seven animals from each group (Kelly *et al.*, 1988). Blood metabolite concentrations will give useful indications of herd nutritional status but there are some caveats to how the tests are made and interpreted. Whitaker *et al.* (1999) and Herdt *et al.* (2000) give useful discussions of sampling methods and techniques and the need to combine blood test data with other herd information.

Valid reference values are essential. Reference levels for some blood constituents are given in Table 14.3 (other values are given by Underwood and Suttle, 1999; Whitaker *et al.*, 1999; Herdt *et al.*, 2000; Hoff *et al.*, 2001). Consider these points when interpreting blood mineral contents:

1. Calcium: Ca levels in blood are maintained by the action of parathyroid hormone on bone – bone Ca thus acts as a reservoir to maintain blood Ca levels (Bronner and Stein, 1995). Ionized Ca in blood does not fall unless there is an intense demand for Ca, e.g. in the dairy cow at calving.
2. Phosphorus: blood P content (blood inorganic P, BIP) is an indicator of variable accuracy of P status (Wise *et al.*, 1963; Little and McMeniman, 1973). The animal's age, the time of day that the sample is collected, sampling stress and exercise all influence BIP levels (Cohen, 1973, 1974). Nevertheless, BIP

Table 14.3. Threshold values for blood minerals in domestic ruminants. (From cattle – Rowlands and Manston, 1976; Payne, 1978; Pelletier *et al.*, 1985; Whitaker, 2004; sheep – Paynter, 1996; deer – Audigé, 1992; Audigé *et al.*, 1995.)

Mineral[a]	Dairy cattle	Sheep	Deer
Phosphorus (inorganic) (mmol/l)	1.4–2.48	1.3–2.7	1.5–4.74
Calcium (mmol/l)	2.35–2.46	2.12–2.87	2–2.83
Magnesium (mmol/l)	0.8–0.99	0.74–1.44	–
Sodium (mmol/100 ml)	135–143		–
Potassium (mmol/100 ml)	4.3–5.7		–
Copper (µmol/l)	7.56–9.4	9.4	3–19
Zinc (µmol/l)	–	9.2	–
Cobalt (nmol vitamin B12/l plasma)	–	0.52	0.08–0.61
Selenium (µmol/l blood)	–	0.76	1–32[b]

[a]In plasma unless otherwise indicated.
[b]Units are kIU/l GSH_{px}.

has been extensively used to indicate the P status of grazing animals.

3. Copper: analysis for total plasma Cu, or the activities of ceruloplasmin or Cu/Zn superoxide dismutase (CuZnSOD) will indicate the animal's Cu status, although less reliably than liver Cu measurements. CuZnSOD levels change about 3 weeks after a change in dietary Cu content (Underwood and Suttle, 1999).

4. Selenium: either total blood Se or the activity of GSH_{px}, an enzyme which contains Se, can be used to identify cattle which are Se-deficient (Ortman *et al.*, 1999). Again, changes in GSH_{px} activity occur about 3 weeks after a change in dietary Se status (Underwood and Suttle, 1999).

Liver

Samples can be obtained post-mortem or from the live animal by making a biopsy with a trocar. Liver Cu and Se values give good indications of the animal's reserves of these nutrients.

Adrenal glands

The ultrastructure of the zona glomerulosa, and the relative thicknesses of the zona glomerulosa and the cortex are both good indicators of Na status. In Na-deficient sheep the zona glomerulosa becomes more mitotically active and the mitochondria become enlarged and rounded (Hill *et al.*, 1983). In Na-replete sheep these cells are shrunken and accumulate lipid. Adrenal glands can only be inspected post-mortem, which reduces the applicability of this method.

Saliva

The ratio of Na to K in saliva falls when the animal becomes Na deficient, and this is a sensitive test for Na status (Murphy and Gartner, 1974). Normal Na and K concentrations for cattle are 139 and 7.4 mmol/l (Murphy and Plasto, 1973). Saliva can be collected by aspiration or by letting the animal chew on a plastic sponge or by aspiration into a pipette placed near the parotid gland (Murphy and Connell, 1970). For the record, pigmy hippopotamuses can be trained to open their mouths for saliva collection (Sharp, 2002). If saliva is too difficult to collect, muzzle secretions can be used instead (Singh and Rani, 1999).

Hair

Some authors claim that the mineral content of hair gives useful information about the animal's mineral status. This is doubtful, as the existing data are contradictory (sometimes within the same experiment or series of experiments) and give little support for dietary effects on hair mineral content. Even in those experiments where changes in mineral contents were observed these were much less than changes in other tissues such as blood and liver (e.g. Fisher *et al.*, 1985). Ca and P in hair are not influenced by the animal's dietary intake of these minerals (Cohen, 1973), although diet may influence the concentrations of Cu, Mg (Fisher *et al.*, 1985) and Se (McDowell *et al.*, 1982). Contamination with dust or soil and the type of cleaning used to prepare the hair for analysis will affect the data. On the other hand, hair analysis is useful in toxicological studies (Tutudaki *et al.*, 2003). The ease of obtaining samples makes this method worth persevering with, but at present it remains one of the less reliable methods of evaluating animal mineral status. For a review, see Combs (1987).

The Metabolic Profile Test

The use of blood metabolite levels to define the nutritional and metabolic status of dairy cows was pioneered in the 1960s. The test is a herd test (i.e. not for use with individual cows). Metabolites such as β-hydroxy butyrate, blood glucose and measures of protein status (Table 14.4) are used. The application and limitations of metabolic profile testing are discussed and illustrated by Lister and Gregory (1978), Payne and Payne (1987), Whitaker *et al.* (1995), van Saun (1997) and Macrae *et al.* (2006). The limitations of the test are related to the insensitivity of some metabolites to changes in nutrition, correct choice of test animals, and the confounding effects of season, sampling time in relation to concentrate feeding and stage of lactation. Caldeira *et al.* (2007) present data which demonstrate comparable responses in BCS and blood metabolites.

Methods for Investigating the Nutritional Status of Grazing Animals

Residual dry matter and the adequacy of pasture supplies

The nutrient intake of grazing animals can sometimes be restricted by overstocking or the overuse of pasture. The concept of allocating pasture supplies by 'residual DM' can help to avoid these

Table 14.4. Threshold values of blood constituents in British dairy cattle. (From Rowlands and Manston, 1976; Pelletier *et al.*, 1985; Whitaker, 2004.)

Blood constituent	Lactating cows	Dry cows
β-hydroxy butyrate (mmol/l)	<1.0	<0.6
Non-esterified fatty acids (mmol/l)	<0.7	<0.4
Glucose (mmol/l)	0.3–3.46	0.3
Urea-N (mmol/l)	1.7–2.8	1.7
Albumin (g/l)	30–33.7	30
Globulin (g/l)	50	50
Packed cell volume (%)	27–38	27–38

problems. The nutrient content (e.g. metabolizable energy or protein contents) of a pasture is not uniformly distributed through the whole height of the sward. The older, stemmier parts of the plant are located towards the base (Fig. 14.2), and they are less digestible and have lower protein contents than young, leafy material. After a period of grazing, animals will encounter material which has less than the required nutrient contents, and their performance will suffer. We can maintain optimum animal performance, if grazing is managed to leave the recommended residual DM or if not more than the recommended proportion of the pasture is utilized (Holmes and Roche, 2007; Morris, 2007).

An alternative definition of residual DM is that it is 'the old plant material left standing or on the ground at the beginning of a new growing season' (Bartolome *et al.*, 2002). This definition is used in North America to manage the grazing pressure applied to rangelands. In other countries, a similar concept, the 'sustainable stocking rate', is used.

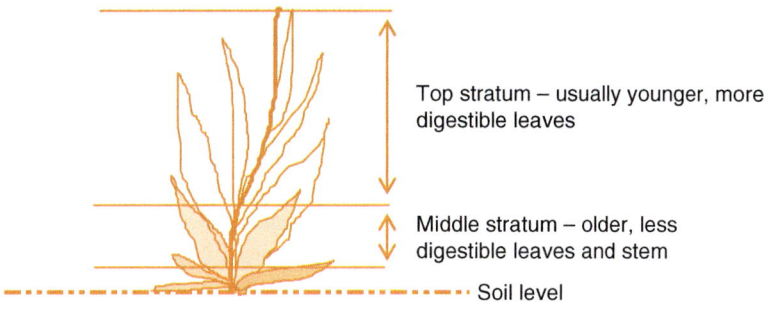

Top stratum – usually younger, more digestible leaves

Middle stratum – older, less digestible leaves and stem

Soil level

Fig. 14.2. Nutrient stratification in temperate grass swards.

Measuring food intake and digestibility

The adequacy of food intake and quality can be gauged from animal condition, but this can be confounded with the effects of non-nutritional health problems.

A rapid method of measuring intake under intensive (controlled) grazing conditions is to do before and after grazing measurements of the DM presentation yield. This works quite well, although there are considerations, about sampling height in relation to grazing height and pasture regrowth rates, which need to be addressed (Smit *et al.*, 2005).

Intake can be estimated by measuring the faecal concentration of an external marker which has been given to the animal in a known amount. From this we can calculate the amount of faeces voided. External markers, e.g. Cr, can be administered using a ruminal controlled release device (commonly called a CRD; Ellis *et al.*, 1981) or in other ways to give a small dose of marker into the animal's digestive tract over several days. The marker mixes with the digesta and is eventually excreted in the faeces. By taking grab (opportunistic) samples of faeces over 7 days and measuring the DM and marker contents, food DM intake can be estimated from the concentration of marker in the faeces provided that we know the digestibility of the food DM.

Alkanes are hydrocarbons which are found in the cuticular waxes of plants. Alkanes with an even number of carbon atoms are present in plants in very low concentrations. We can dose the animal with one of these, e.g. the C36 alkane, to measure intake, and use a naturally occurring odd-numbered alkane as an internal indicator to calculate the food digestibility. By combining this information we can calculate the DM intake (Smit *et al.*, 2005):

$$\text{DM intake (kg/day)} = C_j / [(F_j/F_i) \times (H_i - H_j)] \tag{14.5}$$

where

C_j = amount of the even-numbered alkane (the external indicator) given each day (mg)

F_i = faecal concentration of the odd-numbered alkane (the internal indicator) (mg/kg DM)

F_j = faecal concentration of the even-numbered alkane (mg/kg DM)

H_i = concentration of the odd-numbered alkane in the food (mg/kg DM)

H_j = concentration of the even-numbered alkane in the food (mg/kg DM)

Investigating diet selection

Plant cuticle fragments can be extracted from faeces and compared microscopically with standard examples of the cuticles of those plants which are present in the grazed area (Stewart, 1967). Each plant has differently shaped epithelial cells and the cuticle models this. It is easy to differentiate legumes from grasses. The method has a venerable history but is still used in animal research (Beresford *et al.*, 1989; de Jong *et al.*, 1995).

Plant alkane profiles in faeces can be used to investigate diet selection. The concentrations of long-chain (C25 to C25) alkanes in faeces are determined and compared with the same alkanes in the range of pasture species to which the animal has access (Dove and Mayes, 1991). A greater proportion of longer-chain (e.g. C35) than shorter-chain (e.g. C25) alkanes are recovered in the faeces. We must correct for this, and feeding trials are needed to get this information. Dove and Mayes (2005) have described methods of using alkanes to calculate digestibility, the intakes of supplements and the basal diet of grazing animals.

Faecal Profiling Using Near Infrared Reflectance Spectroscopy

'Faecal profiling' is the use of information from faeces to predict animal performance (for reviews see Dryden, 2003; Stuth *et al.*, 2003; Landau *et al.*, 2006). This can be done using chemical measures, e.g. N and acid detergent fibre (ADF) content to predict the digestible energy and protein contents of herbivore foods. Faecal profiling was developed by workers at Texas A&M University in the 1990s (Stuth and Tolleson, 2000). Near infrared reflectance spectroscopy (NIRS) can be calibrated to identify sex, tick infestation, animal species, pregnancy and diet protein content in wild animals, including deer, elephants and bison (Tolleson and Stuth, 2002). The method has been adopted in northern Australia to predict diet protein content and organic matter digestibility, the content of herbs (dicotyledonous plants) and grasses (monocots) in the diet, and animal growth (Coates, 1999). NIRS faecal profiling is a rapid and robust method, but it requires a substantial amount of work to generate suitable prediction equations and this will inhibit the uptake of the methodology into practical animal management. Nevertheless, where valid prediction equations are available, NIRS faecal profiling may improve farm productivity and profitability. The method is completely non-invasive as faecal samples can be collected after voiding. NIRS is a flexible tool which is potentially attractive for investigating wild animal nutrition status and it has been used, for example, to identify food trees used by Australian possums (McIlwee *et al.*, 2001), for the rapid evaluation of pasture quality for the Australian northern hairy nosed wombat (Woolnough and Foley, 2002) and to determine food characteristics of deer, cattle and goats (Lyons and Stuth, 1992; Leite and Stuth, 1995; Showers *et al.*, 2006).

Additional Techniques for Investigating the Nutritional Status of Wild Animals

Herd dynamics

The nutritional condition of wild animals is revealed in information such as population size, the age and sex structure and reproduction rates. For details of how these can be assessed see Caughley and Sinclair (1994).

Faeces (scats) constituents

Carnivore diets can be characterized from indigestible remains such as bones, teeth, nails, claws and hair, scales, exoskeletons or chetae (Reynolds and Aebischer, 1991; Lobert *et al.*, 2001; Triggs and Brunner, 2002). These data can be quantified by expressing the number of occurrences as proportions of the total number of occurrences (Ciucci *et al.*, 1996). Prey can be identified from DNA recovered from faeces (Kohn and Wayne, 1997; Farrell *et al.*, 2000) or from the types of sterols present (Leeming *et al.*, 1996).

Diet diversity and overlap

The diversity index of Simpson (1949) as modified by Pianka and Pianka (1976) gives a mathematical value to describe the range of food types in an animal's diet:

$$\text{Diversity index} = 1/[(\Sigma p_i^2) \times n] \qquad (14.6)$$

where

p_i = proportion of the ith resource present in the diet

n = number of items in the diet

Dietary overlap, the degree to which two diets share the same ingredients, can be calculated with the formula of Pianka (1973):

$$\text{Dietary overlap} = \Sigma p_{ij}\, p_{ik}/[(\Sigma p_{ij}^2 \times p_{ik}^2)^{0.5}] \qquad (14.7)$$

where

p_{ij} = proportion of the ith component in the diet of the jth animal

p_{ik} = proportion of the ith component in the diet of the kth animal

Digesta collected post-mortem

As the cell wall content (fibre content) of a herbivore's diet increases, we expect the digestibility to fall (e.g. Barnes, 1973). Brown *et al.* (1995) have developed this idea and shown with deer that the ADF content of rumen contents or faeces is an index of diet quality. This is a useful adaptation of the internal marker approach, because it is usually impossible to get an accurate sample of what a grazing animal has actually eaten, and thus get an accurate measure of the internal marker concentration in the consumed food. Forsyth *et al.* (2005) have adapted this method to predict diet selection by deer.

Rumen contents collected post-mortem can be used to estimate diet digestibility using the zero-time fermentation method of Carrol and Hungate (1954). Strained rumen samples are incubated *in vitro* for up to 120 min. Volatile fatty acid (VFA) concentrations determined at intervals during the incubation are plotted and the regression extrapolated to zero time to get the VFA concentration in the living animal and the rate of VFA production. For more recent descriptions of the method's use see Olsen and Mathiesen (1996) and Sørmo *et al.* (1997).

References

Anderson, A.E., Medin, D.E. and Bowden, D.C. (1972) Indices of carcass fat in a Colorado mule deer population. *Journal of Wildlife Management* 36, 579–594.

Arman, P., Hopcraft, D. and McDonald, I. (1975) Nutritional studies on East African herbivores. 2. Losses of nitrogen in the faeces. *British Journal of Nutrition* 33, 265–276.

Audigé, L., Wilson, P.R. and Morris, R.S. (1998) A body condition score system and its use for farmed deer hinds. *New Zealand Journal of Agricultural Research* 41, 545–553.

Audigé, L.J.M. (1992) Serum biochemical values of rusa deer (*Cervus timorensis russa*) in New Caledonia. *Australian Veterinary Journal* 69, 268–271.

Audigé, L.J.M., Wilson, P.R. and Morris, R.S. (1995) Deer herd health productivity and data. *Proceedings of the*

Deer Branch New Zealand Veterinary Association Conference – 1995 12, 31–56.

Barnes, R.F. (1973) Laboratory methods of evaluating feeding value of herbage. In: Butler, G.W. and Bailey, R.W. (eds) Chemistry and Biochemistry of Herbage, Vol. 3. Academic Press, London, pp. 179–214.

Bartolome, J.W., Frost, W.E., McDougald, N.K. and Connor, J.M. (2002) California Guidelines for Residual Dry Matter (RDM) Management on Coastal and Foothill Annual Rangelands. Rangeland Monitoring Series, University of California Agriculture and Natural Resources Publication 8092, Oakland, California. Available at: http://anrcatalog.ucdavis.edu

Beresford, N.A., Lamb, C.S., Mayes, R.W., Howard, B.J. and Colgrove, P.M. (1989) The effect of treating pastures with bentonite on the transfer of ^{137}Cs from grazed herbage to sheep. Journal of Environmental Radioactivity 9, 251–264.

Bredon, R.M., Harker, K.W. and Marshall, B. (1963) The nutritive value of grasses grown in Uganda when fed to Zebu cattle. 1. The relation between the percentage of crude protein and other nutrients. Journal of Agricultural Science, Cambridge 61, 101–104.

Broderick, G.A. and Clayton, M.K. (1997) A statistical evaluation of animal and nutritional factors influencing concentrations of milk urea nitrogen. Journal of Dairy Science 80, 2964–2971.

Bronner, F. and Stein, W.D. (1995) Calcium homeostasis – an old problem revisited. Journal of Nutrition 125, 1987S–1995S.

Brown, R.D., Hellgren, E.C., Abbott, M., Ruthven, D.C. and Bingham, R.L. (1995) Effects of dietary energy and protein restriction on nutritional indices of female white-tailed deer. Journal of Wildlife Management 59, 595–609.

Caldeira, R.M., Belo, A.T., Santos, C.C., Vazques, M.I. and Portugal, A.V. (2007) The effect of body condition score on blood metabolites and hormonal profiles in ewes. Small Ruminant Research 68, 233–241.

Carrol, E.J. and Hungate, R.E. (1954) The magnitude of microbial fermentation in the bovine rumen. Applied Microbiology 2, 205–214.

Caughley, G. and Sinclair, A.R.E. (1994) Wildlife Ecology and Management. Blackwell Science, Cambridge, Massachusetts.

Chapman, C.A., Webb, T., Fronstin, R., Wasserman, M.D. and Santamaria, A.M. (2005) Assessing dietary protein of colobus monkeys through faecal sample analysis: a tool to evaluate habitat quality. African Journal of Ecology 43, 276–278.

Ciucci, P., Boitani, L., Pelliccioni, E.R., Rocco, M. and Guy, I. (1996) A comparison of scat-analysis methods to assess the diet of the wolf Canis lupus. Wildlife Biology 2, 37–48.

Coates, D.B. (1999) Faecal spectroscopy (NIRS) for nutritional profiling of grazing cattle. In: Eldridge, D. and Freudenberger, D. (eds) People and Rangelands Building the Future. Proceedings of the VI International Rangelands Congress, Townsville, VI International Rangeland Congress Inc., Aitkenvale, Queensland, Australia, Vol. 1, pp. 466–467.

Cohen, R.D.H. (1973) Phosphorus nutrition of beef cattle. 2. Relation of pasture phosphorus to phosphorus content of blood, hair and bone of grazing steers. Australian Journal of Experimental Agriculture and Animal Husbandry 13, 5–8.

Cohen, R.D.H. (1974) Phosphorus nutrition of beef cattle. 4. The use of faecal and blood phosphorus for the estimation of phosphorus intake. Australian Journal of Experimental Agriculture and Animal Husbandry 14, 709–714.

Combs, D.K. (1987) Hair analysis as an indicator of mineral status of livestock. Journal of Animal Science 65, 1753–1758.

Crenshaw, T.D. (2001) Calcium, phosphorus, vitamin D, and vitamin K in swine nutrition. In: Lewis, A.J. and Southern, L.L. (eds) Swine Nutrition, 2nd edn. CRC Press, Boca Raton, Florida, pp. 187–212.

de Jong, C.B., Gill, R.M.A., van Wieren, S.E. and Burlton, F.W.E. (1995) Diet selection by roe deer Capreolus capreolus in Kielder Forest in relation to plant cover. Forest Ecology and Management 79, 91–97.

Dou, Z., Knowlton, K.F., Kohn, R.A., Wu, Z., Satter, L.D., Zhang, G., Toth, J.D. and Ferguson, J.D. (2002) Phosphorus characteristics of dairy feces affected by diets. Journal of Environmental Quality 31, 2058–2065.

Dove, H. and Mayes, R.W. (1991) The use of plant wax alkanes as marker substances in studies of the nutrition of herbivores: a review. Australian Journal of Agricultural Research 42, 913–952.

Dove, H. and Mayes, R.W. (2005) Using n-alkanes and other plant wax components to estimate intake, digestibility and diet composition of grazing/browsing sheep and goats. Small Ruminant Research 59, 123–139.

Dryden, G.McL. (2003) Near Infrared Reflectance Spectroscopy: Applications in Deer Nutrition. RIRDC Publication No. W03/007. Rural Industries Research and Development Corporation, Canberra, Australia.

Earle, D.F. (1976) A guide to scoring dairy cow condition. Journal of Agriculture, Victoria 74, 228–231.

East of Scotland Agricultural College (1976) Condition Scoring of Cattle. Bulletin No. 6 (revised edn). East of Scotland Agricultural College, Edinburgh, UK.

Eicher, R., Bouchard, E. and Bigras-Poulin, M. (1999) Factors affecting milk urea nitrogen and protein concentrations in Quebec dairy cows. Preventive Veterinary Medicine 39, 53–63.

Ellis, K.J., Laby, R. and Burns, R.G. (1981) Continuous controlled administration of chromic oxide to sheep. Proceedings of the Nutrition Society of Australia 6, 145.

European Community (1976) EU Directive 76/371/EEC. Official Journal of the European Communities, No. L102/1. Available at http://europa.eu.int/eur-lex/lex/RECH_reference_pub.do

Farrell, L.E., Roman, J. and Sunquist, M.E. (2000) Dietary separation of sympatric carnivores identified by molecular analysis of scats. *Molecular Ecology* 9(10), 1583–1593.

Fearon, P. (2005) *Sow Condition Scoring*. Department of Primary Industries and Fisheries, Brisbane, Queensland, Australia. Available at: http://www2.dpi.qld.gov.au/pigs/4324.html

Ferguson, J.D., Galligan, D.T. and Thomsen, N. (1994) Principal descriptors of body condition score in Holstein cows. *Journal of Dairy Science* 77, 2695–2703.

Finch, N.A. (2000) Perfomance and condition of the wild red deer of South Eastern Queensland. BAppSc (Hons) thesis, University of Queensland, Gatton, Australia.

Fisher, D.D., Wilson, L.L., Leach, R.M. and Scholz, R.W. (1985) Switch hair as an indicator of magnesium and copper status of beef cows. *American Journal of Veterinary Research* 46, 2235–2240.

Forsyth, D.M., Richardson, S.J. and Menchenton, K. (2005) Foliar fibre predicts diet selection by invasive Red Deer (*Cervus elaphus scoticus*) in a temperate New Zealand forest. *Functional Ecology* 19, 495–504.

Grace, N.D. (1981) Phosphorus kinetics in sheep. *British Journal of Nutrition* 45, 367–374.

Grainger, C. and McGowan, A.A. (1982) The significance of pre-calving nutrition of the dairy cow. In: McMillan, K.L. and Taufa, V.K. (eds) *Dairy Production from Pasture*. New Zealand Society of Animal Production, Hamilton, New Zealand, pp. 135–171.

Grainger, C., Wilhelms, G.D. and McGowan, A.A. (1982) Effect of body condition at calving and level of feeding in early lactation on milk production of dairy cows. *Australian Journal of Experimental Agriculture and Animal Husbandry* 22, 9–17.

Herdt, T.H., Rumbeiha, W. and Braselton, W.E. (2000) The use of blood analyses to evaluate mineral status in livestock. *Veterinary Clinics of North America: Food Animal Practice* 16, 423–444.

Hill, G.M. and Spears, J.W. (2001) Trace and ultratrace elements in swine nutrition. In: Lewis, A.J. and Southern, L.L. (eds) *Swine Nutrition*, 2nd edn. CRC Press, Boca Raton, Florida, pp. 229–261.

Hill, P.A., Coghlan, J.P., Butkus, A. and Ryan, G.B. (1983) Structural and functional studies of the adrenal zona glomerulosa in sodium-depleted and sodium-loaded sheep. *Cell and Tissue Research* 229, 515–531.

Hoff, B., Schrier, N., Boermans, H., Faulkner, H. and Hussein, A. (2001) Assessment of trace mineral and vitamin E status beef cows in Ontario. *Canadian Veterinary Journal* 42, 384–385.

Holmes, C.W. and Roche, J.R. (2007) Pastures and supplements in dairy production systems. In: Rattray, P.V., Brookes, I.M. and Nicol, A.M. (eds) *Pasture and Supplements for Grazing Animals*. New Zealand Society of Animal Production, Occasional Publication No. 14, NZSAP, Hamilton, New Zealand, pp. 221–242.

Holst, P.J., Murison, R.D. and Wadsworth, J.C. (2002) Bone mineralization and strength in range cattle. *Australian Journal of Agricultural Research* 53, 947–954.

Jeffries, B.C. (1961) Body condition scoring and its use in management. *Tasmanian Journal of Agriculture* 32, 19–21.

Kelly, J.M., Whitaker, D.A. and Smith, E.J. (1988) A dairy herd health and productivity service. *British Veterinary Journal* 144, 470.

Kohn, M. and Wayne, R.K. (1997) Facts from feces revisited. *Tree* 12, 223–227.

Landau, S., Glasser, T. and Dvash, L. (2006) Monitoring nutrition in small ruminants with the aid of near infrared reflectance spectroscopy (NIRS) technology: a review. *Small Ruminant Research* 61, 1–11.

Leeming, R., Ball, A., Ashbolt, N. and Nichols, P. (1996) Using faecal sterols from humans and animals to distinguish faecal pollution in receiving waters. *Water Research* 30, 2983–2900.

Leite, E.R. and Stuth, J.W. (1995) Fecal NIRS equations to assess diet quality of free-ranging goats. *Small Ruminant Research* 15, 223–230.

Lister, D. and Gregory, N.G. (eds) (1978) *The Use of Blood Metabolites in Animal Production*. Occasional Publication No. 1, British Society of Animal Production, Milton Keynes, UK.

Little, D.A. (1972) Bone biopsy in cattle and sheep for studies of phosphorus status. *Australian Veterinary Journal* 48, 668–670.

Little, D.A. and McMeniman, N.P. (1973) Variation in bone composition of grazing sheep in south-western Queensland, related to lactation and type of country. *Australian Journal of Experimental Agriculture and Animal Husbandry* 13, 229–233.

Lobert, B., Lumsden, L., Brunner, H. and Triggs, B. (2001) An assessment of the accuracy and reliability of hair identification of south-east Australian mammals. *Wildlife Research* 28, 637–641.

Lyons, R.K. and Stuth, J.W. (1992) Fecal NIRS equations for predicting diet quality of free-ranging cattle. *Journal of Range Management* 45, 238–244.

McDowell, L.R., Kiatoko, M., Bertrand, J.E., Chapman, H.L., Pate, F.M., Martin, F.G. and Conrad, J.H. (1982) Evaluating the nutritional status of beef herds from four soil order regions of Florida. 2. Trace minerals. *Journal of Animal Science* 55, 38–47.

McIlwee, A.M., Lawler, I.R. and Cork, S.J. (2001) Coping with chemical complexity in mammal–plant interactions: near-infrared spectroscopy as a predictor of *Eucalyptus* foliar nutrients and of the feeding rates of folivorous marsupials. *Oecologia* 128, 539–548.

Macrae, A.I., Whitaker, D.A., Burrough, E., Dowell, A. and Kelly, J.M. (2006) Use of metabolic profiles for the assessment of dietary adequacy in UK dairy herds. *Veterinary Record* 159, 655–661.

Mahan, D.C. (2001) Selenium and vitamin E in swine nutrition. In: Lewis, A.J. and Southern, L.L. (eds)

Swine Nutrition, 2nd edn. CRC Press, Boca Raton, Florida, pp. 281–314.

Marquez, M. and Coblentz, B.E. (1987) Metatarsal and mandibular marrow fat in black-tailed deer. *Journal of Wildlife Management* 51, 38–40.

Mehrez, A.Z., Ørskov, E.R. and McDonald, I. (1977) Rates of rumen fermentation in relation to ammonia concentration. *British Journal of Nutrition* 38, 437–443.

Menzies, P.I. (1997) Reproductive health management programs. In: Youngquist, R.S. (ed.) *Current Therapy in Large Animal Theriogenology*. Saunders, Philadelphia, Pennsylvania, pp. 643–649.

Miller, E.L. (1982) The nitrogen needs of ruminants. In: Thomson, D.J., Beever, D.E. and Gunn, R.G. (eds) *Forage Protein in Ruminant Animal Production*. British Society of Animal Production, Occasional Publication No. 6, Thames Ditton, UK, pp. 79–88.

Morris, S.T. (2007) Pastures and supplements in beef production systems. In: Rattray, P.V., Brookes, I.M. and Nicol, A.M. (eds) *Pasture and Supplements for Grazing Animals*. New Zealand Society of Animal Production, Occasional Publication No. 14, NZSAP, Hamilton, New Zealand, pp. 243–254.

Mould, E.D. and Robbins, C.T. (1981) Nitrogen metabolism in elk. *Journal of Wildlife Management* 45, 323–334.

Murphy, G. and Connell, J.A. (1970) A simple method of collecting saliva to determine the sodium status of cattle and sheep. *Australian Veterinary Journal* 46, 595–598.

Murphy, G.M. and Gartner, R.J. (1974) Letter: sodium levels in the saliva and faeces of cattle on normal and sodium deficient diets. *Australian Veterinary Journal* 50, 280–281.

Murphy, G.M. and Plasto, A.W. (1973) Liveweight response following sodium chloride supplementation of beef cows and their calves grazing native pasture. *Australian Journal of Experimental Agriculture and Animal Husbandry* 13, 369–374.

Murray, R.M., Heard, R.W., Tiddy, R., Boniface, A.N. and Coates, D.B. (1994) Tail bone densitometry as a technique for measuring bone mineral status in cattle. *Proceedings of the Australian Society of Animal Production* 20, 325–328.

Nagy, K.A. (1987) Field metabolic rate and food requirement scaling in mammals and birds. *Ecological Monographs* 57, 111–128.

National Research Council (NRC) (1998) *Nutrient Requirements of Swine*, 10th rev. edn. National Academy Press, Washington, DC.

National Research Council (NRC) (2000) *Nutrient Requirements of Beef Cattle*, 7th rev. edn. Update. National Academy Press, Washington, DC.

National Research Council (NRC) (2001) *Nutrient Requirements of Dairy Cattle*, 7th rev. edn. National Academy Press, Washington, DC.

National Research Council (NRC) (2007) *Nutrient Requirements of Horses*, 5th rev. edn. National Academy Press, Washington, DC.

Neiland, K.A. (1970) Weight of dried marrow as an indicator of fat in caribou femurs. *Journal of Wildlife Management* 34, 904–907.

Nicholson, M.J. and Butterworth, M.H. (1986) *A Guide to Condition Scoring of Zebu Cattle*. International Livestock Centre for Africa, Addis Ababa, pp. 3–29.

Olsen, M.A. and Mathiesen, S.D. (1996) Production rates of volatile fatty acids in the minke whale (*Balaeoptera acutorostrata*) forestomach. *British Journal of Nutrition* 75, 21–31.

Oosting, S.J., Verdonk, J.M.J.H. and Spinhoven, G.G.B. (1989) Effect of supplementary urea, glucose and minerals on the *in vitro* degradation of low quality feeds. *Asian-Australasian Journal of Animal Science* 2, 583–590.

Ortman, K., Andersson, R. and Holst, H. (1999) The influence of supplements of selenite, selenate and selenium yeast on the selenium status of dairy heifers. *Acta Veterinaria Scandinavica* 40, 23–34.

Payne, J.M. (1978) The Compton metabolic profile test. In: Lister, D. and Gregory, N.G. (eds) *The Use of Blood Metabolites in Animal Production*. Occasional Publication No. 1, British Society of Animal Production, Milton Keynes, UK, pp. 3–12.

Payne, J.M. and Payne, S. (1987) *The Metabolic Profile Test*. Oxford University Press, New York.

Paynter, D.I. (1996) Diagnosis of mineral deficiencies. In: Masters, D.G. and White, C.L. (eds) *Detection and Treatment of Mineral Nutrition Problems in Grazing Sheep*. Australian Centre for International Agricultural Research, Canberra, Australia.

Pianka, E.R. (1973) The structure of lizard communities. *Annual Review of Ecology and Systematics* 4, 53–74.

Pianka, E.R. and Pianka, H.D. (1976) Comparative ecology of twelve species of nocturnal lizards (Gekkonidae) in the Western Australian desert. *Copeia* No. 1, 125–142.

Pelletier, G., Tremblay, A.V. and Hélie, P. (1985) Facteurs influençant le profil métanologie des vaches laitières. *Canadian Veterinary Journal* October 26, 306–311.

Poppi, D.P. and Ternouth, J.H. (1978) Variation in bone phosphorus content of the twelfth rib of sheep. *Proceedings of the Australian Society of Animal Production* 12, 127.

Reynolds, J.C. and Aebischer, N.J. (1991) Comparison and quantification of carnivore diet by faecal analysis: a critique, with recommendations, based on a study of the fox *Vulpes vulpes*. *Mammal Review* 21, 97–122.

Riney, T. (1955) Evaluating condition of free-ranging red deer (*Cervus elaphus*), with special reference to New Zealand. *New Zealand Journal of Science and Technology* 36, 429–463.

Rowlands, G.J. and Manston, R. (1976) The potential uses of metabolic profiles in the management and selection of cattle for milk and beef production. *Livestock Production Science* 3, 239–256.

Russel, A.J.F., Doney, J.M. and Gunn, R.G. (1969) Subjective assessment of body fat in live sheep. *Journal of Agricultural Science, Cambridge* 72, 451–454.

Sannes, R.A., Messman, M.A. and Vagnoni, D.B. (2002) Form of rumen-degradable carbohydrate and nitrogen on microbial protein synthesis and protein efficiency of dairy cows. *Journal of Dairy Science* 85, 900–908.

Satter, L.D. and Slyter, L.L. (1974) Effect of ammonia concentration on rumen microbial protein production *in vitro. British Journal of Nutrition* 32, 199–208.

Sharp, M. (2002) Training of pygmy hippopotamus (*Hexaprotodon liberiensis*) for saliva collection to determine estrus cycles. *Animal Keepers' Forum* 29, 236–238.

Showers, S.E., Tolleson, D.R., Stuth, J.W., Kroll, J.C. and Koerth, B.H. (2006) Predicting diet quality of white-tailed deer via NIRS fecal profiling. *Rangeland Ecology and Management* 59, 300–307.

Simpson, E.H. (1949) Measurement of diversity. *Nature* 163, 688.

Sinclair, A.R.E., Krebs, C.J. and Smith, J.N.M. (1982) Diet quality and food limitation in herbivores: the case of the snowshoe hare. *Canadian Journal of Zoology* 60, 889–897.

Singh, S.P. and Rani, D. (1999) Assessment of sodium status in large ruminants by measuring the sodium-to-potassium ratio in muzzle secretions. *American Journal of Veterinary Research* 60, 1074–1081.

Smit, H.J., Taweel, H.Z., Tas, B.M., Tamminga, S. and Elgersma, A. (2005) Comparison of techniques for estimating herbage intake of grazing dairy cows. *Journal of Dairy Science* 88, 1827–1836.

Sørmo, W., Haga, Ø.E., White, R.G. and Mathiesen, S.D. (1997) Comparative aspects of volatile fatty acid production in the rumen and distal fermentation chamber in Svaalbard reindeer. *Rangifer* 17, 81–95.

Standing Committee on Agriculture (SCA) (1990) *Feeding Standards for Australian Livestock. Ruminants.* CSIRO Publications, Melbourne, Australia.

Stephenson, R.G.A. (1985) Sheep Nutrition – rumen ammonia test. (a) basic application of the test. *Qld. DPI Refnote* R17/Jun 85. Queensland Department of Primary Industries, Brisbane, Australia.

Stephenson, R.G.A. and McGuigan, K.R. (1985) Sheep Nutrition – rumen ammonia test. (b) reagents and procedure. *Qld. DPI Refnote* R18/Jun 85. Queensland Department of Primary Industries, Brisbane, Australia.

Stewart, D.R.M. (1967) Analysis of plant epidermis in faeces: a technique for studying the food preferences of grazing herbivores. *Journal of Applied Ecology* 4, 83–111.

Stuth, J., Jama, A. and Tolleson, D. (2003) Direct and indirect means of predicting forage quality through near infrared reflectance spectroscopy. *Field Crops Research* 84, 45–56.

Stuth, J.W. and Tolleson, D.R. (2000) Managing the nutritional status of grazing animals using near-infrared

spectroscopy. *Compendium on Continuing Education for the Practising Veterinarian* 10, S108–S115.

Ternouth, J.H. (1989) Endogenous losses of phosphorus by sheep. *Journal of Agricultural Science, Cambridge* 113, 291–297.

Tolleson, D.R. and Stuth, J.W. (2002) Exploring grazing animal physiology via fecal NIRS. *Proceedings of an NIRS Workshop,* University of Alberta, Alberta, Canada, October, 2002.

Torbit, S.C., Carpenter, L.H., Batmann, R.M., Alldredge, A.W. and Whilte, G.C. (1988) Calibration of carcass fat indices in wintering mule deer. *Journal of Wildlife Management* 52, 582–588.

Trevaskis, L.M. and Fulkerson, W.J. (1999) The relationship between various animal and management factors and milk urea, and its association with reproductive performance of dairy cows grazing pasture. *Livestock Production Science* 57, 255–265.

Triggs, B. and Brunner, H. (2002) *Hair ID: an Interactive Tool for Identifying Australian Mammalian Hair.* CSIRO Publishing, Collingwood, Victoria, Australia.

Tuckwell, C. (2003) *The Deer Farming Handbook.* Rural Industries Research and Development Corporation, Canberra, Australia, pp. 255–256.

Tutudaki, M., Tsakalof, A.K. and Tsatsakis, A.M. (2003) Hair analysis used to assess chronic exposure to the organophosphate diazinon: a model study with rabbits. *Human and Experimental Toxicology* 22, 159–164.

Tyler, N.J.C. (1987) Body composition and energy balance of pregnant and non-pregnant Svalbard reindeer during winter. *Symposium of the Zoological Society, London,* 57, 203–229.

Underwood, E.J. and Suttle, N.F. (1999) *The Mineral Nutrition of Livestock,* 3rd edn. CAB International, Wallingford, UK.

van Saun, R.J. (1997) Nutritional profiles: a new approach for dairy herds. *Bovine Practitioner* 31(2), 43–50.

Waldner, C., Campbell, J., Jim, G.K., Guichon, P.T. and Booker, C. (1998) Comparison of 3 methods of selenium assessment in cattle. *Canadian Veterinary Journal* 39, 225–231.

Westwood, C.T., Lean, I.J. and Kellaway, R.C. (1998) Indications and implications for testing of milk urea in dairy cattle: a quantitative review. Part 1. Dietary sources and metabolism. *New Zealand Veterinary Journal* 46, 87–96.

Whitaker, D.A. (2004) Metabolic profiles. In: Andrews, A.H., Blowey, R.W., Boyd, H. and Eddy, R.G. (eds) *Bovine Medicine: Diseases and Husbandry of Cattle,* 2nd edn. Blackwell Science, Oxford, UK, pp. 804–817.

Whitaker, D.A., Kelly, J.M. and Eayres, H.F. (1995) *Use and Interpretation of Metabolic Profiles in Dairy Cows.* Department of Veterinary Clinical Studies, University of Edinburgh, Edinburgh, UK.

Whitaker, D.A., Goodger, W.J., Garcia, M., Perera, B. M.A.O. and Wittwer, F. (1999) Use of metabolic profiles in dairy cattle in tropical and subtropical countries on smallholder dairy farms. *Preventive Veterinary Medicine* 38, 119–131.

Wildman, E.E., Jones, G.M., Wagner, P.E., Boman, R.L., Trout, H.F. and Lesch, T.N. (1982) A dairy cow body condition scoring system and its relationship to selected production variables in high producing Holstein dairy cattle. *Journal of Dairy Science* 65, 495–501.

Winks, L., Laing, A.R., O'Rourke, P.K. and Wright, G.S. (1979) Factors affecting response to urea-molasses supplements by yearling cattle in tropical Queensland. *Australian Journal of Experimental Agriculture and Animal Husbandry* 19, 522–529.

Wise, M.B., Ordoveza, A.L. and Barrick, E.R. (1963) Influence of variations in dietary calcium:phosphorus ratio on performance and blood constituents of calves. *Journal of Nutrition* 79, 79–84.

Woolnough, A.P. and Foley, W.J. (2002) Rapid evaluation of pasture quality for a critically endangered mammal, the northern hairy-nosed wombat (*Laseorhinus knefftii*). *Wildlife Research* 29, 91–100.

Wrench, J.M., Meissner, H.H. and Grant, C.C. (1997) Assessing diet quality of African ungulates from faecal analyses: the effect of forage quality, intake and herbivore species. *Koedoe* 40, 125–136.

Wu, Z., Satter, L.D., Blohowiak, A.J., Stauffacher, R.H. and Wilson, J.H. (2001) Milk production, estimated phosphorus excretion, and bone characteristics of dairy cows fed different amounts of phosphorus for two or three years. *Journal of Dairy Science* 84, 1738–1748.

Chapter 14

15 Nutrition and the Environment

Nutrient Excretion and Environmental Pollution

The pollution of waterways and lakes as a result of the runoff of excreted N and P from animal production units is a major environmental threat in some regions. Eutrophication of waterways with N and P can adversely affect aquatic organisms through nitric acid acidification, overgrowth of algae and subsequent hypoxia, poisoning by NH_3, uncontrolled plant growth and sedimentation (Smith, 2003; Camargo and Alonso, 2006). Volatilization of NH_3 from urine and faeces can reduce soil Ca, Mg and K, leading to increased risk of root damage, overgrowth and dieback, and infections in trees (Lee and Dollard, 1994; IROEC, 2001). Faecal material also finds its way into streams and contributes to their microbiological load. Flies and odours associated with beef cattle feedlots are other nuisances which can be related to how these animals are managed and fed (Watts and Tucker, 1993; McGinn et al., 2003). Government regulations and community pressure both work to encourage (or require) producers to reduce N and P losses from farms. This incurs a cost – in Australia, Herath (1996) estimated that it cost lot feeders AUS$27–41/head to comply with the regulations of P runoff.

The main nutritional approach to dealing with the problems of eutrophication and acidification is to improve the efficiencies of nutrient use by careful ration formulation, and so reduce the amounts of N and P excreted from the animal. Chen (2001) gives a general overview of how inefficiencies in animal production are related to issues of global warming and pollution, and suggests some possible solutions.

Both pigs and dairy cattle are inefficient users of nutrients. In The Netherlands, sows and finisher pigs excrete 65–75% of the N and P consumed, and weaners excrete about 35% of P and 45% of N consumed (van Der Peet-Schwering and Den Hartog, 2000). Castillo et al. (2001) have shown that overall N utilization by dairy cows is poor, rarely exceeding 0.30 g milk N/g of dietary N intake. Thus, dairy cattle may excrete more than two-thirds of the N they eat (ignoring maintenance and body weight changes). Techniques to improve the efficiency of nutrient use vary depending on the type of animal. Paik (2001) and Han et al. (2001) have reviewed a number of these, including food processing, various food additives and amino acid supplementation.

A varying proportion of plant P is in the form of phytic acid, especially in seeds and in protein meals made from oilseeds. Only some 20% of this form of P is available to monogastric animals, and pigs excrete large amounts of phytate-P. Phytases can be added to foods and will release P from its combination within the phytic acid molecule and thus make it more available. In some cases, plant phytases naturally occurring in foods, especially rye grain, may be active enough to have an effect on P availability in pig diets (Zimmermann et al., 2002). The effectiveness of added phytase appears to be greater, at low dietary P levels, when the pig's food intake is greater (1.2 versus 2.8 times maintenance; Steiner et al., 2006), possibly because of a slightly higher gastric pH in pigs fed more food. Improved availability of food P allows the manager to reduce the amount of supplementary P which has to be added, as well as reducing the amount of faecal P voided.

Cu and Zn are sometimes used as growth promoters in pig diets. Some users include them at levels well above the nutritional requirements, e.g. 125–250 mg Cu/kg and 1000–3000 mg Zn/kg (Smith et al., 1997; Hill et al., 2000); although current recommendations for pig foods in the EU are to limit Cu and Zn additions to 175 and 250 mg/kg, respectively (European Commission, 2003a,b). A substantial proportion, approaching 100%, of this added mineral is excreted with adverse effects on soils, plants and animals when the manure is subsequently spread on pasture (Jondreville *et al.*, 2003). Inorganic Zn may chelate with diet phytate and reduce the efficacy of added phytase (Augspurger *et al.*, 2004). Chelated minerals, complexed with amino acids (lysine or methionine), partially hydrolysed protein (proteinate) or polysaccharides, are sometimes used in an attempt to improve the efficiency of absorption of these minerals. However, there is no consistent evidence that chelated forms of Cu and Zn improve the absorption of these minerals by pigs (Case and Carlson, 2002; Carlson *et al.*, 2004; Veum *et al.*, 2004). The best way of reducing mineral excretion by animals is to formulate the dietary concentration to more nearly reflect the animals' nutritional requirements.

Amino acid incorporation into body tissues depends on the diet supplying exactly the correct amounts of amino acids at the time that the tissue is synthesized. Animal managers often ensure against deficiencies, and thus reduced performance, by formulating diets with more than the required amounts of essential amino acids, and protein generally. The excess N is excreted. When pigs are fed diets with about 12% protein they excrete less N in urine than if given diets with more conventional protein contents (e.g. 16%). As we would expect, this is achieved at the cost of reduced performance (Panetta *et al.*, 2006). N excretion can be reduced and performance maintained by feeding lower-protein diets with appropriate amino acid supplementation, although there may be some reduction in animal growth rate or increase in carcass fatness (Figueroa *et al.*, 2002; Shriver *et al.*, 2003). If these manipulations are done correctly, there will be less unused amino acid N and less N excreted in, particularly, the urine. This will reduce the amounts of NH_3 released into the environment. To quote Kerr and Easter (1995): 'it seems that for each one percentage unit reduction in dietary crude protein combined with amino acid supplementation, total N losses (fecal plus urinary) can be reduced by approximately 8%'. Zervas and Zijlstra (2002) showed that when young grower pigs were allowed to eat ad lib, a 15% reduction in dietary protein content (from 18.5% to 15.5%) with supplemented essential amino acids gave a 27% reduction in urinary N excretion and a 6% reduction in total N excretion. Kerr and Easter (1995) also reported less urinary N excretion (with no change in faecal N) when pigs were fed a low protein (12%) diet plus the essential amino acids lysine, tryptophan and threonine. There was better N retention if this diet was further supplemented with non-essential amino acids, but at the cost of a nearly doubled urinary N excretion. The question is whether to formulate the diet to maximize growth or to maximize N retention. In one sense, the answer will depend on the relative economic impacts of fast growth compared to the cost of penalties for environmental pollution.

Zervas and Zijlstra (2002) added a fermentable fibre source (beet pulp) to pig diets and increased faecal N excretion and reduced urinary N. A similar result was reported by Kreuzer *et al.* (1998), who also found that NH_3 release from manure (mixed faeces and urine) was less when these diets were fed. Addition of glucanases (enzymes which hydrolyse the non-starch polysaccharides in certain cereal grains) may reduce the amount of volatile N in pig excreta in diets based on wheat, but not barley (Nyachoti *et al.*, 2006; Leek *et al.*, 2007).

Antibiotics and probiotics can also improve pig performance and reduce N and P excretion (Table 15.1). The use of some of these, especially antibiotics and hormones, is regulated by law in some jurisdictions. The data in Table 15.1 should be viewed as an example of what might be achieved, but are not a recommendation for pig management.

Dairy farming is also a potentially important source of water and soil pollution. As an example, in Dutch intensive dairying only some 14% of N inputs and 37% of P inputs are converted into dairy products (van Keulen *et al.*, 2000), with the remainder lost through runoff and leaching, NH_3 volatilization and accumulation of P in the soil. There are several approaches to reducing the amounts of N excreted by dairy cattle. Kebreab *et al.* (2000) suggest that reducing the amount of N fertilizer used to grow pasture and forage crops intended for silage making will reduce the amounts of N excreted by cattle. Less fertilizer gives a lower-N crop and thus less excess N in the diet. Grasses with high N contents may contain a substantial amount of non-protein N, which is largely ruminally degradable and thus likely to contribute to urinary N excretion.

Dutch workers have considered the advantages and disadvantages of formulating rations using a 'least-mineral' rather than a least-cost basis. Under Dutch conditions, current ration formulation methods give excesses of 51.4 kg of N and 17.0 kg of P per heifer per year. Using a 'least-mineral' approach to ration formulation should reduce the amount of P excreted by better matching requirements to supply. In their simulation, Mourits *et al.* (2000)

Table 15.1. Effects of antibiotics, probiotics, organic acids and hormones on excretion of dry matter and N by pigs. (From Han *et al.*, 2001.)

Medication	Inclusion level	Change in performance (%)		
		FCR[a]	DM excretion	N excretion
Antibiotics				
Salinomycin	25 mg/kg	+11.0	−17.3	−18.8
Virginiamycin	40 mg/kg	+5.9	−22.5	−12.7
Flavomycin	10 mg/kg	+5.2	−12.8	−5.1
Probiotics				
Clostrodium butyricum	0.03%	+3.8	−15.8	−25.4
Lactic acid bacteria	0.5%	+8.8	−12.6	−4.2
Organic acids				
Formic acid	2%	+8.1	−0.1	−1.8
Citric acid	2%	+11.0	−0.1	−1.2
Hormones				
β-agonists	0.5 mg/kg	+0.25	−40.8	−37.6
β-agonists	2.75 mg/kg	+0.4	−10.5	−12.4
Porcine somatotrophin	4 mg/kg	+0.4	−15.5	−21.4

[a]FCR = food conversion ratio (i.e. the amount of food needed to produce 1 kg of gain).

showed that P excretion would be reduced, but at the cost of slightly lower heifer growth rates and reduced breeding opportunities.

In ruminants, many of the inefficiencies of N use occur in the rumen, where it is rare for a diet to provide the exact amounts of ruminally degradable protein (RDP) and ruminally undegradable protein needed by the animal, and where hyper-NH_3 producing bacteria reduce the efficiency of diet protein use (Eschenlauer *et al.*, 2002). Also important is the timing of supplies of RDP and ruminally fermentable energy in relation to the needs of microbial metabolism. Simulation studies by Rotz *et al.* (1999) have shown that feeding a less degradable protein source to dairy cattle should reduce volatile N loss, especially when cattle are fed a concentrate-rich diet. Improving the synchronicity of N and energy supplementation (Castillo *et al.*, 2000) improves the efficiency of incorporation of N from RDP into microbial protein and thus reduces the amount of RDP-N which is ultimately excreted in urine.

Faeces from beef feedlots can be a major source of environmental pollution. They contaminate waterways, cause objectionable smells and increased fly populations. Faecal excretion can be reduced by processing the diet to increase organic matter digestibility. Watts and Tucker (1993) tested the effects of several grain-processing methods and the addition of enzymes on faeces excretion. Cattle fed dry-rolled grain voided larger amounts of faeces than those given steam-flaked grain, and the starch content of faeces from dry-rolled sorghum (25%) was much greater than in faeces derived from steam-flaked sorghum or dry-rolled barley (2.4–4.2%). A diet formulated with dry-rolled sorghum gave faeces with more volatile solids than did diets with either steam-flaked sorghum or with either rolled or steam-flaked barley (7.4 versus 4.8 to 5.1 kg/1000 kg animal live weight). Sorghum-based diets are thus likely to generate more faecal odour than barley-based diets. Because it increases starch digestibility, fine milling can help to control unpleasant odours in feedlots by reducing the output of undigested starch in cattle faeces (Reis *et al.*, 2001). Watts and Tucker (1993) also found that faecal moisture content had a major effect on odour production. Saline drinking water can lead to wetter faeces (Sinclair *et al.*, 1996), so it is important that water quality is continuously monitored – for both animal welfare and environmental reasons.

Greenhouse Gases

Global warming exacerbated by biogenic methane, and to a smaller extent N_2O, is expected to have varying, but in some cases quite deleterious, effects on animal production. Crop and forage yields may be reduced in southern Europe because of water shortages and extreme weather events (Olesen and Bindi, 2002). Global warming will reduce the areas and/or seasons of the subtropics in which C3 grasses can be grown, and the associated reduction in pasture nutritive value (C4 grasses are less digestible than C3 species) will potentially reduce animal productivity. There are also cultural and societal impacts, including effects on water, soil and air quality, dangers of fire resulting from increased fuel loads if grazing is reduced, and changes in landscape values (IPCC, 2007).

The greenhouse gases which arise from animal production are methane, N_2O and CO_2. Much attention is paid to methane because it has 21 times the global warming potential of CO_2 (UNFCCC, 2006) and is produced in greater amounts than N_2O. It is estimated that 15–20% of the world's methane is produced as a by-product of microbial fermentation of carbohydrate in animals' digestive tracts. The total methane emission by domestic and wild animals is some 80 Tg/year (10^{12} g; Jensen, 1996). Only 30% of total methane emission from animals originates from natural sources, whereas 70% is linked to human activities, including livestock production (Demeyer and Fievez, 2000). Although methanogens (microorganisms which produce methane) have been isolated from the faeces of a wide variety of non-ruminant species, e.g. rat, horse, pig, monkey, baboon, rhinoceros, hippopotamus, giant panda, goose, turkey and chicken (Jensen, 1996), the contributions of monogastrics to global methane production is not great.

Domestic animals, mainly ruminants, contribute a variable proportion of the total greenhouse gases produced by human activity. As we would expect, the proportion of methane from anthropogenic animal sources, i.e. excluding methane produced by wild animals, varies according to the proportion of GDP derived from industrial sources. Some indicative values for the proportion of anthropogenic animal-derived methane (UNFCCC, 2006) are: Japan 2%, South Africa 9%, the European Community 9% (The Netherlands 8%, UK 7%), Canada 7%, the USA 6%, China 15%, Australia 18%, Indonesia 26%, Argentina 41% and New Zealand 50%.

Vermorel (1997) has calculated from metabolism chamber data that sheep and goats produce from 33 to 45 g methane/day, and emissions from horses are about 45–53 g/day for saddle horses, 82 g/day for draught horses and 21–39 g/day for ponies. Direct measurement of grazing sheep in Western Australia gave 12 g methane/day (Leuning et al., 1999), and values for sheep and lactating dairy cattle grazing New Zealand pastures were 19 g/day and 263 g/day, respectively (Lassey et al., 1997).

There are several ways of manipulating ruminant diets to reduce methane synthesis. In the main, these work by reducing the amount of free H_2 produced by rumen microorganisms, generally by using the longer-chain volatile fatty acids (VFA) or long-chain fatty acids as H sinks (Hegarty, 1999a; Demeyer and Fievez, 2000). Techniques which have been investigated include increasing dietary fat content (Machmüller et al., 2001), using ionophores such as monensin (O'Kelly and Spiers, 1992), adding halogenated methane analogues (bromomethane, chloroform, iodoform, carbon tetrachloride; Lanigan, 1972; McCrabb et al., 1997), defaunation (removal of protozoa from the rumen; Hegarty, 1999b), encouraging the growth of rumen acetogens (organisms which use H as a precursor of acetic acid rather than methane; Joblin, 1999), addition of cysteine (Takahashi et al., 1997) and the inclusion of tanniniferous plants (Carulla et al., 2005) or a food preservative, nisin, in the diet (Klieve and Hegarty, 1999; Sar et al., 2005). Ionophores reduce the numbers of rumen microorganisms which produce methane. Monensin, for example, reduces the numbers of some bacterial species which produce CO_2 and H_2 – these are precursors of methane. Rumen protozoa also produce a large amount of H_2. Defaunation reduces methane synthesis as well as increasing the flow of protein into the hindgut. A survey of 1000 diets by Giger-Reverdin et al. (2000) clearly showed that increasing dietary fat was associated with reduced methane production. Takahashi et al. (1997) have suggested that S compounds may chelate those metals which are needed as cofactors for the enzymes involved in methane synthesis, or that these substances may compete with CO_2 as a H sink. The response to cysteine is dose-dependent – an effective dose may be about $0.5 \, g/kg^{0.75}$/day.

Dietary additives effectively reduce methane production but are impracticable in the management of grazing ruminants. In this situation, the most useful approach may be to increase animal productivity so that the proportion of food energy which is used for maintenance, rather than production, is diminished. This means, of course, that the proportion of methane attributable to each unit of product is similarly reduced. Kirchgessner et al. (1995) have suggested from modelling studies that a dairy cow producing 5000 kg milk/year emits 110 kg methane/year, and that doubling milk production only adds 5 kg to the annual methane production. In beef steers, methane release is higher in low-intensity production than in intensive systems, e.g. 175 versus 135 g methane/kg weight gain.

Although the measures described above appear to be effective, the situation is much more complex than this simple analysis indicates. In the first place, if increased production is achieved by feeding cereal grains, manufactured foods or conserved forages, there is a greenhouse gas cost associated with the production, transport, processing and feeding of these ingredients. Second, if we include N_2O emissions in our calculations, this can alter the balance in favour of less productive, but less N-intensive, extensive production systems (Howden and Reyenga, 1999). There are other, less obvious, effects of managing farming to reduce greenhouse gases. Using trees as carbon sinks (e.g. farm forestry or allowing trees or shrubs to grow on grazing lands) will reduce the area of land available for animal (and crop) production.

References

Augspurger, N.R., Spencer, J.D., Webel, D.M. and Baker, D.H. (2004) Pharmacological zinc levels reduce the phosphorus-releasing efficacy of phytase in young pigs and chickens. *Journal of Animal Science* 82, 1732–1739.

Camargo, J.A. and Alonso, Á. (2006) Ecological and toxicological effects of inorganic nitrogen pollution in aquatic ecosystems: a global assessment. *Environment International* 32, 831–849.

Carlson, M.S., Boren, C.A., Wu, C., Huntington, C.E., Bollinger, D.W., Veum, T.L. (2004) Evaluation of various inclusion rates of organic zinc either as polysaccharide or proteinate complex on the growth performance, plasma, and excretion of nursery pigs. *Journal of Animal Science* 82, 1359–1366.

Carulla, J.E., Kreuzer, M., Machmüller, A. and Hess, H.D. (2005) Supplementation of *Acacia mearnsii* tannins decreases methanogenesis and urinary nitrogen in forage-fed sheep. *Australian Journal of Agricultural Research* 56, 961–970.

Case, C.L. and Carlson, M.S. (2002) Effect of feeding organic and inorganic sources of additional zinc on growth performance and zinc balance in nursery pigs. *Journal of Animal Science* 80, 1917–1924.

Castillo, A.R., Kebreab, E., Beever, D.E. and France, J. (2000) A review of efficiency of nitrogen utilisation in lactating dairy cows and its relationship with environmental pollution. *Journal of Animal and Feed Sciences* 9, 1–32.

Castillo, A.R., Kebreab, E., Beever, D.E., Barbi, J.H., Sutton, J.D., Kirby, H.C. and France, J. (2001) The effect of energy supplementation on nitrogen utilization in lactating dairy cows fed grass silage diets. *Journal of Animal Science* 79, 240–246.

Chen, D. (2001) Environmental challenges of animal agriculture and the role and task of animal nutrition in environmental protection. *Asian-Australasian Journal of Animal Sciences* 14, 423–431.

Demeyer, D. and Fievez, V. (2000) Ruminants et environnement: la methanogenese. *Annales de Zootechnie* 49, 95–112.

Eschenlauer, S.C.P., McKain, N., Walker, N.D., McEwan, N.R., Newbold, C.J. and Wallace, R.J. (2002) Ammonia production by ruminal microorganisms and enumeration, isolation, and characterization of bacteria capable of growth on peptides and amino acids from the sheep rumen. *Applied and Environmental Microbiology* 68, 4925–4931.

European Commission (2003a) *Opinion of the Scientific Committee for Animal Nutrition on the Use of Copper in Feedingstuffs*. Health and Consumer Protection Directorate-General. Available at: http://ec.europa.eu/food/fs/sc/scan/out115_en.pdf

European Commission (2003b) *Opinion of the Scientific Committee for Animal Nutrition on the Use of Zinc in Feedingstuffs*. Health and Consumer Protection Directorate-General. Available at: http://ec.europa.eu/food/fs/sc/scan/out120_en.pdf

Figueroa, J.L., Lewis, A.J., Miller, P.S., Fischer, R.L., Gómez, R.S. and Diedrichsen, R.M. (2002) Nitrogen metabolism and growth performance of gilts fed standard corn-soybean meal diets or low-crude protein, amino acid-supplemented diets. *Journal of Animal Science* 80, 2911–2919.

Giger-Reverdin, S., Sauvant, D., Vermorel, M. and Jouany, J.P. (2000) Modelisation empirique des facteurs de variation des rejets de methane par les ruminants. *7emes Rencontres autour des Recherches sur les Ruminants*, Paris, France, 6–7 Decembre, 2000.

Han, I.K., Lee, J.H., Piao, X.S. and Li, D. (2001) Feeding and management system to reduce environmental pollution in swine production. *Asian-Australasian Journal of Animal Sciences* 14, 432–444.

Hegarty, R.S. (1999a) Mechanisms for competitively reducing ruminal methanogenesis. *Australian Journal of Agricultural Research* 50, 1299–1306.

Hegarty, R.S. (1999b) Reducing rumen methane emissions through elimination of rumen protozoa. *Australian Journal of Agricultural Research* 50, 1321–1328.

Herath, G. (1996) A review of costs of removing phosphorus to control algal blooms in waterways. *Australian Journal of Environmental Management* 3, 189–201.

Hill, G.M., Cromwell, G.L., Crenshaw, T.D., Dove, C.R., Ewan, R.C., Knabe, D.A., Lewis, A.J., Libal, G.W.,

Mahan, D.C., Shurson, G.C., Southern, L.L. and Veum, T.L. (2000) Growth promotion effects and plasma changes from feeding high dietary concentrations of zinc and copper to weanling pigs (regional study). *Journal of Animal Science* 78, 1010–1016.

Howden, S.M. and Reyenga, P.J. (1999) Methane emissions from Australian livestock: implications of the Kyoto Protocol. *Australian Journal of Agricultural Research* 50, 1285–1291.

Indicators and Reporting Office of Environment Canada (IROEC) (2001) *Nutrients in the Canadian Environment*. Available at: http://www.ec.gc.ca/soer-ree/English/SOER/nutrientseng.pdf

Intergovernmental Panel on Climate Change (IPCC) (2007) *Climate Change 2007: Climate Change Impacts, Adaptation and Vulnerability. Summary for Policy Makers*. Working Group II Contribution to the Intergovernmental Panel on Climate Change Fourth Assessment Report. Available at: http://www.ipcc.ch/SPM6avr07.pdf

Jensen, B.B. (1996) Methanogenesis in monogastric animals. *Environmental Monitoring and Assessment* 42, 99–112.

Joblin, K.N. (1999) Ruminal acetogens and their potential to lower ruminant methane emissions. *Australian Journal of Agricultural Research* 50, 1307–1313.

Jondreville, C., Revy, P.S. and Dourmad, J.Y. (2003) Dietary means to better control the environmental impact of copper and zinc by pigs from weaning to slaughter. *Livestock Production Science* 84, 147–156.

Kebreab, E., Castillo, A.R., Beever, D.E., Humphries, D.J. and France, J. (2000) Effects of management practices prior to and during ensiling and concentrate type on nitrogen utilization in dairy cows. *Journal of Dairy Science* 83, 1274–1285.

Kerr, B.J. and Easter, R.A. (1995) Effect of feeding reduced protein, amino acid-supplemented diets on nitrogen and energy balance in grower pigs. *Journal of Animal Science* 73, 3000–3008.

Kirchgessner, M., Windisch, W., Muller, H.L. (1995) In: von Engelhardt, W., Leonhard-Marek, S., Breves, G. and Giesecke, D. (eds) *Ruminant Physiology: Digestion, Metabolism, Growth and Reproduction*. Proceedings of the 8th International Symposium on Ruminant Physiology, Ferdinand Euke Verlag, Stuttgart, Germany, pp. 333–348.

Klieve, A.V. and Hegarty, R.S. (1999) Opportunities for biological control of ruminal methanogenesis. *Australian Journal of Agricultural Research* 50, 1315–1320.

Kreuzer, M., Machmüller, A., Gerdemann, M.M., Hanneken, H. and Wittmann, M. (1998) Reduction of gaseous nitrogen loss from pig manure using feeds rich in easily-fermentable non-starch polysaccharides. *Animal Feed Science and Technology* 73, 1–19.

Lanigan, G.W. (1972) Metabolism of pyrrolizidine alkaloids in the ovine rumen. IV. Effects of chloral hydrate and halogenated methanes on rumen methanogenesis and alkaloid metabolism in fistulated sheep. *Australian Journal of Agricultural Research* 23, 1085–1091.

Lassey, K.R., Ulyatt, M.J., Martin, R.J., Walker, C.F. and Shelton, I.D. (1997) Methane emissions measured directly from grazing livestock in New Zealand. *Atmospheric Environment* 31, 2905–2914.

Lee, D.S. and Dollard, G.J. (1994) Uncertainties in current estimates of emissions of ammonia in the United Kingdom. *Environmental Pollution* 86, 267–277.

Leek, A.B.G., Callan, J.J., Reilly, P., Beattie, V.E. and O'Doherty, J.V. (2007) Apparent component digestibility and manure ammonia emission in finishing pigs fed diets based on barley, maize or wheat prepared without or with exogenous non-starch polysaccharide enzymes. *Animal Feed Science and Technology* 135, 86–99.

Leuning, R., Baker, S.K., Jamie, I.M., Hsu, C.H., Klein, L., Denmead, O.T. and Griffith, D.W.T. (1999) Methane emission from free-ranging sheep: a comparison of two measurement methods. *Atmospheric Environment* 33, 1357–1365.

McCrabb, G.J., Berger, K.T., Magner, T., May, C. and Hunter, R.A. (1997) Inhibiting methane production in Brahman cattle by dietary supplementation with a novel compound and the effects on growth. *Australian Journal of Agricultural Research* 48, 323–329.

McGinn, S.M., Janzen, H.H. and Coates, T. (2003) Atmospheric ammonia, volatile fatty acids, and other odorants near beef feedlots. *Journal of Environmental Quality* 32, 1173–1182.

Machmüller, A., Dohme, F., Soliva, C.R., Wanner, M. and Kreuzer, M. (2001) Diet composition affects the level of ruminal methane suppression by medium-chain fatty acids. *Australian Journal of Agricultural Research* 52, 713–722.

Mourits, M.C.M., Berentsen, P.B.M., Huirne, R.B.M. and Dijkhuizen, A.A. (2000) Environmental impact of heifer management decisions on Dutch dairy farms. *Netherlands Journal of Agricultural Science* 48, 151–164.

Nyachoti, C.M., Arntfield, S.D., Guenter, W., Cenkowski, S. and Opapeju, F.O. (2006) Effect of micronized pea and enzyme supplementation on nutrient utilization and manure output in growing pigs. *Journal of Animal Science* 84, 2150–2156.

O'Kelly, J.C. and Spiers, W.G. (1992) Effect of monensin on methane and heat productions of steers fed lucerne hay either *ad libitum* or at the rate of 250 g/hour. *Australian Journal of Agricultural Research* 43, 1789–1793.

Olesen, J.E. and Bindi, M. (2002) Consequences of climate change for European agricultural productivity, land use and policy. *European Journal of Agronomy* 16, 239–262.

Paik, I.K. (2001) Management of excretion of phosphorus, nitrogen and pharmacological level minerals to reduce environmental pollution from animal production. *Asian-Australasian Journal of Animal Sciences* 14, 384–394.

Panetta, D.M., Powers, W.J., Xin, H., Kerr, B.J. and Stalder, K.J. (2006) Nitrogen excretion and ammonia emissions from pigs fed modified diets. *Journal of Environmental Quality* 35, 1297–1308.

Reis, R.B., Emeterio, F.S., Combs, D.K., Satter, L.D. and Costa, H.N. (2001) Effects of corn particle size and source on performance of lactating cows fed direct-cut grass-legume forage. *Journal of Dairy Science* 84, 429–441.

Rotz, C.A., Satter, L.D., Mertens, D.R. and Muck, R.E. (1999) Feeding strategy, nitrogen cycling, and profitability of dairy farms. *Journal of Dairy Science* 82, 2841–2855.

Sar, C., Mwenya, B., Pen, B., Morikawa, R., Takaura, K., Kobayashi, T. and Takahashi, J. (2005) Effect of nisin on ruminal methane production and nitrate/nitrite reduction *in vitro*. *Australian Journal of Agricultural Research* 56, 803–810.

Shriver, J.A., Carter, S.D., Sutton, A.L., Richert, B.T., Senne, B.W. and Pettey, L.A. (2003) Effects of adding fiber sources to reduced-crude protein, amino acid-supplemented diets on nitrogen excretion, growth performance, and carcass traits of finishing pigs. *Journal of Animal Science* 81, 492–502.

Sinclair, S.E., Clarke, M.R. and Kelly, A.M. (1996) Waste minimisation in the Australian feedlot industry: a perspective with regard to salinity parameters observed in commercial feedlots. *Animal Production in Australia* 21, 278–281.

Smith, J.W., Tokach, M.D., Goodband, R.D., Nelssen, J. L. and Richert, B.T. (1997) Effects of the interrelationship between zinc oxide and copper sulfate on growth performance of early-weaned pigs. *Journal of Animal Science* 75, 1861–1866.

Smith, V.H. (2003) Eutrophication of freshwater and coastal marine ecosystems: a global problem. *Environmental Science and Pollution Research International* 10, 126–139.

Steiner, T., Mosenthin, R., Fundis, A. and Jakob, S. (2006) Influence of feeding level on apparent total tract digestibility of phosphorus and calcium in pigs fed low-phosphorus diets supplemented with microbial or wheat phytase. *Livestock Science* 102, 1–10.

Takahashi, J., Chaudhry, A.S., Beneke, R.G., Suhubdy and Young, B.A. (1997) Modification of methane emission in sheep by cystine and a microbial preparation. *Science of the Total Environment* 204, 117–123.

United Nations Framework Convention on Climate Change (UNFCCC) (2006) *Greenhouse Gas Inventory Data*. Bonn, Germany. Available at: http://unfccc.int/ 2860.php

Watts, P.J. and Tucker, R.W. (1993) The effect of ration on waste management and odour control in feedlots. *Recent Advances in Animal Nutrition in Australia 1993*, University of New England, Armidale, New South Wales, Australia, pp. 117–129.

van Der Peet-Schwering, C.M.C. and Den Hartog, L.A. (2000) Manipulation of pig diets to minimize the environmental impact of pig production in the Netherlands. *Pig News and Information* 21, 53N–58N.

van Keulen, H., Aarts, H.F.M., Habekotté, B., van der Meer, H.G. and Spiertz, J.H.J. (2000) Soil–plant–animal relations in nutrient cycling: the case of dairy farming system 'De Marke'. *European Journal of Agronomy* 13, 245–261.

Veum, T.L., Carlson, M.S., Wu, C.W., Bollinger, D.W. and Ellersieck, M.R. (2004) Copper proteinate in weanling pig diets for enhancing growth performance and reducing fecal copper excretion compared with copper sulfate. *Journal of Animal Science* 82, 1062–1070.

Vermorel, M. (1997) Emissions annuelles de methane d'origine digestive par les ovins, les caprins et les equins en France. *Productions Animales* 10, 153–161.

Zervas, S. and Zijlstra, R.T. (2002) Effects of dietary protein and fermentable fiber on nitrogen excretion patterns and plasma urea in grower pigs. *Journal of Animal Science* 80, 3247–3256.

Zimmermann, B., Lantzsch, H.-J., Mosenthin, R., Schöner, F.-J., Biesalski, H.K., Drochner, W. (2002) Comparative evaluation of the efficacy of cereal and microbial phytases in growing pigs fed diets with marginal phosphorus supply. *Journal of the Science of Food and Agriculture* 82, 1298–1304.

16 Nutritional Genomics

Genotype Differences in Nutrient Use and Requirements

Animal species develop characteristics of nutrient use and requirements which reflect the nutritional environment in which they have evolved. This can lead to consistent differences between animal species. Examples are the differences between tropical and temperate cattle in their ability to use low-quality pastures or respond to concentrate-rich foods (e.g. Moran, 1985), or the ability of intermediate feeders such as the rusa deer (*Cervus timorensis*) to grow without adverse health consequences on concentrate-rich diets (Puttoo *et al.*, 1998) which a grazer herbivore such as the cow (*Bos* spp.) would not easily tolerate. It is useful to know how well particular breeds and species of animals tolerate particular nutritional environments. Then we can tailor our feeding management to take advantage of, or allow for, these differences.

'Nutritional genomics' is the study of the interface of nutrient requirements and metabolism, nutrient supply from foods and the animal's genetic characteristics. The term was coined by DellaPenna (1999). The study of nutritional genomics is proceeding rapidly in human nutrition, where the 'productivity' in question is health status, quality of life or longevity. We can also apply this concept to animal nutrition, where we are interested in more conventional forms of production. Typically, we ask: 'To what extent do the responses of individual boars differ when they are fed the same type of food, and how heritable is this difference?' This is a routine test in pig artificial breeding centres (Robinson and Buhr, 2005). Or: 'Why does my horse suffer repeated episodes of myoglobinuria ("tying-up")?' Perhaps the horse has a genetic predisposition to equine polysaccharide storage myopathy (Valentine, 2005). In these, and in many other, examples of idiosyncratic responses to a nutritional

environment, the underlying reason is the characteristics of the individual animal's genotype and its response to nutrients and foods.

Nutritional genomics subsumes two discipline areas: 'nutrigenomics' and 'nutrigenetics'. To quote Kussmann *et al.* (2006):

> Nutrigenetics asks the question how individual genetic disposition, manifesting as single nucleotide polymorphisms, copy-number polymorphisms and epigenetic phenomena, affects susceptibility to diet. Nutrigenomics addresses the inverse relationship, that is how diet influences gene transcription, protein expression and metabolism.

These concepts are illustrated in Fig. 16.1. The variety of genetic, biochemical and statistical techniques needed to explore nutritional genomics is described by Swanson *et al.* (2003).

Nutrigenetics: Case Studies of Genotype × Nutrient Interactions in Beef Cattle, Dairy Cattle and Pigs

There are substantial and well-recognized differences between *Bos indicus* (tropical) cattle breeds and those cattle which originated in Europe, the *Bos taurus* breeds. These differences are related to the abilities of these breeds to survive in nutrient-deficient environments and to take advantage of nutrient-rich conditions. Some of these differences are described in the following examples.

Genotype × diet quality interactions and the performance of Indonesian cattle

Moran (1985) investigated the growth and nutrient utilization of four Indonesian cattle breeds and the swamp buffalo (*Bubalis bubalis*). The cattle breeds were Ongole (*B. indicus*), Bali (*Bos sondaicus*), Madura (*B. indicus* × *B. sondaicus*) and Grati, a Friesian cross (*B. indicus* × *B. taurus*).

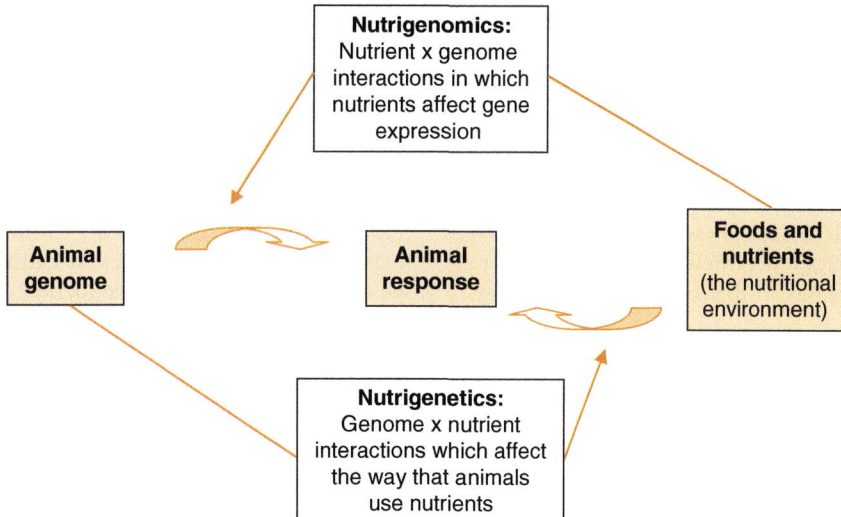

Fig. 16.1. The interrelationships between an animal's genetic characteristics (the genome) and the food that it eats (the nutritional environment).

When the cattle were fed a concentrate-rich diet, *B. sondaicus* genotypes (Bali and Madura cattle) ate the least food dry matter (DM), and consequently had the lowest metabolizable energy (ME) and N intakes and grew the slowest. The performance rankings were altered when a roughage-rich diet was given, as the Madura cattle ate more food and the Grati grew more slowly (Table 16.1). There were no differences in food constituent or energy digestibilities. When the effects of different rates of DM intake were allowed for, diet ME contents (which would reflect any differences in methane production and urine energy excretion) were the same for all breeds. However, N retention was highest in the Grati breed and lowest in the Madura cattle. This may be another characteristic of the *B. sondaicus* genotype but, unfortunately, we cannot be certain of this because these measurements were not made on the Bali cattle.

Experiments with cattle of different breeds

Hunter and Siebert (1986) reported three experiments in which they investigated the effects of forage type on food consumption by Hereford and Brahman cows. High-fibre, low-N forages like

Table 16.1. Dry matter intake and growth of Indonesian cattle and buffalo fed elephant grass with ad libitum concentrate for 154 days or with 30% concentrate for 112 days. (From Moran, 1985.)

Attribute	Diet	Madura	Ongole	Bali	Grati	Buffalo
		\multicolumn{5}{c}{Animal genotype (breed)}				
DM intake						
(kg/day)	ad lib concentrate	5.53[c]	6.50[b]	6.02[b,c]	7.97[a]	5.80[b,c]
	30% concentrate	6.15[b]	6.50[a,b]	–	7.08[a]	6.61[a,b]
(g/kg$^{0.75}$/day)	ad lib concentrate	72.6[b]	73.3[b]	76.8[b]	84.9[a]	76.9[a,b]
	30% concentrate	87.6[a,b]	82.8[b]	–	92.1[a]	89.5[a,b]
Growth (kg/day)	ad lib concentrate	0.59[c]	0.81[a,b]	0.66[b,c]	0.90[a]	0.73[a,b]
	30% concentrate	0.55[b]	0.65[a,b]	–	0.78[a]	0.59[b]
FCR[d]	ad lib concentrate	9.8[a]	8.2[a]	9.7[a]	9.3[a]	8.2[a]
	30% concentrate	11.6[a]	10.3[a]	–	9.5[a]	11.7[a]

[a,b,c] Within diets and attributes, means with similar notations are not different (P < 0.05).
[d] FCR = food conversion ratio (i.e. the amount of food required to produce 1kg of gain).

black speargrass (*Heteropogon contortus*) hay, and to a lesser extent pangola grass (*Digitaria eriantha*) hay, were eaten in smaller amounts by both cattle breeds. Also, both breeds ate similar amounts of these roughages. As the forage quality improved, the amounts of forage DM eaten increased, but the Herefords' intakes increased more than those of the Brahmans (Table 16.2). These differences between breeds and forages persisted when intakes were measured at a series of different live weights (Fig. 16.2).

Vercoe (1970) showed that *B. indicus* cattle have a lower maintenance requirement for energy than *B. taurus* cattle; and Hennessy (1987) demonstrated the greater ability of Brahman steers to maintain rumen NH_3 levels when given low-quality hay.

Monensin is a rumen modifier which increases the amount of propionic acid produced in the rumen. Propionate is a glucogenic volatile fatty acid which, when absorbed in increased amounts from the rumen, should increase the amount of lactose synthesized in the mammary gland, and thus increase milk production. Granzin and Dryden (1999) were interested to see if this effect was expressed to similar degrees in cattle of differing genetic merits for milk production. Low genetic merit Friesian cows produced less milk when given monensin than cows of high genetic merit. They also had higher serum glucose levels; this suggests that they could not use for lactose production the extra glucose that they obtained as a result of monensin administration.

Table 16.2. Intake of dry matter from various forages by Hereford and Brahman steers. (From Hunter and Siebert, 1986.)

Forage	Hereford	Brahman	Advantage to Hereford (%)
Speargrass	11.3	11.8	−4.2
Pangola grass	17.8	16.1	10.6
Pasture	21.5	20.1	7.0
Pasture + lucerne	24.9[a]	21.7[b]	14.7
Lucerne	27.6[a]	23.0[b]	20.0

[a,b]Means in the same row with similar notations are not different (P < 0.05).

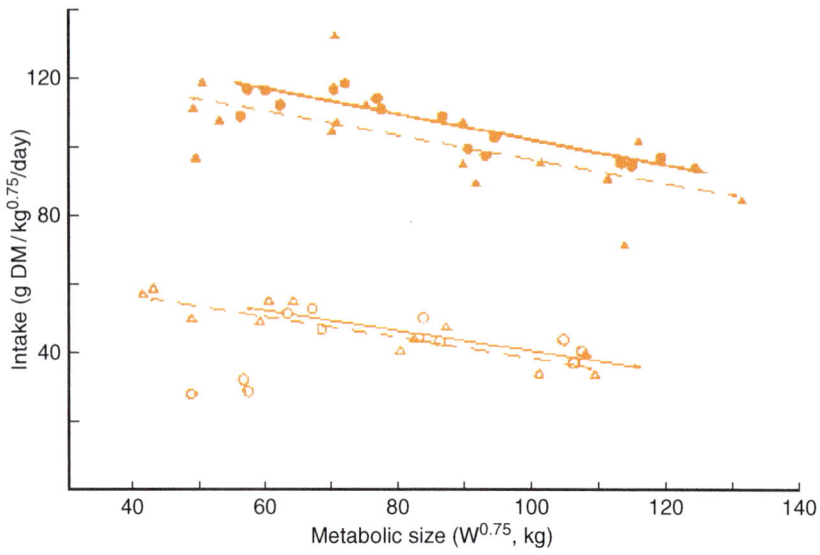

Fig. 16.2. Relationship between increasing metabolic size and dry matter intake by Hereford and Brahman steers. (Redrawn from Hunter and Siebert, 1986.)

Effects of sex and genotype in pigs

The development of pig strains with specific growth and carcass attributes has led to a re-evaluation of the nutrient requirements of these strains. Some of the characteristics of pigs of different sexes and lean growth genotypes are illustrated in the following two experiments.

Two lines of Large White × Landrace pigs (a control line, and one selected for high lean growth) were developed over five generations of breeding by McPhee *et al.* (1991). They were given three diets of varying digestible energy (DE) content (14, 15 and 16 MJ/kg and 0.85 g lysine/MJ DE until week 6; or 13, 14 and 15 MJ/kg and 0.7 g lysine/MJ DE until week 12). The pigs were grown from 25 to 81 kg live weight. Growth of the control pigs (data from males and females pooled) was maximized, irrespective of the diet DE content, at about 17 g lysine/day, whereas growth of the selected pigs was maximized when the DE content was 14 MJ/kg and at about 20 g lysine/day. The selected line pigs grew faster than the controls (0.77 versus 0.69 kg/day, $P < 0.05$) and were leaner. Male and female grower pigs are commonly reported to differ in their ability to respond to high diet nutrient contents, especially when the pigs are greater than 50 kg live weight. In this experiment, males grew faster than females (0.76 versus 0.71 kg/day, $P < 0.05$) and had leaner carcasses.

Noblet *et al.* (1999) examined the maintenance energy requirements of Pietrain, Large White and Meishan pigs. They found that each pig strain and/or type (males, females and castrates) had different maintenance requirements (as indicated by fasting heat production), and very different capacities to deposit tissue protein and fat. Energy requirements for maintenance (kJ/kg$^{0.75}$/day) for the seven genotype × sex groups examined were: Large White castrates, 482; Large White × Meishan castrates, 471; Large White females, 475; synthetic line males, 455; Meishan castrates, 449; Large White males, 439; and Pietrain males, 413. Castrated males had the highest maintenance requirements, followed by the female Large Whites, which is consistent with other published data on the growth of different pig sexes. The maintenance energy requirement rankings were paralleled exactly by the ratio of protein to lipid deposition. Castrates deposited about three times as much fat as protein. The growth data do not allow a comprehensive comparison of the four breeds, but suggest that the Mieshan breed performs more poorly than the others.

These results clearly have genetic origins, but the actual characteristics of the animals' genomes which are responsible for these between-species or between-breed differences are not defined. To do this, we must explore the animals' genomes in more detail.

Genes and Gene Expression

Genes are lengths of deoxyribonucleic acid (DNA) which carry sequences of the purine and pyrimidine bases, adenine, cytosine, guanine and thymine. These sequences code for specific amino acids which, after they have been joined together, form the protein that this gene codes for. Genes have several different regions, including a regulatory area (which is a gene which codes for a repressor molecule), a promoter region (where the two DNA strands can be more easily separated to allow the insertion of the RNA polymerase enzyme), exons (regions which carry the codes for specific amino acids) and introns (which are regions which do not have any protein coding function).

Alberts *et al.* (2002) give a comprehensive description of gene expression, but in brief, it involves the synthesis of a protein (e.g. an enzyme) which has a metabolic function. The process involves the sequence:

1. Transcription: the information (i.e. the sequence of bases) on part of a DNA molecule is used to synthesize a new molecule of messenger RNA (mRNA). This process is initiated by the action of a protein, such as a hormone, which binds to the promoter region of the gene (see Yang, 1998 for a description of transcription sites and factors). RNA polymerase then creates an mRNA molecule which has a base sequence which complements the sequence on the part of the DNA molecule being transcribed.
2. Processing: transcribed introns (regions which do not code for amino acids) are removed from the new mRNA molecule, and the molecule is capped and tailed.
3. Transport: the processed mRNA moves from the nucleus into the cytoplasm, where it binds to a ribosome (either free in the cytoplasm or attached to the endoplasmic reticulum).
4. Translation: the information carried on the mRNA molecule (i.e. the sequence of bases) is

translated into a protein (i.e. a sequence of amino acids) by matching the base sequence of mRNA with that (complementary) sequence on transfer RNA (tRNA) molecules associated with the ribosome.

Gene expression can be regulated by inducer and repressor genes. These are located near the site of the gene whose transcription they influence. Murray *et al.* (2000) describe the actions of the *lac* and *trp* genes in *Escherichia coli*, to illustrate the actions of repression and induction on gene expression. 'Repression' occurs when a repressor gene synthesizes a molecule (a repressor molecule) which binds to the DNA site where the RNA polymerase enzyme would normally begin to transcribe the base sequence of the gene. This binding prevents RNA polymerase from initiating the process of gene expression. The *lac* gene is an example of this type of repression. In some gene systems, the repressor molecule is the product of the gene. This provides a form of feedback inhibition which controls the amount of product produced. 'Induction' occurs when a gene is normally repressed, but inducer molecules bind to the repressor molecules and prevent them from binding to DNA. The *trp* gene is an example of this form of induction.

Single nucleotide polymorphism and animal performance

A single nucleotide polymorphism (SNP) occurs when a single base (adenine, guanine, cytosine or thymine) in a gene is substituted by a different base. There are several examples of SNPs affecting growth, milk composition and carcass composition in domestic animals (e.g. Li *et al.*, 2004; Morris *et al.*, 2007; and review by Laible *et al.*, 2006). Two examples are:

1. Substitution of adenine for guanine in a region of the porcine melanocortin-4 receptor (MC4R) gene: this SNP causes asparagine, instead of aspartic acid, to be inserted into the receptor protein (Kim *et al.*, 2000). The MC4R is located in the paraventricular nucleus and is part of the mechanism which controls food intake. By changing the amino acid the receptor functionality changes and pigs with this SNP grow more quickly, eat more food and are somewhat fatter. These outcomes reflect the role of melanocortin-4 in the brain, which is to inhibit food intake.
2. Cholecystokinin is also involved in the control of food intake, and thus affects growth and fatness:

SNPs have been found at two locations in the gene coding for the type A cholecystokinin receptor (CCKAR) of the pig digestive tract. These involve substitution of adenine for guanine and of guanine for cytosine (Houston *et al.*, 2006). The SNP at the +179 base appears to have the greatest effect on pig performance and pigs with the guanine variant eat about 5% more food and grow 3% more quickly than animals with adenine at this position in the gene.

Nutrigenomics: the Effects of Nutrition on Gene Expression

Nutrigenomics is the study of how nutrients can affect the expression of genes. There are three sub-disciplines: 'transcriptomics', which is the study of how nutrition may affect the way that genetic information is used to synthesize proteins, i.e. the effects of nutrition on mRNA synthesis; 'proteomics', which is the study of the total pool of proteins synthesized as a result of gene expression and the metabolic roles of these proteins; and 'metabolomics', or the study of the total pool of metabolites produced as a result of gene expression.

Examples of nutrient effects on gene expression

If we understand how nutrition can affect the expression of genes we can explain a number of nutritional effects on animal performance, including the genotype × nutritional environment interactions described earlier. Gene expression is usually monitored by measuring the amounts of mRNA encoding for the proteins which are under the influence of the gene in question (Dawson, 2006). There is, as yet, not very much information on animal nutrigenomics but the field is rapidly expanding in human nutrition (Ordovas and Mooser, 2004). Nevertheless, there are some data on nutritional effects on gene expression in animals. Some of this is described below.

Choi *et al.* (1998) investigated the effects of sequential under-feeding and adequate feeding of dairy heifers. They fed an experimental group a sequence of restriction and repletion diets (80% and 125% of the control diet). When cells from the heifers' mammary glands were cultured, those from the experimental group secreted more protein. The concentrations of mRNA coding for αs1 and β-caseins were nearly twice as high in cells from the experimental group than the control group. These results were associated with better growth and feed/gain ratio in the experimental group. A similar regulation of those enzymes which control the storage and mobilization of fat from adipose cells has been demonstrated by Bonnet et al. (1998). These workers found that when previously underfed ewes or cows were given adequate food, the amounts of mRNA coding for lipoprotein lipase and fatty acid synthase enzymes in their adipocytes increased by 19 and 25 times in ewes and six and eight times in cows.

Specific nutrient deficiencies or excesses can also alter gene expression. It is well known that a Zn deficiency reduces the animal's appetite. Sun et al. (2006) fed rats Zn-deficient (3 mg/kg) or Zn-excess (348 mg/kg) diets. In comparison to those given an adequate diet (36 mg/kg) rats given the deficient food responded by down-regulating the expression of genes coding for neuropeptide Y, and those given the Zn excess increased the expression of genes coding for ghrelin and melanin concentrating hormone. The changes in expression of these appetite-enhancing factors were accompanied by corresponding changes in food intake and live weight. As a second example, we can consider metallothionein, a small protein that binds and exchanges metal ions. Its activity in the intestine is influenced by Zn status, and Zn-deficient rats show a down-regulation of the MT-1 gene (Blanchard et al., 2001).

Recently, the effects of maternal nutrition on the developing fetus (also called 'fetal programming') have received attention, particularly in human nutrition. The thesis is that inappropriate feeding, e.g. deficiencies of specific nutrients, or starvation or overfeeding have adverse effects on the developing fetus' genome. The topic has been reviewed by Rees (2002) and Rhind (2004).

With this type of data, we are no longer in the position of having to view the results of our nutritional manipulations as though the animal was a 'black box'. Because nutritional manipulations can alter mRNA synthesis rates, we now have a powerful mechanism to explain, and thus help to further manipulate, animal responses through nutritional management.

The Nutritive Value of Transgenic Foods

Genetically modified (GM) crops are tested for safety against the concept of 'substantial equiva-lence' (or 'comparative assessment'). In essence, this test requires that the GM crop must have the same nutritional characteristics as a comparable crop which has a history of safe use. The substantial equivalence test requires 'molecular characterization, assessment of the safety of expressed proteins, comparative analysis of the macro- and micronutrient composition and contents of antinutrients and toxicants, determination of patterns of use and exposure, determination of toxicologic and nutritional characteristics, studies of safety in animals, and overall evaluation of safety' (Florentino, 2005).

Substantial equivalence in the nutritional characteristics of foods made from GM plants has been demonstrated by several authors. 'Roundup Ready' soybeans gave almost exactly similar growth and carcass responses in pigs as conventional soybean meal (Cromwell et al., 2002). Feeding trials with poultry, pigs and ruminants suggested that the incorporation of the Bacillus thuringiensis (Bt) toxin gene or the glufosinate (phosphinothricin) tolerance (Pat) gene into maize and sugarbeets did not significantly affect their chemical composition (cell wall composition, fatty acid or amino acid profiles) or nutrient utilization (Bohme et al., 2001; Aulrich et al., 2002). There is some limited evidence that transgenic plants may be more easily digested than conventional plants. Bohme et al. (2001) fed transgenic sugarbeet and maize to pigs. Their results indicated that the Pat crops had slightly higher ME contents: 14.1 versus 13.7 MJ/kg DM for sugarbeets, and 16.0 versus 15.8 MJ/kg DM for maize grain. There were no differences in the ME contents of either sugarbeets or ensiled sugarbeet tops fed to ruminants. Pigs fed 'Roundup Ready' soybean meal were slightly fatter than those given normal meal (Cromwell et al., 2002).

There is concern that transgenic DNA from plants used as foods could be incorporated into rumen microorganisms or the animal. Rumen bacteria grown in vitro in a medium which included Bt maize responded differently to challenge with ampicillin (Koch et al., 2006) but that the mechanism involved gene transfer is only speculative. DNA transfer from transgenic organisms has not yet been detected in animal tissues (e.g. Flachowsky and Aulrich, 2001; Aulrich et al., 2002; Flachowsky et al., 2005) although fragments of natural plant DNA have been found in chicken muscle (Aulrich et al., 2002).

Nutrition of Transgenic Animals

Transgenic animals are produced by inserting 'exotic' genetic material into an animal's genome; cloned animals are produced by inserting existing nuclear material from an animal of (usually high) genetic quality. If, as is suggested by Wheeler and Walters (2001) and Vajta and Gjerris (2006), future transgenic and cloned animals will grow faster, use food more efficiently, have more offspring, be more disease resistant and/or have different body or product chemical compositions, then these animals will impose greater nutritional demands on their managers. These demands are likely to be simply quantitative, i.e. 'more of the same thing'. On the other hand, if the animal has a new metabolic characteristic, perhaps the capacity to hydrolyse phytin through the insertion of a phytase gene (Golovan *et al.*, 2001), then their managers will have to design new feeding regimes. This area of animal nutrition is very much in its infancy, but we will have to carefully monitor developments.

References

Alberts, B., Johnson, A., Lewis, J., Raff, M., Roberts, K. and Walter, P. (2002) *Molecular Biology of the Cell*, 4th edn. Garland Science, New York.

Aulrich, K., Bohme, H., Daenicke, R., Halle, I. and Flachowsky, G.T.I. (2002) Novel feeds – a review of experiments at our institute. *Food Research International* 35, 285–293.

Blanchard, R.K., Moore, J.B., Green, C.L. and Cousins, R.J. (2001) Modulation of intestinal gene expression by dietary zinc status: effectiveness of cDNA arrays for expression profiling of a single nutrient deficiency. *Proceedings of the National Academy of Sciences* 98, 13507–13513.

Bohme, H., Aulrich, K., Daenicke, R. and Flachowsky, G. (2001) Genetically modified feeds in animal nutrition 2nd communication: glufosinate tolerant sugar beets (roots and silage) and maize grains for ruminants and pigs. *Archiv fur Tierernahrung* 54, 197–207.

Bonnet, M., Faulconnier, Y., Flechet, J., Hocquette, J.F., Leroux, C., Langin, D., Martin, P. and Chilliard, Y. (1998) Messenger RNAs encoding lipoprotein lipase, fatty acid synthase and hormone-sensitive lipase in the adipose tissue of underfed-refed ewes and cows. *Reproduction, Nutrition, Development* 38, 297–307.

Choi, Y.J., Jang, K., Yim, D.S., Baik, M.G., Myung, K.H., Kim, Y.S., Lee, H.J., Kim, J.S. and Han, I.K. (1998) Effects of compensatory growth on the expression of milk protein gene and biochemical changes of the mammary gland in Holstein cows. *Journal of Nutritional Biochemistry* 9, 380–387.

Cromwell, G.L., Lindemann, M.D., Randolph, J.H., Parker, G.R., Coffey, R.D., Laurent, K.M., Armstrong, C.L., Mikel, W.B., Stanisiewski, E.P. and Hartnell, G.F. (2002) Soybean meal from Roundup Ready or conventional soybeans in diets for growing–finishing swine. *Journal of Animal Science* 80, 708–715.

Dawson, K.A. (2006) Nutrigenomics: feeding the genes for improved fertility. *Animal Reproduction Science* 96, 312–322.

DellaPenna, D. (1999) Nutritional genomics: manipulating plant micronutrients to improve human health. *Science* 285, 375–379.

Flachowsky, G. and Aulrich, K. (2001) Nutritional assessment of feeds from genetically modified organism. *Journal of Animal and Feed Sciences* 10, 181–194.

Flachowsky, G., Chesson, A. and Aulrich, K. (2005) Animal nutrition with feeds from genetically modified plants. *Archives of Animal Nutrition* 59, 1–40.

Florentino, R.F. (ed.) (2005) Executive summary. Proceedings of the symposium and workshop on biotechnology-derived nutritious foods: challenges and opportunities in Asia. *Food and Nutrition Bulletin* 264, 404–408.

Golovan, S.P., Meidinger, R.G., Ajakaiye, A., Cottrill, M., Wiederkehr, M.Z., Barney, D.J., Plante, C., Pollard, J.W., Fan, M.Z., Hayes, M.A., Laursen, J., Hjorth, J.P., Hacker, R.R., Phillips, J.P. and Forsberg, C.W. (2001) Pigs expressing salivary phytase produce low-phosphorus manure. *Nature Biotechnology* 19, 741–745.

Granzin, B.C. and Dryden, G.McL. (1999) The effects of monensin on milk production and levels of metabolites in blood and rumen fluid of Holstein-Friesian cows in early lactation. *Australian Journal of Experimental Agriculture* 39, 933–940.

Hennessy, D.W. (1987) The role of nutrition in the genotype × environment interactions of *Bos taurus* and *Bos indicus* cattle in New South Wales. *Recent Advances in Animal Nutrition in Australia* 1987, University of New England, Armidale, NSW, Australia, pp. 64–71.

Houston, R.D., Haley, C.S., Archibald, A.L., Cameron, N.D., Plastow, G.S. and Rance, K.A. (2006) A polymorphism in the 5'-untranslated region of the porcine cholecystokinin type a receptor gene affects feed intake and growth. *Genetics* 174, 1555–1563.

Hunter, R.A. and Siebert, B.D. (1986) The effects of genotype, age, pregnancy, lactation and rumen characteristics on voluntary intake of roughage diets by cattle. *Australian Journal of Agricultural Research* 37, 549–560.

Kim, K.S., Larsen, N., Short, T., Plastow, G. and Rothschild, M.F. (2000) A missense variant of the porcine melanocortin-4 receptor (*MC4R*) gene is associated with fatness, growth, and feed intake traits. *Mammalian Genome* 11, 131–135.

Koch, M., Strobel, E., Tebbe, C.C., Heritage, J., Breves, G. and Huber, K. (2006) Transgenic maize in the presence of ampicillin modifies the metabolic profile and microbial population structure of bovine rumen fluid *in vitro. British Journal of Nutrition* 96, 820–829.

Kussmann, M., Raymond, F. and Affolter, M. (2006) OMICS-driven biomarker discovery in nutrition and health. *Journal of Biotechnology* 124, 758–787.

Laible, G., Wagner, S. and Alderson, J. (2006) Oligonucleotide-mediated gene modification and its promise for animal agriculture. *Gene* 366, 17–26.

Li, C., Basarab, J., Snelling, W.M., Benkel, B., Murdoch, B., Hansen, C. and Moore, S.S. (2004) Assessment of positional candidate genes *myf*5 and *igf*1 for growth on bovine chromosome 5 in commercial lines of *Bos taurus. Journal of Animal Science* 82, 1–7.

McPhee, C.P., Williams, K.C. and Daniels, L.J. (1991) The effect of selection for rapid lean growth on the dietary lysine and energy requirements of pigs fed to scale. *Livestock Production Science* 27, 185–198.

Moran, J.B. (1985) Comparative performance of five genotypes of Indonesian large ruminants. I. Effect of dietary quality on live weight and feed utilization. *Australian Journal of Agricultural Research* 36, 743–752.

Morris, C.A., Cullen, N.G., Glass, B.C., Hyndman, D.L., Manley, T.R., Hickey, S.M., McEwan, J.C., Pitchford, W.S., Bottema, C.D.K. and Lee, M.A.H. (2007) Fatty acid synthase effects on bovine adipose fat and milk fat. *Mammalian Genome* 18, 64–74.

Murray, R.K., Granner, D.K., Mayes, P.A. and Rodwell, V.W. (2000) *Harper's Biochemistry*, 25th edn. Section 4. Lang Medical Books/McGraw-Hill, New York.

Noblet, J., Karege, C., Dubois, S. and van Milgen, J. (1999) Metabolic utilisation of energy and maintenance requirements in growing pigs: effects of sex and genotype. *Journal of Animal Science* 77, 1208–1216.

Ordovas, J.M. and Mooser, V. (2004) Nutrigenomics and nutrigenetics. *Current Opinion in Lipidology* 15, 101–108.

Puttoo, K., Dryden, G.McL. and McCosker, J.E. (1998) Performance of weaned rusa (*Cervus timorensis*) deer given concentrates of varying protein content with sorghum hay. *Australian Journal of Experimental Agriculture* 38, 33–39.

Rees, W.D. (2002) Manipulating the sulfur amino acid content of the early diet and its implications for long-term health. *Proceedings of the Nutrition Society* 61, 71–77.

Rhind, S.M. (2004) Effects of maternal nutrition on fetal and neonatal reproductive development and function. *Animal Reproduction Science* 82–83, 169–181.

Robinson, J.A.B. and Buhr, M.M. (2005) Impact of genetic selection on management of boar replacement. *Theriogenology* 63, 668–678.

Swanson, K.S., Schook, L.B. and Fahey, G.C. Jr (2003) Nutritional genomics: implications for companion animals. *Journal of Nutrition* 133, 3033–3040.

Sun, J.-Y., Jing, M.-Y., Wang, J.-F., Zi, N.-T., Fu, L.-J., Lu, M.-Q. and Pan, L. (2006) Effect of zinc on biochemical parameters and changes in related gene expression assessed by cDNA microarrays in pituitary of growing rats. *Nutrition* 22, 187–196.

Vajta, G. and Gjerris, M. (2006) Science and technology of farm animal cloning: state of the art. *Animal Reproduction Science* 92, 211–230.

Valentine, B.A. (2005) Diagnosis and treatment of equine polysaccharide storage myopathy. *Journal of Equine Veterinary Science* 25, 52–61.

Vercoe, J.E. (1970) The fasting metabolism of Brahman, Africander and Hereford × Shorthorn cattle. *British Journal of Nutrition* 24, 599–606.

Wheeler, M.B. and Walters, E.M. (2001) Transgenic technology and applications in swine. *Theriogenology* 56, 1345–1369.

Yang, V.W. (1998) Eukaryotic transcription factors: identification, characterisation and functions. *Journal of Nutrition* 128, 2045–2051.

17 Animal Responses to Stock Food Processing

An Overview of Animal Food Processing

Animal food processing can involve particle size reduction by chopping or milling (this is also called 'grinding' or 'cracking'), compaction by wafering or pelleting, the application of heat and/or moisture, application of chemicals such as NH_3 and NaOH and mixing with other ingredients. We process foods to:

1. Increase digestibility or the efficiency of utilization of metabolizable energy (ME);
2. Improve the content of nutrients other than available energy;
3. Increase the amount of food eaten or the rate of consumption by increasing the food's bulk density or by changing its particle size;
4. Destroy pathogens (e.g. *Salmonella*), toxins (especially those which occur naturally in legume seeds) and weed seeds;
5. Improve handling characteristics, so as to reduce waste during feeding or to facilitate the use of automated food handling equipment;
6. Improve storage characteristics by preventing the oxidation of fats, the destruction of vitamins or mould growth;
7. Allow the intimate mixing of ingredients by making particle sizes more uniform;
8. Facilitate other forms of processing (e.g. milling prior to pelleting, chopping before ensiling);
9. Improve the appeal of the food to the (human) purchaser or the (animal) consumer by pelleting or extrusion, or by adding colourants or aromas.

The processing methods commonly used for roughages and concentrates are outlined in Table 17.1, together with summaries of their effects.

Physical and Chemical Processing of Cereal Grains

The intact grain is protected in four ways against digestion by animals. The seedcoat protects the endosperm and the digestibility of whole (unmilled) grain can be quite low, e.g. 50–75% for cattle (Table 17.2). Milling breaks the seed into small pieces and exposes more endosperm to the action of digestive enzymes. A second limit to digestion is imposed by the structure of the grain endosperm – starch granules buried deep within the endosperm are inaccessible to enzymes. Third, the starch granules are embedded in a protein matrix (McAllister *et al.*, 1993; Swan *et al.*, 2006). The fourth factor is the chemistry of starch, as amylose is more resistant to digestion than amylopectin (Gallant *et al.*, 1992).

Milling

This is the most common 'cold' process used in stock food processing; it is the breaking of a grain or other seed or plant part. The only changes caused by milling are physical: the material is broken into small pieces. There is no chemical change other than the possible loss of some food constituents in small particles lost as dust. Most commonly, grain is milled so that it is broken into two or three pieces with a minimum of fines (these are particles less than about 350 µm). Roller mills are sometimes preferred over hammer mills for processing cereal grains, although there is little evidence that there are differences between the two types of mill which reflect on animal performance (Ohh *et al.*, 1983; Stock and Mader, 1987; Laurinen *et al.*, 2000; Choct *et al.*, 2004).

There are between-species differences in the animal response to grain milling. Animals which chew their food effectively during ingestion or rumination (e.g. pigs, horses, sheep, goats and deer) are able to digest whole grain fairly completely (Ørskov, 1976, 1979; Vipond *et al.*, 1980; Wondra *et al.*, 1995; Rapetti and Bava, 2004). Starch digestion by horses is influenced variably by dry milling. Precaecal digestion of starch is improved, especially

Table 17.1. Food processing methods and their uses.

Food class/type	Processing method	Effect of processing
Green roughages	Chopping	Reduces particle size, allows mixing and ensiling (facilitates the exclusion of air)
	Ensiling	Allows indefinite storage
	Ammoniation	Preserves by preventing mould growth
Dry roughages	Chopping	Reduces particle size to allow mixing, facilitates automated food handling and the use of food troughs; improves intake rate
	Chemical treatment	Increases intake and digestibility and (after ammoniation) N content
	Milling	Increases bulk density and intake rate; a preliminary to further processing
	Pelleting	Increases bulk density; reduces the possibility of diet selection
	Wafering	Increases bulk density and intake rate
Cereal grains	Milling	Increases digestibility; allows more intimate mixing with other ingredients; used as a preliminary step to further processing
	Steaming and flaking	Increases digestibility, kills weed seeds
	Chemical treatment	Increases digestibility, destroys pathogenic organisms
	Ensiling	Increases digestibility by activating seed enzymes
Legume seeds (including soybeans)	Heating	Destroys some endogenous poisons
	Milling	Increases digestibility
Mixed foods	Addition of nutrient and non-nutritive additives	Prevents fats oxidizing (e.g. butylated hydroxy toluene) add/mask flavours (e.g. vanilla, caramel), increase mineral/vitamin contents, promote nutrient utilization (e.g. monensin), protect against pathogens (e.g. coccidiostats, antibiotics), etc.
	Mixing	Used in the manufacture of a nutritionally complete diet; reduces diet selection; used as a preliminary step to further processing
	Pelleting	Increases digestibility; prevents ration demixing; prevents diet selection; kills some pathogenic organisms

oat and maize starch (review by Julliand *et al.*, 2006). On the other hand, Julliand *et al.* (2006) found no evidence that starch digestion in the caecum was influenced by milling.

Pigs respond favourably to being given milled grain (Healy *et al.*, 1994; Wondra *et al.*, 1995) and there is some evidence that fine milling can increase grain digestibility over and above the levels usually obtained from coarse milling. When sorghum, wheat or maize are fed to weaner (Healy *et al.*, 1994) and finishing pigs (Wondra *et al.*, 1995), finer milling may increase nutrient digestibility and daily gain, and improve the efficiency of food use, by 5–8%. In these studies, the finely milled grain had particles about 300–400 µm compared to 900 µm to 1 mm in diameter, or 1 versus 2–4 mm diameter. Fine milling of barley grain for pigs is less effective than with sorghum and maize (Laurinen *et al.*, 2000; Medel *et al.*, 2000). This could be due to differences in the endosperm characteristics of the two grain types. Laurinen *et al.* (2000) showed that finer milling (roller milling, the majority of particles 1–2 mm compared to 2–4 mm) improved the digestibilities of energy, total carbohydrate and protein of both wheat and barley fed to growing and finishing pigs. There was no effect on nutrient digestibility between finely and coarsely hammer-milled grain, but in this study the difference in particle size was not large, and the majority of particles in the coarse grain were between 250 µm and 2 mm. Clearly, 'fine' and 'coarse' milling have to be carefully defined. Overall, the data support the contention of several other authors (Wondra *et al.*, 1995; Mavromichalis *et al.*, 2000; Kim *et al.*, 2005) that an optimum particle size for pigs (taking both animal response and the cost of production into account) is about 600 µm.

In most cases, the grain in cattle diets is a component (perhaps less than 50%) of the total diet, and much of the work exploring processing effects has been done with mixed grain/roughage rations. Starch digestibility is a better index of grain digestion than dry matter (DM) digestibility in mixed diets because in these most of the starch

Table 17.2. Digestibility for cattle of the constituents of whole and milled cereal grains.

Grain	Constituent	Digestibility (%)		Reference
		Whole	Milled	
Oats	Dry matter	71.9–74.6	–	Toland, 1978b[b]
	Organic matter	76.7	81.0	Toland, 1976[b]
	Organic matter	51.9	82.9	Nordin and Campling, 1976[a]
	Starch	93.4	99.1	Toland, 1976[c]
Barley	Dry matter	62.0	81.4	Hironaka et al., 1992[a]
	Dry matter	75.2	78.9	Ørskov and Greenhalgh, 1977[a]
	Organic matter	52.5	85.2	Toland, 1976[b]
	Organic matter	62.1	78.0	Nordin and Campling, 1976
	Starch	49.4	98.8	Toland, 1976[c]
	Starch	86.5	94.1	Hironaka et al., 1992[a]
Maize	Dry matter	76.6	79.0	Gorocica-Buenfil and Loerch, 2005[a]
	Starch	92.9	94.7	Gorocica-Buenfil and Loerch, 2005[a]
	Protein	73.5	71.5	Gorocica-Buenfil and Loerch, 2005[a]
	Neutral detergent fibre	61.5	58.0	Gorocica-Buenfil and Loerch, 2005[a]
Wheat	Dry matter	53.8–68.3	–	Toland, 1978a[b,d]
	Dry matter	72.6–77.5	–	Toland, 1978b[b]
	Organic matter	41.0	88.0	Nordin and Campling, 1976[a]
	Organic matter	62.9	87.7	Toland, 1976[b]
	Starch	62.0	99.0	Toland, 1976[c]
	Starch	14.4	93.3	Fulkerson and Michell, 1985
Sorghum	Organic matter	44.3	94.3	Nordin and Campling, 1976[a]

[a]Data are for grain fed as the sole ration constituent or for the whole diet in which grain was >80%.
[b]Digestibility data for grain calculated by difference.
[c]Digestibility data for starch assumes that grain was the only starch source.
[d]Range is for diets fed at 1.5% and 3% of live weight daily.

is from the cereal grain component. There is usually an improvement in grain digestion when it is milled before feeding to cattle (Table 17.2; Moe and Tyrrell, 1977; Axelsen et al., 1979; Fulkerson and Michell, 1985; and reviews by Campling, 1991; Mathison, 1996). Mathison (1996) cites data showing that starch digestion may be improved by milling (compared to whole grains) by as much as 37% for barley-based diets, 22% for oats and 15–20% for wheat. Compared to coarse milling, fine milling may further improve starch digestibility and also increase the proportion which is fermented in the rumen (Ceresnakova et al., 2005).

Maize is an exception to the general rule that milling improves digestibility (NRC, 2000) because even cattle masticate this grain fairly effectively (Simeone et al., 2003). Moe and Tyrrell (1977) and Gorocica-Buenfil and Loerch (2005) report data for maize-based diets which indicates that milling, compared to feeding whole grain, slightly increases energy digestibility and metabolizability, and that

fine milling is more effective than coarse milling (Table 17.3).

Milling of most grains improves food use efficiency by growing cattle (Axelsen et al., 1979; May and Barker, 1984/5; Stock and Mader, 1987) and increases milk yield in lactating cows (Moe and Tyrrell, 1977; Fulkerson and Michell, 1985; San Emeterio et al., 2000). The extent of improvement depends on the grain used and the proportion of grain in the diet, but differences of between 5% and 35% in growth and 0.029 kg milk fat/kg grain have been reported. Because it increases starch digestibility, fine milling can help to control unpleasant odours in feedlots by reducing the output of undigested starch in cattle faeces (Reis et al., 2001).

Final words of caution: although it might improve grain digestibility or increase the ruminal fermentation of starch, fine milling is likely to reduce the efficiency of starch energy use by the ruminant, and more importantly, increase the risk of gastric or oesophageal ulcers in pigs and rumenitis in cattle

Table 17.3. Partitioning of energy in diets containing 45% (lactating cattle) or 55% (dry cattle) maize grain. (From Moe and Tyrrell, 1977.)

Energy transactions (MJ/day)	Dry cows			Lactating cows		
	Whole[a]	Coarsely milled	Finely milled	Whole	Coarsely milled	Finely milled
Total	131.6	131.2	131.7	295.0	311.1	310.7
Digestible	77.2	79.6	84.4	155.4	174.8	189.1
Methane	9.2	8.5	9.1	17.0	17.0	17.9
Metabolizable	64.6	67.5	72.3	129.3	149.9	162.6
Metabolizability[b]	0.84	0.85	0.86	0.83	0.86	0.86

[a]Maize grain processing method; values in the table refer to energy in the whole diet.
[b]Metabolizability = amount of metabolizable energy/amount of digestible energy.

(Reimann *et al.*, 1968; Hedde *et al.*, 1985; Wondra *et al.*, 1995). The effect of grain milling on horse health is less clear. Disorders such as laminitis and enterotoxaemia are associated with the digestion of excessive amounts of starch in the equine large intestine (Garner *et al.*, 1977).

Tempering and ensiling grain

Whole grains can be treated with cold water as a preliminary processing step. This is called 'tempering'. It is the 'softening of the grain kernel that occurs following the addition of 4 to 8% moisture' (Zinn *et al.*, 1998). Tempering for up to an hour, followed by rolling, reduces the amount of fines produced during milling and may improve growth and efficiency of food use (Zinn *et al.*, 1998). A different form of tempering is used in flour milling (Fang and Campbell, 2003). It involves 'conditioning of grain in water for up to 24 h . . . so that the grain reaches a final water content of 15 to 16%' (Estrada-Girón *et al.*, 2005). This form of tempering separates the endosperm, embryo and seedcoat and makes the endosperm break more easily during milling.

Grain can also be ensiled. This is different to forage ensiling which is a preserving process only and does not improve the nutritive value. Ensiled grain (high-moisture grain, or HM grain) is made by harvesting grain at about 30% moisture content, optionally milling it (using a roller mill), then storing it anaerobically for at least 15 days. If the grain is not milled before ensiling it must be rolled before being fed. Dry grain (12.5–14% moisture) can be reconstituted, by spraying water on to the dry grain as it is augered into the silo, to increase the moisture content to about 30%. Seed enzymes are activated and adhering bacteria use the grain constituents as nutrients (Pflugfelder *et al.*, 1986). This increases the solubilities of grain carbohydrates and nitrogenous constituents (Pflugfelder *et al.*, 1986; Balogun *et al.*, 2005). It seems that the processes associated with seed germination have a greater effect on digestibility than those that occur during anaerobic storage (Pflugfelder *et al.*, 1986; Balogun *et al.*, 2005).

There are two caveats about making HM grain: first, rolling the grain prior to reconstitution and ensiling may reduce or prevent the beneficial changes which otherwise occur during these processes (Stock and Mader, 1987); and second, mould growth with accompanying risk of animal poisoning can occur in the moist grain. This can be controlled by adding an inhibitor such as 1% gaseous NH_3 or 0.5–1.5% propionic acid (Sauer and Burroughs, 1974; Kaspersson *et al.*, 1988).

HM grain is a variable commodity, affected by grain type, moisture content at ensiling and the length of the ensiling period. Ensiled grain is more digestible than milled dry grain (Clark and Harshbarger, 1972; Stock and Mader, 1987; Aldrich *et al.*, 1993; Balogun *et al.*, 2005, 2006) when fed to cattle. According to a series of studies reviewed by Stock and Mader (1987) HM sorghum is used for growth in feedlot cattle about as efficiently as steam-flaked (SF) grain and more efficiently than dry-milled grain. However, there are contradictory results. Data from Huck *et al.* (1998), and a meta-analysis by Owens *et al.* (1997), indicate that HM grain gives similar results in growth and the ratio of live weight gain to food intake (FCR) as milled grain. There is no convincing evidence that HM grain improves pig performance (Cole *et al.*, 1970; Goransson and Ogle, 1985).

Heat processing

Moist heat processing changes the grain's chemistry. Starch gelatinization is a physico-chemical change in the structure of starch caused by the forcing of water molecules between adjacent α-glucan strands of amylopectin. Gelatinization of starch involves the swelling and hydration of the starch granule, loss of crystallinity through the irreversible breakage of H bonds in the crystalline regions of the granule (Rooney and Pflugfelder, 1986; Tester *et al.*, 2004; Svihus *et al.*, 2005), the 'leaching' of amylose from the granule and a partial breakdown of amylose and amylopectin molecules to dextrins (Atwell *et al.*, 1988; Jacobs and Delcour, 1998). Starch, especially the amylose, recrystallizes after the heated grain has cooled (Svihus *et al.*, 2005). This is called 'retrogradation'. Retrograded starch is much less digestible than native starch (Zinn *et al.*, 2002).

Some heat processes, e.g. pelleting, will reduce (but not eliminate) contamination by pathogens such as *Salmonella* bacteria provided that the temperature reached in the food is high enough, more than 70°C (Jones and Richardson, 2004).

Responses of pigs to heat-treated cereal grains

Micronizing and roasting of barley and maize increase starch gelatinization by up to 50% units, and this is associated with a tendency for improved starch and energy digestion (Costa *et al.*, 1976; Medel *et al.*, 1999, 2000; Thacker, 1999; Zarkadas and Wiseman, 2002). Medel *et al.* (2000) reported increased gelatinization following expanding of barley, with a small tendency for improved organic matter digestion. Notwithstanding these results, the importance of gelatinization in grains processed for animal foods is debatable. Even when extensive gelatinization has occurred, there may be only a very small, or no, increase in starch digestibility (Fernandes *et al.*, 1975; Hongtrakul *et al.*, 1998; Medel *et al.*, 2000). Lawlor *et al.* (2003) reported a threefold increase in starch gelatinization (about a 40% units increase) following steaming and flaking of both wheat and maize but found no change in animal performance. The relevance of gelatinization remains an unresolved issue in cereal grain processing.

A great deal of the work on the responses of pigs to heat-treated cereal grains has been done with weaners (5–20 kg live weight) and growers (20 kg and more) fed diets based on barley and maize. In general, heat treatments give similar results with both pig types, but there may be differences between grains: e.g. Medel *et al.* (1999) obtained positive responses when barley was extruded or micronized, rather than maize. Lawlor *et al.* (2003) could not demonstrate improvements in any aspect of weaner performance when they fed a mixture of SF maize and wheat. However, the results are mixed. Medel *et al.* (2002, 2004) reported impressive (16–18%) improvements in the growth of weaners fed diets based on SF barley or maize.

Increased energy availability following heat treatment does not always improve growth (Medel *et al.*, 1999, 2000 compared to Thacker, 1999 and Zarkadas and Wiseman, 2002). Instead, the usual animal response is for growth to be similar to that of pigs fed untreated grains but for the FCR to improve (Costa *et al.*, 1976; Skoch *et al.*, 1983a,b; Medel *et al.*, 1999; Thacker, 1999; Medel *et al.*, 2004).

Heat treatment appears to increase starch digestion in the small intestine and to reduce fermentation in the caecum (Van Der Poel *et al.*, 1990a; Medel *et al.*, 2002, 2004). This will increase the amount of glucose absorbed and available for growth and general metabolism. Also, pigs may be more susceptible to swine dysentery when large amounts of non-starch polysaccharides or of resistant starch enter the large intestine (Pluske *et al.*, 1996). Thus, heat processing of cereal grains may lower the incidence of this disease.

Heat processing can affect non-energy nutrients. Heat treatment tends to reduce N utilization by pigs, leading to lower N retention (Costa *et al.*, 1976, pelleted maize) or increased excretion of undigested food N (Laurinen *et al.*, 1998, expanded barley). Van Der Poel *et al.* (1990a) noted that extrusion of maize increased the ileal (but not the whole-tract) digestibility of cystine and proline while not affecting any other amino acid. Heat-labile vitamins (e.g. vitamin A, vitamin D, vitamin C and several B group vitamins) may be damaged by heat treatments (Awuah *et al.*, 2007).

Responses of ruminants to heat-processed cereal grains

The starch in milled maize, oats, barley and wheat is very digestible, with 'whole-tract' digestibilities of 91–100% (Theurer, 1986; Huntington, 1997). Sorghum starch is less completely digested (80–92%). Steam flaking increases the whole-tract digestibility of maize and sorghum starch by 3–11% units, e.g. from 87% to 98% for sorghum, but gives little extra improvement with barley, oats or wheat (Huntington, 1997; Theurer et al., 1999). Very similar results for micronized and SF grains are summarized in Theurer (1986) for both sheep and cattle.

Consistent with their whole-tract digestibilities, wheat, barley and oat starches are almost completely digested in the rumen (about 80–95%, Ørskov, 1986; Huntington, 1997), but those of maize and sorghum are less available (70–90% and 60–80%, respectively, Huntington, 1997; Theurer et al., 1999). The effects of SF and micronizing applied to sorghum and maize on the sites of starch digestion in cattle are shown in Table 17.4. The post-ruminal starch digestibilities of milled barley, oats and wheat are 75–85% (Huntington, 1997). These values are increased, by 2% (oats and wheat) and 13% (barley), by heat processing. When we compare the values for ruminal and post-ruminal starch digestibilities reported between different experiments we find that these are relatively uniform except for milled sorghum. For this grain and processing method there is a substantial variation between experiments, with coefficients of variation (CV) of 20–40%. However, CV values for SF sorghum are only 1.5–3%.

It has been suggested (Ørskov, 1986; Huntington, 1997) that when we give cattle conventional feedlot diets, especially those which are based on sorghum or maize, we may exceed the capacity of the pancreas to synthesize enough amylase to digest the starch entering the small intestine. By moving starch digestion back into the rumen, heat processing can avoid overloading the ruminant's capacity for small intestinal starch digestion. Further, heat processing can increase the amount of microbial protein entering the small intestine (Xiong et al., 1991; Zinn et al., 1998; Theurer et al., 1999) because increasing the amount of ruminally fermentable starch allows the microbial population to increase. However, the ruminal digestion of starch will cause volatile fatty acids (VFA) to be substituted for glucose as the major end product of starch digestion; this will reduce the efficiency of utilization of starch energy, for growth at least.

The meta-analysis of Owens et al. (1997) is a careful review of the effects of grain processing on growing cattle performance. This study suggests that the inclusion of SF maize, sorghum and wheat at 55% or more in feedlot rations, in comparison to milled grain:

- does not change growth rates of cattle given any type of grain;
- reduces food intakes by 12% (maize-based diets), 17% (sorghum) and 10% (wheat); and
- improves the FCR by 11% (maize-based diets), 15% (sorghum) and 10% (wheat).

Other studies with maize give similar results. Data reviewed by Zinn et al. (2002) give an average 6% improvement in growth rate following SF together with a reduction of 7% in DM intake, giving an 14% improvement in FCR. In later work by Huck et al. (1998) SF did not change food intake but obtained 8% increase in growth and FCR. In their review of sorghum processing, Stock and Mader (1987) reported no change in growth rate but a 12% improvement in FCR averaged over all heat processing methods (SF, popping, micronizing and exploding). In contrast to the results obtained with these grains, heat processing barley and oats (compared to milling) has no additional effect on animal performance (Mathison et al., 1991; Owens et al., 1997; Zinn and Barajas, 1997).

Giving SF grain to lactating dairy cattle increases milk yield and the content of protein and solids-not-fat, but generally decreases milk fat content (Theurer et al., 1999; Nikkah et al., 2004).

Steam flaking reduces the amount of undigested starch reaching the large intestine. While this has beneficial nutritional effects, it may improve the gut environment for pathogenic bacteria. Steam flaking of grain has been associated with an

Table 17.4. Ruminal and post-ruminal digestion of starch in maize and sorghum processed in different ways. (From Theurer, 1986; Huntington, 1997.)

Grain and processing method	Starch digestibility (%)		
	Total tract	Rumen	Post-ruminal
Sorghum			
Milled	89.7	58.3	72.1
Steam-flaked	98.7	80.0	93.2
Micronized	96.7	63.7	91.8
Maize			
Milled	93.8	76.1	74.8
Steam-flaked	99.0	85.6	91.6

increased shedding of *Escherichia coli* O157 by lot-fed cattle (Fox *et al.*, 2007).

Responses of horses to heat-treated cereal grains

There are few published data but the indications are that heat treatments give little improvement in whole-tract digestibility over dry milling (Julliand *et al.*, 2006; Särkijärvi and Saastamoinen, 2006). Vervuert *et al.* (2004) could not show any increase in the glycaemic indexes of steamed or popped maize. Data cited by Vervuert *et al.* (2004) suggest that starch digestion in the equine small intestine varies from about 50% to 95%, with barley starch being more extensively digested than that of oats or maize, that the proportion of starch digested declines by 1% unit with each 10g of starch consumed and that there is no consistent effect of heat treatment.

Although there may not be any nutritional advantage in heat processing grains for horses, there may be good health reasons. McLean *et al.* (2000) reported more lactate production in the caeca of ponies fed rolled barley grain than in those given either hay cubes alone or fed with 50% extruded or micronized grain. They pointed out that heat treatment might help to protect horses from disorders stemming from excessive caecal starch fermentation because elevated caecal lactate concentrations in horses are associated with disorders such as laminitis (Garner *et al.*, 1977; Rowe *et al.*, 1994). Of interest is the observation of Al Jassim (2006) that steam flaking of sorghum, compared to milling, increases the production of lactate from bacterial fermentation in the pre-glandular region of the horse's stomach.

A summary of heat treatments for cereal grains

The evidence discussed in the preceding sections suggests that heat treatments:

1. Increase the susceptibility of cereal grain starch to digestion by small intestinal enzymes (except in the equine);
2. May reduce the amount of fermentable starch arriving in the porcine and equine large intestine and thus help in the control of disorders related to excess lactate production;

3. Appear to give only small improvements (up to 5% units) in the digestibility of barley, oat and wheat starch in the rumen, but improve the digestibility of maize and sorghum starch by up to 8% and 20% units, respectively;
4. May avoid overloading the capacity of the ruminant small intestine to digest starch;
5. Reduce the excretion of undigested starch in the faeces, thus reducing feedlot odours;
6. Should improve FCR by 10–15%.

Chemical processing of cereal grains

Grain can be treated with NaOH (3–4% by weight) applied as an aqueous solution and remaining in contact with the grain for several days. The treatment has variable effects on digestibility, food intake and animal performance. Compared to milling, NaOH treatment gives similar (Ørskov and Greenhalgh, 1977; Ørskov *et al.*, 1980) or smaller (Greenhalgh *et al.*, 1980; Vipond *et al.*, 1980; Barnes and Ørskov, 1981; Ørskov *et al.*, 1981; and *in sacco* data from Demetrova and Vajda, 1998) increases in the digestibility of whole grains fed to sheep and cattle. However, NaOH treatments can increase the storage life of HM grain by reducing mould growth, and NaOH-treated grain may have fewer adverse effect on forage digestion than dry-rolled grain (Ørskov and Greenhalgh, 1977) when these are fed in a mixture, although not everyone has obtained this result (Greenhalgh *et al.*, 1980). The optimum NaOH application rate varies between grains (Ørskov *et al.*, 1980) and may be higher when the grains are fed with forages than alone (Barnes and Ørskov, 1981; Ørskov *et al.*, 1981).

Processing of Oilseed Meals

Oilseed meals are the by-products of oil extraction from oilseeds, and processing conditions are designed to maximize the yield of high-quality oil rather than to produce a high-quality animal food. Processing methods (Carr, 1989; Salunkhe and Desai, 1989) vary according to the type of oilseed. Most soybean processing plants remove the oil by solvent extraction. Heat is used in this process (Grieshop *et al.*, 2003), but it is a lower-temperature method (60–88°C, depending on the duration of heating) than expeller extraction, although some plants incorporate a preliminary steam conditioning followed by extrusion to break the oil cells. Oil is extracted by soaking the prepared beans in hexane. This is followed by desolventizing, toasting (105–110°C) to inactivate endogenous toxins, and drying. Solvent-extracted meals have less than 1% oil and more protein than comparable expeller-extracted products. The extruder/expeller method is becoming popular for biodiesel production from soybeans (Karr-Lilienthal *et al.*, 2006). In this process the prepared seed is heated to 130–135°C immediately before pressing. This process produces a protein meal which contains some 4% or more oil and has a slightly lower protein content than that of solvent-extracted meals.

The nutritional quality of oilseed meals depends on the availability of amino acids especially lysine, destruction of toxins and other anti-nutritive factors which are present in most oilseeds, the resistance of its protein to rumen degradability while still being susceptible to intestinal digestion, the presence of residual oil and its effect on the ME content and storage life of the meal, and the presence of seedcoat which will dilute the ME and protein contents. We cannot optimize all these factors in the same set of processing conditions. Heat applied during processing adversely affects amino acid content, especially lysine, arginine and methionine (Chang *et al.*, 1987), and their availability to non-ruminants (Van der Poel *et al.*, 1990b; Marty and Chavez, 1995; Li *et al.*, 2002). Amino acids become unavailable due to the Maillard reaction (where free amino groups react with a reducing sugar), reactions with phenolic substances, fat oxidation products and other amino acids (Bjanason and Carpenter, 1970). On the other hand, heating increases the amount of bypass protein in the meal (Broderick and Craig, 1980) and inactivates some plant toxins (Souffrant *et al.*, 1985; Feng *et al.*, 2003). Removal of the hull gives a more protein- and ME-rich food because the fibre-rich hull has been removed, and also reduces phytate concentrations (Kracht *et al.*, 2004).

Forage Processing

Chopping and milling

Chopping (also called 'chaffing') is used to facilitate forage handling, including mixing with other ingredients and feeding from feed bins or troughs. Eating long pieces of roughage is a wasteful process – when an animal has to pull long pieces of material free from a mass it drops excess pieces on to the ground

where it is often wasted. When it chews long pieces into a bolus suitable for swallowing it loses material from the mouth. Chopping increases the rate at which animals eat and reduces the amount of chewing required to process the food (Kenney *et al.*, 1984; Dulphy and Demarquilly, 1994; Dryden *et al.*, 1995) and may increase DM intake and organic matter digestibility (Tafaj *et al.*, 2001; sheep and cattle). However, chopping does not always improve these characteristics. For example, Shaver *et al.* (1986) could demonstrate no differences between chopped and long lucerne hay fed to dairy cattle in either food intake or intake rate, chewing activity, rumen VFA profile or milk fat content.

When milled roughages are fed to ruminants the processing often increases food intake (McSweeney and Kennedy, 1992; Goetsch *et al.*, 1997; De Vega *et al.*, 2000) and/or the rate at which animals consume food (Shaver *et al.*, 1986; Faichney and Brown, 2004), reduces both the extent and rate of cell wall digestion (Shaver *et al.*, 1986; Uden, 1988; Le Liboux and Peyraud, 1998, 1999; De Vega *et al.*, 2000; Tafaj *et al.*, 2001; Yang *et al.*, 2002) and increases the rate at which undigested food passes from the rumen (Uden, 1988; Le Liboux and Peyraud, 1998; Bernard *et al.*, 2000; Tafaj *et al.*, 2001; Yang *et al.*, 2002). These outcomes can be explained by the way that the ruminant processes its food (Fig. 17.1). Roughage processing which makes it easier for undigested food residues to leave the rumen should also reduce the extent of digestion of those food components (the cell walls) which have to be digested by rumen microbial enzymes. The faster emptying of food particles from the rumen should also allow a greater intake of food. As well as the interactions between critical particle size for leaving the rumen, undigested particle flow and roughage milling, there is some evidence that other factors are involved. Animals fed milled roughages chew (including rumination) less than those fed long roughage (Shaver *et al.*, 1986; Le Liboux and Peyraud, 1998, 1999; Tafaj *et al.*, 1999; Krause *et al.*, 2002). Chewing promotes the adhesion of cellulolytic bacteria, and possibly fungi, to forage particles (Pan *et al.*, 2003). This may explain why milling increases the lag time before digestion starts *in sacco* (Shaver *et al.*, 1986) and why the digestibility of the cell wall of milled forages is reduced.

Milling and pelleting of forages also alter the way that the energy of the digested food is used by the animal (Table 17.5). This can be related to

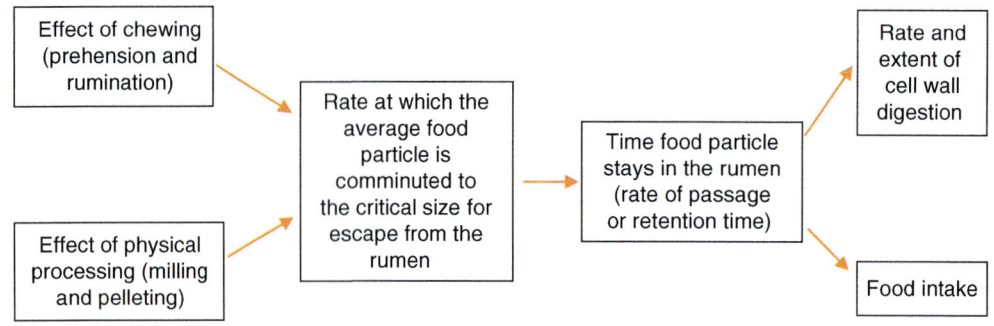

Fig. 17.1. Events in the rumen through which food processing (milling and pelleting) can influence intake and digestion in ruminants.

Table 17.5. Partitioning of the total energy (MJ/100 MJ) of grass hay processed in different ways and fed to sheep. (From Blaxter and Graham, 1956.)

Feeding level	Energy lost in	Chopped	Medium-ground, cubed	Finely ground, cubed
600 (g/day)	Faeces	23.2	29.6	28.1
	Urine	5.1	4.2	4.7
	Methane	8.3	7.4	7.8
	Heat	17.8	13.5	13.4
	Net energy	54.4	54.7	54.0
1500 (g/day)	Faeces	26.8	34.8	37.9
	Urine	5.2	4.9	4.8
	Methane	7.6	5.9	4.6
	Heat	28.8	21.2	21.9
	Net energy	68.4	66.8	69.2

changes in the rumen VFA profile: finely milled roughages are fermented to give more propionic acid and less acetic acid (Shaver *et al.*, 1986; Le Liboux and Peyraud, 1998). This is accompanied by a lower rumen pH (Le Liboux and Peyraud, 1998, 1999; Krause *et al.*, 2002; Krause and Combs, 2003) and, in some cases, more digestion of starch in the rumen than is the case when chopped forages are fed (Yang *et al.*, 2002). Reduced ruminal acetate, or pH, is associated with reduced milk fat content (Shaver *et al.*, 1986; Krause *et al.*, 2002). Apart from a report by Yang *et al.* (2002) that feeding chopped, rather than milled, forage improves it, there is little direct evidence that forage particle size affects rumen microbial protein synthesis.

The net result of the interplay of these effects in the ruminant is that, when ad lib intakes are allowed, the effect of increased digestible organic matter intake (De Vega *et al.*, 2000) outweighs that of reduced digestibility. Consequently, animals may grow faster, although this does not always happen (Tafaj *et al.*, 2001; Yang *et al.*, 2002). Feeding finely milled forages can lead to rumen hyperkeratosis and reduced or depraved appetite; these effects can be alleviated by feeding a small amount of long forage (Nicholson, 1981).

In contrast to ruminants, horses respond little to the milling and pelleting of roughages. Argo *et al.* (2002) found no consistent differences in digestible energy intakes when they fed milled and pelleted forage, even though the chopped forage required more chewing and was consumed more slowly. This is not necessarily a disadvantage – prolonging feeding times by giving hay in a not-easily-prehended form is a well-known way of combating boredom in stabled horses.

In two experiments (Drougol *et al.*, 2000a,b), milling and pelleting a mixture of lucerne and cocksfoot hays fed at equalized intakes to horses gave longer retention of forage particles in the hindgut (caecum and colon), but no change in whole-tract diet constituent digestibilities. A similar lack of difference in apparent digestibilities of neutral detergent fibre (NDF) and DM was also reported by Argo *et al.* (2002). Drougol *et al.* (2000b) examined the extent and rate of digestion in the hindgut. DM and NDF in chopped hay were digested faster, *in sacco*, than those in the milled and pelleted hay, and the effective degradation (i.e. taking account of the effects of both digestion rate and residence time in the hindgut) of these constituents was greater for chopped hay, even though milling gave a longer particle retention time in the hindgut. It is possible that chewing has a similar effect on the colonization of roughage particles by fibrolytic bacteria in both ruminants and equines. The effect of milling and pelleting on VFA production appears to be similar in both horses and ruminants. Drougol *et al.* (2000b) noted that milled hay gave less acetic acid, although there was no effect on the concentrations of either propionic or butyric acids in the hindgut.

Chemical treatment of forages

Very fibrous (low-quality) forages have low digestibilities; this and related effects on resistance to chewing, together with their generally low protein and mineral contents, make them unpromising candidates as animal foods. Nevertheless, they can be improved by chemical treatments which hydrolyse the chemical bonds between lignin and hemicellulose and which may add non-energy nutrients, particularly N.

Mild alkalis solubilize some of the cell wall hemicellulose and non-core lignin (but not core ligin, i.e. the fraction not soluble in sulfuric acid) (Chesson, 1981). Cellulose is largely unaffected by chemical treatments. NaOH is the most effective alkali for improving *in vivo* organic matter digestibility, giving improvements of 10–20% units. Other alkalis, particularly CaO or $Ca(OH)_2$ (Djajanegara and Doyle, 1989; Granzin and Dryden, 2003), Na_2CO_3 (Sriskandarajah and Kellaway, 1984), have been generally found to be less effective than NaOH or to have no additional effect when applied with NaOH (Owen *et al.*, 1982; Meeske *et al.*, 1993; Zaman and Owen, 1995). NaOH applied at 20–50 g/kg straw reduces the amounts of acid detergent fibre (ADF) and NDF in the treated forage (Mishra *et al.*, 2000; Granzin and Dryden, 2003), and substantially increases DM, organic matter (OM) and cell wall constituent digestibility, and DM intake (Higgins, 1981; Sriskandarajah and Kellaway, 1984; Meeske *et al.*, 1993; Granzin and Dryden, 2003).

NaOH increases the amount, but not the efficiency (g/kg ruminally digestible OM), of microbial protein synthesis in the rumen (Sriskandarajah and Kellaway, 1984) and, consistently with this, reduces the rumen NH_3 concentration (Meeske *et al.*, 1993).

Ozone gives good digestibility enhancement, with both monocotyledonous and dicotyledonous straws (Ben-Ghedalia and Miron, 1981; Ben-Ghedalia and Shefet, 1983), SO_2 solubilizes hemicellulose and lignin giving improved digestibility (Ben-Ghedalia and Miron, 1984). NaOCl and H_2O_2 are generally ineffective (Mishra *et al.*, 2000; Granzin and Dryden, 2003). $NaClO_2$ gives good *in sacco* degradability (Ford *et al.*, 1987), but *in vivo* digestibility is not changed possibly because of the loss of rumen fungi. H_2O_2 is sometimes applied with NaOH in the 'alkaline hydrogen peroxide' (AHP) treatment, e.g. 10 kg NaOH for 15 min followed by 50 l of 50% H_2O_2 applied to 100 kg of straw (Meeske *et al.*, 1993). AHP is sometimes more effective than NaOH (Chaudhry, 1998) but it is generally no more useful than NaOH alone (Meeske *et al.*, 1993).

Several acids have been used to treat low-quality forages, including hydrochloric acid (Israelides *et al.*, 1979) and phosphoric acid (Borhami *et al.*, 1982). However, acids per se have little beneficial effect on the nutritive value of low-quality forages. Borhami *et al.* (1982) found that addition of either phosphoric, formic or acetic acids immediately after ammoniation did not further improve the digestibility of barley straw.

NH_3, as anhydrous NH_3, NH_4OH, or urea hydrolysed by food-borne bacteria, is widely used (Sundstol, 1984). NH_3 is generally applied at 4 kg/100 kg straw DM. If the reaction takes place at ambient temperature, NH_3 must react with the straw for about 7 days for the treatment to be effective (Sankat and Lauckner, 1991), or up to 21 days if urea is the NH_3 source (Elangovan *et al.*, 2001). Thermo-ammoniation is a process where the treated straw is held at 90°C for 24 h, with similar effects as the longer period at ambient temperature.

NH_3 treatment increases the digestibility of straws (Dryden and Kempton, 1983; Abou-El-Enin *et al.*, 1999) by 5–15% units, reduces the lag time before *in sacco* digestion starts and increases the potential digestibility and intake of digestible DM (Dryden and Kempton, 1983). It increases the rumen microbial protein yield and reduces the amount of chewing needed (Chermiti *et al.*, 1994).

These effects reflect the solubilization of hemicellulose which weakens the cell wall and promotes colonization by rumen bacteria (Selim *et al.*, 2004).

Ammoniation characteristically increases the N content, by up to 300%. Most of this added N is ruminally available (Dryden and Kempton, 1983). Fifty per cent to 75% of the applied NH_3-N is lost when the stack is aired (Dryden, 1982; Taiwo *et al.*, 1995). The added N can be trapped by addition of organic (acetic and formic) and inorganic (phosphoric) acids (Borhami *et al.*, 1982), molasses (Taiwo *et al.*, 1995), $Ca(OH)_2$ (Zaman and Owen, 1995) and SO_2 gas (Dryden and Leng, 1988). The combined NH_3/SO_2 treatment (e.g. 35 g NH_3/kg DM for 30 days, then 20 g/kg DM SO_2 for 14 days; Naseeven and Kincaid, 1992) gives a greater improvement in N content and *in sacco* digestibility than is obtained from NH_3 alone. Unfortunately the treated straw is not very palatable, and 20 g SO_2/kg DM is near the maximum that sheep will accept (G.McL. Dryden, 1986, Gatton, Australia, unpublished data). However, NH_3/SO_2 straw could be a component of a mixed ration. Naseeven and Kincaid (1992) showed that treated wheat straw mixed with bromegrass hay (as 43% of the mixture) promoted similar growth as bromegrass hay fed alone.

Chemical treatments can have adverse effects. NaOH is caustic and can ulcerate the animal's mouth and load the food and soils (from excretion of excess Na in urine) with Na. It is corrosive and may damage the equipment used to apply it or to handle the treated straw. If NH_3 is used and the reaction takes at temperatures higher than 70°C (e.g. thermo-ammoniation or under black plastic on a hot day), a toxic substance may form from the reaction of NH_3 with reducing sugars. Roughages which have substantial contents of soluble sugars, e.g. failed cereal grain crops harvested as forage are the most likely to form toxins (Perdok and Leng, 1987). The symptoms include hyperexcitability, trembling and 'a sudden stampeding involving galloping in circles and colliding with other animals and fences'. The toxin is excreted in milk and affects suckling calves. The toxin might be 4-methyl-imidazole, but this has not been proven (Weiss *et al.*, 1986).

References

Abou-El-Enin, O.H., Fadel, J.G. and Mackill, D.J. (1999) Differences in chemical composition and fibre diges-tion of rice straw with, and without, anhydrous ammonia from 53 rice varieties. *Animal Feed Science and Technology* 79, 129–136.

Aldrich, J.M., Muller, L.D. and Varga, G.A. (1993) Non-structural carbohydrate and protein effects on rumen fermentation, nutrient flow and performance of dairy cows. *Journal of Dairy Science* 76, 1091–1105.

Al Jassim, R.A.M. (2006) Supplementary feeding of horses with processed sorghum grains and oats. *Animal Feed Science and Technology* 125, 33–44.

Argo, C.M., Cox, J.E., Lockyer, C. and Fuller, Z. (2002) Adaptive changes in the appetite, growth and feeding behaviour of pony mares offered ad libitum access to a complete diet in either a pelleted or chaff-based form. *Animal Science* 74, 517–528.

Atwell, W.A., Hood, L.F., Lineback, D.R., Varriano-Marston, E. and Zobel, H.F. (1988) The terminology and methodology associated with basic starch phenomena. *Cereal Foods World* 33, 306–311.

Awuah, G.B., Ramaswamy, H.S. and Economides, A. (2007) Thermal processing and quality: principles and overview. *Chemical Engineering and Processing* 46, 584–602.

Axelsen, A., Nadin, J.B., Crouch, M. and Edwards, C.B.H. (1979) Feeding whole or cracked wheat or lupins to beef cattle, and a comparison between whole wheat and oats. *Australian Journal of Experimental Agriculture and Animal Husbandry* 19, 539–546.

Balogun, R.O., Rowe, J.B. and Bird, S.H. (2005) Fermentability and degradability of sorghum grain following soaking, aerobic or anaerobic treatment. *Animal Feed Science and Technology* 120, 141–150.

Balogun, R.O., Bird, S.H. and Rowe, J.B. (2006) Germination temperature and time affect *in vitro* fermentability of sorghum grain. *Animal Feed Science and Technology* 127, 125–132.

Barnes, B.J. and Ørskov, E.R. (1981) Utilization of alkali-treated grain. 2. Utilization by steers of diets based on hay or straw and mixed with either NaOH-treated or rolled barley. *Animal Feed Science and Technology* 6, 347–354.

Ben-Ghedalia, D. and Miron, J. (1981) Effect of sodium hydroxide, ozone and sulphur dioxide on the composition and *in vitro* digestion of wheat straw. *Journal of the Science of Food and Agriculture* 32, 224–228.

Ben-Ghedalia, D. and Miron, J. (1984) The response of wheat straw varieties to mild sulphur dioxide treatment. *Animal Feed Science and Technology* 10, 269–276.

Ben-Ghedalia, D. and Shefet, G. (1983) Chemical treatments for increasing the digestibility of cotton straw. 1. Effect of ozone and sodium hydroxide treatments on rumen metabolism and on the digestibility of cell walls and organic matter. *Journal of Agricultural Sciences, Cambridge* 100, 393–400.

Bernard, L., Chaise, J.P., Baumont, R. and Poncet, C. (2000) The effect of physical form of orchardgrass hay on the passage of particulate matter through the rumen of sheep. *Journal of Animal Science* 78, 1338–1354.

Bjanason, J. and Carpenter, K.J. (1970) Mechanisms of heat damage in proteins. 2. Chemical changes in pure proteins. *British Journal of Nutrition* 24, 313–329.

Blaxter, K.L. and Graham, N.M. (1956) The effect of the grinding and cubing process on the utilization of the energy of dried grass. *Journal of Agricultural Science, Cambridge* 47, 207–217.

Borhami, B.E.A., Sundstol, F. and Garmo, T.H. (1982) Studies on ammonia-treated straw. II. Fixation of ammonia in treated straw by spraying with acids. *Animal Feed Science and Technology* 7, 53–59.

Broderick, G.A. and Craig, W.M. (1980) Effect of heat treatment on ruminal degradation and escape, and intestinal digestibility of cotton seed meal protein. *Journal of Nutrition* 119, 2381–2389.

Campling, R.C. (1991) Processing cereal grains for cattle – a review. *Livestock Production Science* 28, 223–234.

Carr, R. (1989) Processing of oilseed crops. In: Robbelen, G., Downey, R.K. and Ashri, A. (eds) *Oil Crops of the World, Their Breeding and Utilization*. McGraw-Hill, New York, pp. 226–259.

Ceresnakova, Z., Chrenkova, M., Kopcekova, J., Sommer, A. and Zitnan, R. (2005) Effect of maize grain treatment on ruminal fermentation and the site and extent of starch digestion in cows. *Journal of Animal and Feed Sciences* 14, 79–91.

Chang, C.J., Tanksley, T.D. Jr, Knabe, D.A. and Zebrowska, T. (1987) Effects of different heat treatments during processing on nutrient digestibility of soybean meal in growing swine. *Journal of Animal Science* 65, 1273–1282.

Chaudhry, A.S. (1998) Nutrient composition, digestion and rumen fermentation in sheep of wheat straw treated with calcium oxide, sodium hydroxide and alkaline hydrogen peroxide. *Animal Feed Science and Technology* 74, 315–328.

Chermiti, A., Teller, E., Vanbelle, M., Collignon, G. and Matatu, B. (1994) Effect of ammonia or urea treatment of straw on chewing behaviour and ruminal digestion processes in non-lactating dairy cows. *Animal Feed Science and Technology* 47, 41–51.

Chesson, A. (1981) Effects of sodium hydroxide on cereal straws in relation to the enhanced degradation of structural polysaccharides by rumen microorganisms. *Journal of the Science of Food and Agriculture* 32, 745–758.

Choct, M., Selby, E.A.D., Cadogan, D.J. and Campbell, R.G. (2004) Effects of particle size, processing, and dry or liquid feeding on performance of piglets. *Australian Journal of Agricultural Research* 55, 237–245.

Clark, J.H. and Harshbarger, K.E. (1972) High moisture corn versus dry corn in combination with either corn silage or hay for lactating cows. *Journal of Dairy Science* 55, 1474–1480.

Cole, D.J.A., Dean, G.W. and Luscombe, J.R. (1970) Single cereal diets for bacon pigs. 2. The effect of methods of storage and preparation of barley on performance and carcass quality. *Animal Production* 12, 1–6.

Costa, P.M.A., Jensen, A.H., Harmon, B.G. and Norton, H.W. (1976) The effects of roasting and roasting temperatures on the nutritive value of corn for swine. *Journal of Animal Science* 42, 365–374.

Demetrova, M. and Vajda, V. (1998) The effect of chemically treated grains on ruminal fermentation. *Czech Journal of Animal Science* 43, 503–509.

De Vega, A., Gasa, J., Guada, J.A. and Castrillo, C. (2000) Frequency of feeding and form of lucerne hay as factors affecting voluntary intake, digestibility, feeding behaviour, and marker kinetics in ewes. *Australian Journal of Agricultural Research* 51, 801–809.

Djajanegara, A. and Doyle, P.T. (1989) Digestion rates in and outflow rates from the rumen of sheep fed untreated or calcium hydroxide-treated wheat straw. *Animal Feed Science and Technology* 25, 179–191.

Drougol, C., Poncet, C. and Tisserand, J.L. (2000a) Feeding ground and pelleted hay rather than chopped hay to ponies. 1. Consequences for *in vivo* digestibility and rate of passage of digesta. *Animal Feed Science and Technology* 87, 117–130.

Drougol, C., Tisserand, J.L. and Poncet, C. (2000b) Feeding ground and pelleted hay rather than chopped hay to ponies. 2. Consequences on fibre degradation in the cecum and the colon. *Animal Feed Science and Technology* 87, 131–145.

Dryden, G.McL. (1982) Loss of ammonia-N from ammoniated barley straw. *Proceedings of the Nutrition Society of Australia* 7, 166.

Dryden, G.McL. and Kempton, T.J. (1983) Digestion of organic matter and nitrogen in ammoniated barley straw. *Animal Feed Science and Technology* 10, 65–75.

Dryden, G.McL. and Leng, R.A. (1988) Effects of ammonia and sulfur dioxide gases on the composition and digestion of barley straw. *Animal Feed Science and Technology* 19, 121–133.

Dryden, G.McL., Stafford, K.J., Waghorn, G.C. and Barry, T.N. (1995) Comminution of roughages by red deer (*Cervus elaphus*) during the prehension of feed. *Journal of Agricultural Science, Cambridge* 125, 407–414.

Dulphy, J.P. and Demarquilly, C. (1994) The regulation and prediction of feed intake in ruminants in relation to feed characteristics. *Livestock Production Science* 39, 1–12.

Elangovan, A.V., Kishan, J. and Sahoo, A. (2001) Fate of ligno-cellulosic components and urea-nitrogen in urea-ammoniated wheat straw. *Animal Nutrition and Feed Technology* 1, 61–68.

Estrada-Girón, Y., Swanson, B.G. and Barbosa-Cánovas, G.V. (2005) Advances in the use of high hydrostatic pressure for processing cereal grains and legumes. *Trends in Food Science and Technology* 16, 194–203.

Faichney, G.J. and Brown, G.H. (2004) The effect of physical form of a lucerne hay on rumination and the passage of particles from the rumen of sheep. *Australian Journal of Agricultural Research* 55, 1263–1270.

Fang, C. and Campbell, G.M. (2003) On predicting roller milling performance V: Effect of moisture content on the particle size distribution from first break milling of wheat. *Journal of Cereal Science* 37, 31–41.

Feng, D.Y., Shen, Y.R. and Chavez, E.R. (2003) Effectiveness of different processing methods in reducing hydrogen cyanide content of flaxseed. *Journal of the Science of Food and Agriculture* 83, 836–841.

Fernandes, T.H., Hutton, K. and Smith, W.C. (1975) A note on the use of micronized barley for growing pigs. *Animal Production* 20, 307–310.

Ford, C.W., Elliott, R. and Maynard, P.J. (1987) The effect of chlorite dilignification on digestibility of some grass forages and on intake and rumen microbial activity in sheep fed barley straw. *Journal of Agricultural Science, Cambridge* 108, 129–136.

Fox, J.T., Depenbusch, B.E., Drouillard, J.S. and Nagaraja, T.G. (2007) Dry-rolled or steam-flaked grain-based diets and fecal shedding of *Escherichia coli* O157 in feedlot cattle. *Journal of Animal Science* 85, 1207–1212.

Fulkerson, W. and Michell, P.J. (1985) Production response to feeding wheat grain to dairy cows. *Australian Journal of Experimental Agriculture* 25, 253–256.

Gallant, D.J., Bouchet, B., Buléon, A. and Pérez, S. (1992) Physical characteristics of starch granules and susceptibility to enzymatic degradation. *European Journal of Clinical Nutrition* 46, 3–16.

Garner, H.E., Hutcheson, D.P., Coffman, J.R., Hann, A.W. and Salem, C. (1977) Lactic acidosis: a factor associated with equine laminitis. *Journal of Animal Science* 45, 1037–1041.

Goetsch, A.L., Patil, A.R., Galloway, D.L., Kouakou, B., Wang, Z.S., Park, K.K. and Rossi, J.E. (1997) Net flux of nutrients across the splanchnic tissues in wethers consuming grasses of different sources and physical forms ad libitum. *British Journal of Nutrition* 77, 769–781.

Goransson, L. and Ogle, R.B. (1985) Anaerobically stored high moisture cereals for growing pigs. *Animal Feed Science and Technology* 12, 159–169.

Gorocica-Buenfil, M.A. and Loerch, S.C. (2005) Effect of cattle age, forage level, and corn processing on diet digestibility and feedlot performance. *Journal of Animal Science* 83, 705–714.

Granzin, B.C. and Dryden, G.M. (2003) Effects of alkalis, oxidants and urea on the nutritive value of rhodes grass (*Chloris gayana* cv. Callide). *Animal Feed Science and Technology* 103, 113–122.

Greenhalgh, J.F.D., Petchey, A.M., Hinks, C.E., Parkinson, H., Laird, R., Rees, E.D. and Fraser, C. (1980) Alkali-treated barley as a supplement to silage for fattening cattle. *Animal Production* 30, 488.

Grieshop, C.M., Kadzere, C.T., Clapper, G.M., Flickinger, E.A., Bauer, L.L., Frazier, R.L. and Fahey, G.C. Jr (2003) Chemical and nutritional characteristics of United States soybeans and soybean meals. *Journal of Agricultural and Food Chemistry* 51, 7684–7691.

Healy, B.J., Hancock, J.D., Kennedy, G.A., Bramel-Cox, P.J., Behnkes, K.C. and Hines, R.H. (1994) Optimum particle size of corn and hard and soft sorghum for nursery pigs. *Journal of Animal Science* 72, 2227–2236.

Hedde, R.D., Lindsey, T.O., Parish, R.C., Daniels, H.D., Morgenthien, E.A. and Lewis, H.B. (1985) Effect of diet particle size and feeding of H2-receptor antagonists on gastric ulcers in swine. *Journal of Animal Science* 61, 179–186.

Higgins, A.J. (1981) The effect of a sodium hydroxide spray treatment on digestibility of barley straw in sheep and goats. *Agricultural Wastes* 3, 145–155.

Hironaka, R., Beauchemin, K.A. and Lysyk, T.J. (1992) The effect of thickness of steam-rolled barley on its utilization by beef cattle. *Canadian Journal of Animal Science* 72, 279–286.

Hongtrakul, K., Goodbrand, R.D., Behnke, K.C., Nelssen, J.L., Tokach, M.D., Bergstrom, J.R., Nessmith, W.B. Jr and Kim, I.H. (1998) The effects of extrusion processing of carbohydrate sources on weanling pig performance. *Journal of Animal Science* 76, 3034–3042.

Huck, G.L., Kreikemeier, K.K., Kuhl, G.L., Eck, T.P. and Bolsen, K.K. (1998) Effects of feeding combinations of steam-flaked grain sorghum and steam-flaked, high-moisture, or dry-rolled corn on growth performance and carcass characteristics in feedlot cattle. *Journal of Animal Science* 76, 2984–2990.

Huntington, G.B. (1997) Starch utilization by ruminants: from basics to the bunk. *Journal of Animal Science* 75, 852–867.

Israelides, C.J., Grant, G.A. and Han, Y.W. (1979) Acid hydrolysis of grass straw for yeast fermentation. *Developments in Industrial Microbiology* 20, 603–608.

Jacobs, H. and Delcour, J.A. (1998) Hydrothermal modifications of granular starch, with retention of the granular structure: a review. *Journal of Agricultural and Food Chemistry* 46, 2896–2905.

Jones, F.T. and Richardson, K.E. (2004) *Salmonella* in commercially manufactured feeds. *Poultry Science* 83, 384–391.

Julliand, V., De Fombelle, A. and Varloud, M. (2006) Starch digestion in horses: the impact of feed processing. *Livestock Science* 100, 44–52.

Karr-Lilienthal, L.K., Bauer, L.L., Utterback, P.L., Zinn, K.E., Frazier, R.L., Parsons, C.M. and Fahey, G.C. Jr (2006) Chemical composition and nutritional quality of soybean meals prepared by extruder/expeller processing for use in poultry diets. *Journal of Agricultural and Food Chemistry* 54, 8108–8114.

Kaspersson, A., Lindgren, S. and Ekstrom, N. (1988) Microbial dynamics in barley grain stored under controlled atmosphere. *Animal Feed Science and Technology* 19, 299–312.

Kenney, P.A., Black, J.L. and Colebrook, W.F. (1984) Factors affecting diet selection by sheep. 3. Dry matter content and particle length of forage. *Australian Journal of Agricultural Research* 35, 831–838.

Kim, J.C., Simmins, P.H., Mullan, B.P. and Pluske, J.R. (2005) The digestible energy value of wheat for pigs, with special reference to the post-weaned animal. *Animal Feed Science and Technology* 122, 257–287.

Kracht, W., Dänicke, S., Kluge, H., Keller, K., Matzke, W., Hennig, U. and Schumann, W. (2004) Effect of dehulling of rapeseed on feed value and nutrient digestibility of rape products in pigs. *Archiv für Tierernährung* 58, 389–404.

Krause, K.M. and Combs, D.K. (2003) Effects of particle size, forage source, and grain fermentability on performance and ruminal pH in midlactation cows. *Journal of Dairy Science* 86, 1382–1397.

Krause, K.M., Combs, D.K. and Beauchemin, K.A. (2002) Effects of forage particle size and grain fermentability in midlactation cows. II. Ruminal pH and chewing activity. *Journal of Dairy Science* 86, 1382–1397.

Laurinen, P., Valaja, J., Näsi, M. and Smeds, K. (1998) Effects of different expander processing conditions on the nutritive value of barley and wheat by-products in pig diets. *Animal Feed Science and Technology* 74, 213–227.

Laurinen, P., Siljander-Rasi, H., Karhunen, J., Alaviuhkola, T., Näsi, M. and Tuppi, K. (2000) Effects of different grinding methods and particle size of barley and wheat on pig performance and digestibility. *Animal Feed Science and Technology* 83, 1–16.

Lawlor, P.G., Lynch, P.B., Caffrey, P.J. and O'Doherty, J.V. (2003) Effect of cooking wheat and maize on the performance of newly weaned pigs. 1. Age and weight at weaning. *Animal Science* 76, 251–261.

Le Liboux, S. and Peyraud, J.L. (1998) Effect of forage particle size and intake level on fermentation patterns and sites and extent of digestion in dairy cows fed mixed diets. *Animal Feed Science and Technology* 73, 131–150.

Le Liboux, S. and Peyraud, J.L. (1999) Effect of forage particle size and feeding frequency on fermentation patterns and sites and extent of digestion in dairy cows fed mixed diets. *Animal Feed Science and Technology* 76, 297–319.

Li, D., Pengbin, X., Liming, G., Shijun, F. and Canghai, H. (2002) Determination of apparent ileal amino acid digestibility in rapeseed meal and cake processed at different temperatures using the direct and difference method with growing pigs. *Archiv für Tierernährung* 56, 339–349.

McAllister, T.A., Phillippe, R.C., Rode, L.M. and Cheng, K.J. (1993) Effect of the protein matrix on the digestion of cereal grains by ruminal microorganisms. *Journal of Animal Science* 71, 205–212.

McLean, B.M.L., Hyslop, J.J., Longland, A.C., Cuddeford, D. and Hollands, T. (2000) Physical processing of barley and its effects on intra-caecal fermentation parameters in ponies. *Animal Feed Science and Technology* 85, 79–87.

McSweeney, C.S. and Kennedy, P.M. (1992) Influence of dietary particle-size on chewing activity and reticuloruminal motility in goats and sheep fed wheaten (*Triticum aestivum*) hay. *Small Ruminant Research* 9, 107–115.

Marty, B.J. and Chavez, E.R. (1995) Ileal digestibilities and urinary losses of amino acids in pigs fed heat processed soybean products. *Livestock Production Science* 43, 37–48.

Mathison, G.W. (1996) Effects of processing on the utilization of grain by cattle. *Animal Feed Science and Technology* 58, 113–125.

Mathison, G.W., Hironaka, R., Kerrigan, B.K., Vlach, I., Milligan, L.P. and Weisenburger, R.D. (1991) Rate of starch degradation, apparent digestibility and rate and efficiency of steer gain as influenced by barley grain volume-weight and processing method. *Canadian Journal of Animal Science* 71, 867–878.

Mavromichalis, I., Hancock, J.D., Senne, B.W., Gugle, T.L., Kennedy, G.A., Hines, R.H. and Wyatt, C.L. (2000) Enzyme supplementation and particle size of wheat in diets for nursery and finishing pigs. *Journal of Animal Science* 78, 3086–3095.

May, P.J. and Barker, D.J. (1984/5) Milling barley and lupin grain in diets for cattle. *Animal Feed Science and Technology* 12, 57–64.

Medel, P., Salado, S., de Blas, J.C. and Mateos, G.G. (1999) Processed cereals in diets for early-weaned piglets. *Animal Feed Science and Technology* 82, 145–156.

Medel, P., Garcia, M., Lazaro, R., de Blas, C. and Mateos, G.G. (2000) Particle size and heat treatment of barley in diets for early-weaned piglets. *Animal Feed Science and Technology* 84, 13–21.

Medel, P., Baucells, F., Gracia, M.I., de Blas, C. and Mateos, G.G. (2002) Processing of barley and enzyme supplementation in diets for young pigs. *Animal Feed Science and Technology* 95, 113–122.

Medel, P., Latorre, M.A., de Blas, C. and Lázaro, R. (2004) Heat processing of cereals in mash or pellet diets for young pigs. *Animal Feed Science and Technology* 113, 127–140.

Meeske, R., Meissner, H.H. and Pienaar, J.P. (1993) The upgrading of wheat straw by alkaline hydrogen peroxide treatment: the effect of NaOH and H_2O_2 on the site and extent of digestion in sheep. *Animal Feed Science and Technology* 40, 121–133.

Mishra, A.S., Chaturvedi, O.H., Khali, A., Prasad, R., Santra, A., Misra, A.K., Parthasarathy, S. and

Jakhmola, R.C. (2000) Effect of sodium hydroxide and alkaline hydrogen peroxide treatment on physical and chemical characteristics and IVOMD of mustard straw. *Animal Feed Science and Technology* 84, 257–264.

Moe, P.W. and Tyrrell, H.F. (1977) Effects of feed intake and physical form on energy value of corn in timothy hay diets for lactating cows. *Journal of Dairy Science* 60, 754–758.

Naseeven, M.R. and Kincaid, R.L. (1992) Ammonia and sulfur dioxide treated wheat straw as a feedstuff for cattle. *Animal Feed Science and Technology* 37, 111–128.

National Research Council (NRC) (2000) *Nutrient Requirements of Beef Cattle*, 7th rev. edn. Update. National Academy Press, Washington, DC.

Nicholson, J.W.G. (1981) Nutrition and feeding aspects of the utilization of processed lignocellulosic waste materials by animals. *Agriculture and Environment* 6, 205–228.

Nikkah, A., Alikhani, M. and Amanlou, H. (2004) Effects of feeding ground or steam-flaked broom sorghum and ground barley on performance of dairy cows in midlactation. *Journal of Dairy Science* 87, 122–130.

Nordin, M. and Campling, R.C. (1976) Digestibility studies with cows given whole and rolled cereal grains. *Animal Production* 23, 305–315.

Ohh, S.J., Allee, G.L., Behnke, K.C. and Deyoe, C.W. (1983) Effects of particle size of corn and sorghum grain on performance and digestibility of nutrients for weaned pigs. *Journal of Animal Science* 57(Suppl. 1), 260–261.

Ørskov, E.R. (1976) The effect of processing on digestion and utilisation of cereals by ruminants. *Proceedings of the Nutrition Society* 35, 245–252.

Ørskov, E.R. (1979) Recent information on processing of grain for ruminants. *Livestock Production Science* 6, 335–347.

Ørskov, E.R. (1986) Starch digestion and utilization in ruminants. *Journal of Animal Science* 63, 1624–1633.

Ørskov, E.R. and Greenhalgh, J.F.D. (1977) Alkali treatment as a method of processing whole grain for cattle. *Journal of Agricultural Science, Cambridge* 89, 253–255.

Ørskov, E.R., Barnes, B.J. and Lukins, B.A. (1980) A note on the effect of different amounts of NaOH application on digestibility by cattle of barley, oats, wheat and maize. *Journal of Agricultural Science, Cambridge* 94, 271–273.

Ørskov, E.R., Barnes, B.J., Macdearmid, A., Williams, P.E.V. and Innes, G.M. (1981) Utilization of alkali-treated grain. 3. Utilization by steers of NaOH-treated and rolled barley in silage-based diets. *Animal Feed Science and Technology* 6, 355–365.

Owen, E., Klopfenstein, T.J., Britton, R.A., Rump, K. and McDonnell, M.L. (1982) Treatment of wheat straw with different alkalies. *Journal of Animal Science* 55(Suppl. 1), 448.

Owens, F.N., Secrist, D.S., Hill, W.J. and Gill, D.R. (1997) The effect of grain source and grain processing on performance of feedlot cattle: a review. *Journal of Animal Science* 75, 868–879.

Pan, J., Koike, S., Suzuki, T., Ueda, K., Kobayashi, Y., Tanaka, K. and Okubo, M. (2003) Effect of mastication on degradation of orchardgrass hay stem by rumen microbes: fibrolytic enzyme activities and microbial attachment. *Animal Feed Science and Technology* 106, 69–79.

Perdok, H.B. and Leng, R.A. (1987) Hyperexcitability in cattle fed ammoniated roughages. *Animal Feed Science and Technology* 17, 121–143.

Pflugfelder, R.L., Rooney, L.W. and Schake, L.M. (1986) The role of germination in sorghum reconstitution. *Animal Feed Science and Technology* 14, 243–254.

Pluske, J.R., Siba, P.M., Pethick, D.W., Durmic, Z., Mullan, B.P. and Hampson, D.J. (1996) The incidence of swine dysentery in pigs can be reduced by feeding diets which limit the amount of fermentable substrate entering the large intestine. *Journal of Nutrition* 126, 2920–2933.

Rapetti, L. and Bava, L. (2004) Effect of grinding of maize and level of starch on digestibility and lactation performance of Saanen goats. *South African Journal of Animal Science* 34(Suppl. 1), 85–88.

Reis, R.B., Emeterio, F.S., Combs, D.K., Satter, L.D. and Costa, H.N. (2001) Effects of corn particle size and source on performance of lactating cows fed direct-cut grass-legume forage. *Journal of Dairy Science* 84, 429–441.

Reimann, E.M., Maxwell, C.V., Kowalczyk, T., Benevenga, N.J., Grummer, R.H. and Hoekstra, W.G. (1968) Effect of fineness of grind of corn on gastric lesions and contents of swine. *Journal of Animal Science* 27, 992–998.

Rooney, L.W. and Pflugfelder, R.L. (1986) Factors affecting starch digestibility with special emphasis on sorghum and corn. *Journal of Animal Science* 63, 1607–1623.

Rowe, J.B., Lees, M.J. and Pethick, D.W. (1994) Prevention of acidosis and laminitis associated with grain feeding in horses. *Journal of Nutrition* 124, 2742S–2744S.

Salunkhe, D.K. and Desai, B.B. (1989) *Postharvest Technology of Oilseeds*. CRC Press, Boca Raton, Florida, pp. 14–15.

San Emeterio, F., Reis, R.B., Campos, W.E. and Satter, L.D. (2000) Effect of coarse or fine grinding on utilization of dry or ensiled corn by lactating dairy cows. *Journal of Dairy Science* 83, 2839–2848.

Sankat, C. and Lauckner, B. (1991) The effect of ammonia treatment on the digestibility of rice straw under various process conditions. *Canadian Agricultural Engineering* 33, 309–313.

Särkijärvi, S. and Saastamoinen, M. (2006) Feeding value of various processed oat grains in equine diets. *Livestock Science* 100, 3–9.

Sauer, D.B. and Burroughs, R. (1974) Efficacy of various chemicals as grain mold inhibitors. *Transactions of the ASAE* 17, 557–559.

Selim, A.S.M., Pan, J., Takano, T., Suzuki, T., Koike, S., Kobayashi, Y. and Tanaka, K. (2004) Effect of ammonia treatment on physical strength of rice straw, distribution of straw particles and particle-associated bacteria in sheep rumen. *Animal Feed Science and Technology* 115, 117–128.

Shaver, R.D., Nytes, A.J., Satter, L.D. and Jorgensen, N.A. (1986) Influence of amount of feed intake and forage physical form on digestion and passage of prebloom alfalfa hay in dairy cows. *Journal of Dairy Science* 69, 1545–1559.

Simeone, A., Beretta, V., Rowe, J.B., Nolan, J.V. and Elizalde, J.C. (2003) Is mastication enough processing for maize grain? *Recent Advances in Animal Nutrition in Australia* 14, 16A.

Skoch, E.R., Binder, S.F., Deyoe, C.W., Allee, G.L. and Behnke, K.C. (1983a) Effects of pelleting conditions on performance of pigs fed a corn-soybean meal diet. *Journal of Animal Science* 57, 922–928.

Skoch, E.R., Binder, S.F., Deyoe, C.W., Allee, G.L. and Behnke, K.C. (1983b) Effects of steam pelleting conditions and extrusion cooking on a swine diet containing wheat middlings. *Journal of Animal Science* 57, 929–935.

Souffrant, W.B., Schumann, B., Gebhardt, G. and Matkowitz, R. (1985) Exocrine pancreas secretion in swine after feeding soybean extract meal. *Archiv für Tierernährung* 35, 383–389.

Sriskandarajah, N. and Kellaway, R.C. (1984) Effects of alkali treatment of wheat straw on intake and microbial protein synthesis in cattle. *British Journal of Nutrition* 51, 289–296.

Stock, R. and Mader, T. (1987) *Grain sorghum processing for beef cattle*. Co-operative Extension, Institute of Agriculture and Natural Resources, University of Nebraska-Lincoln, NebGuide G74-136-A, Lincoln, Nebraska.

Sundstol, F. (1984) Ammonia treatment of straw: methods for treatment anf feeding experience in Norway. *Animal Feed Science and Technology* 10, 173–187.

Svihus, B., Uhlen, A.K. and Harstad, O.M. (2005) Effect of starch granule structure, associated components and processing on nutritive value of cereal starch: a review. *Animal Feed Science and Technology* 122, 303–320.

Swan, C.G., Bowman, J.G.P., Martin, J.M. and Giroux, M.J. (2006) Increased puroindoline levels slow ruminal *digestion* of wheat (*Triticum aestivum* L.) starch by cattle. *Journal of Animal Science* 84, 641–650.

Tafaj, M., Steingass, H., Susenbeth, A., Lang, G.U. and Drochner, W. (1999) Influence of hay particle size at different concentrate and feeding levels on digestive processes and feed intake in ruminants. *Archiv fur Tierernahrung* 52, 167–184.

Tafaj, M., Steingass, H. and Drochner, W. (2001) Influence of hay particle size at different concentrate and feeding levels on digestive processes and feed intake in ruminants. 2. Passage, digestibility and feed intake. *Archiv fur Tierernahrung* 54, 243–259.

Taiwo, A.A., Adebowale, E.A., Greenhalgh, J.F.D. and Akinsoyinu, A.O. (1995) Techniques for trapping ammonia generated from urea treatment of barley straw. *Animal Feed Science and Technology* 56, 133–141.

Tester, R.F., Karkalas, J. and Qi, X. (2004) Starch structure and digestibility. Enzyme–substrate relationship. *World's Poultry Science Journal* 60, 186–195.

Thacker, P.A. (1999) Effect of micronization on the performance of growing/finishing pigs fed diets based on hulled and hulless barley. *Animal Feed Science and Technology* 79, 29–41.

Theurer, C.B. (1986) Grain processing effects on starch utilization by ruminants. *Journal of Animal Science* 63, 1649–1662.

Theurer, C.B., Huber, J.T., Delgardo-Elorduy, A. and Wanderley, R. (1999) Invited review: summary of steam-flaking corn or sorghum grain for lactating dairy cows. *Journal of Dairy Science* 82, 1950–1959.

Toland, P.C. (1976) The digestibility of wheat, barley and oat grain fed either whole or rolled at restricted levels with hay to steers. *Australian Journal of Experimental Agriculture and Animal Husbandry* 16, 71–75.

Toland, P.C. (1978a) Effect of level of feeding a mixed ration of whole wheat and hay on the digestion of wheat by steers. *Australian Journal of Experimental Agriculture and Animal Husbandry* 18, 25–28.

Toland, P.C. (1978b) Influence of some digestive processes on the digestion by cattle of cereal grain fed whole. *Australian Journal of Experimental Agriculture and Animal Husbandry* 18, 29–33.

Uden, P. (1988) The effect of grinding and pelleting hay on digestibility, fermentation rate, digesta passage and rumen and faecal particle size in cows. *Animal Feed Science and Technology* 19, 145–157.

Van Der Poel, A.F.B., Den Hartog, L.A., van Stiphout, W.A.A., Bremmers, R. and Huisman, J. (1990a) Effects of extrusion of maize on ileal and faecal digestibility of nutrients and performance of young piglets. *Animal Feed Science and Technology* 29, 309–320.

Van der Poel, A.F.B., Mollee, P.W., Huisman, J. and Liener, I.E. (1990b) Variations among species of animals in response to the feeding of heat-processed beans. 1. Bean processing and effects on growth, digestibility and organ weights in piglets. *Livestock Production Science* 25, 121–135.

Vervuert, I., Coenen, M. and Bothe, C. (2004) Effects of corn processing on the glycaemic and insulinaemic responses in horses. *Journal of Animal Physiology and Animal Nutrition* 88, 348–355.

Vipond, J.E., Abdalla, S.A. and King, M.E. (1980) The value of processing cereal grains for sheep. *Animal Production* 30, 487–488.

Weiss, W.P., Conrad, H.R., Martin, C.M., Cross, R.F. and Shockey, W.L. (1986) Etiology of ammoniated hay toxicosis. *Journal of Animal Science* 63, 525–532.

Wondra, K.J., Hancock, J.D., Behnke, K.C., Hines, R.H. and Stark, C.R. (1995) Effects of particle size and pelleting on growth performance, nutrient digestibility, and stomach morphology in finishing pigs. *Journal of Animal Science* 73, 757–763.

Xiong, Y., Bartle, S.J. and Preston, R.L. (1991) Density of steam-flaked sorghum grain, roughage level, and feeding regimen for feedlot steers. *Journal of Animal Science* 69, 1707–1718.

Yang, W.Z., Beauchemin, K.A. and Rode, L.M. (2002) Effects of particle size of alfalfa-based dairy cow diets on site and extent of digestion. *Journal of Dairy Science* 85, 1958–1968.

Zaman, M.S. and Owen, E. (1995) The effect of calcium hydroxide and urea treatment of barley straw on chemical composition and digestibility *in vitro*. *Animal Feed Science and Technology* 51, 165–171.

Zarkadas, L.N. and Wiseman, J. (2002) Influence of micronization temperature and pre-conditioning on performance and digestibility in piglets fed barley-based diets. *Animal Feed Science and Technology* 95, 73–82.

Zinn, R.A. and Barajas, R. (1997) Influence of flake density on the comparative feeding value of a barley-corn blend for feedlot cattle. *Journal of Animal Science* 75, 904–909.

Zinn, R.A., Alvarez, E.G., Montano, M.F., Plascencia, A. and Ramirez, J.E. (1998) Influence of tempering on the feeding value of rolled corn in finishing diets for feedlot cattle. *Journal of Animal Science* 76, 2239–2246.

Zinn, R.A., Owens, F.N. and Ware, R.A. (2002) Flaking corn: processing mechanics, quality standards, and impacts on energy availability and performance of feedlot cattle. *Journal of Animal Science* 80, 1145–1156.

18 Feed Mill Design and Management: an Introduction

Basics of Stock Food Processing Facility Design

Stock food manufacturing encompasses the reception and storage of ingredients, ingredient (principally cereal grain) milling, methods of mixing ingredients, further processing of ingredients (e.g. pelleting or extrusion) and the warehousing techniques that ensure quality control. The design and operation of commercial stock food manufacturing facilities, including consideration of pelleting, extrusion and other grain processing methods, is described in detail in McEllhiney (1985, 1994).

Ingredient storage

Most ingredients are stored in bulk, in bins and silos. Capacities can vary from 1 or 2 t to several hundred tonnes, and they are made from wood, concrete, steel or steel mesh and plastic film lining, among other materials (Andrews, 1996; Viljoen, 2001; Bullen, 2006). There are some important safety issues related to the design, construction and use of both temporary and permanent grain storages. Users must be aware of the risks associated with falling grain, the build-up of noxious gases in silos, silo structural failure and injuries caused by augers and falls from ladders, among others (Bahlmann *et al.*, 2002).

Characteristics of ingredients which are relevant to the design of storages include the angle of repose (Fig. 18.1, Table 18.1), the coefficient of friction, sphericity, bulk density and the extent to which it will take up water in storage. Examples of these and the ways in which they can be measured are given by Oje and Ugbor (1991) and Mwithiga and Sifuna (2006).

Storages should empty quickly and completely. We can either use a sweep auger in a flat-floored silo or design the base of the silo or bin so that it empties by gravity. If the base of a bin or the cone at the bottom of a silo is angled at less than the angle of repose of the material stored in it, then the bin will not empty completely. This leads to contamination of material subsequently put into the storage and possible infestation by insects and mould due to the accumulation of old food.

Ingredient bulk densities are used to design storages of adequate size. Representative values for cereals and other grains are given in Table 18.1. These are indicative values for Australian and North American products. Bulk densities vary with moisture content, growing conditions and grain variety, and the angle formed in a silo is shallower during filling, and steeper when emptying.

Moving stock food ingredients

Bulk materials are most commonly moved around the feed mill by chutes under the influence of gravity (which is the cheapest form of transport), augers and bucket elevators. Augers have a continuous metal blade (the 'flight') wound in a corkscrew-like fashion around a central shaft. This rotates in the auger casing. Grain or other material is drawn into the auger by a short section of flight protruding from the casing, becomes trapped between the flight and the casing and is then forced to travel along the auger. Augers of different diameters and rotational speeds transport material at rates of 5–250 t/h. Bucket elevators consist of a series of containers (the 'buckets') attached to an endless belt which rotates inside a casing. Grain fills the containers as they move around into the bottom of the 'up' leg of the elevator, is moved to the top of the leg and is then spilt out of the bucket as it turns over into the 'down' leg. Boumans (1985) and Fairchild (1994) give detailed descriptions of the design and operation of augers, bucket elevators and other conveying equipment.

Stock food ingredients will flow under gravity provided that the chute angle is steeper than the material's angle of repose. Grains which are

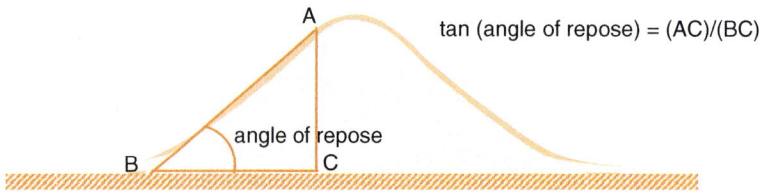

$$\tan (\text{angle of repose}) = (AC)/(BC)$$

angle of repose

A

B C

Fig. 18.1. The angle of repose of a foodstuff.

Table 18.1. Bulk densities and angles of repose of examples of cereal grains, oilseeds and grain legumes. (From Andrews, 1996; Wilcke, 1999; Worley, 2002; Bullen, 2005.)

Grain or seed	Bulk density (kg/m³)	Angle of repose(°)
Barley	62–69	16–30
Canary	70	
Chickpea	60	
Faba bean	75	
Linseed	73	14–25
Lupin	75	
Maize	72–79	15–27
Oats	44–56	18–32
Peas	75	
Rapeseed	67	
Rye	71	17–26
Safflower	53	24
Sorghum	73	30
Soybean	74–77	16–29
Triticale	69	
Wheat	75–88	16–28

approximately spherical and have not retained a fibrous external husk after threshing, e.g. sorghum, wheat, maize (corn), triticale, millet, safflower and linseed, have relatively low angles of repose between 20° and 30° (Bullen, 2005; Mwithiga and Sifuna, 2006) and flow easily. Grains with higher than normal moisture contents have steeper angles of repose (Amin *et al.*, 2004; Mwithiga and Sifuna, 2006). Foodstuffs that are very fibrous or fluffy, have high oil or molasses contents, or have angular or large particles may form piles with almost vertical sides. Oilseed meals, meat meals, whole cottonseed, chopped forages and some granular minerals have large angles of repose and do not move readily under gravity. These ingredients are often better handled by auger, bucket elevator, belt conveyer or drag conveyer.

Pest control

Grain should be stored at moisture contents of no more than 9–13% (Bullen, 2005) so that the growth of insects and microorganisms is inhibited. Grain can be cooled and dried in the silo by aeration, i.e. blowing ambient air through perforated tubes set in the base of the silo. This usually reduces insect growth (Thorpe, 1998). However, to be fully effective this air should be dry and cool, and it may be necessary to aerate at night only, and/or to use dried or refrigerated air (Thorpe, 1998). Although aeration is the cheapest and most ecologically friendly method of grain preservation it is not as reliable in the tropics as it is in temperate regions. Mathematical models are available to predict the stability of grain stored under aeration (Thorpe, 1997; Chawla, 2002; Lopes *et al.*, 2006).

Aeration is usually accompanied by treatment with insecticidal dusts, gases or sprays (Driscoll *et al.*, 2000; Bartholomaeus and Haritos, 2005; Bullen, 2005) or by flooding the silo with CO_2 if organically preserved grain is required (Driscoll *et al.*, 2000). Insecticide use is often controlled by regulation, and in some cases the penalty for breaching these regulations is severe because inappropriate insecticide use can cause residues in animal tissues.

The Design of Feed Mills

'Elementary process' flow diagrams

These diagrams are flowcharts which show the sequence of processes that are used to produce a given type of stock food, from the placing of ingredients into storage to the packaging or storage of the finished products. An elementary process flow diagram simply lists the processes involved in the order in which they are carried out; it does not give any indication of the nature of these processes or details of the equipment used.

Consider the production of a beef cattle feedlot ration in a small stock food processing facility (a 'feed mill'). The material flow and processes described below are typical of many small commercial or on-farm feed mills (Fig. 18.2):

A. Cereal grains and protein meals are delivered and stored in bulk; minerals and vitamins are delivered in bags or boxes usually packed in approximately 1 t lots on pallets.
B. Whole grain is milled through a hammer mill or roller mill and the milled grain is stored in a milled grain supply bin.
C. When a batch of food is prepared, the main ingredients (grain and protein meals, and possibly Na, Ca and P sources) are augered into the mixer; this is equipped with load cells and the required amounts of ingredients are accumulated before they are mixed together.
D. Minor ingredients, i.e. those which are included in amounts of 5 kg/t or less such as trace mineral and vitamin premixes, are weighed using a small electronic platform scale and hand-tipped into the batching bin.
E. The ingredients are mixed.
F. The mixed ration is discharged to bulk product storage.
G. The ration is transported to the user.

The process flow described above and in Fig. 18.2 is as simple as we would expect to find in a commercial feed mill and most are more complex than this. In Fig. 18.2H an alternative batching and mixing sequence is shown. Use of a separate batching bin, mixer and surge bin allows the processes of batching (weighing out and assembling the ration ingredients), mixing and transporting the mixed ration to storage or bagging it, to be carried out simultaneously. This can reduce the time needed to produce a batch of food from about 15 min to perhaps 5 min (or increase processing rates from 8 to 24 t/h, assuming a 2 t capacity mixer). Increased complexity can improve the speed of material flow and allows more processing options, but increases the cost of the unit. Complexity of design should match the required performance.

Many commercial feed mills use additional processing methods and equipment so that a wider range of products is made than just meals. Liquid ingredients such as molasses, or fats or oils, may be added. Rations may be presented as pellets (Fig. 18.2I) or crumbles (a 'crumble' is a pellet which has been coarsely rolled so as to break it into

smaller pieces). Some facilities process grains by steaming and flaking, expansion or extrusion. This introduces requirements to hold the mixture to be processed in a supply bin, to sieve pelleter output to remove unpelleted particles, to cool and store the product and possibly to dry the product.

'Specific process' flow diagrams

So far, we have considered what happens in preparing a mixed ration, but we have not said anything about the equipment that is used to make these things happen. A flowchart in which the nature and capacity of the equipment used is described is called a 'specific process' flow diagram. We shall consider a specific process flow diagram (Fig. 18.3) of a feed mill which is able to produce about 10 t/h, or about 20,000 t annually, using the types of processes which were described in Fig. 18.2.

A. Receiving ingredients: bulk ingredients (e.g. cereal grains, protein meals, limestone and salt) are delivered by road in approximately 30 t capacity trucks. Ingredients are unloaded into a 10 t capacity drive-over hopper (roadpit) (A1). The hopper does not have to be able to accommodate the whole 30 t because the conveying system removes material as it is being delivered. The roadpit auger (A2) and bucket elevator (A3) are rated at 60 t grain/h. The auger has a close-fitting casing so as to be essentially self-cleaning, thus reducing cross-contamination between different deliveries. Material emptying from the bucket elevator is directed by a distributor head into chutes (A4) through which it slides by gravity into one of the silos or ingredient bins. Bagged and boxed ingredients (e.g. trace mineral and vitamin premixes, medications, other minerals) are usually delivered as 1–1.5 t pallet loads (depending on the density of the ingredient and size of bag) or in 1 t bulk bags and are unloaded by forklift truck.
B. Ingredient storage: cereal grains are stored in ten 250 t silos (B1), while protein meals are stored in 12 30 t bins (B2). The amount of bulk ingredient storage needed depends on the seasonality of ingredient supply, the availability and reliability of on-farm or other storage and the likelihood of ingredients spoiling during storage. In this example, grain is stored for 3 months and protein meals for 1 month. Bagged and boxed ingredients are stored in stacks or on their pallets. Premixes sufficient for 2 or 3 weeks' production are purchased

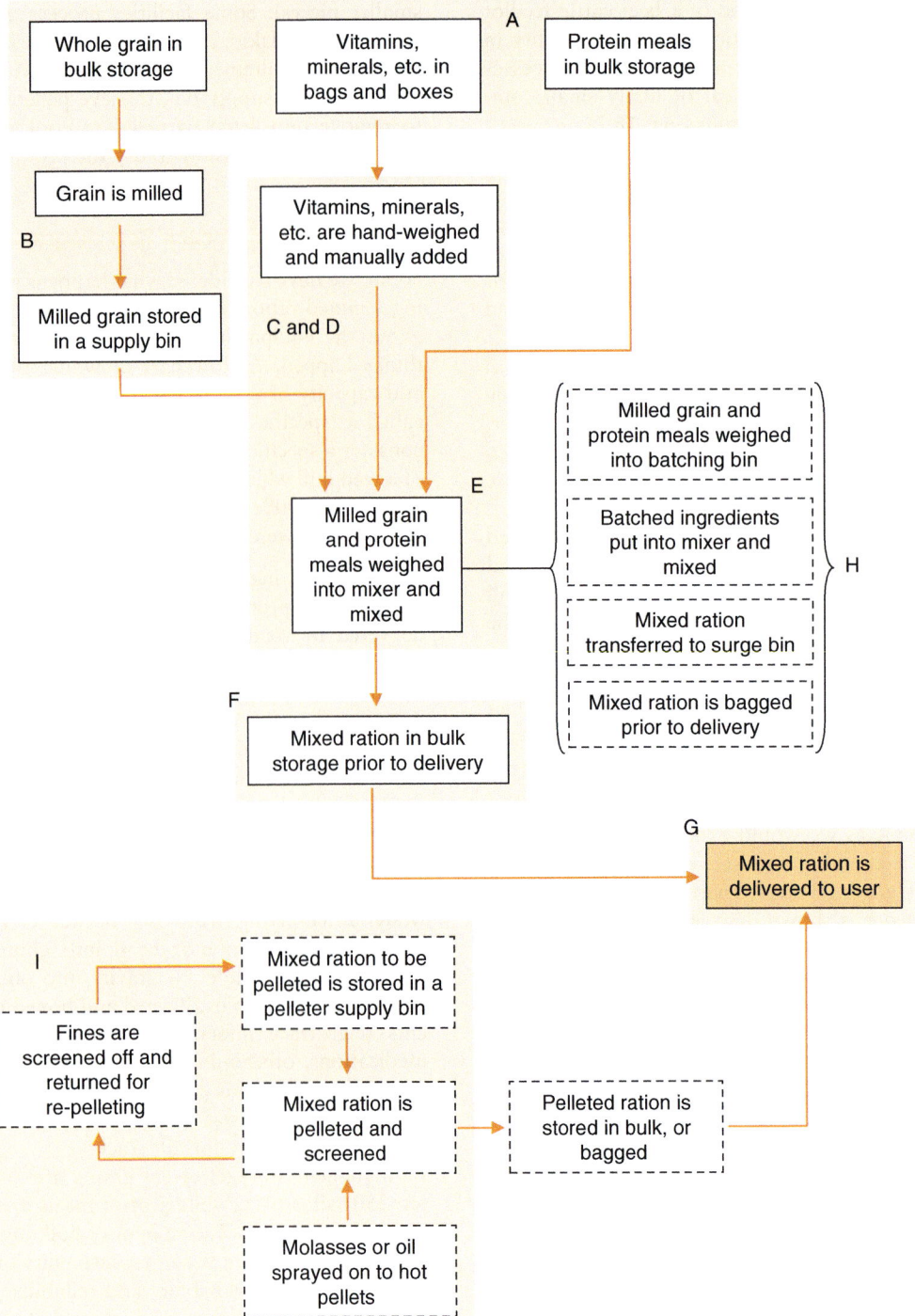

Fig. 18.2. Elementary process flow for a small feed mill. Dashed boxes indicate alternative material flows and processes. See text for the key and explanations.

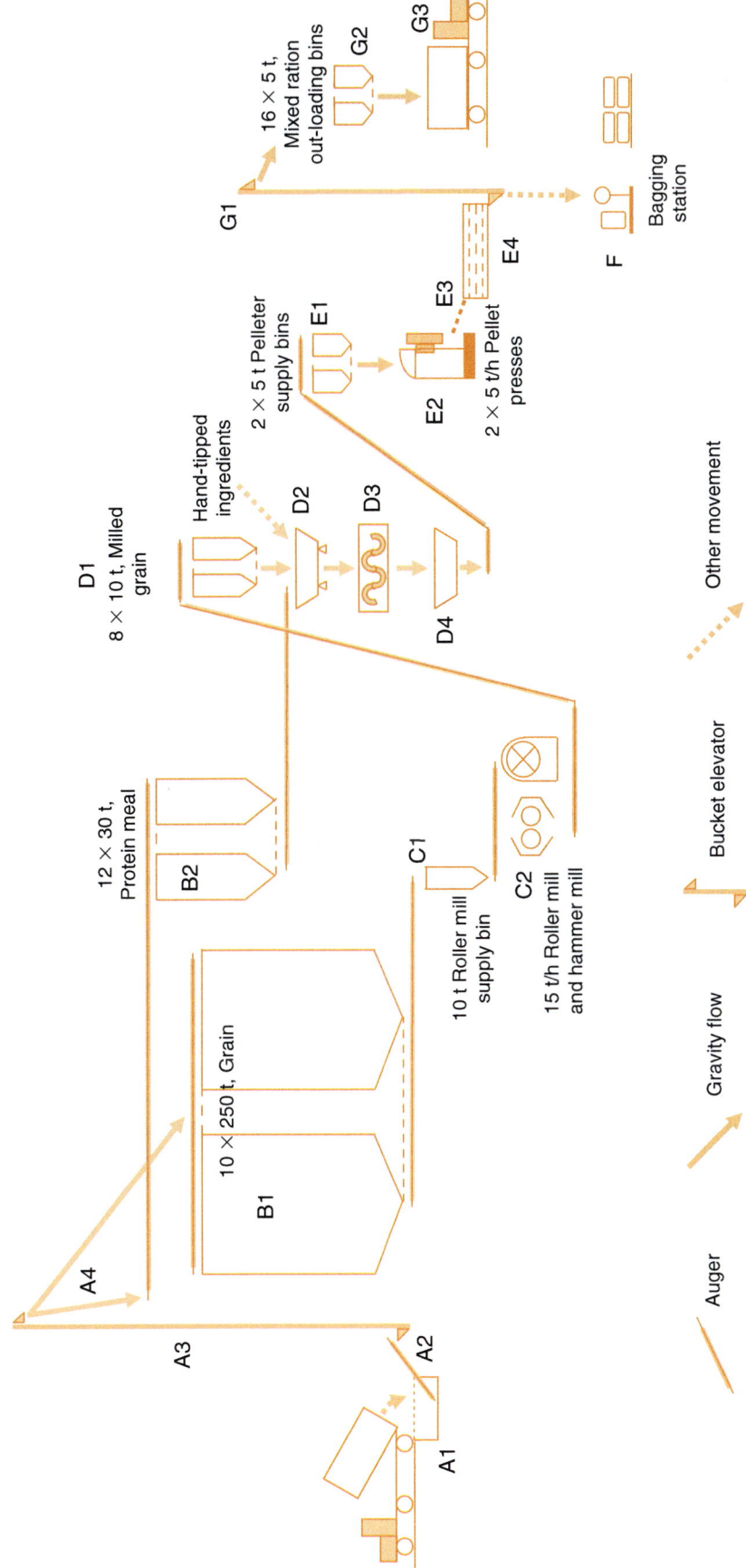

Fig. 18.3. Specific process flowchart for a feed mill with production capacity of 20,000 t/year. See text for key.

from a speciality supplier. Vitamin premixes are unstable at high temperatures (e.g. in a warm summer) and should be stored in a refrigerator. It may be essential to separate animal protein meals from other ingredients.

C. Grain processing: cereal grains are stored whole and are milled shortly before being used. They are stored temporarily in a 10 t supply bin (C1) before being processed through a roller mill or a hammer mill (C2) both rated at 15 t/h. The machines are protected by permanent magnets in the chutes through which the grain enters the mills. Milled grains are augered into eight 10 t batching supply bins (D1), enough for 1 day's production.

D. Batching and mixing: grains and protein meals are augered into a batching bin (D2; 2 t capacity). This is suspended on load cells and the ingredient weights are electronically recorded. The batching bin has a hatch for hand-tipping minor ingredients. These are weighed out on an electronic scale of appropriate capacity and scale sensitivity. The batched ingredients are fed by gravity into a horizontal ribbon mixer (D3; 2 t capacity). The mixed batch is discharged into a 2 t surge bin (D4) below the mixer.

E. Pelleting: the mill runs two pellet presses (E2; each rated at 5–6 t/h and capable of pelleting the whole mill output). Each is supplied with the mixed meal to be pelleted from a supply bin (E1; 10 t capacity) located above the pellet press. The mixture to be pelleted is conditioned (exposed to steam) before pelleting. Pellets are passed over a scalper (E3) which sieves off unpelleted meal and small pieces of broken pellet before the finished pellets are conveyed through cooling towers (E4) where they cool and harden (some mills use cooling bins).

F. Bagging: a bagging station (F) has a supply bin, a scale and a bag sewer. The bags are stacked on pallets ready for shipment.

G. Storage and out-loading of bulk product: the finished product is conveyed by bucket elevator (G1) from the surge bin or the cooling towers to a distributor head from where it flows by gravity into one of sixteen 5 t out-loading bins (G2). The product is loaded by gravity into delivery trucks (G3) which have a capacity of 14 t and are divided into as many as six compartments. The trucks are unloaded by truck-mounted augers, which can be raised and swivelled to reach the client's bin hatches, or by pneumatic conveyers.

Milling

Hammer mills and roller mills are commonly used to mill food ingredients. Their particular advantages and disadvantages are outlined below. Hammer mills can process both grains and forages, but roller mills can process grains only. Information on the design and operation of hammer mills and roller mills is given by Rexroat (1985), Heimann (1994) and Behnke (1996).

Hammer mills

Hammer mills (Fig. 18.4) contain a set of metal 'hammers' which move at up to 3600 rpm, or 65–100 m/s (Goodband *et al.*, 1991; Ajayi and Clark, 1997) mounted on a shaft and surrounded by a screen. When food particles become small enough they fall through the holes in the screen into the plenum chamber and are removed from the hammer mill by either an auger or a pneumatic conveying system.

Hammers can be made in several different ways; the hammer in Fig. 18.4B is reversible (can be exchanged end-for-end) and rotatable (exchanged side-for-side) giving four usable corners. The screen is a steel plate which is drilled with holes and curved so as to fit around the circumference of the arc described by the hammers. The hammers do not touch the screen, but too wide a gap reduces milling performance (Goodband *et al.*, 1991). Holes of 4–8 mm diameter are used in screens for milling grains. Roughages which are to be incorporated in pellets may be milled through screens as small as 4 mm with an accompanying very high power requirement and slow processing rate. The required particle size depends on the roughage inclusion rate, the die size and the extent of pre-pelleting conditioning. For other purposes, roughages are usually milled through screens up to some 50 mm in diameter. This chops the forage into pieces about 3–4 cm long.

Hammer mills work mainly by impact; the grains or other material are broken (milled) when they are struck by the hammers or when they are thrown by this impact against the mill casing or the screen. Impacts in the hammer mill occur randomly and there is no control over the minimum size of the particles which are produced. Hammer-milled grain typically contains fines, although the amount depends on several factors including grain type and the grain moisture content. Shear, which occurs

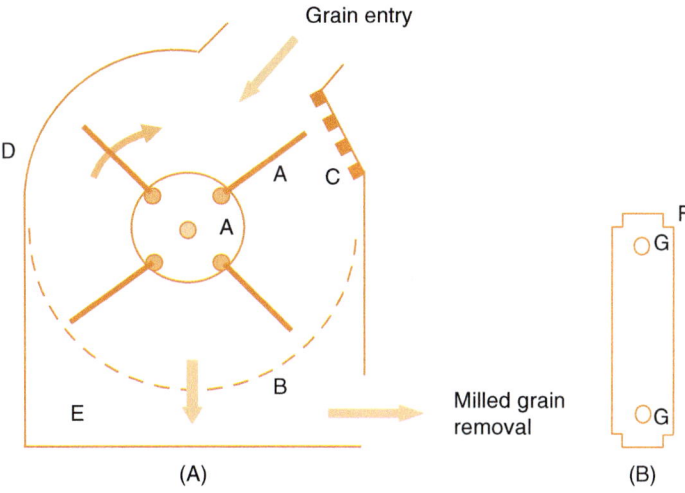

Fig. 18.4. A hammer mill. (A) Construction: A = hammer assembly, B = screen, C = breaker plates, D = casing, E = plenum chamber; (B) hammer design: F = notched corner, G = mounting holes.

when food particles are struck by the sharp edges of the hammers, becomes increasingly important as grain moisture or oil contents increase because grains become more plastic at higher moisture contents (Ajayi and Clark, 1997). Shear is less costly in energetic terms than impact milling (Ajayi and Clark, 1997) and these authors suggest that greater economy of milling would be achieved by using thinner hammers when the desired product is a coarse grind, rather than a fine one.

Factors which influence product quality and processing rate include:

- Grain moisture content: the energy requirement increases with increasing moisture content (Islam and Matzen, 1988) and milling becomes very difficult at more than 16% moisture; grain moisture content is influenced by air humidity.
- Grain type: more fibrous grains mill more slowly and require more power; grain mixtures process more quickly than unmixed grain; oil released from grain legumes during milling may form a seal across the hammer mill screen (B. Hosking, 2007, Oakey, Australia, unpublished data).
- Screen characteristics: smaller holes (i.e. a finer grind), greater area of metal between holes, or blocked holes, reduce processing speed and increase the energy requirement (Islam and Matzen, 1988).
- Required particle size: as the particle size becomes smaller it takes more energy and

greater hammer speeds to achieve this (Ajayi and Clark, 1997) because small particles have less inertia than large ones.

- Air-flow through the hammer mill.
- The degree of wear: the sharp edges of unworn components, especially hammer tips and screen holes, tend to fracture grain by shear, rather than by impact, and this gives less fines.

Roller mills

Roller mills consist of a pair of steel cylinders (or 'rolls') between 250 and 1500 mm long, mounted parallel to each other and separated by a gap which can be adjusted by moving one of the rolls; the other rotates on a fixed axle (Fig. 18.5). The rolls rotate in opposite directions so that they enter the roll gap in the same direction. Grains fall into the gap, are caught between the counter-rotating rolls and are cracked as they pass between them. One of the pair of rolls usually rotates faster than the other (e.g. a differential of up to 2.5:1). This speed differential imparts a shearing action on the grain. The roll gap is the main factor which determines the product particle size (Campbell *et al.*, 2001; Handreck *et al.*, 2004) and the amount of unmilled grain, but the nature of the product is also influenced by grain characteristics such as density, hardness, grain size and moisture content (Fang and Campbell, 2003b). If a fine end product

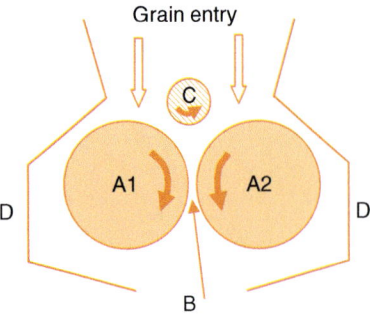

Fig. 18.5. A roller mill: A = rolls (A1 is fixed, A2 is mounted on a spring assembly), B = roll gap, C = feeder roll (not present in all roller mills), D = casing.

is required, pairs of rolls can be mounted in series, each pair giving a successively finer grind. Gwirtz (2005) has described recent advances in roller mill design. Energy consumption per unit of product declines as roll speed increases, but increases as the speed differential is reduced (Guritno and Haque, 1994).

The rolls are generally corrugated, i.e. they have grooves (or 'flutes') cut into their surfaces. The orientation of the fluting can vary from longitudi-

nal (i.e. parallel to the roll axis) to a very shallow spiral (slope of 4–6%) in which the groove runs around and along the roll. Spirally fluted rolls are paired so that the spirals on each roll run in different directions, i.e. the flutes cross each other at an angle twice the angle each flute makes with the long axis of the roll (Matz, 1991).

The flute shapes shown in Fig. 18.6B are suitable for many general milling purposes. Optimum milling action is obtained from rolls which have 4–5 flutes/cm (Goodband *et al.*, 1991) and where the flutes have relatively sharp edges. In many commonly used flutes (e.g. the Le Page, Dawson or saw-tooth shapes; Heimann, 1994) the edges are at different angles. This gives a groove which has a steep or 'sharp' side and a shallow or 'dull' side. A roller mill is running 'sharp to dull' if the steeper sides of the flutes on the faster roll and the shallower sides of the flutes on the slower roll enter the roll gap first (Fig. 18.6C). This is called the roller mill 'disposition'. Rolls operating sharp to sharp give a uniform particle size distribution, the other dispositions give increased proportions of large and fine particles and less intermediate sizes (Fang and Campbell, 2003a). Milling action increases as the absolute speed of the roll surface increases, and

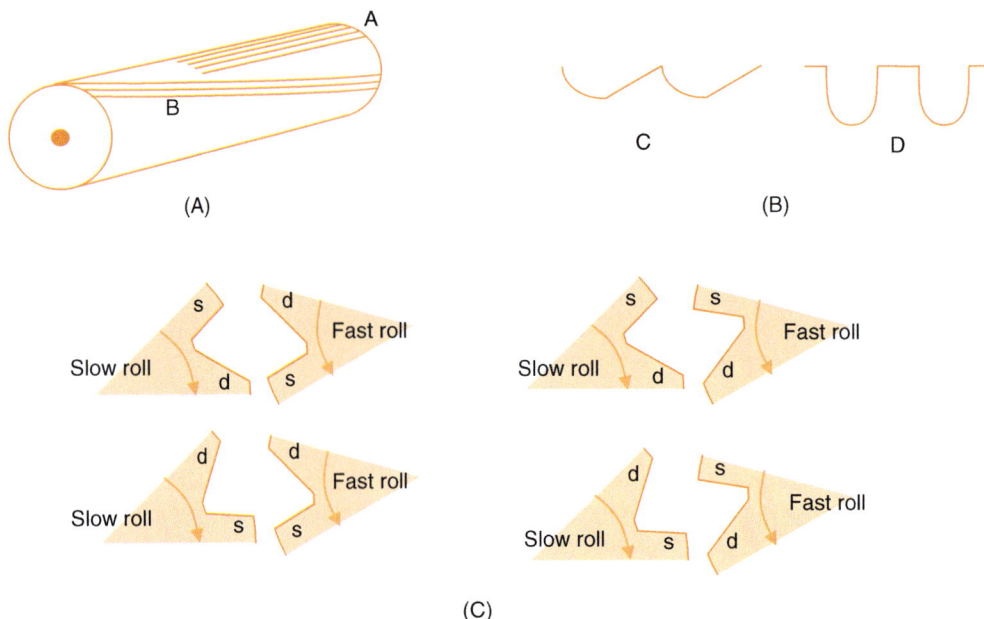

Fig. 18.6. Details of roller mill construction. (A) Roll corrugation: A = longitudinal fluting; B = spiral fluting (exaggerated); (B) types of fluting: C and D = typical flute cross sections; (C) roll disposition: rolls rotating sharp (s) to dull (d), sharp to sharp, dull to dull and dull to sharp.

with increasing differential between the speeds of the fast and slow rolls (Heimann, 1994).

Comparison of roller mills and hammer mills

Roller mills can process cereal grain with as much as 30% moisture (e.g. high-moisture grain), although at this moisture content material may adhere to the rolls (Goodband *et al.*, 1991). Roller-milled product tends to have less fines than hammer-milled material, and with suitable choices of roller fluting, rotation speed and gap setting, grain processing can vary from just removing the seed-coat, to flattening the grain or to milling to a fine flour (Wang *et al.*, 2007). Hammer mills cannot process foods which contain more than about 16% moisture, and their working surfaces wear readily – e.g. the hammers, screens, breaker plates, internal cotter pins and extraction fan components. On the other hand, roller mills are liable to severe and expensive damage if stones or tramp metal are present in the food and are caught in the gap between the rolls. Hammer mills tolerate choking and can run empty and are not susceptible to serious damage from stones or tramp metal; they are easy to maintain – screen and hammer replacement is simple. Very importantly for some on-farm applications, they can process roughages as well as grains. When hammer mills and roller mills are set to give similar end-product characteristics, they use similar amounts of energy (Fang *et al.*, 1997). Energy use is most affected by the gap size in roller milling, and by the screen hole size in hammer milling (Fang *et al.*, 1997). Hammer-milled grain tends to have a larger proportion of fines, i.e. dusty particles which pass a 350 μm screen (Fig. 18.7).

Mixing

Principles of mixing

Mixing is a process which disperses the ingredients of a ration so that, ideally, each small sub-sample has the same proportion of each ingredient as is specified in the ration formulation. In animal food processing the objective of mixing is to scatter particles of dissimilar ingredients until they constitute a blended mass. During mixing, food particles move with respect to each other from their initial position in the mixer (i.e. immediately after the ingredients have been batched) to their final position. During this process, we reach a state which we define as a 'mixture'.

Food mixtures include many different solids and liquids. Each of these has different physical properties: solids differ in particle size, shape, bulk density, coefficient of friction, elasticity, electrostatic charge and hygroscopicity; liquids differ in viscosity and density (Lindley, 1991b). Particle trajectories are influenced by mechanisms which cause them to clump together (agglomeration) or to separate into pools of similar particles (segregation) (Lindley, 1991a). Both agglomeration and segregation interfere with mixing. Agglomeration is favoured by:

- Mechanical interlocking, e.g. between particles of millrun or oilseed meals, and this is facilitated by fibrosity and angularity.
- Surface attraction especially in fine powders of less than 100 μm diameter (e.g. the 'fines' in milled grain); electrostatic attraction is important at low humidities and the effects of water adsorption at high humidities (Karra and Fuerstenau, 1977).

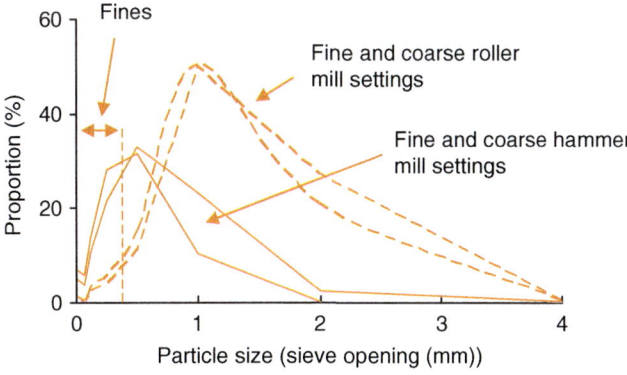

Fig. 18.7. Particle size distribution of hammer-milled and roller-milled barley grain. (Drawn from data in Laurinen *et al.*, 2000.)

- Moisture (e.g. molasses, oil or water) in the mixture affects particle behaviour according to whether the particles are hydrophobic or hydrophilic, the range in particle sizes and the mixer speed (Li and McCarthy, 2006).
- Moisture may dissolve readily soluble ingredients (e.g. urea) which then form a 'glue' between adjacent particles when the moisture evaporates.

Segregation is facilitated (Lindley, 1991a):

- When particles of different sizes, shapes or densities take different paths through an ingredient mass as it is mixed.
- By percolation, i.e. the movement of fine particles between coarse ones – this may cause the fine particles to move to the bottom and large ones to rise to the top of the bulk of material.
- By shear between layers of particles of different sizes, as may occur in a horizontal mixer.

Mixing of solids (Lindley, 1991a) occurs through convection, which is the movement of groups of particles from one place to another within the ingredient mass; diffusion, which is the movement of individual particles throughout the ingredient mass; and shear, which is the slipping of layers of different ingredients past each other. Viscous liquids, such as molasses and oils, are mixed into solids mainly by shear. It can be difficult to mix solid particles which have the same size and shape but different densities, and dense, smooth, round particles tend to move to the bottom of an ingredient mass while irregular particles tend to move to the top (Lindley, 1991b). Incomplete mixing may also occur because of inadequate or excessive mixing times, worn equipment and the use of ingredients which are difficult to mix (e.g. molasses, large fibrous or small dense particles).

Over-mixing can be a problem in stock food processing. This is because the more a mass of ingredients is moved and vibrated, the greater are the opportunities for particles with similar characteristics to move along similar trajectories through the mass. Particle movement is necessary for mixing to occur at all, but it also is responsible for the aggregation of similar materials in pools within the food mass if the mixing process is continued for too long (Li and McCarthy, 2006).

Mixing equipment

Batch mode mixing, rather than continuous mixing, is commonly used in stock food mills (Niranjan

et al., 1994) because it conveniently allows for changing recipes between batches. In batch mixing, the ingredients in a given quantity of mix, determined by the capacity of the mixer, are accumulated before mixing. The ingredients are mixed and then moved to further processing or storage. The process is then repeated.

Vertical and horizontal mixers (Lindley, 1991b; Rielly *et al.*, 1994) are the main types used in stock food processing. Horizontal mixers are the most common in feed mills because they generally produce a more homogeneous mixture in a shorter time. The small footprint of vertical mixers suits them for certain types of on-farm mobile stock food processors.

The simplest form of vertical mixer is the 'single-screw' mixer (Fig. 18.8). The auger lifts the ingredients from the bottom of the mixer and distributes them over the mass of ingredients at the top thus, after a time, mixing them. The mixer is filled through an auger at the bottom of the mixer or through a filling hatch in the top of the mixer. These mixers usually discharge by gravity through a chute or by elevating the material via the auger through the top of the mixer.

Vertical mixers work by a combination of convection, shear and diffusion. The main mixing zones are at the top and bottom of the central auger (Goodband *et al.*, 1991). Ingredients are picked up at the base of the mixer and then discharged from the top of the auger in a spray of material which is distributed over the top of the food mass. The material discharged from the auger top is subjected to centripetal force and moves as individual particles through the air (i.e. by diffusion). Large, denser particles may be thrown further and move to near the mixer wall, while smaller and/or less dense particles may remain near the auger. Convection is probably the main process operating on the material as it enters the bottom of the auger. Shear may occur as the mass of food moves downwards through the mixer body. Percolation of small particles through the ingredient mass probably also occurs because the mass vibrates as the auger works.

Mixing in a vertical mixer may not be necessarily complete. There is a tendency for the layers of different ingredients to be simply moved around the mixer without actually intermixing. The material near the auger will be collected and circulated, but that nearer the wall may not be moved; these tendencies are greater at larger mixer diameters. Overloading reduces mixer efficiency (Goodband

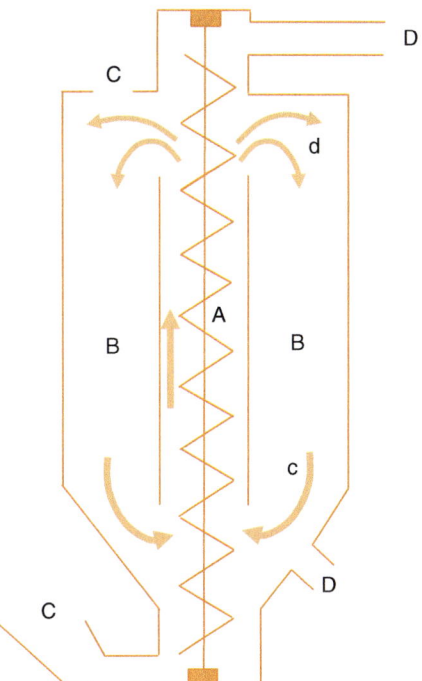

Fig. 18.8. Single-screw vertical mixer. A = central auger, B = ingredient mass, C = alternative ingredient entry ports, D = alternative product discharge ports. Arrows show ingredient flows (c = convection, d = diffusion).

along the length of the mixer body (Fig. 18.9). Many mixers have a second ribbon, or else small paddles, mounted close to the central axle to give additional mixing action in the core of the batch (Goodband *et al.*, 1991). Ribbon mixers act almost exclusively by convection, i.e. the material to be mixed is conveyed from end to end, top to bottom and side to side in the mixer (Rielly *et al.*, 1994). Although low mixing speeds (20–25 rpm) are used there may be some turbulence created in the wake of the ribbon or in voids which develop near the shaft (Laurent *et al.*, 2000), which may cause some diffusion mixing. Percolation (of small, dense particles through the mass of larger, lighter particles) may also occur, as in the vertical mixer.

Discharge hatches are located at the base of the mixer casing. In 'drop-bottom' mixers, the whole base of the casing opens to discharge the load. Horizontal mixers are fairly compact, with capacities of 1–5 t. They are suitable for mounting in a sequence with buffer and surge bins (Fig. 18.3). Horizontal mixers may have hatches for adding pre-weighed amounts of minor ingredients and premixes by hand. These hatches must be guarded so that hands cannot be introduced into the mixer. See Farenholz (1994) for a description of mixer operations.

Mixer calibration

Mixing times must be long enough to achieve a mixture, but not so long that they cause demixing. The optimum time varies with mixer type and with different ration formulations. Mixers should be 'calibrated'. This can be done by assaying the variability

et al., 1991). Liquid ingredients (water, molasses or oils) should be prepared as a premix with milled grain or some similar ingredient before being introduced to a vertical mixer.

Horizontal ribbon mixers have a helical flight (the 'ribbon') which rotates on an axle running

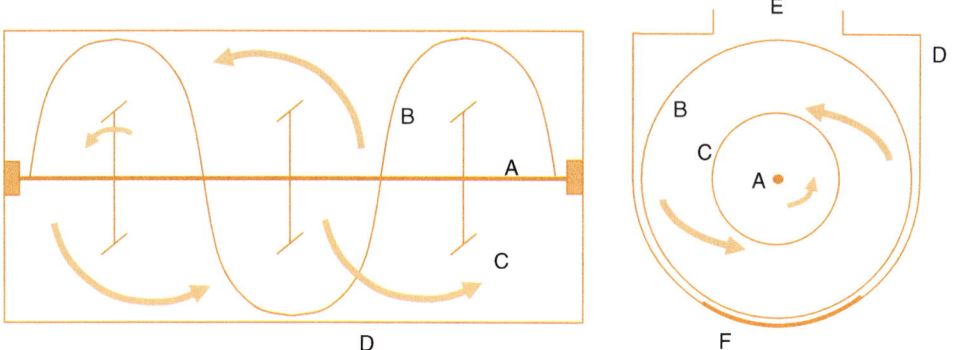

Fig. 18.9. Horizontal ribbon mixer. A = axle, B and C = ribbon blades, D = casing, E = ingredient entry port, F = product discharge hatch. Arrows show ingredient flows.

in the N or Na content, or the amount of coloured iron particles, between samples taken from different places within the mass of ingredient in the mixer (Herrman and Behnke, 1994; Boloh and Gill, 2006). A suitable mixing protocol will limit the variation between these samples to a coefficient of variance (CV) of 10%. This should include the variation caused by errors inherent in the laboratory assay as well as the between-position sampling variation (Wicker and Poole, 1991). Tracers such as iron particles, methyl violet and ytterbium, which are not usual components of food mixtures, can be used to detect cross-contamination between mixes (Boloh and Gill, 2006).

Premixes

These are mixtures of a micro-ingredient and a carrier, e.g. bran, millrun, finely ground grain, lime or silica. The choice of carrier is influenced by the nature of the active ingredients, e.g. their hygroscopicity or deliquescence. Micro-ingredients are ingredients which are only a small proportion of the ration, e.g. trace minerals and vitamins; these are difficult to mix evenly throughout the ration. The value of preparing micro-ingredients as premixes is that the ratio of major to minor ingredient is vastly changed. A vitamin may be incorporated into a stock food at a concentration of 0.1 mg/kg, but incorporated into a premix at 100 mg/kg, i.e. a 1000-fold change in the ratio. This makes the problem of the even distribution of the vitamin one which can be easily handled, rather than one which approaches impossibility.

Heat Processing

Moist heat processes, such as pelleting, steam flaking and extrusion, are widely used in manufacturing foods for domestic production animals and in pet food manufacture. Pelleting involves forcing a previously steamed or moistened ('conditioned') mixture through small holes in a thick steel ring (a 'die ring') (Fairfield, 1994). High temperatures (75–135°C, caused by friction as the food passes through the die ring) and the presence of moisture cause some starch gelatinization as well as changes in soluble sugars, cell wall polysaccharides and proteins. The food particles adhere forming a pellet. The size of the pellet is determined by the shape and size of the die ring holes and knives which are set outside the die ring.

Steam flaking involves conditioning dry grain in steam prior to rolling it between close-set, heavy rolls (Heimann, 1994). Maize is sometimes milled before conditioning. Exposure to steam may vary from 30 to 60 min at ambient pressure to reach a temperature of 100–110°C and a final moisture content of 18–20% (Hale et al., 1966; Theurer et al., 1999) or for 50–120 s at 3.5 kg/cm² pressure to reach a temperature of 130–150°C and a final moisture content of 15–25% (Osman et al., 1970; Theurer, 1986). A less intensive process (steam rolling) involves steaming for up to 15 min to give a final water content of 15% (Theurer et al., 1999). Surfactants may be used to improve the penetration of steam into the starch granules (Malcolm and Kiesling, 1993; Zinn et al., 1998). Steam flaking gives variable results unless the process is carefully controlled. Flake density is a commonly used index of processing effectiveness, and the optimum appears to be about 350 g/l (Swingle et al., 1999). Zinn et al. (2002) have reviewed the processing characteristics needed to produce high-quality steam-flaked (SF) grain.

When grain, or a food mixture, is extruded it is prepared by steam conditioning, and then the hot, moist material is forced through the extruder. This has a barrel enclosing an auger, a plate at the end of the barrel which is pierced by holes of different shapes and sizes, and a knife which cuts the food mixture as it is extruded through the die plate (Hauk et al., 1994). The food mixture reaches 125–170°C for 15–30 s (Rowe et al., 1999).

Other heat processes such as micronizing, popping, exploding and roasting are described by Hale and Theurer (1972). Micronized grain is fed on to a belt in a one-grain thick layer, passed under gasheated ceramic plates (150–200°C for 30–90 s) and then rolled (Fernandes et al., 1975). Grain is popped by exposing it to dry heat at 150°C, causing the rapid conversion of water in the grain to steam and thus 'popping' or expanding the grain endosperm (Parker et al., 1999). In roasting, grain is exposed to direct heat for 30–35 s to raise the grain temperature to between 100°C and 160°C (Costa et al., 1976).

Chopping and Milling of Forages

Chopping and milling are similar physical processes but which differ in the particle size of the processed forage. We will define chopping as a process which gives a minimum particle size of about 10 mm, and milling as giving an average particle size of 2–5 mm. Forage processing equipment may use blades, flails (i.e. thin metal hammers), conventional hammer mills or augers to cut forage into small pieces. Some common forage processing equipment includes:

1. Chaff cutters: forage is squeezed between two belts and is presented to knives which rotate vertically to the plane of the forage. This gives a uniform cut length. Chaff cutters will not process easily very wet or very dry material (such as straw) and bales of hay must be broken up and teased out before the forage is presented to the machine.
2. Counter-running augers: these are used in some mobile feeder-mixers. This gives some degree of processing but the product is coarse. If other ingredients are added, chopping and mixing are combined in one operation.
3. Flails: these are used in 'bale breakers' and some forage harvesters. They rotate at high speed and break the forage when they impact against it. Flail choppers can give excessive variation in particle length when they are used to process dry forages. For example, when lucerne (alfalfa) hay is chopped the particles can be from 1 or 2 mm to about 5 cm long, and excessive dust may be produced when the dry leaves are broken by the flails.
4. Hammer mills: these are commonly used to mill roughages. The hammer mill may be a stand-alone device or incorporated into a multi-purpose machine. Roughage processing can be a very slow process with a high power requirement if a finely milled product is required, but can also be quite rapid if the roughage is processed through a screen with large holes (e.g. 5 cm diameter) or if no screen at all is used. The choice of equipment is determined by the required end product. If a coarsely chopped product is required, then machinery similar to the tub grinder described below is suitable and processes the material rapidly. Hammer milling through a small screen (e.g. 6 mm diameter holes) can produce a finely milled material useful for adding to concentrate-based meals or pelleted foods. Hammer mills used for roughage processing are often set up with a cyclone to collect the dust produced during milling. Tub grinders have a circular container (the 'tub'), a hammer mill and a moving belt assembly to discharge the chopped forage. The tub can be loaded with quantities of roughage, either loose or in intact small or large (round or rectangular) bales. It rotates in a horizontal plane above a hammer mill set into the base of the tub and which has spaced thin hammers and can be fitted with screens of different sizes. The chopped forage falls on to the belt conveyer which can throw the material several metres.

The Principles of Warehousing

The person in control of a warehouse has one of the key tasks in a feed mill, as he or she is responsible for:

- ensuring that the quality of raw materials accepted into the mill conforms to the required standards;
- obtaining and correctly processing the documentation which must accompany all raw material and finished product deliveries;
- allocating storage space for the newly received ingredients and ensuring that they are used in proper rotation and do not deteriorate while in storage, separating meat meals from other ingredients;
- maintaining records of stock usage and notifying the person responsible for purchasing ingredients when new stock is needed;
- allocating finished product awaiting delivery to storage and documenting those deliveries;
- maintaining ingredient stocks in a clean and hygienic (i.e. insect- and mould-free) condition; and
- collecting and storing samples of ingredients received and of finished products for quality assurance purposes.

These warehousing functions normally occur as a sequence of events which are shown in the flowcharts in Figs 18.10 and 18.11. Although the two principal functions (inspecting and documenting incoming ingredients and stock control) are shown in two different flowcharts, they actually occur simultaneously. The process of ingredient reception is summarized in Fig. 18.10 and the procedures involved in stock control, i.e. the storage of ingredients and products, allocation of ingredients and record keeping, are shown in Fig. 18.11. One of the essential aspects of ingredient reception is to identify infestation by live insects. Visual inspection is the first step, but may have to be supplemented with sieving or a flotation test. These and other methods including near infrared reflectance spectroscopy, acoustic detection of insects and chemical detection methods, have been evaluated by Neethirajan *et al.* (2007).

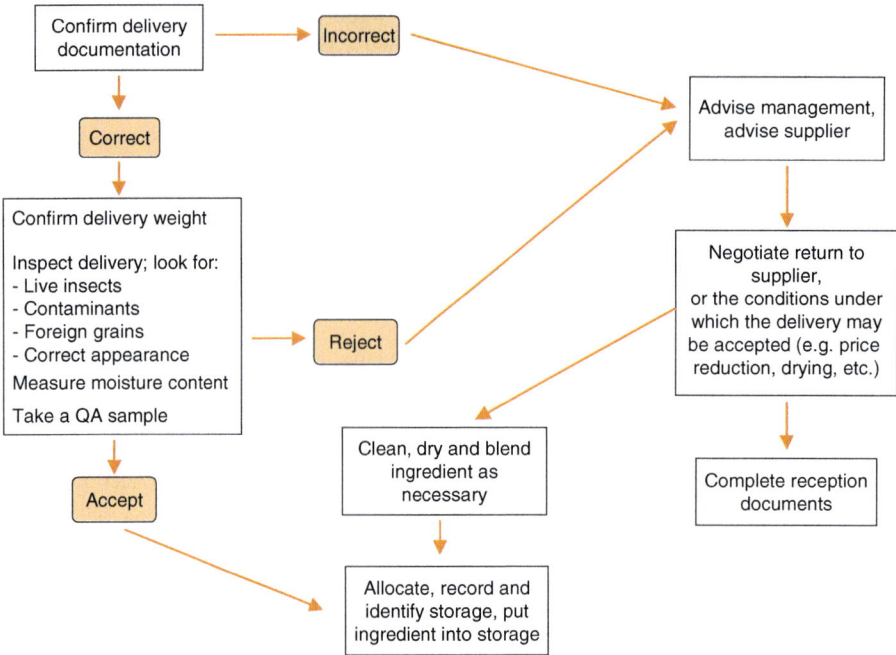

Fig. 18.10. Flowchart of events which occur during ingredient reception.

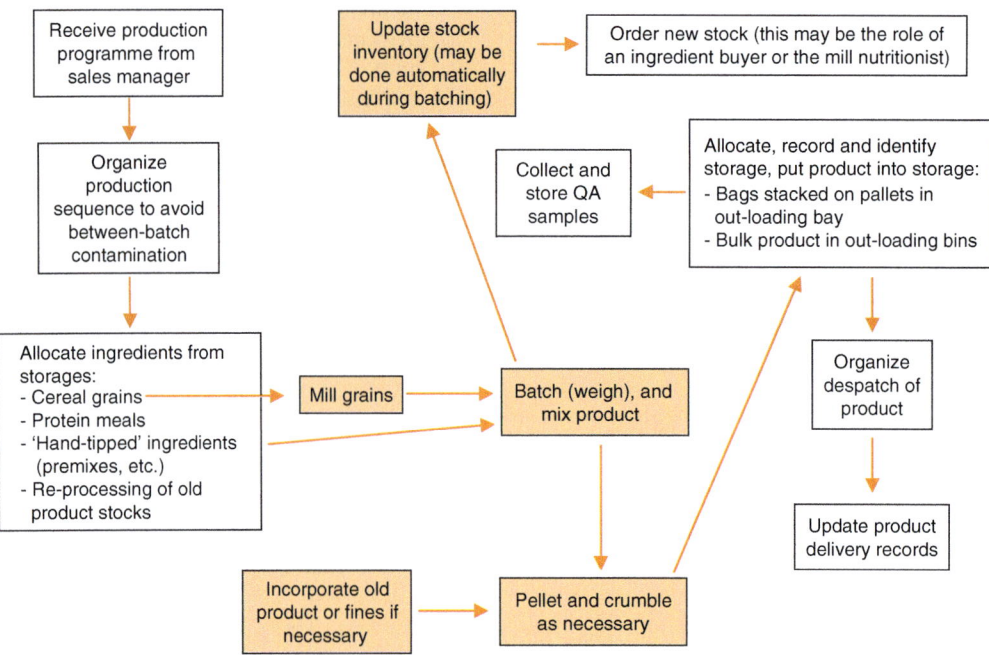

Fig. 18.11. Flowchart of events related to the production stream and stock control (actions in orange boxes are carried out by the production staff, those in clear boxes are done by the warehouse manager).

Mill Hygiene

The warehouse manager must maintain good mill hygiene and safe ingredient storage practices. Mill hygiene includes general cleanliness and dust control and the control of insects and other pests in stored ingredients. Food spillages and dust allow insect and spider populations to become established and encourage rodents and other vermin.

Feed mill personnel can be exposed to poisonous levels of trace minerals, vitamins and other food additives. Examples (but not a complete list) of potentially harmful substances which are used in stock foods are (ILO-CIS, 2004): manganese oxide dust can cause neural disorders after long-term exposure; methionine dust is potentially explosive; 4-hydroxy-3-nitro-phenylarsonic acid (a growth promotor used in poultry diets) is poisonous and flammable; propionic acid (used as a mould inhibitor) is pungent, corrosive and flammable; and sodium selenite (used as a source of the mineral Se) is potentially poisonous. Sometimes the hazard derives from the carrier in the premix rather than the active ingredient. Silicosis may be a risk from working with premixes based on silica, for example. Premixes are more hazardous than the compounded food containing them because of the great difference in concentration of the active ingredients. Users should consult the appropriate material safety data sheets for each premix ingredient.

Dust accumulation in a feed mill increases the risk of fire and explosion. Grain and stock food ingredient dusts are very reactive compared to coal dust. Dusts are hazardous if they have an ignition sensitivity index of 1.0–5.0 and an explosion severity index of between 1.0 and 2.0. In comparison to Pittsburgh coal dust which has an index of 1 for these characteristics, grain dust has indices of 2.8 and 3.3, respectively; wheat flour, 2.1 and 1.8; DL-methionine, 6.2 and 1.5; and thiamine, 2.7 and 3.1 (NMAB, 1982). Dust explosions need an ignition source as well as fuel. These can be smouldering cigarette butts, sparks caused by grinding or welding during plant maintenance or electrical faults. Others are due to equipment failure: hot surfaces (e.g. tramp metal caught in rollers, seized bearings, bags caught around auger flights) are responsible for about one-third of all grain store and feed mill fires and explosions; static electricity is also an ignition source (House, 1991). Dust explosions can occur in silos and other storages (about 20% of explosions), dust filters and separators (14%), conveying systems (10%) and mixers (5%). Read Eckhoff (1991, 2005) for detailed reviews of the causes and nature of dust explosions and Ribereau-Gayon (2000) for descriptions of their human effects.

Quality Assurance in Stock Food Processing

Stock food manufacturers must ensure that their products are adequately nutritious and are safe. Either or both of these characteristics may be regulated by government or industrial bodies. Even if these controls do not exist, the marketplace will ensure that manufacturers who make a substandard product will eventually lose their customers. A sound and properly implemented quality assurance programme will materially assist in maintaining product quality and safety.

Quality assurance procedures which are the particular responsibility of the warehouse manager include ingredient reception and storage, the selection and delivery of the correct ingredients into the processing area and the sampling of finished products. The correct use and maintenance of milling and other processing equipment and the correct measurement of ingredients for batching and appropriate mixing procedures are the responsibility of the staff who operate this equipment.

Dealing with incoming ingredients is the most critical of all quality control procedures because mistakes made here affect all the other activities within the feed mill. Control of incoming ingredient quality involves specifying in the purchase contract the required physical, chemical and nutritional quality characteristics; and then testing the ingredients on arrival to ensure that they comply with these specifications. Some ingredient quality characteristics are controlled by law (e.g. animal protein meal residues in foods intended for ruminants, insecticide residues, contamination with poisonous materials or inclusion of antibiotics; e.g. Europa, undated; USDA, 2006). Others are covered by industry codes of practice (e.g. NACMA, 2006). Others must be explicitly negotiated when the ingredient is ordered. Deliveries which fail to meet allowed tolerances should be returned to the supplier, especially if the breach concerns insect infestations or the presence of banned or poisonous substances. Some deliveries may exceed the purchaser's own warning limits. These deliveries may be accepted, perhaps subject to further negotiations on price, if this will not cause a serious problem for the feed mill. For example, cereal grain which has more than the allowed moisture content may be accepted if the grain can be dried; if grain exceeds a purity tolerance, e.g. more than 5% off-type grain, it may be accepted if this will not greatly influence its nutritional or processing values.

Large feed mills generally operate a quality control laboratory. The most important uses of this analytical data are to:

1. Identify batches of ingredient or product which differ from the usual composition for that material;
2. Identify batches of ingredient which fail to meet the contractual or statutory composition and/or safety requirements;
3. Develop a nutrient content data base for the ingredients commonly used in the mill;
4. Monitor the operation of processing equipment.

Analytical data are often used to generate statistical process control charts (Bender *et al.*, 1982; StatSoft Inc., 2006). Stock food production systems often fail because of wear to equipment, rather than from a single incident, and this occurs imperceptibly. Gradual failure can be identified by plotting data such as pellet hardness, grain particle size or some indicator of inadequate mixing (e.g. protein or a trace mineral content), against time. If the graph moves outside predetermined control limits, the problem can be identified and rectified.

Many quality control procedures depend on examining samples taken from a mass of material. The reliability of conclusions depends on the degree to which the small sample represents the bulk of the material sampled. Good sampling procedure relies on choosing randomly the places to sample so that there is no unintentional bias, and taking enough samples to adequately represent the bulk of substance sampled (European Community, 1976). Samples should be retained from every batch of ingredient received and stock food made, and these should be kept for 3 months.

Rations can be contaminated by residues which remain in batching bins, mixers, augers, storage bins and delivery vehicles. The warehouse manager and the person who determines the mill's manufacturing and product delivery schedules should take care to avoid cross-contamination. The problems caused by cross-contamination go beyond the effects on nutrient content, although these can be

serious enough. A grower pig ration has very different nutritional characteristics to working horse or poultry layer rations, for example. Some additives used in stock food manufacturing have withholding periods – they must not be fed for specified periods before slaughter or at all if the animal produces milk. Further, some ingredients are poisonous to some livestock while being safe with others. For example, monensin can be used in ruminant rations but it is toxic to horses (Hall, 2001); meat meals may not be permitted in ruminant foods.

Quality control systems

There are two world-recognized quality control assurance systems, the ISO 9000:2000 standards and the Hazard Analysis Critical Control Point (HACCP) system. HACCP was introduced by NASA to safeguard the production of food for its astronauts. It is principally a food safety assurance system which was developed to provide a way of identifying and correcting problems of microbial contamination of foods, but it can be adapted to any food production use including ensuring that stock foods are nutritious and safe. The seven principles of HACCP are hazard analysis, critical control point identification, establishment of critical limits, monitoring procedures, corrective actions, record keeping and verification procedures. These are described by various authorities, including FSIS (1998). For detailed information on how to implement a HACCP system consult the New South Wales Food Authority (2005).

ISO9000:2000 is a family of standards which 'specifies requirements for a quality management system for any organization that needs to demonstrate its ability to consistently provide product that meets customer and applicable regulatory requirements and aims to enhance customer satisfaction' (ISO, undated). Implementation of the ISO9000:2000 standards gives stock food manufacturers a management system which facilitates the application of a quality assurance system.

References

Ajayi, O.A. and Clark, B. (1997) High velocity impact of maize kernels. *Journal of Agricultural Engineering Research* 67, 97–104.

Amin, M.N., Hossain, M.A. and Roy, K.C. (2004) Effects of moisture content on some physical properties of lentil seeds. *Journal of Food Engineering* 65, 83–87.

Andrews, A. (1996) *Storing, Handling and Drying Grain. A Management Guide for Farms.* Queensland Department of Primary Industries, Brisbane, Australia.

Bahlmann, L., Klaus, S., Heringlake, M., Baumeier, W., Schmucker, P. and Wagner, K.F. (2002) Rescue of a patient out of a grain container: the quicksand effect of grain. *Resuscitation* 53, 101–104.

Bartholomaeus, A.R. and Haritos, V.S. (2005) Review of the toxicology of carbonyl sulfide, a new grain fumigant. *Food and Chemical Toxicology* 43, 1687–1701.

Behnke, K.C. (1996) Feed manufacturing technology: current issues and challenges. *Animal Feed Science and Technology* 62, 49–57.

Bender, F.E., Douglass, L.W. and Kramer, A. (1982) *Statistical Methods for Food and Agriculture.* AVI Publishing Company, Westport, Connecticut.

Boloh, Y. and Gill, C. (2006) Measuring cross-contamination in feed manufacturing. *Feed International* 27 (4), 28–29.

Boumans, G. (1985) *Grain Handling and Storage.* Elsevier, Amsterdam, The Netherlands, pp. 87–206.

Bullen, K. (2005) *Grain storage – Basic design principles.* Available at: http://www2.dpi.qld.gov.au/fieldcrops/15052.html

Bullen, K. (2006) *Grain Storage – Steel Mesh Silos for On-farm Grain Storage.* Available at: http://www2.dpi.qld.gov.au/fieldcrops/14196.html

Campbell, G.M., Bunn, P.J., Webb, C. and Hook, S.C.W. (2001) On predicting roller milling performance. Part II. The breakage function. *Powder Technology* 115, 243–255.

Chawla, K. (2002) *Management of Cereal Grain in Storage.* Available at: http://www1.agric.gov.ab.ca/$department/deptdocs.nsf/all/agdex4509

Costa, P.M.A., Jensen, A.H., Harmon, B.G. and Norton, H.W. (1976) The effects of roasting and roasting temperatures on the nutritive value of corn for swine. *Journal of Animal Science* 42, 365–374.

Driscoll, R., Longstaff, B.C. and Beckett, S. (2000) Prediction of insect populations in grain storage. *Journal of Stored Products Research* 36, 131–151.

Eckhoff, R.K. (1991) *Dust Explosions in the Process Industries.* Butterworth/Heinemann, Oxford, UK.

Eckhoff, R.K. (2005) Current status and expected future trends in dust explosion research. *Journal of Loss Prevention in the Process Industries* 18, 225–237.

Europa (undated) *Activities of the European Union. Summaries of Legislation. Food Safety. Animal Nutrition.* Available at: http://europa.eu/scadplus/leg/en/s81000.htm

European Community (1976) EU Directive 76/371/EEC. *Official Journal of the European Communities*, No. L102/1. Available at: http://europa.eu.int/eur-lex/lex/RECH_reference_pub.do

Fairchild, F.J. (1994) Bucket elevators. In: McIllheny, R.R. (tech. ed.) *Feed Manufacturing Technology IV.* American Feed Industry Association, Arlington, Virginia, pp. 174–178.

Fairfield, D. (1994) Pelleting cost center. In: McIllheny, R.R. (tech. ed.) *Feed Manufacturing Technology IV.* American Feed Industry Association, Arlington, Virginia, pp. 111–130.

Fang, C. and Campbell, G.M. (2003a) On predicting roller milling performance IV: effect of roll disposition on the particle size distribution from first break milling of wheat. *Journal of Cereal Science* 37, 21–29.

Fang, C. and Campbell, G.M. (2003b) On predicting roller milling performance V: effect of moisture content on the particle size distribution from first break milling of wheat. *Journal of Cereal Science* 37, 31–41.

Fang, Q., Boloni, I., Haque, E. and Spillman, C.K. (1997) Comparison of energy efficiency between a roller mill and a hammer mill. *Applied Engineering in Agriculture* 13, 631–635.

Farenholz, C. (1994) Proportioning and mixing cost center. In: McIllheny, R.R. (tech. ed.) *Feed Manufacturing Technology IV.* American Feed Industry Association, Arlington, Virginia, pp. 98–102.

Fernandes, T.H., Hutton, K. and Smith, W.C. (1975) A note on the use of micronized barley for growing pigs. *Animal Production* 20, 307–310.

Food Safety and Inspection Service (FSIS) (1998) *Key Facts: The Seven HACCP Principles.* United States Department of Agriculture, Washington, DC. Available at: http://www.fsis.usda.gov/OA/background/keyhaccp.htm

Goodband, R.D., Murphy, J.P., Behnke, K.C. and Harner, J.P. (1991) Selection of equipment critical in on-farm mixing. *Feedstuffs* 63, 16–18, 29.

Guritno, P. and Haque, E. (1994) Relationship between energy and size-reduction of grains using a 3-roller mill. *Transactions of the ASAE* 37, 1243–1248.

Gwirtz, J. (2005) Advances in roller mill design. *Muhle + Mischfutter* 142, 315–316.

Hale, W.H. and Theurer, C.B. (1972) Feed preparation and feed processing. In: Church, D.C. (ed.) *Digestive Physiology and Nutrition of Ruminants, Vol. 3 Practical Nutrition.* O&B Books, Corvallis, Oregon, pp. 49–76.

Hale, W.H., Cuitin, L., Saba, W.J., Taylor, B. and Theurer, B. (1966) Effect of steam processing and flaking milo and barley on performance and digestion by steers. *Journal of Animal Science* 25, 392–396.

Hall, J.O. (2001) Toxic feed constituents in the horse. *Veterinary Clinics of North America – Equine Practice* 17, 479.

Handreck, B., Poetschke, L. and Senge, C. (2004) Effect of size reducing parameters within the milling process of wheat middlings. *Getreidetechnologie* 58, 211–216.

Hauk, B., Rokey, G., Smith, O., Herbster, J. and Sunderland, R. (1994) Extrusion cooking systems. In: McIllheny, R.R. (tech. ed.) *Feed Manufacturing Technology IV.* American Feed Industry Association, Arlington, Virginia, pp. 131–139.

Heimann, M. (1994) Roller mill systems. In: McEllhiney, R.R. (tech. ed.) *Feed Manufacturing Technology IV.*

American Feed Industry Association, Arlington, Virginia, pp. 91–97.

Herrman, T. and Behnke, K. (1994) *Testing Mixer Performance*. Kansas State University Agricultural Experiment Station and Cooperative Extension Service. Available at: http://www.oznet.ksu.edu/library/grsci2/MF1172.pdf

House, C. (1991) US records 15 grain dust explosions in '90. *Feedstuffs* 63, 21–23.

International Occupational Safety and Health Information Centre (ILO-CIS) (2004) *International Chemical Safety Cards (ICSC)*. Available at: http://www.ilo.org/public/ english/protection/safework/cis/products/icsc/dtasht/index.htm

International Organization for Standardization (ISO) (undated) *Selection and Use of the ISO 9000:2000 Family of Standards*. Available at: http://www.iso.org/iso/en/iso9000–14000/understand/selection_use/selection_use.html

Islam, M.N. and Matzen, R. (1988) Size distribution analysis of ground wheat by hammer mill. *Powder Technology* 54, 235–241.

Karra, V.K. and Fuerstenau, D.W. (1977) The effect of humidity on the trace mixing kinetics in fine powders. *Powder Technology* 16, 97–105.

Laurent, B.F.C., Bridgwater, J. and Parker, D.J. (2000) Motion in a particle bed agitated by a single blade. *AIChE Journal* 46, 1723–1734.

Laurinen, P., Siljander-Rasi, H., Karhunen, J., Alaviuhkola, T., Näsi, M. and Tuppi, K. (2000) Effects of different grinding methods and particle size of barley and wheat on pig performance and digestibility. *Animal Feed Science and Technology* 83, 1–16.

Li, H. and McCarthy, J.J. (2006) Cohesive particle mixing and segregation under shear. *Powder Technology* 164, 58–64.

Lindley, J.A. (1991a) Mixing processes for agricultural and food materials: 1. Fundamentals of mixing. *Journal of Agricultural Engineering Research* 48, 153–170.

Lindley, J.A. (1991b) Mixing processes for agricultural and food materials: 3. Powders and particulates. *Journal of Agricultural Engineering Research* 49, 1–19.

Lopes, D.deC., Martins, J.H., Melo, E.deC. and Monteiro, P.M.deB. (2006) Aeration simulation of stored grain under variable air ambient conditions. *Postharvest Biology and Technology* 42, 115–120.

McEllhiney, R.R. (tech. ed.) (1985) *Feed Manufacturing Technology III*. American Feed Industry Association, Arlington, Virginia.

McEllhiney, R.R. (tech. ed.) (1994) *Feed Manufacturing Technology IV*. American Feed Industry Association, Arlington, Virginia.

Malcolm, K.J. and Kiesling, H.E. (1993) Dry matter disappearance and gelatinization of grains as influenced by processing and conditioning. *Animal Feed Science and Technology* 40, 321–330.

Matz, S.A. (1991) *The Chemistry and Technology of Cereals as Food and Feed*, 2nd edn. Van Nostrand Reinhold, New York.

Mwithiga, G. and Sifuna, M.M. (2006) Effect of moisture content on the physical properties of three varieties of sorghum seeds. *Journal of Food Engineering* 75, 480–486.

National Agricultural Commodities Marketing Association (NACMA) (2006) *Commodity Standards*. Available at: http://www.nacma.com.au/grain_specifications

National Materials Advisory Board (NMAB) (1982) *Classification of Dusts Relative to Electrical Equipment in Class II Hazardous Locations*. National Academy Press, Washington, DC.

Neethirajan, S., Karunakaran, C., Jayas, D.S. and White, N.D.G. (2007) Detection techniques for stored-product insects in grain. *Food Control* 18, 157–162.

New South Wales Food Authority (2005) *General Guidelines for the Development and Implementation of a Food Safety Program*. Available at: http://www.foodauthority.nsw.gov.au/industry/pdf/Part%201%20-%20General%20Guidelines%20Food%20Safety %20Program.pdf

Niranjan, K., Smith, D.L.O., Rielly, C.D., Lindley, J.A. and Phillips, V.R. (1994) Mixing processes for agricultural and food materials: Part 5, review of mixer types. *Journal of Agricultural Engineering Research* 59, 145–161.

Oje, K. and Ugbor, E.C. (1991) Some physical properties of oilbean seed. *Journal of Agricultural Engineering Research* 50, 305–313.

Osman, H.F., Theurer, B., Hale, W.H. and Mehen, S.M. (1970) Influence of grain processing on *in vitro* enzymatic starch digestion of barley and sorghum grain. *Journal of Nutrition* 100, 1133–1140.

Parker, M.L., Grant, A., Rigby, N.M., Belton, P.S. and Taylor, J.R.N. (1999) Effects of popping on the endosperm cell walls of sorghum and maize. *Journal of Cereal Science* 30, 209–216.

Rexroat, D.W. (1985) Material processing cost center. In: McEllhiney, R.R. (tech. ed.) *Feed Manufacturing Technology III*. American Feed Industry Association, Arlington, Virginia, pp. 137–150.

Ribereau-Gayon, R. (2000) Les explosions de poussières dans le stockage agroalimentaire: les risques d'accident catastrophique et leurs conséquences médicales. *Médecine de Catastrophe – Urgences Collectives* 3, 13–20.

Rielly, C.D., Smith, D.L.O., Lindley, J.A., Niranjan, K. and Phillips, V.R. (1994) Mixing processes for agricultural and food materials: Part 4, assessment and monitoring of mixing systems. *Journal of Agricultural Engineering Research* 59, 1–18.

Rowe, J.B., Choct, M. and Pethick, D.W. (1999) Processing cereal grains for animal feeding. *Australian Journal of Agricultural Research* 50, 721–736.

StatSoft, Inc. (2006) *Quality control charts*. In: *Electronic Statistics Textbook*. StatSoft, Tulsa, Oklahoma. Available at: http://www.statsoft.com/textbook/stathome.html

Swingle, R.S., Eck, T.P., Theurer, C.B., De la Llata, M., Poore, M.H. and Moore, J.A. (1999) Flake density of steam-processed sorghum grain alters performance and sites of digestibility by growing–finishing steers. *Journal of Animal Science* 77, 1055–1065.

Theurer, C.B. (1986) Grain processing effects on starch utilization by ruminants. *Journal of Animal Science* 63, 1649–1662.

Theurer, C.B., Huber, J.T., Delgardo-Elorduy, A. and Wanderley, R. (1999) Invited review: summary of steam-flaking corn or sorghum grain for lactating dairy cows. *Journal of Dairy Science* 82, 1950–1959.

Thorpe, G.R. (1997) Modelling ecosystems in ventilated conical bottomed farm grain silos. *Ecological Modelling* 94, 255–286.

Thorpe, G.R. (1998) The modelling and potential applications of a simple solar regenerated grain cooling device. *Postharvest Biology and Technology* 13, 151–168.

United States Department of Agriculture (USDA) (2006) *Grain, Rice and Pulses. Official US Standards.* Available at: http://www.gipsa.usda.gov/GIPSA/webapp?area=home&subject=grpi&topic=sq-ous

Viljoen, J. (2001) Select grain stores carefully to control costs. *Farming Ahead 119.* pp. 44–45. Available at: http://sgrl.csiro.au/comm/farm_storage.pdf

Wang, R., Koutinas, A.A. and Campbell, G.M. (2007) Dry processing of oats – application of dry milling. *Journal of Food Engineering* 82, 559–567.

Wicker, D.L. and Poole, D.R. (1991) How is your mixer performing? *Feed Management* 42, 40–44.

Wilcke, W. (1999) *Calculating Bushels.* University of Minnesota Extension Service. Available at: http://www.bae.umn.edu/extens/postharvest/bushels.pdf

Worley, J. (2002) *Aerating Grain in Storage.* Bulletin 712. Cooperative Extension Service, University of Georgia College of Agricultural and Environmental Sciences. Available at: http://pubs.caes.uga.edu/caespubs/pubcd/B712.htm

Zinn, R.A., Alvarez, E.G., Montano, M.F., Plascencia, A. and Ramirez, J.E. (1998) Influence of tempering on the feeding value of rolled corn in finishing diets for feedlot cattle. *Journal of Animal Science* 76, 2239–2246.

Zinn, R.A., Owens, F.N. and Ware, R.A. (2002) Flaking corn: processing mechanics, quality standards, and impacts on energy availability and performance of feedlot cattle. *Journal of Animal Science* 80, 1145–1156.

This page intentionally left blank

Index

Acid detergent fibre *see* food analysis
Acidosis 60
Allergens 75
Allometry of nutrient use 181–183
    see also energy, water
Amino acids
    arginine 20, 34, 195, 203, 204, 263
        see also specific foods
    availability 34–36, 68, 263
        FDNB binding (for lysine) 35
        ileal digestibility 34
        indicator amino acid oxidation 35
        mobile nylon bag digestibility 35
        slope-ratio assay 35
    bypass 202
    cystine, cysteine 20, 34, 59, 62, 63, 131, 194, 195, 203, 260
    efficiency of use of absorbed amino acids 197, 198, 203
    environmental pollution 242
    essential 18–19, 20, 33, 34, 61, 64, 65, 67, 68, 89, 194–195, 203, 221, 242
    first-limiting 20, 194, 203
    functions as nutrients 20
    gene expression 251, 252
    histidine 20, 34, 63, 195, 203
    ideal amino acid mixture 197, 198, 204–205, 206
    isoleucine 20, 172, 195, 203
    leucine 20, 195, 203
    lysine 19, 20, 34, 35, 59, 64–65, 158, 172, 192, 194, 195, 203, 204, 205, 206–207
        see also specific foods
    methionine 20, 34, 195, 203, 263
    phenylalanine 20, 195, 203
    requirements
        horses 207
        pigs 204, 205, 206
        ruminants 198–200
    taurine 12, 20, 195
    threonine 20, 172, 195, 203, 242
    tryptophane 20, 59, 63, 195, 203
    tyrosine 20, 195, 203
    valine 20, 34, 195, 203
α-Amylase 52, 75, 85, 91–92, 261
Animal production level *see* net energy
Animal protein meals 19, 67, 278, 287
    see also specific foods

Anti-nutritional factors *see* secondary substances
Archaea 1, 3, 6, 96, 99, 102
Arsenic 20, 124, 131, 142
Ascorbic acid *see* vitamin C
Ash *see* food analysis
Autotrophs 1

B group vitamins 8, 21, 59, 61, 65, 88, 101, 102, 151, 152, 155, 221, 260
    deficiencies 157–158
    functions 156
    measurement units 151
    sources 156–157
    supplementation 158
    see also specific vitamins
Bacteria 1, 3, 4, 12, 29, 36, 49, 50, 60, 75, 76, 79, 87, 88, 89, 90–91, 93, 155, 157, 198, 244, 261, 262, 263, 265
    see also rumen microbial population
Basal metabolic rate (BMR) 180, 183
Basal metabolism 180–181
    measurement 183
    standard conditions 181, 183
    thermoneutral environment 183
Batching of stockfood ingredients 275, 276, 278, 286, 287, 288
Biotin 20, 59, 101, 152, 155, 156, 157, 158
Blood analysis *see* nutritional status assessment
Blood constituents *see* nutritional status assessment
Blood meal 67, 68
Body measurements, relationships with energy use 181–183
    see also allometry of nutrient use
Boron 20, 124, 131, 142
Bromine 20, 131, 142
Bypass protein 64, 90, 173, 198, 263
    see also undegradable dietary protein

Calcitriol *see* vitamin D
Calcium 19, 20, 21, 124, 131–134
    availability 133
    blood levels 231–232
    calcium:phosphorus ratio 132

Calcium (*continued*)
  deficiency disorders (arthritis, hypocalcaemia,
      osteochondrosis, osteodystrophia fibrosa,
      osteomalacia, osteoporosis, rickets) 133–134
  oxalic acid (oxalate) 79–80, 173
Calorimetry 183–184
  direct and indirect 184, 194
  heat equivalence of respired oxygen 185
  open and closed circuit 184
  oxygen consumption related to energy use
      184, 185
  respiratory quotient (RQ) 184, 185, 194
Canola meal *see* rapeseed meal
Carnivores 1, 2, 3, 4–6, 12, 28, 118, 152, 227, 235
  anatomical and physiological adaptations
      8, 85–86
Carotenoids *see* vitamin A
Cellulose *see* plant cell wall
Cereal grain by-products 19, 57, 60–62, 130, 133
  bran 61
  brewers' grains 61–62
  millrun (middlings) 61
  pollard 61
Cereal grain processing
  endosperm 256, 257, 259, 284
  exploding 261, 284
  extrusion 260, 262, 275, 284
  heat-labile nutrients 157, 260
  heat treatments, responses 262
      grains 260, 261, 262
      horses 262
      pigs 260
      ruminants 260–262
      starch 260
  high moisture (HM) grain, ensiling 259, 281
  micronizing 75, 260, 261, 262, 284
  milling 60, 61, 243, 256–259
      animal disorders (gastric and oesophageal
          ulcers, enterotoxaemia, laminitis,
          rumenitis) 258–259
      comparison of roller mills and hammer
          mills 281
      digestibility 256–258, 259, 261
      hammermills 256, 257, 278–279
      maize-based diets 257–258, 259
      particle size of cereal grains, optimum for
          pigs 257
      roll corrugation and disposition 280
      roller mills 256, 259, 279–281
  pelleting 284
  popping 261, 262, 284
  roasting 260, 284
  steam flaking, steam flaked (SF) grain 60, 243, 257,
      259, 260, 261, 262, 275, 284
  tempering 259
  see also chemical processing of foods, extrusion,
      pelleting, starch, stockfood processing

Cereal grains
  amino acid and protein content relationships 57,
      59, 64
  angle of repose 273–274
  bulk densities 274
  digestion and digestibility 52, 101, 258
  effects on animal health 60, 134
  global use 57, 58
  minerals 130, 132–133, 135, 143
  non-starch polysaccharides 78, 242
  nutrient contents 57–60
      variability 59–60, 223
  physico-chemical structure 50–52
      aleurone layer 50, 51, 61
      endosperm 28, 50, 51–52, 59, 60, 61,
          78, 101
      β-glucan 25, 51–52, 60, 78, 87, 101, 242
      puroindoline 51, 52
      seedcoat 50–51, 52, 59, 61, 64, 156,
          256, 259
      testa 51
  phytin (phytate) 74, 75, 76–77
  starch *see* starch
  tannins 78
  vitamins 155, 156
  see also cereal grain processing
Chemical processing of foods 262, 265–266
  cereal grains 262
  forages 265–266
      acids 265
      alkalis, sodium hydroxide (NaOH) 265, 266
      ammonia 265–266
      hydrogen peroxide 265
      ozone 265
      thermo-ammoniation 266
Cholecaliferol *see* vitamin D
Choline 20, 65, 152, 156, 157, 158
Chromium 20, 124, 130, 131, 137, 138, 234
Cobalt 20, 21, 62, 100, 124, 130, 131, 137, 138, 139, 232
  see also vitamin B12
Coconut (copra) meal 19, 64, 65, 66, 200
Concentrate foods
  food intake 162, 164, 207–208
  nutritive value 57
  rumen function 99, 100, 105
  thiamin status of ruminants 157
  use in animal feeding 17, 19, 21, 34, 90,
      243, 249
  see also cereal grains, oilseed meals, secondary
      substances *and specific foods*
Copper 20, 21, 62, 68, 77, 124, 130, 131, 137, 138,
      139–140, 143, 221, 241, 242
  ceruloplasmin, Cu/Zn superoxide dismutase 232
  deficiencies (anaemia, arthritis, diarrhoea,
      impaired fertility, ataxia, osteochondrosis,
      swayback) 133, 139–140, 230
  supplements 140

Copra meal *see* coconut meal

Cottonseed meal, whole cottonseed 16, 19, 64, 66, 75, 76, 77, 90, 274

Critical temperature 183, 191

Crude fibre *see* food analysis

Crude protein *see* food analysis

Cyanocobalamin *see* vitamin B12

Cyanogens (dhurrin, linamarin, lotaustralin) 74, 78

Cyclopropenoid fatty acids (malvalic acid, sterculic acid) 77

Deficiency symptoms 229–230

Diet selection 2, 10, 11, 12, 88, 89, 172–174

Dietary cation anion difference (DCAD) 135–136

Digestibility 11, 12, 25, 34, 40, 99, 122, 135, 170, 201, 229, 234, 260
   apparent, true 28, 29
   associative 31
   cereal grains 52, 59, 60, 61
   chemical treatment 265–276
   definition 28
   effective (digestibility, degradability) 32–33
   forages 41–43, 50, 263, 264
   ileal 34, 205
   lignin 47–49
   measurement 29–30, 234
      difference 31
      gas production 31
      in vitro and *in sacco* 30–31, 35
      in vivo 29–30
      marker ('indicator') 30, 234
   oilseed meals 64
   potential 31–32
   protein fractions in digesta 201, 206
   rate of digestion 31–32
   see also plant cell wall, starch digestion

Digestible energy (DE) 28–30, 32, 33, 100, 122, 162, 234, 258
   contents 57, 65
      see also specific foods
   definition 29
   efficiency of conversion to metabolizable energy 33, 191, 193
   measurement 29–30
   see also digestibility, energy, energy requirements, metabolizable energy, net energy

Digestion
   foregut fermentation 7, 8, 10
   functional diagrams 85–87
   hindgut fermentation 7, 10
   limitations to mammalian digestion 88, 93
   mammalian (oral, gastric and intestinal) 89–94
      α-amylase 52, 60, 75, 85, 91–92, 261
      brush border peptidases 90–91
      carbohydrate 91–92

      chylomicrons 94
      esterase 85, 89, 93
      fat emulsification 93, 94
      intestinal enzymes 92
      limit dextrin 92
      lipase, colipase 85, 93, 94, 253
      lipid digestion and absorption 93–94
      pancreatic enzymes 90–91
      pepsin 30, 35, 90, 91
      pro-enzymes (zymogens) 89–90
      proteases 35, 52, 74, 75, 89
      protein and nucleic acid digestion 89–91
      trypsin 63, 74, 90, 91
   sequence of digestion 85, 86–87, 91, 92, 94, 104, 107
      effect on nutrient supply in herbivores 88–89
   rumen (microbial) digestion 103–108
      bacterial cellulosome 104
      cellulose and hemicellulose digestion 93, 103–104
      deaminases 106–107
      endo- and exo-glucanases 104
      fat digestion 107
      hyper-ammonia-producing bacteria 107
      methane synthesis 87, 95, 99, 102, 104, 105, 106, 243–245, 259, 264
      protein digestion 105–107
      rate limiters to rumen digestion 99, 100, 105–106
      saturated and unsaturated fatty acids 107, 108
      Stickland reaction 106
      trans and *cis* fatty acids 107, 108
   see also digestive physiology

Digestive physiology
   caecotrophy, coprophagy 88
   comminution of food 1, 6, 88, 95, 170
   comparison of carnivores, herbivores and omnivores 85–89
   digesta flow control (mucus-trap, wash-back) 88
   digesta passage rate 6, 8, 11, 28, 30, 33, 36, 43, 75, 88, 171–172, 198, 200, 263
   digestion kinetics 11, 31–33, 47, 164, 170–171, 200, 264–265
   digestive tract morphology 85–88
   foregut and hindgut fermenters 7, 8, 10
   flow of digesta from the rumen 32, 33, 170, 200
   herbivores 28, 86, 87–89, 90, 97
   mouth and teeth morphology indexes (hypsodontic index, incisor arcade breadth, incisor width ratio, palatal width, relative muzzle width) 170
   non-ruminants 85–87
   ruminants 1, 2, 8, 10, 11, 31, 49, 76, 79, 85, 86, 87–89, 90, 92, 93, 94–95

Digestive physiology (*continued*)
    rumination 1, 7, 8, 10, 87, 88, 89, 170, 171, 187,
        188, 226, 256, 263, 264
    see also digestion
Digestive tracts, functional diagrams *see* digestion
1, 25-dihydroxy cholecalciferol *see* vitamin D

Effective NDF 34
Efficiencies of nutrient use
    absorbed amino acids 34, 197–198, 205
    digestible energy 33, 191, 193
    ideal protein (e_p) 197–198, 204
    metabolizable energy (k) 187–189, 191, 193
    metabolizable protein (k_N) 201
    uptake of RAN by rumen microorganisms
        (k_{RAN}) 201–202, 203
Electrolytes 135–136
Endogenous N loss 195–197
    endogenous urinary N (EUN), metabolic faecal
        N (MFN) 36, 195–197, 202
    protein turnover rates of mammalian tissues and
        enzymes 196
Energy
    digestibility 29–30
    digestible energy (DE) 28–30
    metabolizability 33, 258, 259
    metabolizable energy (ME) 29, 33
    net energy (NE) 29, 33
    partitioning 29
Energy requirements 189–195
    horses 193–194, 195
        efficiency of energy use 193
        growth 193–194
        l'unité fourragère cheval (UFC) 193
        maintenance 193
        work 194, 195
    pigs 191–193
        breeding sow 192–193
        efficiencies of energy use 191
        growth 192
        effects of sex and genotype 192
        maintenance 191–192
        pregnancy anabolism 193
    ruminants 189–191
        growth 190
        maintenance 189–190
        milk production 190–191
        pregnancy 191
Erucic acid 65, 75, 77
Essential amino acids *see* amino acids
Essential nutrients 18–21, 34, 35, 36, 101
    see also minerals, vitamins, *and specific foods*
Ether extract *see* food analysis
Evolution
    artiodactyls 3, 7
    mammalian carnivory and herbivory 3, 4–5, 6–8

methane production and methanogens 6
perissodactyls 3, 7
primitive feeding behaviours 3–4
symbiotic digestive microorganisms 6–7
teeth and jaws 4, 6
Extrusion 64, 157, 256, 260, 262, 263, 273, 275, 284

Fat-soluble vitamins 151, 152–155
    see also specific vitamins
Fatty acids 8, 12, 18, 20, 25, 86–87, 88–89, 93, 94,
        101, 107, 108, 116, 135, 154, 155, 156, 158,
        185, 233, 244, 253
    see also specific foods
Feeding behaviour 1–3, 8–9
    evolution 1, 3–8
    plasticity 11–12
    primitive 3–5
Feeding methods 1
Feeding standards 28, 37, 189, 204, 206
    definition 208
    nutritional models 208–209
    requirements, allowances 208
Feeding styles *see* nutritional classification
Feedmilling
    design 273–278
        elementary process flow diagrams 274–275, 276
        specific process flow diagrams 275, 277, 277
    hygiene and safety 273, 287
        dust explosions 287
    see also stockfood processing
Field metabolic rate (FMR) 181, 182, 183, 186
Fish meals 19, 67, 68
Fluorine 20, 130, 131, 132, 134, 142
Folate 151, 152, 155, 156, 157, 158
Food analysis 22–25
    detergent-based systems 23, 24
    proximate (Weende) analysis 22–23
    Uppsala method 23
Food classification 16–17, 18, 19
Food constituents *see* food analysis, nutrients, *and
        specific constituents*
Food evaluation systems 28–29, 29–30, 33, 34, 36–37
Food intake
    diet selection and food intake 172–174
        competition for preferred foods 164
        finding and prehending foods 173
        food nutrient content 172
        other available foods 172
        palatability (taste) 172–173
        social factors 164, 172–173
    digestive tract-level regulation 169–172
        digesta flow 169–170, 171–172
        receptors 169
        rumen digestion 170–171
    grazing behaviour 164, 174
    learning and food intake 164, 173–174

neophobia 173–174
    post-ingestive feedback 173
    self-medication 173
neuro-endocrine control 165, 166–169
    arcuate nucleus 165, 166, 167, 168
    ghrelin 166, 167, 168, 174, 253
    hypothalamus 162, 164, 166, 168
    insulin 164, 165, 166, 167, 168–169, 174
    leptin 164, 165, 166, 167, 168, 169, 174
    neuropeptide Y 165, 166, 167, 168, 253
    neuropeptide Y/agouti related protein (NPY/
        AGRP) neurones 166, 167, 168, 174, 175
    olfactory bulb 166
    paraventricular nucleus 165, 166, 167, 168, 252
    peptide YY 164, 167
    propiomelanocortin/cocaine and amphetamine
        regulated transcript (POMC/CART)
        neurones 166, 167, 168, 174, 175
    vagus nerve 165, 166, 167, 168, 174
    predicting food intake 189, 207–208
    regulation 162
        examples 163–164
        motivation or compulsion 162–164, 166
        overview 164, 165
        redundancy 174–175
        set point 175
        short-term and long-term control 163, 164,
            166, 168
Food prehension 1, 4, 95, 170, 188, 264
Forage processing see chemical processing of foods,
    roughage processing
Foregut fermenters 7, 8, 10

Genotypic differences see nutritional genomics
Glucosinolates 65, 75, 77
Glycaemic index 33–34, 61, 262
Gossypol 66, 74, 75, 76, 90
Grain legumes (pulses) 19, 57, 63, 74, 274, 279
Grasses
    cell wall amount and digestibility 42, 43
    cell walls (type I and type II) 44
    leaves 40–42
        anatomy 40, 41
        bundle sheath 40, 42, 170
        digestibility 41, 42, 170
        epidermis 40, 42, 170
        mesophyll 40, 42, 170
        phloem 40, 42, 170
        schlerenchyma 40, 42, 170
        vascular tissue 40, 42, 103
        xylem 40, 41, 42, 43, 170
    photosynthetic types (C3, tropical, panicoid; C4,
        temperate, festucoid) 40, 244
    stems 41–43
        tissues 41, 43
        digestibility 43

Grazing animals, assessment of nutritional status
    alkanes 30, 227, 234
    diet selection 234
    food intake and digestibility 234
    nutrient stratification 233
    residual dry matter 233
Greenhouse gases 243–245
    biogenic methane 243–244
    global warming 243–244
    methane production, methods of reduction
        243–244
Groundnut (peanut) meal 64, 65, 66

Herbivores, nutritional classification
    browsers 3
    concentrate selectors 2, 9, 10, 11
    foregut and hindgut fermenters 7, 8, 10
    grazers 2, 3, 9, 10, 11, 87, 89, 248
    mixed (intermediate) feeders 2, 3, 89
Heterotrophs 1
Hind-gut fermenters 7, 8, 10
Hypsodontic index 8, 10

Incisor width ratio 8, 9, 11
Ideal protein 198, 204–205, 206
    efficiency of use (e_p) 198, 204
    see also protein
Ileal digestibility 16, 34, 204–205, 206
Ingredient handling, receival and storage
    aeration of silos 274
    angle of repose 273–274
    bulk densities 273–274
    handling equipment (augers, bucket elevators) 273,
        274, 275, 277
    insect infestation, pest control 274
    storage equipment (silos, bins) 273, 275, 277
Intake (of food) see food intake
Intestinal digestion see digestion
Iodine 20, 77, 131, 140
Iron 20, 62, 68, 75, 76, 77, 90, 124, 130, 131, 135,
    137, 138, 140–141, 142, 230
Isoflavones 74, 75, 76

Lectins 74–75
Lignin
    analysis 23–24
    chemistry 46–47, 48
    digestibility of forages 10, 28, 41, 43, 47–50,
        101, 265
    plant cell wall 10, 24, 25, 44, 45, 46–47
Limiting nutrient 19, 20, 130, 134
Lipoic acid 131, 152, 156
Lithium 20, 131, 142
L'unité fourragère cheval (UFC) 193

Magnesium 20, 21, 77, 122, 124, 125, 130, 132, 134–135, 136–137, 138, 140, 230, 232, 241
Maintenance energy requirement 180, 181, 184–187, 195
    factors influencing (activity, temperature regulation) 186–187
    horses 193
    measurement (change in live weight, comparative slaughter) 185–186
    pigs 191–192
    ruminants 189–190
    see also calorimetry
Maintenance protein requirement 195–196
    horses 206
    pigs 204
    ruminants 202, 203
    see also endogenous N loss
Mammalian digestion *see* digestion
Manganese 20, 76, 124, 131, 137, 138, 287
Matières azotées digestibles cheval (MADC) 206
Maximum metabolic rate (MMR) 181
Meat meals, meat and bone meals 19, 62, 67, 68, 133, 134, 274, 285, 288
Menadione *see* vitamin K
Menaquinone *see* vitamin K
Metabolic size 181–183
    biological meaning 182
Metabolizable energy (ME) 29, 33, 89, 189, 190, 191, 228, 256
    bypass starch 201
    contents 57–58, 65
        also see specific foods
    efficiency of use (k) 33, 187, 224
        differences between foods 187–189
        growth (k_g) 190, 191, 194
        lactation (k_l) 190, 191
        maintenance (k_m) 190, 191, 193
        pregnancy (k_p) 191
    fermentable 200, 201
    see also digestible energy, energy requirements, net energy
Metabolizable protein (MP) 194, 200–201, 203
    efficiency of use 203
    see also net protein
Methanogens 1, 6, 96, 97, 99, 102, 104, 105, 106, 244
Microbial (crude) protein (MCP) 105, 198, 200, 201, 202, 229, 243, 261, 265
Microbial digestion *see* digestion
Milk by-products 19, 20, 67, 68
Mimosine 78, 79, 98
Minerals 19, 20, 21, 23, 25, 77, 130–142
    absorption coefficients (Ca and P) 133
    bioavailability 36–37, 77, 141
    deficiencies 229–230
        see also specific minerals
    drinking water 120, 122, 124

    essential 20, 130–131
    major (macrominerals) 131–137
        requirements 136–137
    metabolic roles 138
    pharmacological effects 140
    toxic 130–131, 138, 140, 141, 142, 287
    trace (microminerals) 137–141
    ultra-trace 130, 133
    see also specific minerals and specific foods
Minimal metabolism 180
    see also basal metabolism
Mixing 275, 276, 278
    equipment 282–283, 285
    ingredient segregation 282
    mixer calibration 283–284
    premixes 284
    principles 281–282
Molasses 19, 62–63, 90, 130, 135, 173, 266, 274, 275, 282–283
Molybdenum 20, 124, 129, 131, 137, 138, 141, 142

Near infrared reflectance spectroscopy 234–235
Net energy (NE) 28, 29, 33, 187–189, 192, 194, 264
    requirements 186–187, 189–191, 191–192, 193–194
    variable net energy 189
    see also digestible energy, energy, metabolizable energy
Net protein (NP) 196–197, 202–203, 206–207
    requirements
        basal endogenous N excretion, maintenance (Pr_M) 202, 206
        growth (Pr_G) 202, 207
        milk production (Pr_L) 202, 206
        pregnancy (Pr_P) 203, 206
        wool growth (Pr_W) 202
    see also metabolizable protein, protein
Neuro-endocrine control of food intake *see* food intake
Neutral detergent fibre *see* food analysis
Niacin 20, 59, 101, 102, 152, 155, 156, 157
Nickel 20, 124, 131, 142
Nitrate/nitrite 79, 96, 122, 124, 125
Nitrogen-free extract *see* food analysis
Non-starch polysaccharides 25, 60, 78, 191, 242, 260
Non-structural polysaccharides *see* food analysis
Nutrient absorption 1, 32, 37, 60, 68, 91, 92, 93, 94, 131, 157, 242
    see also specific minerals
Nutrient data, methods of expression 21–22
Nutrient pollution 241–243
Nutrient requirements 180
    factorial and empirical approaches 180
    maintenance and production requirements, definition 180
    see also energy requirements, protein requirements, minerals, vitamins

Nutrients 17–22
   deficiency symptoms 229–230
   definition 17–18
   essential nutrient 18–20, 194, 195, 203
   functions 20
   limiting nutrient 19–20, 130, 131, 194, 203
Nutrigenetics *see* nutritional genomics
Nutrigenomics *see* nutritional genomics
Nutritional classification 1–3
   practical uses 12
   see also carnivores, herbivores, omnivores
Nutritional ecology 1–12
Nutritional genomics 248–254
   fetal programming 253
   forage intake by cattle 249–250
   genes, gene expression 251–253
   genotype × diet quality interactions in cattle 248–249
   nutrient effects on gene expression, examples 252–253
   response by dairy cattle to monensin 250
   sex, breed and lean growth genotype in pigs 251
   single nucleotide polymorphisms (SNP) and animal performance 252
Nutritional investigative methods *see* nutritional status assessment
Nutritional status assessment
   blood analysis, minerals 231–232
      ceruloplasmin 232
      Cu/Zn superoxide dismutase 232
      glutathione peroxidase 232
      inorganic P 231
      threshold values 232
   bone analyses, Ca and P 229, 231
   comparison with performance norms 226–227
   deficiency symptoms 230
      see also nutrients, *and specific minerals and vitamins*
   energy status indices 227–228
      bone marrow fat content 227, 228
      body condition scores 227–228
      kidney fat index 227, 228
      muscle index 228
   faecal analyses 229, 231
   faecal profiling 234–235
   food analysis 227
   general approaches 226
   metabolic profile test 233
   protein status indices (faecal N, milk urea, rumen ammonia) 229
   tissue analyses (liver, adrenal glands, saliva, hair) 232–233
   see also grazing animals assessment of nutritional status, wild animals assessment of nutritional status
Nutritive value of foods 16, 17
   measurement 28–37
      expression (air-dry, as-received, as-fed, dry matter basis) 21
      units 21, 151–152

   see also food analysis, food evaluation systems, *and specific foods*

Oilseeds, oilseed meals 17, 19, 63–67
   available amino acids 64, 65
   bypass protein content 64, 65
   full-fat oilseed meals 66–67
   global use 63–64
   oil extraction methods (extruder/expeller, solvent) 262–263
   toxins 65, 74, 75, 76
      see also secondary substances
   see also specific foods
Omnivores 2, 3, 6, 85, 87, 89
Osteochondrosis 63, 133
Osteodystrophia fibrosa (ODF, 'bighead') 80, 133, 230
Osteomalacia, osteoporosis 133, 155, 230
Oxalate 65, 79–80, 133, 173, 230

Palatal width 8, 9
Palm kernel meal 19, 64, 66
Pantothenic acid (vitamin B5) 20, 59, 65, 101, 152, 155, 156, 157
Peanut meal *see* groundnut meal
Pelleted foods 119, 157, 187, 256, 257, 260, 263–264, 265, 275, 276, 277
   see also cereal grain processing
Pelleting 278, 284
Phosphorus 19, 20, 21, 23, 36, 76–77, 107, 122, 130, 131–134, 136–137, 139, 140, 172, 230, 231, 232, 241, 242
   see also nutritional status assessment, *and specific foods*
Phylloquinone *see* vitamin K
Phytase 75, 77, 132, 133, 241, 242, 254
Phytin (phytic acid, phytate) 37, 59, 65, 74, 75, 76–77, 132, 133, 143, 241, 254
Plant cell wall 44–50
   architecture 44–45
   cellulose 6, 10, 23, 24, 25, 44, 45, 46, 47, 60, 78
      amorphous and crystalline 45
      microfibrils 45
   cutin 50
   hemicelluloses 6, 10, 23, 24, 25, 44, 45, 46, 47, 49, 78, 103
      arabinoxylan 46, 60, 78
      glucomannan 46
      glucuronoarabinoxylan 46, 47
      xyloglucan 46, 47
   lignin 6, 10, 24, 25, 44, 46–50
      chemistry 46–47
      negative effect on digestibility 47–50
   matrix carbohydrates 45–46, 47
   pectins 23, 24, 25, 44, 45–46, 47, 78, 103
   protein 50
   silica 50

Plant secondary metabolites *see* secondary substances
Pollution by excess nutrients 241–243
Potassium 20, 62, 129, 131, 134, 135, 136–137, 140, 232, 241
Prehension 1, 4, 6, 95, 170, 188, 264
Protease inhibitors (Bowman-Birk, Kunitz-type) 74, 75
Protein
    biological value (BV) 35–36, 197–198
    bypass protein 64, 90, 173, 198, 263
        see also undegradable dietary protein
    degradability (dg) 32, 36, 59, 64, 65, 200
        see also ruminally degradable protein
    degradability fractions
        indigestible (fraction c) 200
        quickly degradable protein (QDP)
            (fraction a) 200, 202
        slowly degraded protein (SDP)
            (fraction b) 200, 202
    degradable intake protein (DIP) 198
        see also ruminally degradable protein
    effective degradability (Edg) 32, 36, 200, 201
    metabolizable protein (MP) 194, 197, 198, 200, 202
        efficiency of use 203
        predicting the supply 200–201
        requirements 203
    microbial (crude) protein (MCP) 198, 200, 201, 202
    protein truly digestible in the intestine (PDIA) 198
    ruminally available N (RAN) 198, 200, 201, 202
        efficiency of uptake by rumen microorganisms (k_{RAN}) 202
    ruminally degradable protein (RDP) 36, 198
    undegradable dietary protein (UDP) 198, 200, 201
        see also bypass protein
    undegradable intake protein (UIP) 198
        see also bypass protein, undegradable dietary protein
Protein quality indices *see* protein
Protein requirements
    horses 206–207
        available protein (AP) 206
        growth 207
        maintenance 206
        matières azotées digestibles cheval (MADC) 206
        reproduction 206
        work 207
    pigs *see* amino acids, protein retention
    ruminants *see* metabolizable protein, net protein
Protein retention (deposition), growing pigs 192, 198, 203, 204, 205
Protozoa 1, 4, 60, 87, 244
    see also rumen microbial population
Pulses *see* grain legumes
Pyridoxine *see* vitamin B6

Quality assurance in stockfood manufacture 287–288
    quality control systems 288
    statistical process control 288
    see also warehousing

Rapeseed (canola) meal 19, 64, 65, 90
Ration formulation 218–224
    basic concepts 218–220
        choosing nutrients and ingredients 221–222
    calculating the nutrient content of a ration 219
    ingredient buying strategies 222
    least cost 220–221
    linear programming 219–221
        multiple optimization 223–224
        nutrient constraints (nutrient, quantity) 219–220, 221
    multiblending 223
    non-cost objectives (least-mineral) 224
    non-feasible problem 222
    nutrient requirements, concentration units 218–219
    nutritional modelling 180, 186, 208–209
        nutritional management models 224
    parametric ranging 222–223
    Pearson square 221, 222
Relative muzzle width 8, 9, 10
Retinoic acid *see* vitamin A
Riboflavin (vitamin B2) 20, 59, 152, 156, 157
Rickets 133, 151, 155, 230
Roughage processing 256, 257, 263, 284–285
    chopping (chaffing) 256, 257, 263, 264, 284–285
    milling 99, 255–256, 257, 284–285
        colonization of particles by rumen bacteria 263
        depraved appetite 264
        energy use 263–265
        food particle retention time 169–170, 264
        rumen hyperkeratosis 264
        rumen volatile fatty acids 263–264
    see also chemical processing of foods, stockfood processing
Rubidium 20, 131, 142
Rumen digestion *see* digestion, rumen microbial fermentation
Rumen function *see* rumen microbial ecology, rumen microbial fermentation, rumen microbial population
Rumen heat of fermentation 29, 33
Rumen microbial ecology 94–103
    consortia (biofilms, microcolonies) 96, 98, 101
    interactions between microorganisms 101–103
Rumen microbial fermentation
    fibre, cellulose digestion 103–104
    hydrogen sinks 104–105
    monosaccharides metabolism (primary fermentation) 104, 105

volatile fatty acid synthesis (secondary fementation) 97, 104, 106

see also digestion

Rumen microbial population

bacteria and archaea 95, 96, 97, 99

acetogenic bacteria 96, 97, 105, 106, 244

ammonia requirements 36, 100, 198, 200, 229, 250, 265

Butyrivibrio fibrisolvens 97, 99, 101, 103, 106, 107

carbohydrate substrates 97, 99–100, 101, 103

Fibrobacter succinogenes 96, 97, 98, 99, 101, 103, 104, 106

fibrolytic bacteria 97, 98, 99, 103, 104

hyper-ammonia producing bacteria 107, 243

Lactobacillus spp. 97, 100, 101, 106

Megasphaera elsdenii 97, 99, 102, 106, 107

methanogens 1, 6, 96, 97, 99, 102, 104, 105, 106, 244

Prevotella ruminicola 97, 99, 103, 105, 107

Ruminoccocus albus 96, 97, 103, 104, 106

Ruminococcus flavefaciens 96, 97, 101, 103, 104, 105, 106

Selenomonas ruminantium 97, 99, 102, 105, 106, 107

Streptococcus bovis 97, 99, 103, 106

bacteriophages 96

conditions for growth (pH, anaerobiosis, temperature, supply of substrate) 95

development in neonates 97

diet-induced changes 99–100

fungi 96, 101, 103

geographical differences 98

nutrient requirements 101

protozoa 96, 97, 98, 99–100, 101, 102, 103, 105–106, 107

cellulose digestion 103

defaunation, effects on protein and fibre digestion, methane production 100, 244

substrates 99–100

seasonal and diurnal changes 98–99

yeasts 97

Rumen N metabolism 199

Ruminally available N (RAN) 198, 200, 201–202

Ruminally degradable protein (RDP) 36, 198

Rumination 7, 8, 10, 87, 88, 89, 170, 171, 188, 263, 264

Secondary substances (anti-nutritional factors, plant secondary metabolites) 74–80

see also specific substances

Selenium 20, 65, 68, 124, 130, 131, 137, 138, 141, 230, 232, 287

Silicon 20, 131, 142

Sodium 20, 58, 59, 117, 122, 125, 130, 131, 134, 135, 136–137, 140, 172, 221, 230, 231, 232, 266

see also specific foods

Soybeans, soybean meal 16, 19, 31, 60, 63–64, 65, 66–67, 74, 75, 76, 197, 253, 257, 262–263, 274

Standard metabolic rate (SMR) 181, 182, 183, 185, 186, 187

see also basal metabolic rate

Standard reference weight 190, 202, 207

Starch 20, 25, 28, 51–52, 57, 60, 63, 173, 256, 259

amylopectin 51, 91, 92, 256, 260

amylose 51, 52, 60, 63, 91, 92, 256, 260

bypass starch 201

digestion 91–92, 97, 99, 100, 101, 105, 243, 256, 257, 258, 259–261, 262, 264

gelatinization 260, 284

resistant starch 12, 24, 60, 201

retrogradation 260

Stockfood processing

animal responses 256–266

methods 75, 256–260, 263, 265

overview 256

reasons for processing foods 256, 257

see also cereal grain processing, chemical processing of foods, oilseeds processing, feedmillng, roughage processing

Sugarbeet pulp 62–63

Sugarcane bagasse 17, 63

Sugar refining by-products *see specific products*

Sulfur 20, 62, 78, 124, 130, 131, 138, 141, 244

Summit metabolism 181, 183

Sunflower, sunflower meal 16, 19, 64, 65–66, 67, 200

Surface area law 182

Tannins 28, 60, 65, 74, 75, 78–79, 88, 173, 227, 229, 244

Thermoneutral environment 183

Thiamin (vitamin B1) 20, 59, 65, 101, 151, 152, 155, 156, 157, 158, 287

Tin 20, 131, 142

Tocopherols *see* vitamin E

Trace minerals *see* minerals

Transgenic animals 254

Transgenic foods, nutritive value 253–254

substantial equivalence 253

Undegradable dietary protein (UDP) 198, 200, 201

see also bypass protein

Vanadium 20, 124, 130, 131, 142

Vitamin A (retinol) 19, 20, 22, 59, 65, 66, 79, 88, 151, 152–154, 221, 260

anti-oxidant role 154

carotenoids 59, 152, 153, 154

night blindness 154

synthesis of visual pigment 154

Vitamin B1 *see* thiamin

Vitamin B2 *see* riboflavin
Vitamin B5 *see* pantothenic acid
Vitamin B6 (pyridoxamine) 20, 59, 101, 152, 155, 156, 157
Vitamin B12 20, 65, 12, 138, 141, 151, 152, 155, 156,
      157–158, 232
Vitamin C (ascorbic acid) 20, 151, 152, 155, 158, 260
Vitamin D (calciferol) 20, 132, 133, 136, 140, 151, 152,
      154–155, 230
Vitamin E (tocopherol) 20, 59, 65, 76, 151, 152, 155
Vitamin K 20, 152, 155, 156
Vitamins 151–158
    classification 152
    deficiency and toxicity symptoms and effects 154,
        155, 157–158, 230
    definition 151
    fat-soluble 152–155
    history 151
    nomenclature 151, 152
    units of measurement 21, 151–152
        definitions (international units, IU; retinol
            activity equivalent, RAE) 22, 152
    water-soluble 155–158
    see also specific vitamins
Volatile fatty acids 87, 90, 104, 105, 106, 188,
      264, 265
    see also rumen microbial fermentation
Voluntary food intake *see* food intake

Warehousing 285–286
    principles 285
    stock control 285–286
    see also feedmilling
Water 115–125
    functions in animals 115
    physiological parameters of water use 117–119

      isotope dilution studies 117–118
      total body water 117, 118
      water turnover rate (fractional, total)
          117–118, 119
      water balance 115, 116, 117, 119, 121
    requirements for drinking water 117, 119–122
      animal productivity 120–121
      environmental effects (humidity,
        temperature) 119
      estimates 121–122
      food composition effects (protein, fibre,
        minerals) 120
    sources 115–117
      drinking water 116, 117
      food water 115–116
      metabolic water 116–117
    water:dry matter ratio 119–120
Water salinity 121, 122–125
    effects of drinking saline water 122–124
    measurement, electrical conductivity 124
    maximum levels for dissolved solids 124–125
    tolerances for saline water 121
    urine osmolarity 122–123
Water-soluble carbohydrates *see* food analysis
Water-soluble vitamins 155–158
    see also specific vitamins
Wild animals, assessment of nutritional status 180, 181,
      227, 229, 234–235
    digesta collected post mortem 235
    dietary diversity and overlap 235
    faecal (scats) analyses 234–235
    herd dynamics 235

Zinc 20, 21, 62, 68, 77, 124, 130, 131, 133, 134, 137,
    138, 139, 141, 142, 232, 241–242, 253

Printed and bound by CPI Group (UK) Ltd, Croydon, CR0 4YY

11/01/2026

14804825-0004